ELECTRON BEAMS, LENSES, AND OPTICS

Volume 1

ELECTRON BEAMS, LENSES, AND OPTICS

Volume 1

A. B. EL-KAREH

DEPARTMENT OF ELECTRICAL ENGINEERING
SYRACUSE UNIVERSITY
SYRACUSE, NEW YORK

J. C. J. EL-KAREH

SYRACUSE, NEW YORK

1970

ACADEMIC PRESS New York and London

ACADEMIC PRESS, INC.
111 Fifth Avenue, New York, New York 10003

United Kingdom Edition published by
ACADEMIC PRESS, INC. (LONDON) LTD.
Berkeley Square House, London W1X 6BA

LIBRARY OF CONGRESS CATALOG CARD NUMBER: 71-107546

PRINTED IN THE UNITED STATES OF AMERICA

Contents

Preface

This book developed during the course of several years. Draft after draft of the manuscript benefited from the give and take between one of the authors (A.B.E.) and the graduate students who participated in his university lectures as well as the industrial scientists and engineers who attended his summer "crash course." As was the case with these students, the authors expect that the audience for this book will have a non-uniform background in electron optics. As such, this book, may be many things to many people.

Our foremost plan was to make the material accessible to newcomers to the field. To this end, every equation, with only one exception, has been derived from first principles. To emphasize the physics of our discussions, we have tried to use elementary mathematics as much as possible with the obvious consequence of sometimes sacrificing elegance and brevity for the sake of clarity.

The bibliography is not a complete one; readily available sources fill this need. We limit ourselves to citing those references that, in our opinion, have a direct bearing on the material covered, or that we have drawn upon in preparing the book.

For workers active in the field, the advanced topics are discussed in depth and the most complete data tables available on electrostatic and magnetic lenses are included. A good deal of these data are original and appear in print for the first time.

This work appears in two volumes. Volume 1 deals with the ideal case; Volume 2 deals in detail with aberrations. Since we have confined ourselves to rotationally symmetrical systems, many important subjects such as quadrupole lenses and deflection have not been included. The authors hope to cover this material and include further data based upon their analysis of a great number of electrostatic and magnetic lenses in a future publication.

The mks system has been used throughout the book.

We are indebted to Dr. Harry Gruenberg for his derivation of Eq. (A5.1) in Chapter V. We also wish to thank Mrs. Louise Naylor for her patience and understanding while she typed the manuscript.

List of Symbols

	all subscripts a refer to aperture plane
	all subscripts i refer to image side
	all subscripts o refer to object side
	all primes are derivatives with respect to z
	all dots are derivatives with respect to time
$\mathbf{A}$	magnetic vector potential
A	a factor
$\mathbf{a}$	acceleration
a	parameter occurring in the formula for the Glaser bell-shaped field; in this text $a = d$ is the half width of the field
B	magnetic flux density; also a factor
B_0	maximum flux density along the axis of a magnetic lens; in the symmetrical case, B_0 is the flux density at the center of the lens
B_p	magnetic flux density in the core of the magnetic lens
C_c	coefficient of chromatic aberration for object or image at infinity
C_s	coefficient of spherical aberration for object or image at infinity
$\bar{C}_s$	coefficient of spherical aberration for finite object and image distances
C_{si}	coefficient of spherical aberration, image side, object at infinity
$\bar{C}_{si}$	coefficient of spherical aberration, image side, object not at infinity
C_{so}	coefficient of spherical aberration, object side, image at infinity
$\bar{C}_{so}$	coefficient of spherical aberration, object side, image not at infinity
c	velocity of light
D	diameter of tube or aperture of an electrostatic lens; also bore diameter of a magnetic lens
D_1	diameter of central electrode of an *einzel* lens
d	half width of central electrode in an *einzel* lens; also half width of magnetic field
E	electric field

e	charge of the electron
F	position of focal point; also a function
$\bar{F}$	reduced focal length
f	focal length
F_{obj}	objective focal length
F_{proj}	projective focal length
G	coil form factor; also center of gravity of an electrostatic lens
H	magnetic field intensity
h	Planck's constant
I	electric current; also a factor (Chapter VI)
J	current density; also a factor (Chapter VI)
K^2	Glaser's factor equal to $eB_0{}^2a^2/8mV$ (Chapter VII)
k^2	factor equal to $eB_0{}^2R^2/8mV_r$ (Chapter VIII)
k	Boltzmann's constant
l	distance from the geometrical center of an *einzel* lens to the outer electrode
log	logarithm to base e
M	magnification; also a factor (Chapter VI)
m	mass of the electron
N	factor (Chapter VI)
NI	ampere turns
n	index of refraction
P	position of principal plane
p	momentum
R	reduced height of ray from axis in electrostatic lenses; also radius of tube or aperture of an electrostatic lens; also bore radius of a magnetic lens
R_1	radius of central electrode of an *einzel* lens
R_2	radius of outer electrode of an *einzel* lens
R_m	mean radius of a torus (Chapter VIII)
r	height of electron from axis in polar coordinates
r_m	minimum radius of beam
r_α	height of the first paraxial ray in the aberration expressions
r_α'	slope of the first paraxial ray in the aberration expressions
r_γ	height of the second paraxial ray in the aberration expressions
r_γ'	slope of the second paraxial ray in the aberration expressions
S	action integral, ray function; also gap width of a magnetic lens
s	spacing between electrodes in electrostatic lens
T	temperature in degrees Kelvin
t	time
u	ratio of potentials at the center of the central and outer electrodes of an *einzel* lens

u_i	image distance
u_o	object distance
V	fixed voltage of electrode in electrostatic lenses; also accelerating potential in magnetic lenses
V_r	accelerating potential in magnetic lenses corrected for relativity
v	velocity of the electron
W	energy
x_i	distance from image to image focal point
x_o	distance from object to object focal point
z	coordinate of axis of rotational symmetry
z_m	midfocal length
z_p	distance from principal plane to plane of reference
α	angle of incidence; also half angle of arrival or departure of electron trajectory; also a factor (Chapter VI)
β	angle; also a factor (Chapter VI); also a factor equal to $k^2 V_r/(NI)^2$ of a magnetic lens
Δl	longitudinal spherical aberration
θ	half angle of arrival or departure of electron trajectory; also angle of rotation of the path in a magnetic lens
$\dot{\theta}_0$	Larmor angular velocity
κ	wave number
λ	wavelength; also a factor (Chapter VI)
μ_0	magnetic permeability in vacuum
ν	frequency
ρ	radius of curvature; also radius in cylindrical coordinates
σ	ratio of voltages (V_2/V_1) in electrostatic immersion lens
τ	period
ϕ	variable voltage in electrostatic lenses
ϕ_i	voltage corresponding to initial velocity of emitted electrons
ψ	magnetic scalar potential

ELECTRON BEAMS, LENSES, AND OPTICS

Volume 1

I

Electron Beams and Light Rays

As we will show in this chapter and in subsequent chapters, many of the laws that govern light rays are applicable to electron beams. Because of this close analogy, the term "electron optics" is used when dealing with the behavior of electrons that are under the influence of electric and magnetic fields. The material covered under the heading "electron optics" does not necessarily mean that it only applies to electrons. The same laws are applicable when dealing with any other charged particle, the only difference being the change in charge, mass, and polarity of the particle. Since most of the common instruments use electron beams rather than other beams, it seems quite normal to use the term electron optics in the general case.

Electron optics and light optics have a lot in common. In view of the fact that light optics has been studied for the past 400 years while the study of electron optics dates back only about 40 years, it seems logical that one should be in a position to exploit every analogy that exists between the two cases. Unfortunately, the analogy is not always valid. As we shall see later on in this chapter, there exist some fundamental limitations to the analogy which must be taken into consideration when we try to apply the laws of light optics in electron optics. In what follows, we shall make a comparison and relate the two cases whenever this is possible.

1.1 Snell's Law

The basic law of geometrical optics is the law of refraction named after Willebrord Snell. This law for light optics is

$$\frac{\sin \alpha_1}{\sin \alpha_2} = \frac{n_2}{n_1} \tag{1.1}$$

Here, n_1 and n_2 are the indices of refraction on two sides of a plane, and α_1 and α_2 are the angles of incidence and refraction with respect to the normal to the plane. The incident ray, the normal to the plane of incidence, and the refracted ray are in one plane. The corresponding situation in electron optics is shown in Fig. 1.1.

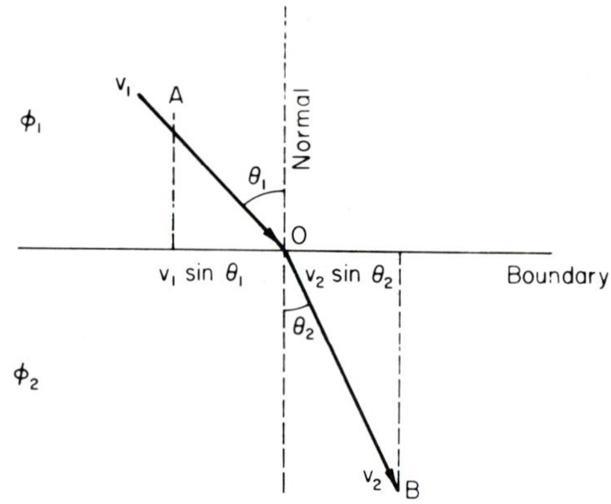

Fig. 1.1. Snell's law in optics, $v_1 \sin \theta_1 = v_2 \sin \theta_2$. The boundary separates two regions of uniform potentials ϕ_1 and ϕ_2.

Let us assume that an electron is moving in a region of uniform potential ϕ_1. Obviously, it will travel along a straight line. At the boundary it changes to a region of uniform potential ϕ_2. When the electron changes from one potential region to another, the component of the velocity normal to the boundary changes accordingly, but the tangential component of the velocity is unchanged. Thus, we have

$$v_1 \sin \theta_1 = v_2 \sin \theta_2 \tag{1.2}$$

or

$$v_1/v_2 = \sin \theta_2/\sin \theta_1 \tag{1.3}$$

If we compare Eq. (1.1) with Eq. (1.3), we can see immediately that the quantity in electron optics which corresponds to the index of refraction in light optics is the electron velocity. But, as is well known, the electron velocity is proportional to the square root of the potential (for slow electrons, of course), hence we have

$$\sin \theta_2/\sin \theta_1 = (\phi_1)^{1/2}/(\phi_2)^{1/2} \tag{1.4}$$

In general, therefore, for slow electrons, we have

$$n = kv = k\sqrt{\phi} \tag{1.5}$$

where k is an arbitrary constant. It is clear from the foregoing that the validity of Snell's law of refraction in the case of the motion of material particles is a direct result of the increase in velocity, or the acceleration, having the same direction as the force causing it.

In the general case, Eq. (1.5) should be written as follows:

$$n = kv/(1 - v^2/c^2)^{1/2} \tag{1.6}$$

in which c is the velocity of light $\approx 3 \times 10^8$ m/sec. For slow electrons this will reduce to Eq. (1.5).

1.2 General Properties of the Two Optical Media

All optical systems must have some kind of carrier of radiant energy. In light optics, the energy carriers are photons, whereas in electron optics they are charged particles, generally electrons. Table 1.1 lists the most important properties of the two types of energy carriers, those of light optics and those most often used in electron optics.

From Table 1.1 we see that both types of carriers obey wave equations of their own. It may therefore be inferred that one of the primary phenomena governing the image formation in both cases is diffraction. Thus Abbe's theory, which we will deal with in a subsequent chapter, applies in both cases under certain conditions. Also, the resolving power can be derived when we use similar approaches in devices which have any of the types of energy carriers, mentioned in Table 1.1.

TABLE 1.1[a]

	Photon	Electron
Charge	0	$-e$
Mass	$h\nu/c^2$	$m = m_0[1 - (v/c)^2]^{-1/2}$
Rest mass	0	m_0
Momentum	$p = h\nu/c$	$p = mv$
Energy	$W = h\nu = cp$	$W = mv^2/2 = c(m_0{}^2c^2 + p^2)^{1/2}$
Wavelength	$\lambda = c/n\nu$	$\lambda = h/mv$
Wave equation	Solution real	Solution complex
Statistics	Bose	Fermi
Index of refraction	$n = c/c' = c/\lambda\nu$	$n = \lambda_0/\lambda = (m/m_0)(v/c)$
Shape of refractive elements	Arbitrary	Limited by Laplace's equation
Optical components	Static	Static and dynamic

[a] Here, h is Planck's constant, 6.624×10^{-34} J-sec; ν is the frequency.

1.3 Fermat's Principle

The analogy between electron and light optics can be most effectively dramatized by considering Fermat's principle of least action. Although in its historical form it is a principle of fastest arrival, it is now expressed as follows:

In any optical medium a scalar quantity n, finite everywhere in space, can be defined so that the line integral in the three-dimensional space taken between any two fixed points is an extremum for the ray which passes through these points. In other words,

$$\delta \int_{P_0}^{P_1} n\, ds = 0 \tag{1.7}$$

Here, ds is the element of length of ray, n denotes the refractive index, which is, in general, a function of position, direction, and frequency, $\int n\, ds$ is the line integral over any line between the limits indicated as subscript and superscript. In other words, Fermat's principle can be expressed as follows: If the value of $\int_{P_0}^{P_1} n\, ds$ along a particular line joining any two points P_0 and P_1 of an optical system is stationary with respect to weak variations that leave the end points unchanged, then the line is a ray of the optical system. As an obvious corollary, we can say that if no line can be drawn joining P_0 and P_1 such that $\int n\, ds$ taken along this line is stationary for weak variations, no ray can pass according to geometrical optics from P_0 to P_1. Fermat's principle can be stated in still another way:

The time required for light to travel along the actual path and the time required for light to travel along any neighboring path differ only by terms of the second order. In other words, for the actual path the first variation of the optical path length is zero.

We shall apply Fermat's principle to the case of refraction at a plane surface. In Fig. 1.2, let AOB be the actual path from A in medium n_1 to B in medium n_2. The time it takes the ray to cover the path AOB is

$$t = \frac{AO}{v_1} + \frac{OB}{v_2} \tag{1.8}$$

where v_1 and v_2 are the velocities of the particle in mediums n_1 and n_2, respectively. Or

$$t = \frac{[a^2 + (c-x)^2]^{1/2}}{v_1} + \frac{(b^2 + x^2)^{1/2}}{v_2} \tag{1.9}$$

By Fermat's principle, the first derivative of t with respect to x is zero.

$$\frac{dt}{dx} = -\frac{(c-x)}{v_1[a^2 + (c-x)^2]^{1/2}} + \frac{x}{v_2(b^2 + x^2)^{1/2}} = 0 \tag{1.10}$$

or

$$\frac{\sin \theta_2}{v_2} - \frac{\sin \theta_1}{v_1} = 0 \tag{1.11}$$

or

$$\frac{\sin \theta_1}{\sin \theta_2} = \frac{v_1}{v_2} \tag{1.12}$$

The velocity in light optics is inversely proportional to the index of refraction. Equation (1.12) is therefore identical to Eq. (1.1). This is Snell's law as stated earlier in this chapter.

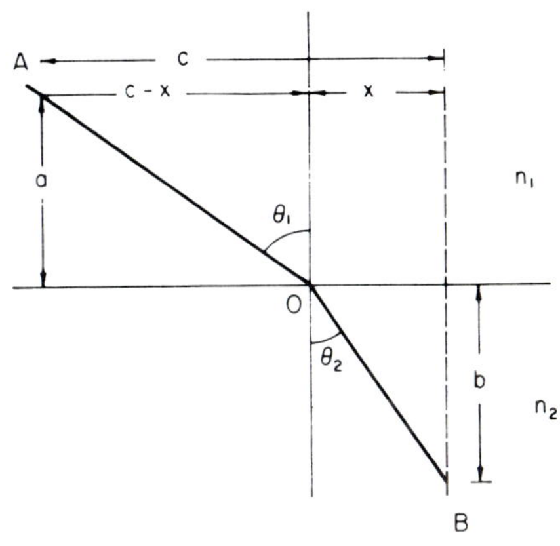

Fig. 1.2. Application of Fermat's principle in obtaining Snell's law.

1.4 Hamilton's Principle

In dynamics, Hamilton showed that all the properties of the motion of a particle of mass m between two points P_0 and P_1 in a potential field, depend on a function of the velocity and position of the particle. Hamilton defined the action as the integral of the momentum over the path, expressed as follows:

$$S = \int_{P_0}^{P_1} mv\, ds = \int_{P_0}^{P_1} mv^2\, dt \tag{1.13}$$

For a constant total energy, the natural path of the particle, the one which is dynamically possible, is selected by the condition which makes the action integral a minimum. In other words

$$\delta S = 0$$

This determines the trajectory.

The curves of equal action together with the paths of the particles which are orthogonal to them have corresponding analogies in the electric case. They are the equipotential surfaces and the lines of force. The field and its distribution determine the motion of a charged particle through it, and as with dynamics, the action integral must fulfil the requirement that it has to be a minimum for any possible trajectory.

If we compare Eq. (1.7) with Eq. (1.13), we will see that the refractive index of a medium in the optical analogy can be considered the equivalent of a corresponding momentum in the dynamical one. This means that when the total energy is constant, the path taken by a massive particle is the same as that taken by a light ray, in a medium whose refractive index at every point of the motion is proportional to the velocity of the particle. This principle shall be used extensively later on.

We must remember that if the electric potential varies appreciably over a distance of the order of magnitude of the wavelength of the wave associated with the electrons, $\lambda = h/mv$, the laws of electron optics which have been presented above are no longer valid, and the phenomena which appear can be described only in terms of the wave nature of the electron. This is the case, for example, when an electron passes through a solid.

1.5 Limitations of the Analogy

One can quite rightly conclude from the preceding section that the analogy with light optics is of great value in electron optics. However, there are some fundamental limitations which should be taken into account. Most of these limitations are listed below.

(a) In light optics, the index of refraction of lenses changes abruptly between materials of different indices of refraction. In electron optics, the changes are continuous.

(b) The fact that an electron carries a charge, whereas a photon does not, presents some problems in electron optics unknown in light optics. Due to the inherent charge of the energy carrier, there exists a mutual repulsion of the individual particles resulting in a tendency of the electron beam to spread. This phenomenon, known as space charge, will be treated in a subsequent chapter.

(c) There is a difference between mass and rest mass of the carriers in light and electron optics. Even at rest, the electron has a mass and the interaction between the electron and matter will have to be different from that of light.

(d) Both the energy and the momentum of the electron are continuously variable and can be changed arbitrarily. This is not the case in light optics.

(e) In light optics, there is almost an infinite choice of refracting media. One can select any transparent substance to build a lens to a given specification. In electron optics, on the other hand, we only have electric and magnetic fields. Light, under normal circumstances, is not influenced by either.

(f) The refractive index for light is usually contained between the limits $n = 1$ and 2.5. In electron optics, the range of refractive indices is enormous. Values as high as 1000 are often used.

(g) By pulsing the electron beam at prescribed intervals, it is possible to obtain some phenomena unknown in light optics. For example, the bunching of electrons in klystrons.

(h) Electric and magnetic lenses are very flexible. The refractive index in electron optics is continuously variable; the refractive index in light optics is practically immutable.

(i) Due to the fact that electric and magnetic lenses have certain important aberrations of the same sign (this will be shown in a subsequent chapter), it is not generally possible to correct for the errors in one lens by using additional lenses. Strictly speaking, there are certain arrangements which have been devised in which errors can be canceled to a certain extent; however, these are highly complex and very involved. On the other hand, it is well known that aberrations can be avoided in light optics by the superposition of various lenses.

(j) Electron beams must travel in good vacuum due to the rapid absorption and scattering of electrons by gases. Light rays are free, although there are light optical instruments which require the use of a reasonably good vacuum for the very far ultraviolet.

(k) If there is interaction of light with matter, it will more often have the character of a reversible reaction with practically no change in entropy of the system. With electrons, the interaction with matter results in changes which will more often have the character of an irreversible reaction with a resulting change in entropy.

(l) As indicated in Table 1.1, the solution of the wave equation in the case of light optics is real, whereas with electron optics it is complex. In addition, there is no single equation that is valid for the two cases in which only the constants differ.

(m) If an ideal specimen which does not absorb any energy is observed in monochromatic light, the whole image formation can be described in terms of refraction and phase differences. If no energy of the illuminating beam is lost to the substance of the specimen, the energy of the transmitted beam will remain unchanged. This is not the case in electron optics. A substantial proportion of an electron beam passing through matter will be scattered incoherently with resulting energy loss to the specimen.

(n) Almost all lenses in electron optics are convergent. In light optics, convergent as well as divergent lenses are used.

These are some of the limitations of the analogy. There are some more of far less significance. However, despite the above limitations, there are enough fundamental similarities to permit the application of the relatively vast experience gained in light optics when learning electron optics.

Further Reading

Marton, L. (1950). *J. Opt. Soc. Am.* **40**, 269.

II

The Paraxial Ray in Symmetrical Electric Fields

If we assume that the inclination of the trajectory of an electron beam with respect to the axis of an electron optical system will always be small enough to allow us to replace its tangent by its sine or its arc, our analysis will be greatly simplified. More specifically, we are neglecting second and higher order powers of the inclination with respect to the axis. Also, if the electron beam is bounded by a cylinder of radius R and length L, a similar simplification will result when we confine our analysis to the rays within a cylinder for which $R \ll L$ so that all second and higher order terms of R/L can be neglected.

Rays which have the properties outlined above are referred to as "paraxial rays," i.e., rays that are close to the axis.

Since in paraxial rays we neglect second and higher order terms, the theory dealing with these rays is referred to as "first-order theory."

An ideal optical lens forms a true optical image in which every point of the image space corresponds to a conjugate point of the object space. As we shall see, subsequently, electron lenses can image objects ideally, provided we confine ourselves to paraxial rays. Hence, the first order theory is also referred to as "the ideal case."

The point-for-point conjugacy which is referred to above is also called "stigmatic imaging." (This should not be confused with astigmatism, a lens error with which we shall deal later on.)

The first laws governing paraxial rays have been systematically developed by Carl Friedrich Gauss. Such a study is therefore appropriately referred to as "Gaussian dioptrics."

We see, therefore, that the literature refers to the case where second and higher order terms are neglected as: (a) the ideal case, (b) paraxial ray

(c) stigmatic imaging, (d) first order theory, and (e) Gaussian dioptrics. They all refer to the same condition.

In what follows we will make a study of electron rays in plane symmetrical as well as in rotationally symmetrical fields. We shall see that to a certain extent the discussion of the Gaussian dioptrics in both cases is very similar. These two cases are very often encountered in electron optical devices. Due to the symmetry a considerable simplification results.

The rotationally symmetrical electric fields are very similar to the concentrical lens systems of light optics. The analogy between light optics and electron optics which was indicated in Chapter I can be extended to the study of optical image formation. We will show that in this case most of the laws of light optics are applicable in electron optics.

2.1 Plane Symmetrical Fields

Plane symmetry is found in various electron optical devices, such as, secondary emission multipliers, deflecting systems, and collimating slit systems. Let us consider the case of a series of parallel slits as shown in Fig. 2.1. In this case the yz plane is a plane of symmetry. If we assume that the

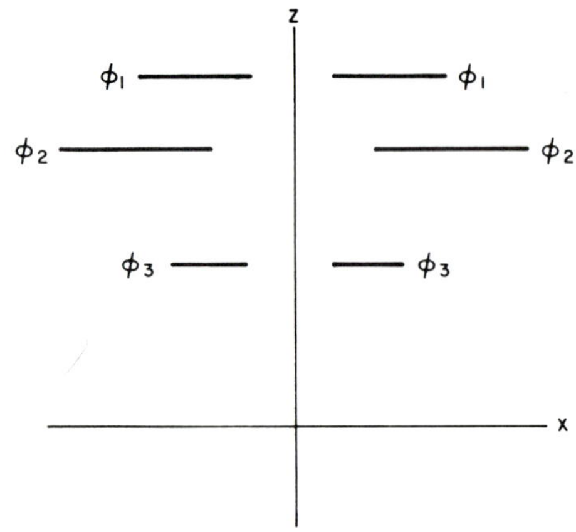

Fig. 2.1. Series of parallel slits forming a system with plane or two-dimensional symmetry.

slits are infinitely long along the y axis, there is no variation of the potential in the y direction if the x and z coordinates are kept constant.

In what follows an expression is going to be derived from which the entire potential distribution can be obtained in terms of the potential along the z axis. Laplace's equation in Cartesian coordinates reduces to:

$$(\partial^2\phi/\partial x^2) + (\partial^2\phi/\partial z^2) = 0 \tag{2.1}$$

where $\phi(x, y, z)$ is the potential. Let us express $\phi(x, z)$ as a power series in x. Since the yz plane is a plane of symmetry, any potential at x will be the same as the one at $-x$. Hence, all odd powers of x must vanish.

$$\phi(x, z) = \sum_{n=0}^{\infty} A_n(z)x^{2n} \tag{2.2}$$

Let us now substitute Eq. (2.2) in Eq. (2.1). We get

$$\sum_{n=0}^{\infty} [A_n''(z)x^{2n} + 2n(2n-1)A_n(z)x^{2n-2}] = 0 \tag{2.3}$$

where $A_n''(z)$ is the second derivative of $A_n(z)$ with respect to z. Since the sum of the coefficients of like powers of x must be equal to zero, we have

$$A_{n-1}''(z) = -2n(2n-1)A_n(z) \qquad n = 1, 2, 3, \quad \text{etc.} \tag{2.4}$$

At $x = 0$, we define

$$A_0(z) = \phi(z) \tag{2.5}$$

where $\phi(z)$ is the potential along the z axis. Then

$$A_1(z) = -\frac{A_0''(z)}{(2)(1)} = -\frac{1}{2}\phi''(z) \tag{2.6}$$

$$A_2(z) = -\frac{A_1''(z)}{4.3} = \frac{1}{4!}\phi^{\text{IV}}(z) \tag{2.7}$$

or, in general,

$$A_n(z) = (-1)^n(2n!)^{-1}\phi^{(2n)}(z) \tag{2.8}$$

and $\phi(x, z)$ can now be written as

$$\phi(x, z) = \sum_{n=0}^{\infty} (-1)^n(2n!)^{-1}\phi^{(2n)}(z)x^{2n} \tag{2.9}$$

or

$$\phi(x, z) - \phi(z) - \tfrac{1}{2}\phi''(z)x^2 + \tfrac{1}{24}\phi^{\text{IV}}(z)x^4 - \cdots \tag{2.10}$$

This expression will be used very often when we analyze electron optical systems having plane symmetry.

2.2 Paraxial Rays in Plane Symmetrical Electric Fields

Let us consider a ray with initial direction along the xz plane. Let the yz plane be a plane of symmetry. If x and z are the coordinates of the electron at a certain time, then its motion in a field plane is given by

$$m(d^2x/dt^2) = -eE_x \qquad \text{and} \qquad m(d^2z/dt^2) = -eE_z \tag{2.11}$$

Here, d^2x/dt^2 and d^2z/dt^2 are the accelerations, and E_x and E_z are the components of the electric field strength in the x and z directions, respectively. If we limit our analysis to first-order terms, the x component of the field can be obtained from Eq. (2.10)

$$E_x = -[\partial\phi(x, z)/\partial x] = \phi''(z) \cdot x \tag{2.12}$$

Substituting this relation in Eq. (2.11), we have

$$m(d^2x/dt^2) = -e\,\phi''(z) \cdot x \tag{2.13}$$

This relation shows that for paraxial rays the force is proportional to the distance of the electron from the axis.

It will be very useful to obtain a relation as a function of z only. We should therefore try to eliminate the time variable in Eq. (2.13).

Equating the kinetic and potential energies, we have

$$\tfrac{1}{2}mv^2 = \tfrac{1}{2}m[(dx/dt)^2 + (dz/dt)^2] = e\phi \tag{2.14}$$

where ϕ is the potential.

We can rewrite Eq. (2.14) as follows:

$$\tfrac{1}{2}m[1 + (dx/dz)^2](dz/dt)^2 = e\phi \tag{2.15}$$

Here, dx/dz is the tangent of the trajectory. Since we are dealing with paraxial rays, $(dx/dz)^2 \ll 1$ and reference to Eq. (2.10) will immediately show that to a first order of approximation

$$\phi(x, z) \approx \phi(0, z) \tag{2.16}$$

Therefore

$$\frac{dz}{dt} = \left[\frac{2e\phi(z)}{m}\right]^{1/2} \tag{2.17}$$

This relation could also have been obtained by integrating Eq. (2.11).

The following transformation will now be used:

$$\begin{aligned} m\frac{d^2x}{dt^2} &= m\left[\frac{2e\phi(z)}{m}\right]^{1/2} \frac{d}{dz}\left(\frac{dx}{dz}\frac{dz}{dt}\right) \\ &= m\left[\frac{2e\phi(z)}{m}\right]^{1/2} \left(\frac{2e}{m}\right)^{1/2} \frac{d}{dz}\left\{[\phi(z)]^{1/2}\frac{dx}{dz}\right\} \end{aligned} \tag{2.18}$$

We can now substitute Eq. (2.18) in Eq. (2.13). We write

$$2e[\phi(z)]^{1/2}\frac{d}{dz}\left\{[\phi(z)]^{1/2}\frac{dx}{dz}\right\} + e\phi''(z) \cdot x = 0 \tag{2.19}$$

By differentiating, we obtain

$$[\phi(z)]^{1/2}\left\{[\phi(z)]^{1/2}\frac{d^2x}{dz^2}+\frac{1}{2[\phi(z)]^{1/2}}\frac{d\phi(z)}{dz}\frac{dx}{dz}\right\}+\frac{1}{2}\phi''(z)\cdot x=0 \tag{2.20}$$

or

$$\phi(z)\frac{d^2x}{dz^2}+\frac{1}{2}\phi'(z)\frac{dx}{dz}+\frac{1}{2}\phi''(z)\cdot x=0 \tag{2.21}$$

and finally

$$\frac{d^2x}{dz^2}+\frac{\phi'(z)}{2\phi(z)}\frac{dx}{dz}+\frac{\phi''(z)}{2\phi(z)}x=0 \tag{2.22}$$

2.3 Rotationally Symmetrical Systems

This kind of symmetry is by far the most common in electron optical devices. Consider the case of two coaxial cylinders of potentials V_1 and V_2 as shown in Fig. 2.2. If we choose a point of radius r at any position along

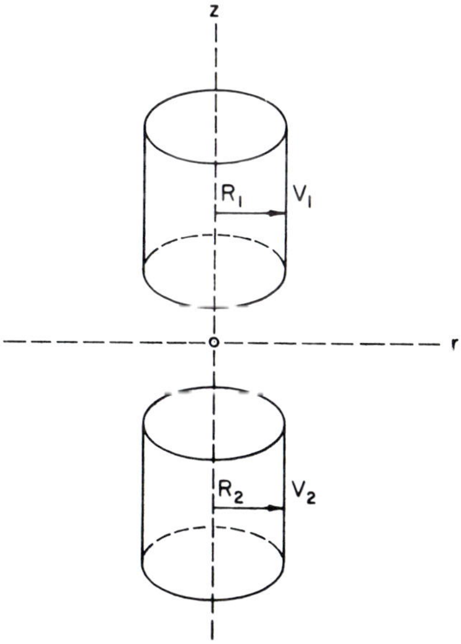

Fig. 2.2. Coaxial cylinders forming a rotationally symmetrical system. R_1 need not be equal to R_2.

the axis z, the potential will not vary as we rotate this point about the axis. In other words, the potential does not vary with θ. It is therefore advantageous to use cylindrical coordinates in this case.

In determining the differential equations for paraxial electron trajectories in rotationally symmetrical electric fields, we shall follow a procedure similar to the case of plane symmetry. In this case Laplace's equation (in cylindrical coordinates) reduces to

$$\frac{\partial^2 \phi}{\partial z^2} + \frac{1}{r}\frac{\partial}{\partial r}\left(r\frac{\partial \phi}{\partial r}\right) = 0 \tag{2.23}$$

Let us express $\phi(z, r)$ as a power series in r. Since rotating θ by any angle, say by π, does not change the potential, all odd powers of r must vanish. Thus,

$$\phi(r, z) = \sum_{n=0}^{\infty} A_n(z) r^{2n} \tag{2.24}$$

$$\frac{1}{r}\frac{\partial}{\partial r}\left(r\frac{\partial \phi}{\partial r}\right) = \sum_{n=0}^{\infty} 4n^2 A_n(z) r^{2n-2} \tag{2.25}$$

$$\frac{\partial^2 \phi}{\partial z^2} = \sum_{n=0}^{\infty} A_n''(z) r^{2n} \tag{2.26}$$

The sum of the coefficients of any power of r in Eq. (2.23) must vanish. Hence

$$A_{n+1}(z) = -\frac{A_n''(z)}{4(n+1)^2} \tag{2.27}$$

Thus, the series is

$$\phi(r, z) = A_0(z) r^0 - \frac{A_0''(z) r^2}{2^2} + \frac{A_0^{\text{IV}}(z) r^4}{2^2 \cdot 4^2} + \cdots + \frac{(-1)^n}{(n!)^2} A_0^{(2n)}(z)\left(\frac{r}{2}\right)^{2n} \tag{2.28}$$

When $r = 0$, let

$$\phi(0, z) = A_0(z) = \phi(z) \tag{2.29}$$

Then

$$\phi(z, r) = \phi(z) - \frac{r^2}{4}\phi''(z) + \frac{r^4}{64}\phi^{\text{IV}}(z) \cdots \tag{2.30}$$

We can see that this expression is similar to the one derived for plane symmetrical systems.

2.4 Paraxial Rays in Rotationally Symmetrical Electric Fields

If the initial direction of the electrons is in a plane which passes through the axis, a so-called meridional plane, then, since there is no potential variation as one rotates along the θ direction, the trajectory of the electrons will remain in this plane. In the following analysis, we will deal with meridional rays only. If the initial velocity of an electron has a circular component $[r(d\theta/dt)]_{z=0}$, its trajectory will not lie along the meridional plane and is referred to as a "skew" electron trajectory.

The motion in a meridional plane can be described by the two coordinates r and z. If E_r is the radial component of the field strength and E_z its axial component, then the motion is given by

$$m(d^2r/dt^2) = -eE_r \qquad \text{and} \qquad m(d^2z/dt^2) = -eE_z \tag{2.31}$$

Again, since we are dealing with only paraxial rays, E_r can be obtained from Eq. (2.30) to give

$$E_r = -\partial\phi(r, z)/\partial r = \tfrac{1}{2}\phi''(z) \cdot r \tag{2.32}$$

As in Eq. (2.14), we have

$$\tfrac{1}{2}mv^2 = \tfrac{1}{2}m[(dz/dt)^2 + (dr/dt)^2] = e\phi \tag{2.33}$$

where ϕ is the potential. If we substitute Eq. (2.32) in Eq. (2.31) we obtain

$$m(d^2r/dt^2) = -(e/2)\phi''(z)r \tag{2.34}$$

This is an important relation since it shows that the force toward the axis is proportional to the distance of the electrons from the axis. This is essential for the existence of optical image formation.

Next we shall eliminate the time variable, and following the same procedure as in the plane symmetrical case, we obtain

$$2e[\phi(z)]^{1/2}\frac{d}{dz}\left\{[\phi(z)]^{1/2}\frac{dr}{dz}\right\} + \frac{e}{2}\phi''(z) \cdot r = 0 \tag{2.35}$$

which finally reduces to

$$\frac{d^2r}{dz^2} + \frac{\phi'(z)}{2\phi(z)}\frac{dr}{dz} + \frac{\phi''(z)}{4\phi(z)}r = 0 \tag{2.36}$$

We have now obtained a linear homogeneous second-order differential equation for the distance of the electron from the axis in rotationally symmetrical electric fields.

Equation (2.36) was first derived by Busch, in 1926. We can easily see that Eq. (2.36) differs from the corresponding equation for plane symmetry Eq. (2.22) only by a factor of 1/2 in the coefficient of the unknown function. Because of the great similarity between these two equations, all quantitative conclusions arrived at through the use of these differential equations are valid for both the plane and rotationally symmetrical electric fields.

Many important deductions can be made from Eq. (2.36). First, the ratio e/m (charge to mass) of the particles does not appear, indicating that the trajectory is the same for any charged particle, no matter what may be its ratio of charge to mass provided it enters the field with the same initial kinetic energy. The particles with different charges come to the same focus but arrive there at different times. Second, the equation is homogeneous in ϕ so that if the voltage is increased proportionately for all the electrodes, the trajectory remains unaltered. Thus, theoretically, alternating currents can be used with no effect on the focus. Finally, the equation is homogeneous in r and z. Any increase in the dimensions of the whole system produces a corresponding increase in the dimensions of the trajectory, since the equipotentials, although of the same form, are now enlarged. If the object is doubled in size, the image will be doubled likewise, keeping the ratio between the two unchanged.

2.5 Picht's Reduced Formula

Equation (2.36) is difficult to use because of the presence of ϕ'' since for most lenses the measurement of this value is very often inaccurate. Picht has introduced a very useful modification by using a new variable

$$R = r\phi^{1/4} \tag{2.37}$$

which after substitution in Eq. (2.36) and simplification finally yields

$$\frac{d^2R}{dz^2} = -\frac{3}{16}\left(\frac{\phi'}{\phi}\right)^2 R \tag{2.38}$$

Here, we have eliminated dR/dz and obtained an equation possessing a single expression characteristic of the electron lens. We can now determine all the Gaussian optical properties of a lens in terms of a characteristic function

$$Q(z) = (\phi'/\phi)^2 \tag{2.39}$$

2.6 Image Formation by Electric Fields

So far, we have tacitly assumed that electric fields have the capability of forming images of various objects. This was already expected since, as we have seen, the equations for the paraxial electron trajectories are homogeneous linear second-order differential equations. In what follows, we shall show beyond doubt that electric fields in which the equation for the trajectories is given by Eq. (2.36) have the property of stigmatic image formation. In other words, the paraxial electron trajectories which emanate from the points of a surface element will come together in corresponding points in such a way that the collection of these points represents a similar image of the object surface. The images can be either enlarged or reduced depending on the geometry of the system.

A homogeneous linear differential equation of the second order has always two independent solutions, and the general solution is a linear combination of the two particular solutions. Let us consider two particular solutions $m(z)$ and $n(z)$, and let us assume that at the point $z = z_0$ along the axis, we have

$$\begin{aligned} m(z_0) &= 1, \qquad & n(z_0) &= 0 \\ m'(z_0) &= 0, \qquad & n'(z_0) &= 1 \end{aligned} \tag{2.40}$$

where m' and n' are derivatives with respect to z.

Any paraxial trajectory $r(z)$ can be represented by a linear combination of the two particular solutions as follows:

$$r(z) = K_1 m(z) + K_2 n(z) \tag{2.41}$$

where K_1 and K_2 are two constants. Let us now consider the trajectories which pass through an object point H_0 of coordinates r_0, z_0. We then have

$$r_0 = K_1 m(z_0) + K_2 n(z_0) = K_1 \tag{2.42}$$

This determines the constant K_1. Next, we differentiate Eq. (2.41), and let $z = z_0$ to obtain

$$r'(z_0) = K_1 m'(z_0) + K_2 n'(z_0) = K_2 \tag{2.43}$$

Let us select a plane $z = z_i$ where the function $n(z)$ has its first zero, then

$$r(z_i) = r_0 m(z_i) \tag{2.44}$$

which is independent of the parameter $r'(z_0)$. This clearly shows that a collection of trajectories which start from H_0 will come together in the image point H_i no matter what their initial slopes are as long as they are small. The plane $z = z_i$ is the image plane or the plane conjugated to the object plane $z = z_0$. The ratio of image magnitude r_i to object magnitude r_0

$$r_i/r_0 = m(z_i) = M \tag{2.45}$$

Here, M is the same for all pairs of points and is called the magnification.

2.7 Image Formation by Paraxial Rays in Rotationally Symmetrical Fields with Superimposed Deflection

We shall show in what follows that the stigmatic image formation in a rotationally symmetrical field will not be disturbed if we deflect the image forming paraxial rays by an external field. It goes without saying that this field should not be so strong as to violate the paraxial properties of the trajectories. From Eq. (2.30), for paraxial rays, we have

$$\phi(z, r) = \phi(z) - \tfrac{1}{4}r^2\,\phi''(z) = \phi(z) - \tfrac{1}{4}(x^2 + y^2)\,\phi''(z) \tag{2.46}$$

The relations for the field strength of the axially symmetrical field are

$$E_{x_a} = -\frac{\partial\phi}{\partial x} = \frac{\phi''}{2}x, \qquad E_{y_a} = -\frac{\partial\phi}{\partial y} = \frac{\phi''}{2}y \tag{2.47}$$

Let the components of the deflecting field be $E_1(z)$ and $E_2(z)$ in the x and y direction, respectively. If we superimpose the deflecting fields to the lens field we have

$$E_x = E_1(z) + \tfrac{1}{2}\phi''(z)x, \qquad E_y = E_2(z) + \tfrac{1}{2}\phi''(z)y \tag{2.48}$$

Integrating Eq. (2.48) and using Eq. (2.46), we obtain

$$\phi(x, y, z) = \phi(z) - E_1(z)x - E_2(z)y - \tfrac{1}{4}(x^2 + y^2)\phi''(z) \tag{2.49}$$

The equation of motion can now be written as

$$m\ddot{x} = -eE_x = -e[E_1(z) + \tfrac{1}{2}\phi''(z)x] \tag{2.50}$$

$$m\ddot{y} = -eE_y = -e[E_2(z) + \tfrac{1}{2}\phi''(z)y] \tag{2.51}$$

To eliminate the time variable, we use the energy equation

$$\tfrac{1}{2}m(\dot{x}^2 + \dot{y}^2 + \dot{z}^2) = e[\phi(z) - E_1(z)x - E_2(z)y - \tfrac{1}{4}(x^2 + y^2)\,\phi''(z)] \tag{2.52}$$

For paraxial trajectories the axial component $\dot{z}$ of the velocity is far larger than the transverse components $\dot{x}$ and $\dot{y}$. Therefore, $\dot{x}^2 \ll \dot{z}^2$ and $\dot{y}^2 \ll \dot{z}^2$. We can then write

$$\tfrac{1}{2}m\dot{z}^2 = e(\phi - E_1x - E_2y) \tag{2.53}$$

While attempting to eliminate the time variable, we obtain higher order terms of E_1x and E_2y, which for the paraxial case can be neglected. Therefore, we can use

$$dz/dt = [2e\phi(z)/m]^{1/2} \tag{2.54}$$

and we follow exactly the same procedure we used in obtaining Eq. (2.36).

We finally get

$$x'' + \frac{\phi'}{2\phi}x' + \frac{\phi''}{4\phi}x = -\frac{E_1(z)}{2\phi} \tag{2.55}$$

$$y'' + \frac{\phi'}{2\phi}y' + \frac{\phi''}{4\phi}y = -\frac{E_2(z)}{2\phi} \tag{2.56}$$

These two differential equations describe the projections of the trajectory on the x-z and the y-z planes. For the special case in which the deflecting field disappears, $E_1 = E_2 = 0$ and if we select one of the coordinates as coinciding with the plane of the trajectory, Eqs. (2.55) and (2.56) should correspond to Eq. (2.36).

As we saw earlier, if we write the solution to the differential equation as

$$x = K_1 m(z) + K_2 n(z) \tag{2.57}$$

in which $m(z)$ and $n(z)$ satisfy the boundary conditions as given in Eq .(2.40), the first constant K_1 represents the initial value of x at $z = z_0$, and the second constant K_2 represents the initial slope.

If x_0 and y_0 are the coordinates of the object point, and α and β are the initial angles of inclination where $\alpha = dx/dz$ and $\beta = dy/dz$ as shown in in Fig. 2.3, we then have

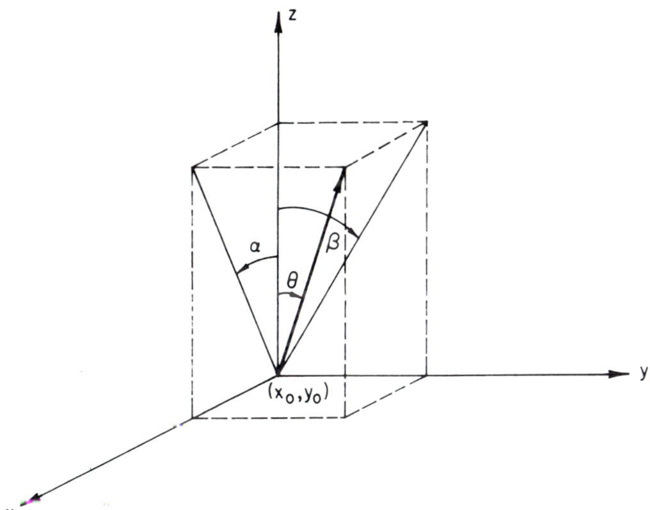

Fig. 2.3. Original direction of trajectory, leaving object point (x_0, y_0), is along arrow; α and β are projections of original angle with the z axis on xOz and yOz planes.

$$x = x_0 m(z) + \alpha n(z) \tag{2.58}$$

$$y = y_0 m(z) + \beta n(z) \tag{2.59}$$

if there is no deflection, that is, if $E_1 = E_2 = 0$. As we pointed out earlier, we have stigmatic image formation in this case.

If only a deflecting field is present, then $\phi = V =$ constant, and we have

$$x'' = -E_1/2V \qquad \text{and} \qquad y'' = -E_2/2V \tag{2.60}$$

These are the well-known equations for deflection in an electric field.

If now we consider the general case in which a deflection is superimposed on a rotationally symmetrical imaging field, the general solution of the system of differential equations is given by

$$x = x_0 m(z) + \alpha n(z) + X \tag{2.61}$$

$$y = y_0 m(z) + \beta n(z) + Y \tag{2.62}$$

Here, X and Y are particular solutions to the corresponding inhomogeneous differential equations (2.55) and (2.56); X and Y depend on E_1 and E_2, respectively. In the image plane we have the magnification of the object plus the deflection. Thus,

$$x_i = X_i + Mx_0 \tag{2.63}$$

$$y_i = Y_i + My_0 \tag{2.64}$$

Therefore, a point on the axis will not be imaged on a point on the axis. We can say that the axis changes to a curve $x = X(z)$ and $y = Y(z)$. This same curvature is experienced by all rays of the imaging bundle. Therefore, the stigmatic character of the imaging formation is preserved.

A practical application of this analysis can be found in cathode-ray tubes. As long as the deflecting field is not too strong as to violate the paraxial properties of the trajectories, cathode-ray tubes can be developed so that deflection is proportional to the deflecting field and the focusing is maintained throughout.

2.8 The Radius of Curvature as a Function of the Potential

Let us consider a rotationally symmetrical electric field about an axis Oz. The equipotentials can be represented by $\phi(r, z) = \text{constant}$. We select the Oz axis so that it passes through point S as shown in Fig. 2.4

$$(dz/dr)_S = 0 \tag{2.65}$$

we have

$$dz/dr = -(\partial\phi/\partial r)/(\partial\phi/\partial z) \tag{2.66}$$

Therefore,

$$(\partial\phi/\partial r)_S = 0 \tag{2.67}$$

The radius of curvature of the equipotential at S is given by

$$\rho_S = \frac{[1 + (dz/dr)^2]_S^{3/2}}{[d^2z/dr^2]_S} \tag{2.68}$$

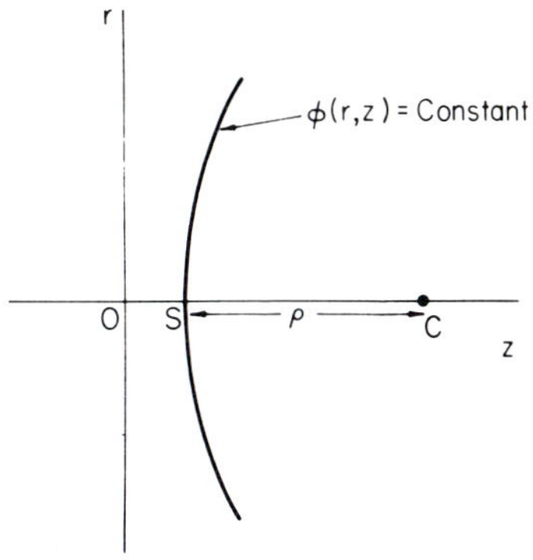

Fig. 2.4. Determination of the radius of curvature as a function of the potential.

which in this case reduces to

$$\rho_S = \left(\frac{d^2z}{dr^2}\right)_S^{-1} \tag{2.69}$$

Let us take the derivative with respect to r of Eq. (2.66). Using Eq. (2.67), we obtain

$$(d^2z/dr^2)_S = -\frac{(\partial^2\phi/\partial r^2)_S}{(\partial\phi/\partial z)_S} \tag{2.70}$$

or

$$(\rho_S)^{-1} = -(\phi')^{-1}(\partial^2\phi/\partial r^2)_S \tag{2.71}$$

where ϕ' is the derivative of ϕ with respect to z. Let us now write Laplace's equation at S

$$(\partial^2\phi/\partial z^2)_S + (\partial^2\phi/\partial r^2)_S + (r_S)^{-1}(\partial\phi/\partial r)_S = 0 \tag{2.72}$$

At S, r, and $\partial\phi/\partial r$ are both zero. By applying l'Hôpital's rule, we obtain

$$(r^{-1}\partial\phi/\partial r)_S = (\partial^2\phi/\partial r^2)_S \tag{2.73}$$

Hence, from Eq. (2.72), we have

$$(\partial^2\phi/\partial r^2)_S = -\tfrac{1}{2}(\partial^2\phi/\partial z^2)_S = -\tfrac{1}{2}\phi'' \tag{2.74}$$

Finally, we can write

$$(\rho_S)^{-1} = \phi''/2\phi' \tag{2.75}$$

2.9 Determination of the Second Derivative of the Potential

When the potential along the axis is found experimentally, the first derivative with respect to z of the potential along the axis can be obtained with sufficient accuracy. For example, graphical differentiation has been used in numerous cases. If an attempt is made to obtain the second derivative by differentiating graphically with respect to z, significant errors appear. In order to avoid this, the following method can be used.

In Eq. (2.30), let $r^2 = \omega$. We obtain

$$\phi(z, r) = \phi(z) - (1/4)\phi''(z)\omega + (1/64)\phi^{\mathrm{IV}}(z)\omega^2 \cdots \tag{2.76}$$

Now differentiate with respect to ω.

$$\partial\phi/\partial\omega = -(1/4)\phi''(z) + (1/32)\phi^{\mathrm{IV}}(z)\omega \tag{2.77}$$

Along the axis, $r = 0$ and $\omega = 0$; therefore,

$$\left(\frac{\partial\phi}{\partial\omega}\right)_{\omega=0} = -\frac{1}{4}\phi''(z) \tag{2.78}$$

or

$$\phi''(z) = -4\left(\frac{\partial\phi}{\partial\omega}\right)_{\substack{\omega=0\\ r=0}} \tag{2.79}$$

If the potential ϕ as a function of r is measured at any point z along the axis and is plotted as a function of ω, the second derivative of ϕ with respect to z, $d^2\phi/dz^2$, can be obtained from the slope of the curve at $\omega = 0$.

III

Analytical Determination of Electrostatic Fields

We have already seen that the index of refraction in electron optics is expressed by $k\sqrt{\phi}$ where k is a constant and ϕ is the potential. It is therefore necessary to determine the potential distribution in any device in order to obtain the trajectories of the electrons. Unfortunately, the problem of determining equipotentials is a very hard one to solve if a large accuracy is desired, which is often the case. Even with the simplest configurations, analytical solutions are very complex, if at all possible. Consequently, several experimental procedures were developed. Nevertheless, we shall attempt to study this problem analytically and see how far we can get in some cases.

3.1 Laplace's Equation

It is well known that an electric field E can be derived from a potential ϕ by differentiation

$$E = -\operatorname{grad}\phi \tag{3.1}$$

This potential ϕ satisfies Laplace's equation

$$\nabla^2\phi = 0 \tag{3.2}$$

and takes predetermined values at the boundaries. Obviously, all electrodes are equipotential surfaces. Any potential distribution which satisfies Laplace's equation and the boundary conditions is unique.

We have seen in the previous chapter how most electron optical devices have either rotational symmetry or plane symmetry. This is a needed blessing in this case since Laplace's equation then reduces to two dimensions. The selection of coordinates depends on the type of symmetry available. In Cartesian coordinates, we have

$$\frac{\partial^2 \phi}{\partial x^2} + \frac{\partial^2 \phi}{\partial y^2} = 0 \tag{3.3}$$

and in cylindrical or polar coordinates, we have

$$\frac{\partial^2 \phi}{\partial z^2} + \frac{\partial^2 \phi}{\partial r^2} + \frac{1}{r}\frac{\partial \phi}{\partial r} = 0 \tag{3.4}$$

3.2 Potential Distribution by Means of Complex Functions

Let

$$w = x + jy = re^{j\varphi} \tag{3.5}$$

Then

$$F(w) = \phi(x, y) + j\psi(x, y) \tag{3.6}$$

Using the Cauchy–Riemann relations

$$\frac{\partial \phi}{\partial x} = \frac{\partial \psi}{\partial y} \quad \text{and} \quad \frac{\partial \phi}{\partial y} = -\frac{\partial \psi}{\partial x} \tag{3.7}$$

and differentiating with respect to x and y, we obtain

$$\frac{\partial^2 \phi}{\partial x^2} + \frac{\partial^2 \phi}{\partial y^2} = 0 \tag{3.8}$$

which is Laplace's equation as given by Eq. (3.3). A similar derivation can lead to Laplace's equation in cylindrical coordinates.

3.3 Method of Separation of Variables

Let

$$\phi(x, y) = F(x)G(y) \tag{3.9}$$

If we insert this in Laplace's equation, we obtain

$$G(y)\frac{\partial^2 F(x)}{\partial x^2} + F(x)\frac{\partial^2 G(y)}{\partial y^2} = 0 \tag{3.10}$$

Let us now divide throughout by $F(x)G(y)$. We get

$$\frac{1}{F(x)}\frac{\partial^2 F(x)}{\partial x^2}+\frac{1}{G(y)}\frac{\partial^2 G(y)}{\partial y^2}=0 \tag{3.11}$$

Since the two variables are independent, each term of the equation must be equal to a constant.

$$\frac{1}{F(x)}\frac{\partial^2 F(x)}{\partial x^2}=k^2 \tag{3.12}$$

$$\frac{1}{G(y)}\frac{\partial^2 G(y)}{\partial y^2}=-k^2 \tag{3.13}$$

The solutions to these equations are

$$F(x)=A\sinh kx+B\cosh kx \tag{3.14}$$

$$G(x)=C\sin ky+D\cos ky \tag{3.15}$$

The general solution is given by

$$\phi(x,y)=\sum_{k=0}^{\infty}[(A_k\sinh kx+B_k\cosh kx)(\sin ky+D_k\cos ky)] \tag{3.16}$$

We have been able to eliminate C_k without loss of generality.

Similarly, a relation can be cbtained in polar coordinates by assuming a function

$$\phi(r,\theta)=R(r)\,\Theta(\theta) \tag{3.17}$$

The general solution is

$$\phi(r,\theta)=\sum_{n=0}^{\infty}[r^n(A_n\sin n\theta+B_n\cos n\theta)+r^{-n}(\sin n\theta+D_n\cos n\theta)] \tag{3.18}$$

3.4 Conformal Transformation

Consider the transformation

$$W=w^n \tag{3.19}$$

where $W=\phi+j\psi$ and $w=x+jy$. Then

$$\phi+j\psi=(x+jy)^n=r^n(\cos n\theta+j\sin n\theta) \tag{3.20}$$

Then

$$\psi=r^n\sin n\theta \tag{3.21}$$

We can assume that the surfaces $r^n \sin n\theta = \text{constant}$ are equipotentials including the special case

$$r^n \sin n\theta = 0 \tag{3.22}$$

in which the equipotential consists of two planes cutting at an angle π/n.

Now suppose $n = 1$. Then $\psi = y$, and we have a uniform field of force. If $n = 2$, then $\psi = 2xy$ and the equipotentials are rectangular hyperbolic cylinders, including as a special case two planes intersecting at right angles (see Fig. 3.1). Hence, this transformation gives the field in the immediate

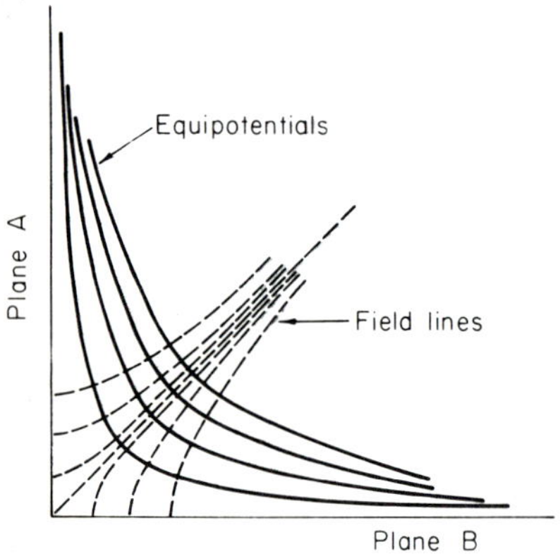

Fig. 3.1. Field lines and equipotentials of two planes intersecting at right angles.

neighborhood of two conducting planes meeting at right angles in any field of force. It will also give the field between two coaxial rectangular hyperbolas.

If $n = 1/2$, we have

$$x + jy = (\phi + j\psi)^2 \tag{3.23}$$

and

$$x = \phi^2 - \psi^2, \qquad y = 2\phi\psi \tag{3.24}$$

or

$$y^2 = 4\psi^2(x + \psi^2) \tag{3.25}$$

In this case the equipotentials are confocal and coaxial parabolic cylinders, including as a special case $\psi = 0$, a semi-infinite plane bounded by the line of foci. This transformation clearly gives the field in the immediate neighborhood of a conducting sharp straight edge in any field of force (see Fig. 3.2).

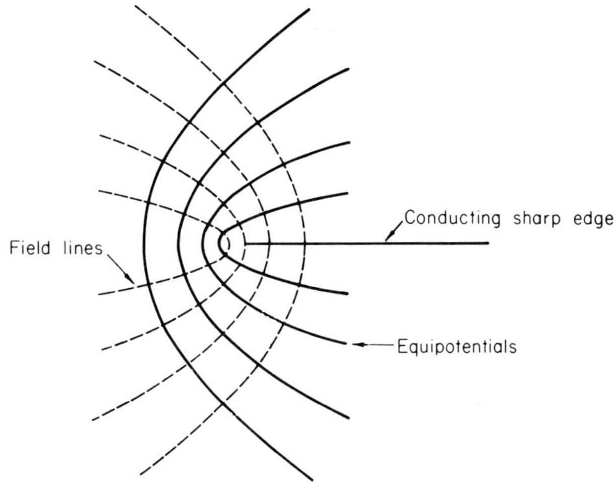

Fig. 3.2. Field lines and equipotentials in the vicinity of a conducting sharp edge.

There are numerous examples of this kind. We shall next develop mathematical expressions for special cases in detail. This will serve two purposes. It will show the reader how the above-mentioned methods of solving Laplace's equation are actually applied. In addition, the examples chosen are of great importance in electrostatic electron lenses and will frequently be used later on.

3.5 Axial Potential of Two Equidiameter Cylinders with Negligible Separation

In this section we shall derive an expression for the potential along the axis of two equidiameter cylinders with negligible separation between them.

Since we have rotational symmetry, Laplace's equation in cylindrical coordinates is

$$\frac{\partial^2\phi}{\partial r^2} + \frac{1}{r}\frac{\partial\phi}{\partial r} + \frac{\partial^2\phi}{\partial z^2} = 0 \tag{3.26}$$

We shall use the method of separation of variables. Let

$$\phi = R(r)Z(z) \tag{3.27}$$

We then have

$$\frac{1}{R}\frac{d^2R}{dr^2} + \frac{1}{rR}\frac{dR}{dr} = -\frac{1}{Z}\frac{d^2Z}{dz^2} \tag{3.28}$$

Since the left- and right-hand members of Eq. (3.28) are functions only of r and z, respectively, the equality is only possible if each member is equal to the same constant, say k^2. This leads to the general solution

$$\phi(r, z) = U + \sum_k [C_1(k)J_0(jkr) + C_2(k)N_0(jkr)] \times [b_1(k)\cos kz + b_2(k)\sin kz] \tag{3.29}$$

where U, C_1, C_2, b_1, and b_2 are arbitrary constants and $J_0(jkr)$ and $N_0(jkr)$ are Bessel's and Neumann's functions of order zero, respectively. The sum is for all possible values of k, real as well as imaginary.

The midplane between the two cylinders is a plane of symmetry for the electric field intensity and is therefore an equipotential surface. We take the origin of the z axis in that plane and assume that the potential there is $\frac{1}{2}(V_1 + V_2)$. We thus have the following boundary conditions at $r = R$ (radius of the cylinders):

$$\phi(R, z) = V_1 \qquad \text{for} \quad z < 0$$
$$\phi(R, z) = V_2 \qquad \text{for} \quad z > 0$$
$$\phi(R, 0) = \tfrac{1}{2}(V_1 + V_2) \qquad \text{for} \quad z = 0$$

Now for $r = 0$, $z \neq 0$, the potential should still be finite, and since $N_0(0) \to -\infty$ we must choose $C_2 = 0$ for all values of k. Also, $\phi(r, z)$ must remain finite for all values of z. Since $\phi(r, z)$ becomes infinitely large for large z if $k^2 < 0$, we eliminate this case by choosing $k^2 > 0$, in other words we assume only real values for k. If we let k change continuously rather than discretely, we can write

$$\phi(r, z) = U + \int_0^\infty C_1(k)J_0(jkr)[b_1(k)\cos kz + b_2(k)\sin kz]\, dk \tag{3.30}$$

In order to satisfy the boundary condition at $\phi(r, 0)$ where the potential is $\frac{1}{2}(V_1 + V_2)$, we find that the only way to achieve this is to let $U = \frac{1}{2}(V_1 + V_2)$ and $b_1(k) = 0$. The solution then becomes

$$\phi(r, z) = \tfrac{1}{2}(V_1 + V_2) + \int_0^\infty C(k)\, J_0\,(jkr)\sin kz\, dk \tag{3.31}$$

According to Fourier's integral theorem, a function

$$f(x) = \pi^{-1} \int_0^\infty du \int_{-\infty}^{+\infty} f(t)\cos u(t - x)\, dt \tag{3.32}$$

if odd, reduces to

$$f(x) = (2/\pi) \int_0^\infty \sin ux\, du \int_0^\infty f(t)\sin ut\, dt \tag{3.33}$$

If we let $x = k$, $u = z$, and $f(x) = C(x)J_0(jxr)$, we can write

$$f(k) = (2/\pi)\int_0^\infty \sin kz\, dz \int_0^\infty C(t)J_0(jtr)\sin tz\, dt \tag{3.34}$$

Now

$$\phi(r, z) - \tfrac{1}{2}(V_1 + V_2) = \int_0^\infty C(t)J_0(jtr)\sin tz\, dt \tag{3.35}$$

Therefore,

$$C(k)J_0(jkr) = (2/\pi)\int_0^\infty (\sin kz)\, dz \cdot [\phi(r, z) - \tfrac{1}{2}(V_1 + V_2)] \tag{3.36}$$

or

$$C(k) = \frac{2}{\pi J_0(jkr)}\int_0^\infty \sin kz\left[\phi(r, z) - \frac{V_1 + V_2}{2}\right] dz \tag{3.37}$$

Now for $r = R$, the radius of the cylinders,

$$C(k) = \frac{2}{\pi J_0(jkR)}\int_0^\infty \sin kz\left[\phi(R, z) - \frac{V_1 + V_2}{2}\right] dz \tag{3.38}$$

For $z > 0$, $\phi(R, z) = V_2$, hence

$$C(k) = \frac{2}{\pi J_0(jkR)}\int_0^\infty \sin kz\left(V_2 - \frac{V_1 + V_2}{2}\right) dz \tag{3.39}$$

The integral $\int_0^\infty \sin kz\, dz$ diverges. We should take

$$\lim_{\alpha\to 0}\int_0^\infty e^{-\alpha z}\sin kz\, dz = \lim_{\alpha\to 0}\frac{k}{\alpha^2 + k^2} = \frac{1}{k} \tag{3.40}$$

We then have

$$G(k) = \frac{V_2 - V_1}{\pi k J_0(jkR)} \tag{3.41}$$

The same value for $G(k)$ can be obtained by considering the condition for $z < 0$. We finally obtain

$$\phi(r, z) = \frac{V_1 + V_2}{2} + \frac{V_2 - V_1}{\pi}\int_0^\infty \frac{\sin kz}{k}\frac{J_0(jkr)}{J_0(jkR)}\, dk \tag{3.42}$$

On the axis, the potential is

$$\phi(0, z) = \frac{V_1 + V_2}{2} + \frac{V_2 - V_1}{\pi}\int_0^\infty \frac{\sin kz}{k}\frac{1}{J_0(jkR)}\, dk \tag{3.43}$$

We can readily see that these relations satisfy the boundary conditions, at $z = 0$ and at $r = R$ by remembering that

$$\frac{2}{\pi}\int_0^\infty \frac{\sin kz}{k}\,dk = \begin{cases} 1 & \text{for} \quad z > 0 \\ 0 & \text{for} \quad z = 0 \\ -1 & \text{for} \quad z < 0 \end{cases} \tag{3.44}$$

The axial potential distribution is as shown in Fig. 3.3. It resembles a tanh ωz distribution, where ω is a constant.

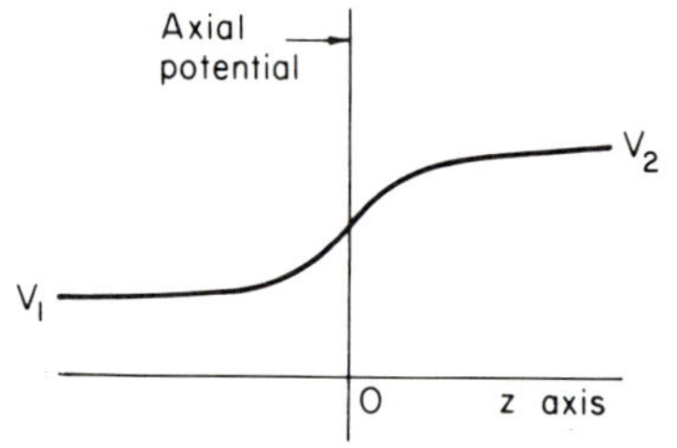

Fig. 3.3. The axial potential distribution of two coaxial cylinders separated by a negligibly small distance.

It seems reasonable, therefore, to assume that the following expression will give the potential distribution to a good approximation

$$\phi(0, z) = \frac{V_1 + V_2}{2} + \frac{V_2 - V_1}{2} \tanh \frac{\omega z}{R} \tag{3.45}$$

In order to determine the value of ω, we request that the derivatives with respect to z of Eqs. (3.45) and (3.43) be equal at the point $r = 0$, $z = 0$. Thus

$$\frac{V_2 - V_1}{\pi R}\int_0^\infty \frac{dk}{J_0(jk)} = \frac{V_2 - V_1}{2}\frac{\omega}{R} \tag{3.46}$$

or

$$\omega = \frac{2}{\pi}\int_0^\infty \frac{dk}{J_0(jk)} = 1.31835 \approx 1.32 \tag{3.47}$$

Using a digital computer, the right-hand and left-hand sides of the following equation were evaluated for various values of z.

$$\int_0^\infty \frac{\sin kz}{k}\frac{1}{J_0(jkR)}\,dk \approx \frac{\pi}{2}\tanh \frac{1.32}{R} z \tag{3.48}$$

Table 3.1 lists these values, their differences, and the percentage error. In this case $R = 1$ and z is taken in units of R. The following equation therefore gives the potential along the axis with a reasonable degree of accuracy.

$$\phi(0, z) = \frac{V_1 + V_2}{2} + \frac{V_2 - V_1}{2} \tanh \frac{1.32}{R} z \tag{3.49}$$

TABLE 3.1

z	Eq. (3.48) left-hand	Eq. (3.48) right-hand	Difference column 2 − column 3	Error (%)
0.2	0.406502	0.405318	0.001184	0.29
0.4	0.759237	0.760033	−0.000795	−0.10
0.6	1.03068	1.03601	−0.005325	−0.52
0.8	1.22227	1.23171	−0.009442	−0.77
1.0	1.35011	1.36154	−0.011437	−0.85
1.2	1.43253	1.44392	−0.011392	−0.80
1.4	1.4847	1.49471	−0.010004	−0.67
1.6	1.51738	1.52547	−0.008089	−0.53
1.8	1.53769	1.54391	−0.006216	−0.40
2.0	1.55031	1.55488	−0.004572	−0.29
2.2	1.55814	1.56139	−0.003256	−0.21
2.4	1.56295	1.56524	−0.002291	−0.15
2.6	1.56595	1.56752	−0.001572	−0.10
2.8	1.56781	1.56887	−0.001055	−0.07
3.0	1.56894	1.56966	−0.000719	−0.05

In Tables 3.2 and 3.3 a comparison is also made between the first derivatives of the left-hand and right-hand sides of Eq. (3.48) and between their second derivatives. We can see that wherever the error is large, the values are negligibly small.

TABLE 3.2

z	First derivative of left-hand Eq. (3.48)	First derivative of right-hand Eq. (3.48)	Difference column 2 − column 3	Error (%)
0	2.08326	2.07345	0.009808	0.47
0.2	1.93454	1.93540	−0.000855	−0.04
0.4	1.56953	1.58803	−0.018503	−1.18
0.6	1.14775	1.17151	−0.023762	−2.07
0.8	0.782622	0.798575	−0.015953	−2.04
1.0	0.511255	0.515637	−0.004382	−0.86
1.2	0.325820	0.321440	0.004380	1.34
1.4	0.204783	0.196022	0.008762	4.28
1.6	0.127728	0.117942	0.009786	7.66
1.8	0.079345	0.070391	0.008953	11.28
2.0	0.049174	0.041809	0.007365	14.98
2.2	0.030444	0.024761	0.005683	18.67
2.4	0.018832	0.014640	0.004192	22.26
2.6	0.011648	0.008647	0.000300	25.77
2.8	0.007200	0.005104	0.002096	29.16
3.0	0.004453	0.003012	0.001442	32.37

TABLE 3.3

z	Second derivative of right-hand Eq. (3.48)	Second derivative of left-hand Eq. (3.48)	Difference column 2 − column 3	Error (%)
0	0	0	0	0
0.2	−1.41007	−1.31840	−0.091661	6.50
0.4	−2.09473	−2.02849	−0.066238	3.16
0.6	−2.02696	−2.03982	−0.012861	−0.63
0.8	−1.59680	−1.65313	0.056328	−3.53
1.0	−1.12603	−1.17994	0.053907	−4.79
1.2	−0.747482	−0.780054	0.032572	−4.36
1.4	−0.480192	−0.492427	0.012235	−2.55
1.6	−0.303040	−0.302382	−0.000658	0.22
1.8	−0.189425	−0.182651	−0.006773	3.58
2.0	−0.117803	−0.109256	−0.008547	7.26
2.2	−0.073050	−0.064977	−0.008072	11.05
2.4	−0.045247	−0.038512	−0.006735	14.89
2.6	−0.027983	−0.022780	−0.005203	18.59
2.8	−0.017323	−0.013458	−0.003865	22.31
3.0	−0.010694	−0.000795	−0.002748	25.70

3.6 Axial Potential of Two Coaxial Equidiameter Cylinders Separated by a Distance

We shall now develop an expression that will give the potential along the axis of two coaxial equidiameter cylinders separated by a distance s.

In what follows, we shall show that

$$\phi(r, z) = \frac{1}{2\pi} \int_{-\infty}^{+\infty} a_k J_0(jkr) e^{jkz} \, dk \tag{3.50}$$

in which J_0 is the normal Bessel function of the first kind of zero order, is a solution to Laplace's equation: $\nabla^2 \phi = 0$.

Proof: The Taylor expansion for the potential in a rotationally symmetrical field is

$$\phi = \phi(r, z) = \phi(0, z) + \frac{r^2}{2!} \frac{\partial^2 \phi}{\partial r^2}\bigg|_{r=0} + \frac{r^4}{4!} \frac{\partial^4 \phi}{\partial r^4}\bigg|_{r=0} + \cdots \tag{3.51}$$

The following relations for Bessel's function of order zero are known.

$$J_0(x) = \sum_{n=0}^{\infty} \frac{(-1)^n}{(n!)^2} \left(\frac{x}{2}\right)^{2n} = 1 - \frac{x^2}{2^2} + \frac{x^4}{(2!)^2 2^4} - \frac{x^6}{(3!)^2 2^6} + \frac{x^8}{(4!)^2 2^8} - \cdots \tag{3.52}$$

$$\left.\frac{d^{2n}J_0(x)}{dx^{2n}}\right|_{x=0} = \frac{(-1)^n}{(n!)^2}\frac{(2n)!}{2^{2n}} \tag{3.53}$$

Let

$$x = jkr \tag{3.54}$$

then

$$\left.\frac{d^{2n}J_0(jkr)}{dr^{2n}}\right|_{r=0} = (-1)^n k^{2n}\frac{(-1)^n}{(n!)^2}\frac{(2n)!}{2^{2n}} = \frac{k^{2n}(2n)!}{(n!)^2 2^{2n}} \tag{3.55}$$

Now, $J_0(0) = 1$, therefore

$$\phi(0, z) = \frac{1}{2\pi}\int_{-\infty}^{+\infty} a_k e^{jkz}\, dk = f(z) \tag{3.56}$$

$$\left.\frac{\partial^{2n}\phi(r, z)}{\partial r^{2n}}\right|_{r=0} = \frac{1}{2\pi}\int_{-\infty}^{+\infty} a_k \frac{k^{2n}(2n)!}{(n!)^2 2^{2n}} e^{jkz}\, dk \tag{3.57}$$

$$\frac{d^{2n}f(z)}{dz^{2n}} = \frac{1}{2\pi}\int_{-\infty}^{+\infty}(jk)^{2n} a_k e^{jkz}\, dk = (-1)^n \frac{1}{2\pi}\int_{-\infty}^{+\infty} k^{2n} a_k e^{jkz}\, dk \tag{3.58}$$

Thus

$$\left.\frac{\partial^{2n}\phi(r, z)}{\partial r^{2n}}\right|_{r=0} = \frac{1}{(-1)^n}\cdot\frac{(2n)!}{(n!)^2 2^{2n}}\cdot\frac{d^{2n}f(z)}{dz^{2n}} = (-1)^n\frac{(2n)!}{(n!)^2 2^{2n}}\cdot\frac{d^{2n}f(z)}{dz^{2n}} \tag{3.59}$$

and

$$\begin{aligned}\phi(r, z) = f(z) &+ \frac{r^2}{2!}(-1)^1\frac{2!}{1^2 2^2}\frac{d^2 f(z)}{dz^2} + \frac{r^4}{4!}(-1)^2\frac{4!}{(2!)^2 2^4}\frac{d^4 f(z)}{dz^4} \\ &+ \cdots + (-1)^n\frac{(2n)!}{(n!)^2 2^{2n}}\frac{r^{2n}}{(2n)!}\frac{d^{2n}f(z)}{dz^{2n}}\end{aligned} \tag{3.60}$$

or

$$\begin{aligned}\phi(r, z) = f(z) &- \left(\frac{r}{2}\right)^2\frac{d^2 f(z)}{dz^2} + \frac{1}{(2!)^2}\left(\frac{r}{2}\right)^4\frac{d^4 f(z)}{dz^4} \\ &+ \cdots + (-1)^n\frac{1}{(n!)^2}\left(\frac{r}{2}\right)^{2n}\frac{d^{2n}f(z)}{dz^{2n}}\end{aligned} \tag{3.61}$$

This equation is identical to Eq. (2.30). Hence, we have shown that Eq. (3.50) is indeed a solution to Laplace's equation.

To determine the solution we have to choose $a(k)$ so that $\phi(r, z)$ satisfies the boundary conditions.

Let us consider a particular case in which the potential ϕ is known along a cylinder coaxial with the z axis in our coordinate system, and with radius R. Thus

$$\phi(R, z) = \frac{1}{2\pi}\int_{-\infty}^{+\infty} a(k)J_0(jkR)e^{jkz}\,dk \tag{3.62}$$

is known.

Fourier's integral theorem states: If a function $G(z)$ is given, we can form the function

$$F(jk) = \int_{-\infty}^{+\infty} G(z)e^{-jkz}\,dz \tag{3.63}$$

which is called its Fourier transform, and inversely we have

$$G(z) = \phi(R, z) = \frac{1}{2\pi}\int_{-\infty}^{+\infty} F(jk)e^{jkz}\,dk \tag{3.64}$$

Thus, with

$$F(jk) = a(k)J_0(jkR) \tag{3.65}$$

we have

$$a(k)J_0(jkR) = \int_{-\infty}^{+\infty} \phi(R, z)e^{-jkz}\,dz \tag{3.66}$$

If R is taken as unit of length

$$a(k)J_0(jk) = \int_{-\infty}^{+\infty} \phi(1, z)e^{-jkz}\,dz \tag{3.67}$$

from which

$$a(k) = [J_0(jk)]^{-1}\int_{-\infty}^{+\infty} \phi(1, z)e^{-jkz}\,dz \tag{3.68}$$

We then have

$$\phi(r, z) = (2\pi)^{-1}\int_{-\infty}^{+\infty} [J_0(jk)]^{-1}\int_{-\infty}^{+\infty} \phi(1, \zeta)e^{-jk\zeta}\,d\zeta\, J_0(jkr)e^{jkz}\,dk \tag{3.69}$$

or

$$\phi(r, z) = (2\pi)^{-1}\int_{-\infty}^{+\infty}\int_{-\infty}^{+\infty} \phi(1, \zeta)[J_0(jkr)/J_0(jk)]e^{jk(z-\zeta)}\,dk\,d\zeta \tag{3.70}$$

The following boundary conditions are considered in this case. For $r = R$,

$$\text{potential} = 0 \qquad \text{for} \quad z < \zeta \quad \text{and for} \quad z > \zeta + d\zeta$$
$$\text{potential} = (d\zeta)^{-1} \qquad \text{for} \quad \zeta < z < \zeta + d\zeta.$$

This is shown in Fig. 3.4. Using these boundary conditions in Eq. (3.66), we have

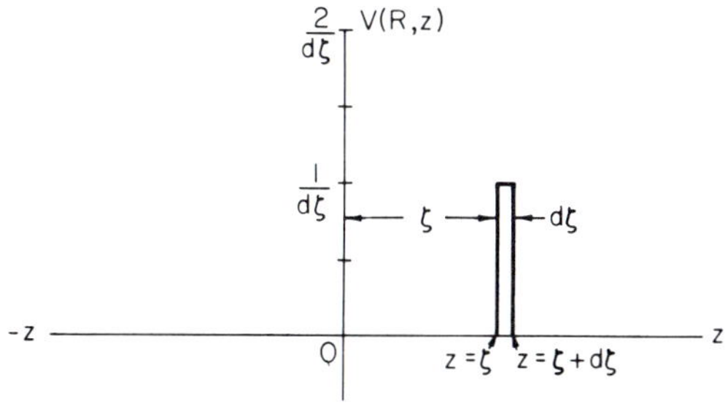

Fig. 3.4. The boundary condition at $r = R$. Potential $= 0$ at $z < \zeta$ and $z > \zeta + d\zeta$. Potential $= 1/d\zeta$ for $\zeta < z < \zeta + d\zeta$.

$$a(k)J_0(jkR) = \int_{\zeta}^{\zeta + d\zeta} \frac{1}{d\zeta} e^{-jkz}\, dz = -\frac{1}{jk}\frac{1}{d\zeta}(e^{-jkz})\Big|_{\zeta}^{\zeta + d\zeta}$$
$$= \frac{1}{d\zeta}\frac{j}{k}(e^{-jk(\zeta + d\zeta)} - e^{-jk\zeta}) \tag{3.71}$$

or

$$a(k)J_0(jkR) = \frac{j}{k\, d\zeta} e^{-jk\zeta}(e^{-jkd\zeta} - 1) \tag{3.72}$$

Expanding the exponential power into a series and neglecting second and higher order terms, we have

$$a(k)J_0(jkR) \cong -\frac{j}{k}\frac{jk\, d\zeta}{d\zeta} e^{-jk\zeta} = e^{-jk\zeta} \tag{3.73}$$

Thus

$$a(k) = \frac{e^{-jk\zeta}}{J_0(jkR)} \tag{3.74}$$

For this particular case the solution is therefore

$$\phi(r, z) = \frac{1}{2\pi}\int_{-\infty}^{+\infty} \frac{e^{-jk\zeta}}{J_0(jkR)} J_0(jkr)e^{jkz}\, dk \tag{3.75}$$

Since we are expressing everything in terms of $R = 1$

$$\phi(r, z) = \frac{1}{2\pi} \int_{-\infty}^{+\infty} \frac{e^{-jk\zeta}}{J_0(jk)} J_0(jkr) e^{jkz} \, dk \tag{3.76}$$

Along the z axis, $r = 0$

$$\phi(0, z) = \frac{1}{2\pi} \int_{-\infty}^{+\infty} \frac{e^{-jk\zeta}}{J_0(jk)} J_0(0) e^{jkz} \, dk = \frac{1}{2\pi} \int_{-\infty}^{+\infty} \frac{e^{jk(z-\zeta)}}{J_0(jk)} \, dk \tag{3.77}$$

since $J_0(0) = 1$.

We introduce a new variable $\xi = z - \zeta$. We then have

$$\phi(0, z) = (2\pi)^{-1} \int_{-\infty}^{+\infty} e^{jk\xi} / J_0(jk) \, dk \tag{3.78}$$

The modified Bessel function can be expanded as

$$J_0(jk) = \sum_{\nu=0}^{\infty} \frac{k^{2\nu}}{2^{2\nu}(\nu!)^2} = 1 + \frac{k^2}{4} + \frac{k^4}{64} + \frac{k^6}{2304} + \cdots \tag{3.79}$$

and $J_0(jk)$ has its minimum value at $k = 0$. Thus $[J_0(jk)]^{-1}$ has its maximum value there. Also, $[J_0(jk)]^{-1}$ is symmetrical with respect to $k = 0$. Therefore, it can be expressed in a series of cosine terms

$$[J_0(jk)]^{-1} = \sum_{n=0}^{\infty} A_n \cos \tfrac{1}{2} nk \tag{3.80}$$

in the interval from -2π to $+2\pi$. At $k = \pm 2\pi$, $J_0(jk) \approx 87$. Therefore, $[J_0(\pm 2\pi)]^{-1}$ is sufficiently small to be neglected.

We are thus justified in using $\pm 2\pi$ instead of $\pm\infty$ in the limits of integration.

Next, we expand $[J_0(jk)^{-1}]$ into a Fourier series. In Eq. (3.80),

$$A_0 = (4\pi)^{-1} \int_{-2\pi}^{+2\pi} [J_0(jk)]^{-1} \, dk \tag{3.81}$$

and

$$A_n = (2\pi)^{-1} \int_{-2\pi}^{+2\pi} [J_0(jk)]^{-1} \cos \tfrac{1}{2} nk \, dk \tag{3.82}$$

Equation (3.78) can now be written as

$$\phi(0, z) = (2\pi)^{-1} \int_{-2\pi}^{+2\pi} e^{jk\xi} \sum_{n=0}^{\infty} A_n \cos \tfrac{1}{2} nk \, dk \tag{3.83}$$

and since n is independent of k, we can write

$$\phi(0, z) = (2\pi)^{-1} \sum_{n=0}^{\infty} \int_{-2\pi}^{+2\pi} A_n (\cos \tfrac{1}{2}nk)(\cos k\xi + j \sin k\xi)\, dk$$

$$= (2\pi)^{-1} \sum_{n=0}^{\infty} A_n \int_{-2\pi}^{+2\pi} \{\tfrac{1}{2}[\cos(\xi + \tfrac{1}{2}n)k + \cos(\xi - \tfrac{1}{2}n)k]$$

$$+ \tfrac{1}{2}j[\sin(\xi + \tfrac{1}{2}n)k + \sin(\xi - \tfrac{1}{2}n)k]\}\, dk \tag{3.84}$$

Since the sine function is odd it does not contribute to the integral, and we have

$$\begin{aligned}\phi(0, z) &= \frac{1}{4\pi} \sum_{n=0}^{\infty} A_n \int_{-2\pi}^{+2\pi} \left[\cos\left(\xi + \frac{n}{2}\right)k + \cos\left(\xi - \frac{n}{2}\right)k\right] dk \\ &= \frac{1}{4\pi} \sum_{n=0}^{\infty} A_n \left[\frac{\sin(\xi + \frac{1}{2}n)k}{\xi + \frac{1}{2}n} + \frac{\sin(\xi - \frac{1}{2}n)k}{\xi - \frac{1}{2}n}\right]_{-2\pi}^{+2\pi}\end{aligned} \tag{3.85}$$

which after simplification yields

$$\phi(0, z) = (\xi/\pi)(\sin 2\pi\xi) \sum_{n=0}^{\infty} A_n(-1)^n(\xi^2 - \tfrac{1}{4}n^2)^{-1} \tag{3.86}$$

For $\xi = 0$, and $n = 0$ we have

$$\phi(0, z) = \lim_{\xi \to 0} A_0\xi (\sin 2\pi\xi)/\pi\xi^2 = 2A_0 \tag{3.87}$$

and for $\xi = \pm\frac{1}{2}m$, $m = 1, \ldots,$ we have

$$\phi(0, z) = \lim_{\xi \to \pm\frac{1}{2}m} \frac{\xi}{\pi}(-1)^m A_m \frac{\sin 2\pi\, \xi}{\xi^2 - \frac{1}{4}m^2} = A_m \tag{3.88}$$

L'Hôpital's rule was used in both cases. Numerical values for the coefficients A_0, A_1, and so forth, were obtained using Eqs. (3.81) and (3.82) with the help of a digital computer. Table 3.4 lists some of these values. Beyond A_6 the coefficients were negligible. Using Eq. (3.86), $\phi(0, z)$ was obtained for

TABLE 3.4

A_0	A_1	A_2	A_3	A_4	A_5	A_6
0.330387	0.436502	0.162508	0.054193	0.016555	0.006774	0.002673

various values of $\xi = z - \zeta$ with the help of a digital computer. The results are plotted in Fig. 3.5 and are also listed in Table 3.5.

We must remember that this represents the axial potential due to a differential cylinder element at $z = \zeta$, and $r = R = 1$. In other words, for a potential element at $z = \zeta$, the value of the potential along the z axis due to the element shown in Fig. 3.4 is given by Fig. 3.5. As we mentioned earlier, this distribution is symmetrical about the ordinate $z = \zeta$, and therefore only the values of $\xi = 0$ to ∞ are plotted.

TABLE 3.5

ξ	$\phi_1(0, z)$ according to Eq. (3.86)	$\phi_2(0, z)$ according to Eq. (3.89)	$\phi_2 - \phi_1$	$100(\phi_2 - \phi_1)/\phi_1$
0	0.660774	0.660000	−0.000774	−0.12
0.1	0.649494	0.648632	−0.000862	−0.13
0.2	0.616971	0.616057	−0.000914	−0.15
0.3	0.566908	0.566433	−0.000475	−0.08
0.4	0.504703	0.505486	+0.000783	+0.16
0.5	0.436502	0.439227	+0.002725	+0.63
0.6	0.368146	0.372904	+0.004758	+1.29
0.7	0.304349	0.310418	+0.006069	+1.99
0.8	0.248188	0.254194	+0.006006	+2.42
0.9	0.200971	0.205373	+0.004401	+2.19
1.0	0.162508	0.164133	+0.001625	+1.00
1.1	0.131592	0.130038	−0.001554	−1.18
1.2	0.106606	0.102318	−0.004288	−4.02
1.3	0.086035	0.080071	−0.005964	−6.93
1.4	0.068766	0.062396	−0.006370	−9.26
1.5	0.054193	0.048461	−0.005732	−10.58
1.6	0.042098	0.037542	−0.004559	−10.82
1.7	0.032448	0.029025	−0.003423	−10.54
1.8	0.025177	0.022406	−0.002771	−11.00
1.9	0.020032	0.017276	−0.002756	−13.75
2.0	0.016555	0.013308	−0.003247	−19.61
2.1	0.014158	0.010244	−0.003914	−27.64
2.2	0.012274	0.007882	−0.004392	−35.78
2.3	0.010493	0.006061	−0.004432	−42.23
2.4	0.008641	0.004660	−0.003982	−46.07
2.5	0.006774	0.003582	−0.003192	−47.12
2.6	0.005087	0.002752	−0.002335	−45.89
2.7	0.003796	0.002115	−0.001681	−44.28
2.8	0.003016	0.001625	−0.001392	−46.14
2.9	0.002706	0.001248	−0.001458	−53.87
3.0	0.002673	0.000959	−0.001714	−64.12

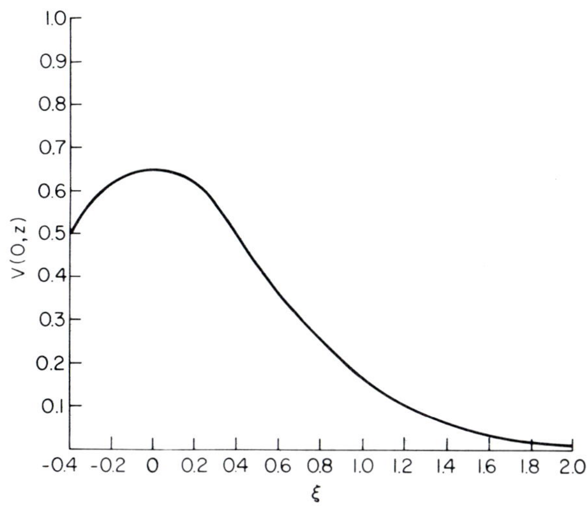

Fig. 3.5. Distribution of the potential due to one element shown in Fig. 3.4. This is also closely approximated by Eq. (3.89).

Using the same value for ω as obtained earlier, we found that these values are very closely matched by the function

$$\phi(0, z) = (\tfrac{1}{2}\omega) \operatorname{sech}^2 \omega\xi \qquad \text{for} \quad \omega = 1.32 \tag{3.89}$$

Table 3.5 lists the values of $\phi(0, z)$ as obtained from Eqs. (3.86) and (3.89). We can easily see that the match is sufficiently close for many practical applications.

If we know the potential at $r = 1$, $\phi(1, z)$, the potential on the axis can be obtained by multiplying the differential cylinder function corresponding to a given value $z = \zeta$ at $r = 1$ by the potential for this value, $\phi(1, \zeta)$ and summing over all values of ζ. We then have

$$\phi(0, z) = \tfrac{1}{2}\omega \int_{-\infty}^{+\infty} \phi(1, \zeta) \operatorname{sech}^2 \omega\xi \, d\zeta \tag{3.90}$$

where $\omega = 1.32$.

A direct application of this method is the evaluation of the axial potential distribution for a system consisting of two electrodes of equal diameter, infinitely long and separated by a distance s as shown in Fig. 3.6.

The potentials of the two cylinders are V_1 and V_2, and the radius is chosen as unity. We shall assume that the potential in the gap varies linearly at $r = 1$.

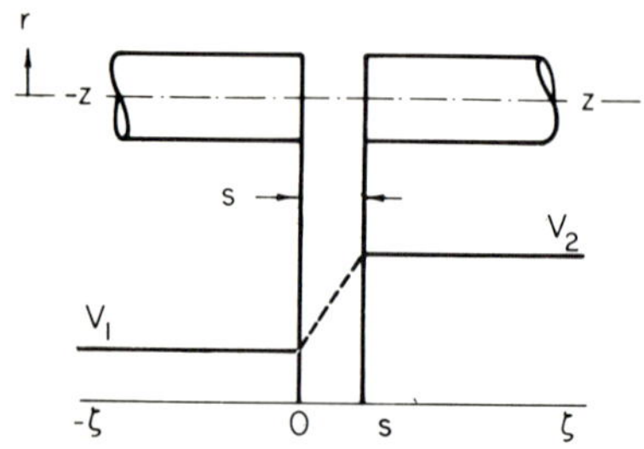

Fig. 3.6. Two concentric cylinders of equal diameter separated by a distance s. It is assumed that the potential varies linearly between the two cylinders at $r = R$.

Thus, we have the following boundary conditions:

$$\phi(1, \zeta) = V_1 \quad \text{at} \quad \zeta < 0 \tag{3.91}$$

$$\phi(1, \zeta) = V_1 + [(V_2 - V_1)/s]\zeta \quad \text{at} \quad 0 < \zeta < s \tag{3.92}$$

$$\phi(1, \zeta) = V_2 \quad \text{at} \quad \zeta > s \tag{3.93}$$

Also $z - \zeta = \xi$, and $d\zeta = -d\xi$.

Substituting these values in Eq. (3.90), we obtain

$$\phi(0, z) = \frac{\omega}{2}\left\{V_1 \int_{-\infty}^{0} \operatorname{sech}^2 \omega\xi \, d\zeta + \int_0^s \left[V_1 + \left(\frac{V_2 - V_1}{s}\right)\zeta\right] \operatorname{sech}^2 \omega\xi \, d\zeta \right.$$
$$\left. + V_2 \int_s^{+\infty} \operatorname{sech}^2 \omega\xi \, d\zeta\right\} \tag{3.94}$$

or

$$\phi(0, z) = \frac{-\omega}{2}\left\{V_1 \int_{\infty}^{z} \operatorname{sech}^2 \omega\xi \, d\xi \right.$$
$$+ \int_z^{z-s} \left[V_1 + \left(\frac{V_2 - V_1}{s}\right)(z - \xi)\right] \operatorname{sech}^2 \omega\xi \, d\xi$$
$$\left. + V_2 \int_{z-s}^{-\infty} \operatorname{sech}^2 \omega\xi \, d\xi\right\} \tag{3.95}$$

which after integration yields

$$\phi(0, z) = -\frac{1}{2}\left\{(\tanh \omega\xi)\, V_1 \Big|_{\infty}^{z} + (\tanh \omega\xi) \left[V_1 + \left(\frac{V_2 - V_1}{s}\right)z\right] \Big|_z^{z-s} \right.$$
$$\left. + (\tanh \omega\xi) V_2 \Big|_{z-s}^{-\infty} - \frac{V_2 - V_1}{\omega s}\left(\omega\xi \tanh \omega\xi - \log\cosh \omega\xi\right)\Big|_s^{z-s}\right\} \tag{3.96}$$

for which we finally obtain

$$\phi(0, z) = \frac{V_1 + V_2}{2} + \left(\frac{V_2 - V_1}{2\omega s}\right) \log \frac{\cosh \omega z}{\cosh \omega(z - s)} \tag{3.97}$$

If the separation s is very small, we take the limit of Eq. (3.97) for $s \to 0$ and obtain

$$\phi(0, z) = \frac{V_1 + V_2}{2} + \left(\frac{V_2 - V_1}{2}\right) \tanh \omega z \tag{3.98}$$

For $\omega = 1.32$ this is identical to Eq. (3.49) with $R = 1$.

Equation (3.97) is accurate despite the fact that the analysis was made assuming a linear distribution at $r = R$ between the cylinders. Even for a separation s equal to the diameter of the cylinders the results are accurate enough for many applications. This was shown to be true by comparing the values obtained with this formula with those obtained using the electrolytic tank and the digital computer. In the latter, Laplace's equation was solved by the relaxation method.

3.7 An Empirical Relation

The following empirical formula for the potential along the axis has been suggested

$$\phi(0, z) = \frac{V_1 + V_2}{2} + \frac{V_2 - V_1}{b} \int_0^z \exp(-(\pi/b^2)z^2)\, dz \tag{3.99}$$

where

$$b = 2R[0.73 + 0.53(s/2R)^2] \tag{3.100}$$

Here s is again the spacing, and R the radius of the cylinders. The above relation is valid as long as $s/2R < 1.75$. The accuracy is of the order of 5%.

3.8 Potential Distribution Due to a Circular Hole Separating Two Uniform Fields

In what follows, we shall derive an expression for the potential for a particular case which has wide applications. We shall be using this expression later on when analyzing certain types of lenses. The solution in this case is mathematically possible if an appropriate selection of coordinates is made.

In Fig. 3.7, we have a circular opening in the xOy plane. The electrode

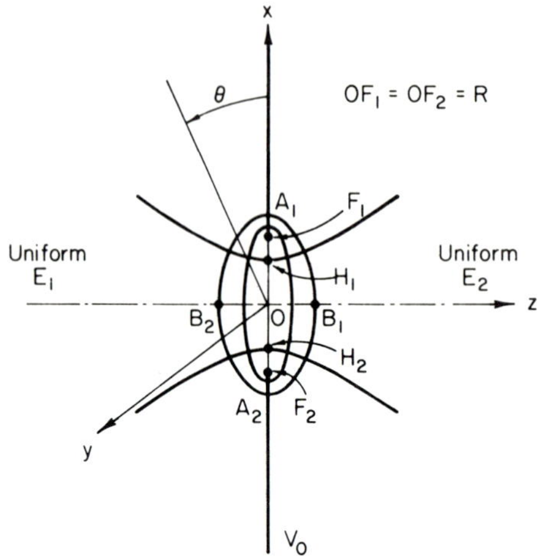

Fig. 3.7. A circular aperture separating two uniform fields. The electrode is at potential V_0 and is assumed very thin.

in that plane is assumed to be very thin and at a potential V_0. This electrode separates two uniform fields. At large distances from the left and the right of the center of the opening, the fields are assumed to be equal to E_1 and E_2, respectively.

In the xOz plane, the equation of an ellipse with foci F_1 and F_2 on the x axis is

$$\frac{x^2}{a^2} + \frac{z^2}{b^2} = 1, \qquad y = 0 \tag{3.101}$$

in which $a > b$ and $a^2 - b^2 = R^2$ if $OF_1 = OF_2 = R$. Therefore, R is the distance of the foci from the origin O. In our diagram, Fig. 3.7, $OA_1 = OA_2 = a$ and $OB_1 = OB_2 = b$. If we let $b = uR$, then $a^2 - u^2R^2 = R^2$ or $a^2 = R^2(1 + u^2)$. The equation of the ellipse can then be written as

$$\frac{x^2}{R^2(1+u^2)} + \frac{z^2}{R^2u^2} = 1, \qquad y = 0 \tag{3.102}$$

The ellipse is real for $u^2 \geq 0$.

If we let this ellipse rotate with respect to the z axis, we obtain an ellipsoid with the following equation:

$$\frac{x^2}{R^2(1+u^2)} + \frac{y^2}{R^2(1+u^2)} + \frac{z^2}{R^2u^2} = 1 \tag{3.103}$$

which in cylindrical coordinates can be written as

$$\frac{r^2}{R^2(1+u^2)} + \frac{z^2}{R^2u^2} = 1 \tag{3.104}$$

The equation of a hyperbola in the xOz plane with foci F_1 and F_2 is

$$\frac{x^2}{c^2} - \frac{z^2}{d^2} = 1, \qquad y = 0 \tag{3.105}$$

Here, $OH_1 = OH_2 = c$, and $d^2 = R^2 - c^2$. Let $d = vR$, then $c^2 = (1 - v^2)R^2$, and Eq. (3.105) can be written as

$$\frac{x^2}{R^2(1-v^2)} - \frac{z^2}{R^2v^2} = 1, \qquad y = 0 \tag{3.106}$$

If we let the hyperbola rotate with respect to the z axis, we obtain a rotational hyperboloid of one sheet. The equation of this hyperboloid is

$$\frac{x^2 + y^2}{R^2(1-v^2)} - \frac{z^2}{R^2v^2} = 1 \tag{3.107}$$

which in cylindrical coordinates is

$$\frac{r^2}{R^2(1-v^2)} - \frac{z^2}{R^2v^2} = 1 \tag{3.108}$$

If $v^2 \geq 0$ and $1 - v^2 \geq 0$, or $v^2 \leq 1$, the hyperbola has the x axis as its real axis.

We now use a coordinate system with u, v, and θ as variables. The surfaces u = constant are the rotational ellipsoids, and the surfaces v = constant are the rotational hyperboloids of one sheet. These were discussed earlier. The surfaces θ = constant are planes containing the z axis. The three sets of surfaces are mutually orthogonal. In Eqs. (3.104) and (3.108), if we let $u = 0$, we get

$$z^2 = 0 \qquad \text{and} \qquad r^2v^2 = R^2v^2(1 - v^2) \tag{3.109}$$

or

$$r^2 = R^2(1 - v^2) \tag{3.110}$$

and since $0 \leq v^2 \leq 1$, then $r^2 \leq R^2$ and $r \leq R$. Therefore, the surface $u = 0$ represents the disc in the $z = 0$ plane which covers the opening in the electrode.

Similarly, if $v = 0$, we have

$$r^2 = R^2(1 + u^2) \tag{3.111}$$

and $r^2 \geq R^2$ or $r \geq R$.

Therefore, $v = 0$ represents the surface of the electrode, excluding the aperture. This is the reason why we have selected these sets of coordinates.

The potential obeys Laplace's equation.

$$\frac{\partial^2 \phi}{\partial r^2} + \frac{1}{r}\frac{\partial \phi}{\partial r} + \frac{\partial^2 \phi}{\partial z^2} = 0 \tag{3.112}$$

We shall now express Laplace's equation in terms of u and v. From Eq. (3.104) and Eq. (3.108), we can derive r and z as expressions of u and v. We write the two equations in the following form:

$$r^2u^2 + z^2(1 + u^2) = R^2u^2(1 + u^2) \tag{3.113}$$

$$r^2v^2 - z^2(1 - v^2) = R^2v^2(1 - v^2) \tag{3.114}$$

from which we obtain

$$z = Ruv \tag{3.115}$$

$$r = R[(1 + u^2)(1 - v^2)]^{1/2} \tag{3.116}$$

Now

$$\frac{\partial \phi}{\partial r} = \frac{\partial \phi}{\partial u}\frac{\partial u}{\partial r} + \frac{\partial \phi}{\partial v}\frac{\partial v}{\partial r} \tag{3.117}$$

and

$$\begin{aligned}\frac{\partial^2 \phi}{\partial r^2} &= \frac{\partial}{\partial r}\left(\frac{\partial \phi}{\partial u}\frac{\partial u}{\partial r} + \frac{\partial \phi}{\partial v}\frac{\partial v}{\partial r}\right) \\ &= \frac{\partial^2 \phi}{\partial u^2}\left(\frac{\partial u}{\partial r}\right)^2 + 2\frac{\partial^2 \phi}{\partial u\,\partial v}\frac{\partial u}{\partial r}\frac{\partial v}{\partial r} \\ &\quad + \frac{\partial^2 \phi}{\partial v^2}\left(\frac{\partial v}{\partial r}\right)^2 + \frac{\partial \phi}{\partial u}\frac{\partial^2 u}{\partial r^2} + \frac{\partial \phi}{\partial v}\frac{\partial^2 v}{\partial r^2}\end{aligned} \tag{3.118}$$

and similarly for $\partial\phi/\partial z$ and $\partial^2\phi/\partial z^2$. Laplace's equation can now be written as follows:

$$\begin{aligned}&\frac{\partial^2 \phi}{\partial u^2}\left[\left(\frac{\partial u}{\partial r}\right)^2 + \left(\frac{\partial u}{\partial z}\right)^2\right] + 2\frac{\partial^2 \phi}{\partial u\,\partial v}\left(\frac{\partial u}{\partial r}\frac{\partial v}{\partial r} + \frac{\partial u}{\partial z}\frac{\partial v}{\partial z}\right) + \frac{\partial^2 \phi}{\partial v^2}\left[\left(\frac{\partial v}{\partial r}\right)^2 + \left(\frac{\partial v}{\partial z}\right)^2\right] \\ &\quad + \frac{1}{r}\left[\frac{\partial \phi}{\partial u}\frac{\partial u}{\partial r} + \frac{\partial \phi}{\partial v}\frac{\partial v}{\partial r}\right] + \frac{\partial \phi}{\partial u}\left(\frac{\partial^2 u}{\partial r^2} + \frac{\partial^2 u}{\partial z^2}\right) + \frac{\partial \phi}{\partial v}\left(\frac{\partial^2 v}{\partial r^2} + \frac{\partial^2 v}{\partial z^2}\right) = 0\end{aligned} \tag{3.119}$$

We solve for $\partial u/\partial r$, $\partial v/\partial r$, $\partial u/\partial z$, and $\partial v/\partial z$, and obtain

$$\frac{\partial u}{\partial r} = \frac{ru}{R^2(u^2 + v^2)} \tag{3.120}$$

$$\frac{\partial v}{\partial r} = \frac{-rv}{R^2(u^2 + v^2)} \tag{3.121}$$

$$\frac{\partial u}{\partial z} = \frac{(1 + u^2)v}{R(u^2 + v^2)} \tag{3.122}$$

and

$$\frac{\partial v}{\partial z} = \frac{u(1 - v^2)}{R(u^2 + v^2)} \tag{3.123}$$

We also obtain $\partial^2 u/\partial r^2$, $\partial^2 u/\partial z^2$, $\partial^2 v/\partial r^2$, and $\partial^2 v/\partial z^2$

$$\frac{\partial^2 u}{\partial r^2} = \frac{u}{R^2(u^2 + v^2)} + \frac{u(1 + u^2)(1 - v^2)(-u^2 + 3v^2)}{R^2(u^2 + v^2)^3} \tag{3.124}$$

$$\frac{\partial^2 u}{\partial z^2} = \frac{u(1 + u^2)}{R^2(u^2 + v^2)^2}\left[1 + v^2 - \frac{2v^2(2 + u^2 - v^2)}{u^2 + v^2}\right] \tag{3.124a}$$

$$\frac{\partial^2 v}{\partial r^2} = \frac{-v}{R^2(u^2 + v^2)} + \frac{v(1 + u^2)(1 - v^2)(3u^2 - v^2)}{R^2(u^2 + v^2)^3} \tag{3.125}$$

$$\frac{\partial^2 v}{\partial z^2} = \frac{v(1 - v^2)(1 - u^2)}{R^2(u^2 + v^2)^2} - \frac{2u^2 v(1 - v^2)(2 + u^2 - v^2)}{R^2(u^2 + v^2)^3} \tag{3.125a}$$

Substituting all these values and that of r from Eq. (3.116) in Laplace's equation, and after some manipulation, we finally obtain

$$(1 + u^2)\frac{\partial^2 \phi}{\partial u^2} + (1 - v^2)\frac{\partial^2 \phi}{\partial v^2} + 2u\frac{\partial \phi}{\partial u} - 2v\frac{\partial \phi}{\partial v} = 0 \tag{3.126}$$

To solve this equation, we shall use the method of separation of variables. We let

$$\phi = U(u)V(v) \tag{3.127}$$

then

$$\frac{1}{U}\frac{d}{du}\left\{(1 + u^2)\frac{dU}{du}\right\} + \frac{1}{V}\frac{d}{dv}\left\{(1 - v^2)\frac{dV}{dv}\right\} = 0 \tag{3.128}$$

The first term is a function of u only, the second a function of v. Each term is therefore equal to a constant which we select as k. Then

$$(1 + u^2)\frac{d^2 U}{du^2} + 2u\frac{dU}{du} - kU = 0 \tag{3.129}$$

and

$$(1 - v^2)\frac{d^2 V}{dv^2} - 2v\frac{dV}{dv} + kV = 0 \tag{3.130}$$

The value of k can be obtained from the boundary conditions. These are:

for $v = 0$, $\phi = V_0$, the potential on the electrode;

for $z \ll 0$, the field is uniform, and equal to E_1;

for $z \gg 0$, the field is uniform, equal to E_2.

In that case we must have

$$\lim_{z \to -\infty} \frac{\partial \phi}{\partial z} = -E_1 \qquad \text{and} \qquad \lim_{z \to \infty} \frac{\partial \phi}{\partial z} = -E_2$$

since $E = -\text{grad}\,\phi$.

We, therefore, require that for large values of z, the potential UV should be proportional to z. If we refer to Fig. 3.7, we can see that large values of z correspond to large values of u. Since $z = Ruv$ from Eq. (3.115), the potential UV, for large values of u, should be proportional to uv for arbitrary values of v. This means that V should be proportional to v. If we let $V = Av$ in Eq. (3.130), we obtain

$$(1 - v^2) \cdot 0 - 2vA + kAv = 0 \tag{3.131}$$

This is valid for $k = 2$. Hence, Eq. (3.129) can be written as

$$(1 + u^2)\frac{d^2U}{du^2} + 2u\frac{dU}{du} - 2U = 0 \tag{3.132}$$

A solution is $U = u$.

Since Eq. (3.129) is linear, we can now assume that the general solution is

$$U = uf(u) \tag{3.133}$$

Then

$$(1 + u^2)u\frac{d^2f(u)}{du^2} + 2(1 + 2u^2)\frac{df(u)}{du} = 0 \tag{3.134}$$

or

$$\frac{d^2f(u)/du^2}{df(u)/du} = -2\frac{1 + 2u^2}{(1 + u^2)u} \tag{3.135}$$

or

$$\frac{d}{du}\left(\log\frac{df(u)}{du}\right) = -2\frac{1 + 2u^2}{(1 + u^2)u} = -2\left(\frac{1}{u} + \frac{u}{1 + u^2}\right) \tag{3.136}$$

or

$$\log\frac{df(u)}{du} = -2\int\frac{du}{u} - 2\int\frac{u\,du}{1 + u^2} = -2\log u - \log(1 + u^2) + \log C_1 \tag{3.137}$$

where C_1 is a constant, and

$$\frac{df(u)}{du} = \frac{C_1}{u^2(1 + u^2)} \tag{3.138}$$

Thus

$$f(u) = \int \frac{C_1\,du}{u^2(1 + u^2)} = -\frac{C_1}{u} - C_1 \arctan u + C_2 \tag{3.139}$$

where C_2 is again a constant of integration.

$$U = uf(u) = -C_1 - C_1 u \arctan u + C_2 u \tag{3.140}$$

or

$$U = K_1 u + K_2(u \arctan u + 1) \tag{3.141}$$

Here, K_1 and K_2 are new constants.

The general solution is, therefore

$$\phi = UV = av[K_1 u + K_2(u \arctan u + 1)] + C \tag{3.142}$$

or

$$\phi = Auv + Bv(u \arctan u + 1) + C \tag{3.143}$$

where $B = aK_2$, $A = aK_1$, and C are all arbitrary constants. Their values depend on the boundary conditions. Now for $v = 0$, we must have $\phi = V_0$, thus $C = V_0$. From Eq. (3.115), we have

$$z/R = uv \tag{3.144}$$

and therefore

$$\phi = \frac{A}{R} z + \frac{B}{R} z\left(\arctan u + \frac{1}{u}\right) + V_0 \tag{3.145}$$

We still have to express u as a function of r and z. If we solve for v from Eq. (3.115) and substitute this in Eq. (3.116), we get

$$r^2 = R^2(1 + u^2)[1 - (z^2/R^2u^2)] \tag{3.146}$$

or

$$u^2 = \frac{r^2 + z^2 - R^2 \pm [(r^2 + z^2 - R^2)^2 + 4R^2z^2]^{1/2}}{2R^2} \tag{3.147}$$

For real values of R, r, and z the term under the radical is positive. In order to get real values for u, we only use the positive sign.

$$u = \pm\left\{\frac{1}{2}\left(\frac{r^2}{R^2} + \frac{z^2}{R^2} - 1\right) + \frac{1}{2}\left[\left(\frac{r^2}{R^2} + \frac{z^2}{R^2} - 1\right)^2 + \frac{4z^2}{R^2}\right]^{1/2}\right\}^{1/2} \tag{3.148}$$

The positive sign of u corresponds to the positive values of z, and the negative sign of u corresponds to the negative sign of z.

We can eliminate the $\pm$ sign in Eq. (3.148) by writing

$$\phi = \frac{A}{R} z + \frac{B}{R} |z| \left(\arctan u + \frac{1}{u} \right) + V_0 \tag{3.149}$$

For

$$z > 0, z = |z| \qquad \text{and} \qquad \lim_{z \to \infty} \phi = \lim_{z \to \infty} \left(\frac{A}{R} z + \frac{B}{R} z \frac{\pi}{2} + V_0 \right) \tag{3.150}$$

and for

$$z < 0, z = -|z| \qquad \text{and} \qquad \lim_{z \to -\infty} \phi = \lim_{z \to -\infty} \left(\frac{A}{R} z - \frac{B}{R} z \frac{\pi}{2} + V_0 \right) \tag{3.151}$$

Thus

$$\lim_{z \to \infty} \frac{\partial \phi}{\partial z} = \frac{A}{R} + \frac{B}{R} \frac{\pi}{2} = -E_2 \tag{3.152}$$

and

$$\lim_{z \to -\infty} \frac{\partial \phi}{\partial z} = \frac{A}{R} - \frac{B}{R} \frac{\pi}{2} = -E_1 \tag{3.153}$$

Solving for A and B, we obtain

$$\frac{A}{R} = -\frac{1}{2}(E_1 + E_2) \qquad \text{and} \qquad \frac{B}{R} = +\frac{1}{\pi}(E_1 - E_2) \tag{3.154}$$

Hence,

$$\phi(r, z) = -\frac{1}{2}(E_1 + E_2) z + \frac{1}{\pi}(E_1 - E_2)|z| \left(\arctan u + \frac{1}{u} \right) + V_0 \tag{3.155}$$

On the axis, Eq. (3.148) reduces to

$$u = |z|/R \tag{3.156}$$

and

$$\phi(0, z) = -\frac{1}{2}(E_1 + E_2) z + \frac{1}{\pi}(E_1 - E_2)\, |z| \left(\arctan \frac{|z|}{R} + \frac{R}{|z|} \right) + V_0 \tag{3.157}$$

For all values of z, Eq. (3.157) can be written as follows:

$$\phi(0, z) = -\frac{1}{2}(E_1 + E_2)z + \frac{R}{\pi}(E_1 - E_2)\left(\frac{z}{R}\arctan\frac{z}{R} + 1\right) + V_0 \quad (3.158)$$

At the center of the aperture

$$\phi(0, 0) = +(R/\pi)(E_1 - E_2) + V_0 \quad (3.159)$$

Figures 3.8 and 3.9 show some typical results. Some values are listed in Tables 3.6 and 3.7.

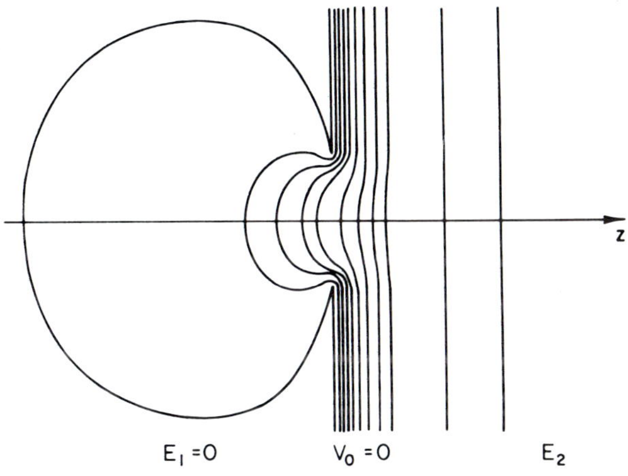

Fig. 3.8. Equipotentials of a single aperture separated by two uniform fields. Here, $E_1 = 0$.

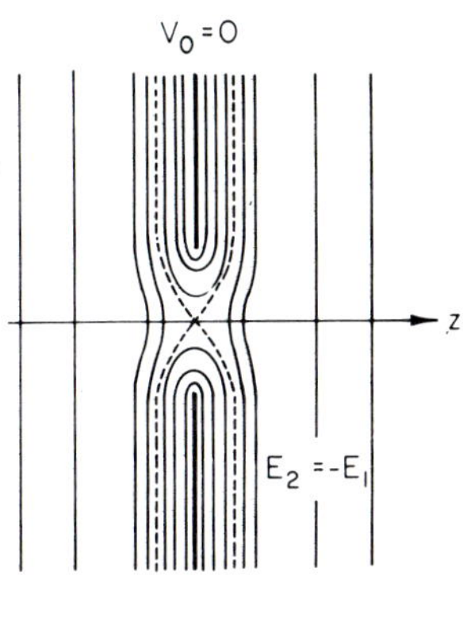

Fig. 3.9. Equipotentials of a single aperture separated by two uniform fields. Here, $E_2 = -E_1$.

TABLE 3.6[a]

r \ z	−5	−4	−3	−2	−1	0	0.5	1	1.5	2.0
0	0.004145	0.006393	0.011061	0.023143	0.068310	0.318310	0.642102	1.06831	1.53756	2.02314
0.25	0.004130	0.006359	0.010962	0.022749	0.065862	0.308202	0.635754	1.06586	1.53663	2.02275
0.50	0.004087	0.006258	0.010674	0.021624	0.058988	0.275665	0.617053	1.05899	1.53401	2.02162
0.75	0.004016	0.006095	0.010218	0.019916	0.049074	0.210542	0.588147	1.04907	1.53015	2.01992
1.00	0.003920	0.005878	0.009629	0.017832	0.038169	0	0.556295	1.03817	1.52566	2.01783
1.25	0.003802	0.005617	0.008946	0.015592	0.028248	0	0.532243	1.02825	1.52115	2.01559
1.50	0.003665	0.005322	0.008209	0.013382	0.020392	0	0.518603	1.02039	1.51703	2.01338
1.75	0.003514	0.005004	0.007454	0.011334	0.014682	0	0.511406	1.01468	1.51355	2.0113
2.00	0.003351	0.004673	0.006712	0.009520	0.010692	0	0.507446	1.01069	1.51074	2.00952

[a] Potentials for $E_1 = 0$ and $E_2 = -1$ and $R = 1$; z and r in units of R. All values are normalized with respect to radius of the aperture and field E_2.

TABLE 3.7[a]

r \ z	0	0.1	0.2	0.3	0.4	0.5	1	1.25	1.5	2
0	0.636620	0.642965	0.651753	0.692284	0.733515	0.784203	1.13662	1.34968	1.57512	2.04629
0.25	0.616405	0.623171	0.643143	0.675419	0.718700	0.771507	1.13172	1.34670	1.57326	2.04550
0.50	0.551329	0.559834	0.534559	0.623487	0.674126	0.734106	1.11798	1.33833	1.56803	2.04325
0.75	0.421084	0.436355	0.477147	0.534531	0.602109	0.676294	1.09815	1.32616	1.56031	2.03983
1.00	0	0.216309	0.326794	0.425419	0.519823	0.612590	1.07634	1.31241	1.55133	2.03566
1.25	0	0.124603	0.243390	0.355248	0.461700	0.564486	1.05650	1.29917	1.54229	2.03118
1.50	0	0.110307	0.219625	0.327305	0.433130	0.537205	1.04078	1.28781	1.53407	2.02676
1.75	0	0.105555	0.210812	0.315529	0.419551	0.522812	1.02936	1.27879	1.52710	2.02267
2.00	0	0.103402	0.206684	0.309743	0.412496	0.514892	1.02138	1.27195	1.52147	2.01904

[a] Potentials for $E_1 = 1$, $E_2 = -1$, $V_{\bullet} = 0$, $R = 1$; z and r in units of R. All values normalized with respect to radius of the aperture and field E_2. The same values apply for negative values of z. We have symmetry in this case.

3.9 Condition at the Aperture

For $z = 0$ and $r^2 > R^2$, $\phi = V_0$, the potential of the electrode. For $z = 0$ and $r^2 < R^2$, $u = 0$, and according to Eq. (3.143), we have

$$\phi(r, 0) = Bv + C \tag{3.160}$$

If we substitute the value of B from Eq. (3.154) and remember that $C = V_0$ we obtain

$$\phi(r, 0) = (R/\pi)(E_1 - E_2)v + V_0 \tag{3.161}$$

Reference to Fig. 3.7 and to the relations preceding Eq. (3.106) shows that

$$r = OH_1 = c = R(1 - v^2)^{1/2} \tag{3.162}$$

or

$$v = (1 - (r^2/R^2))^{1/2} \tag{3.163}$$

and therefore

$$\phi(r, 0) = \frac{E_1 - E_2}{\pi}(R^2 - r^2)^{1/2} + V_0 \tag{3.164}$$

which for $r = 0$ reduces to Eq. (3.159) as expected.

3.10 Condition at the Saddle Point

If $E_1E_2 < 0$ and $|E_1| = |E_2|$, the equipotentials become straight lines intersecting at the center of the aperture. This branch point is referred to as a "saddle point."

Let us consider a point along the axis at $z = z_1$, very close to the center of the aperture at $z = z_0$. Expanding the potential in a series, we obtain

$$\phi_0(z_1) = \phi_0(z_0) + z_1\, \phi_0'(z_0) + \tfrac{1}{2}z_1^{\,2}\, \phi_0''(z_0) + \cdots \tag{3.165}$$

If we now substitute this value for $\phi_0(z_1)$ in Eq. (2.30) we get

$$\phi(r, z) = \phi_0(z_0) + z_1\, \phi_0'(z_0) + \tfrac{1}{2}z_1^{\,2}\, \phi_0''(z_0) - \tfrac{1}{4}r^2\phi_0''(z_0) + \cdots \tag{3.166}$$

At the saddle point, the equipotential $\phi(r, z) = \phi_0(z_0)$. Hence

$$\tfrac{1}{4}r^2\phi_0'' = z_1\phi_0' + \tfrac{1}{2}z_1^{\,2}\phi_0'' \tag{3.167}$$

We have already seen from Eq. (2.75) that the radius of curvature of the equipotential on the axis is given by

$$\rho = 2\phi_0'/\phi_0'' \tag{3.168}$$

The radius of curvature at the saddle point tends to zero. Hence $\phi_0' = 0$, and we obtain

$$\tan^2 \gamma = r^2/z_1{}^2 = 2 \qquad \text{or} \qquad \gamma = 54^\circ\, 44' \tag{3.169}$$

where γ is the angle between the straight line equipotential and the axis.

Further Reading

Bertram, S. (1940). *Proc. I.R.E.* **28**, 418–420.
Bertram, S. (1942). *J. Appl. Phys.* **13,** 496–502.
Ollendorff, F. (1932). "Die Potentialfelder der Elektrotechnik." Springer, Berlin.

IV

General Properties of Electrostatic Lenses

In Chapter II we saw how axially symmetrical electric fields have the capability of forming images. A device which has the ability of forming images of objects irradiated by electrons, or of emitting such particles, is called an "electron lens." There are two types of lenses, the electrostatic and the magnetic. Before starting a detailed analysis of the electrostatic lens, let us first review certain properties of lenses already known from light optics.

4.1 Cardinal Points of a Lens

(a) *Focal Points.* A focal point is the image of a bundle of rays incident on a lens parallel to the axis. Figure 4.1a shows the focal point F. The plane perpendicular to the axis and passing through F is known as the "focal plane."

(b) *Principal Points.* Figure 4.1b shows how the incident ray and the refracted ray are extended until they intersect. The planes passing through the intersections and perpendicular to the axis are called "principal planes." The intersections of the principal planes with the axis are called "principal points."

The main property of the principal planes is that they have unit-positive lateral magnification. The distance from the focal point to the principal point is called the "focal length." The distance from the geometrical center of the lens to the focal point is called the "midfocal length."

(c) *Nodal Points.* If we consider all the rays that pass through a lens from an off-axis point to its corresponding image point, we find that there always exists one ray which emerges in the image space parallel to its direction in the object space. Figure 4.1c shows the points where the extensions of the segments of this particular ray meet the axis. They are called "nodal points." The planes perpendicular to the axis which pass through them are called "nodal planes."

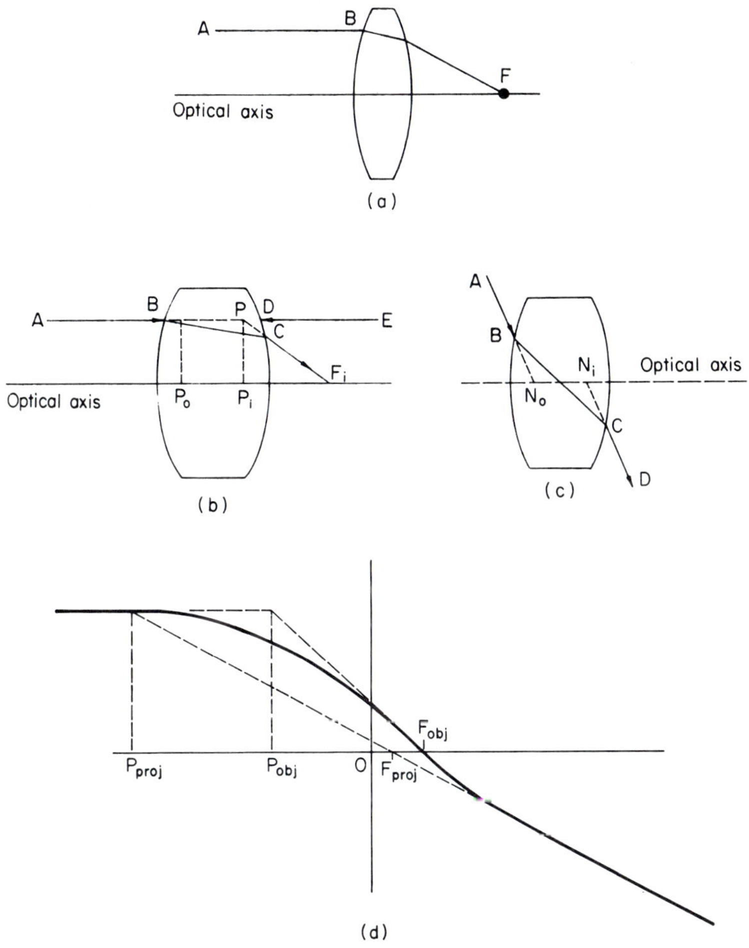

Fig. 4.1. (a) Definition of focal point: Incident ray AB is parallel to the axis. F is the focal point. (b) Definition of principal points: Ray AB, which is parallel to the optical axis, is extended to meet CF at P; PP_i is the principal plane; P_i is the principal point. Similarly, we obtain P_o for ray ED. (c) Definition of nodal points: AB and CD are parallel; N_o and N_i are the nodal points obtained by extending AB and CD to meet the optical axis. (d) Definition of objective and projective cardinal points in a strong lens: F_{obj} and F_{proj} are the objective and projective focal points, respectively; P_{obj} and P_{proj} are the objective and projective principal points, respectively.

The focal points, principal points, and nodal points of a lens are called the "cardinal points" of the lens.

(d) *Objective and Projective Cardinal Points.* In many instances the electron beam trajectory crosses the axis inside the lens. This takes place in very strong lenses. We cannot assume here that the path of the trajectory will be rectilinear after crossing the axis because it will still be modified by the field until it emerges from the lens. This condition is illustrated in Fig. 4.1d where the trajectory first crosses the axis at the point F_{obj}. This is referred to as the objective focal point; $F_{obj}P_{obj}$ is the objective focal length; OF_{obj} is called the objective midfocal length, and P_{obj} is the objective principal point.

If we extend the trajectory as it leaves the lens back to the axis and to the parallel incoming beam, we obtain the projective focal point F_{proj} and the projective principal point P_{proj}, and OF_{proj} is the projective midfocal length. It is clear that in a relatively weak lens where the intersection occurs outside the lens, the projective and objective cardinal points are identical.

4.2 Important Lens Relations

If the refractive indices of the media on the left and on the right of a lens are n_o and n_i, respectively, and the corresponding focal lengths are f_o and f_i, then

$$n_o/n_i = f_o/f_i \tag{4.1}$$

If u_o is the object distance measured from the object to its principal plane, and u_i the image distance from its corresponding principal plane, then

$$(n_o/u_o) + (n_i/u_i) = n_o/f_o \tag{4.2}$$

and

$$(n_o/u_o) + (n_i/u_i) = n_i/f_i \tag{4.3}$$

In general, when the media on the left and right of a lens have the same index of refraction, we have

$$f_o = f_i = f \qquad \text{and} \qquad (1/u_o) + (1/u_i) = 1/f \tag{4.4}$$

4.3 Newton's Formula

If x_o is the distance from the projection on the axis of an object point to F_o and x_i is the distance along the axis from its corresponding image point to F_i, then

$$x_o x_i = f_o f_i \tag{4.5}$$

If $f_o = f_i = f$, then

$$x_o x_i = f^2 \tag{4.6}$$

4.4 Sign Convention

Generally speaking, any sign convention will be satisfactory so long as it is consistent. The following is the convention we shall use in this text:

(a) A plane of reference is chosen near or at the geometrical center of the lens. All distances measured on the left of this reference plane are negative, and those on the right are positive. This reference plane will be called the " midplane of the lens."

(b) The image-side focal length is measured from the image-side principal plane and is positive. The object-side focal length is negative.

(c) Transverse dimensions, such as object and image sizes, are positive above the axis, and negative below the axis.

(d) Angles are positive if the slope of the ray with respect to the axis is positive; otherwise, they are negative.

4.5 Weak and Strong Lenses

In light optics, if the thickness of a lens is small compared with other dimensions associated with its optical properties (such as focal lengths, object and image distance, etc.), then such a lens is referred to as a "thin lens." Otherwise, the lens is considered to be "thick." In thin lenses, the principal planes, nodal planes, and geometric center of the lens are all confused in one plane at the center. All characteristic dimensions are then measured from the center of the lens.

In electron optics, lenses are referred to as "weak" or "strong" depending on whether the height of the ray changes appreciably inside the lens. We will show that, in electron optics, the principal planes of a weak lens are not confused as in light optics. Although the distinction in electron optics and light optics between the two types of lenses appears analogous, it is basically different.

4.6 The Law of Helmholtz–Lagrange

We shall use Fermat's principle in deriving an important relation in optics. Refer to Fig. 4.2. By Fermat's principle, we have the following relations for the two optical paths, defined by the product of length and index of refraction,

$$H_o B H_i = H_o P_o H_i \tag{4.7}$$

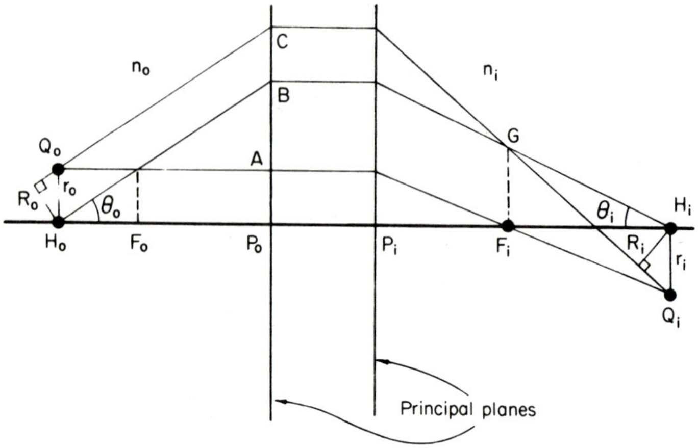

Fig. 4.2. The law of Helmholtz–Lagrange obtained from Fermat's principle.

and

$$Q_o CQ_i = Q_o AQ_i \tag{4.8}$$

Here, $H_o BH_i$, and so forth, are optical paths, not geometrical lengths.

Since we assume paraxial rays, or first order theory,

$$F_i H_i \approx F_i Q_i \qquad \text{and} \qquad GR_i \approx GH_i \tag{4.9}$$

Since G and F_i lie in the focal plane, we also have

$$Q_o AF_i = H_o P_o F_i \qquad \text{and} \qquad R_o CG = H_o BG \tag{4.10}$$

or, combining Eqs. (4.9) and (4.10), we have

$$Q_o AQ_i = H_o P_o H_i \qquad \text{and} \qquad R_o CR_i = H_o BH_i \tag{4.11}$$

Using Eqs. (4.7) and (4.8), we have

$$R_o Q_o = R_i Q_i \tag{4.12}$$

The optical paths $R_o Q_o$ and $R_i Q_i$ can now be expressed as follows:

$$n_o r_o \theta_o = n_i r_i \theta_i \tag{4.13}$$

where we have assumed that $\sin\theta \approx \theta$ for small angles. This relation is generally known as Lagrange's relation, Helmholtz–Lagrange's equation, or the theorem of Clausius. Whatever its name may be, it is a very useful relation in electron optics. Equation (4.13) is valid only for paraxial rays.

If we consider a system which is perfect, i.e., one in which the magnification is constant for all object positions, no matter what their height from the axis may be, then the angle θ can be as large as we want, and we obtain

$$n_o r_o \sin \theta_o = n_i r_i \sin \theta_i \tag{4.14}$$

This is Abbe's relation, generally referred to as the "sine law." It is valid for all rays, whether or not paraxial, so long as the system is free from aberrations. In electron optics, an ideal system is obtained only when θ is very small, and in this case $\sin \theta \approx \theta$. Abbe's relation then reduces to the Helmholtz–Lagrange equation.

4.7 Lens Relations in Electron Optics

All the above relations obtained in light optics are valid in electron optics. For the sake of comparison, we shall derive some of the above relations using the paraxial-ray equation in electron optics.

Equation (2.35) can be written in the following form:

$$[\phi(z)]^{1/2} \frac{d}{dz} \left\{ [\phi(z)]^{1/2} \frac{dr}{dz} \right\} = \frac{-r\phi''(z)}{4} \tag{4.15}$$

From Chapter II we know that we can choose any two rays that satisfy this equation to obtain the general solution. We then select two rays r_1 and r_2 such that r_1 leaves the lens parallel to the axis and r_2 enters the lens parallel to the axis. See Fig. 4.3. We then have the following initial conditions:

$$r_1 = r_i \qquad r_1' = 0 \qquad \text{at} \quad z = \infty$$

$$r_2 = r_o \qquad r_2' = 0 \qquad \text{at} \quad z = -\infty$$

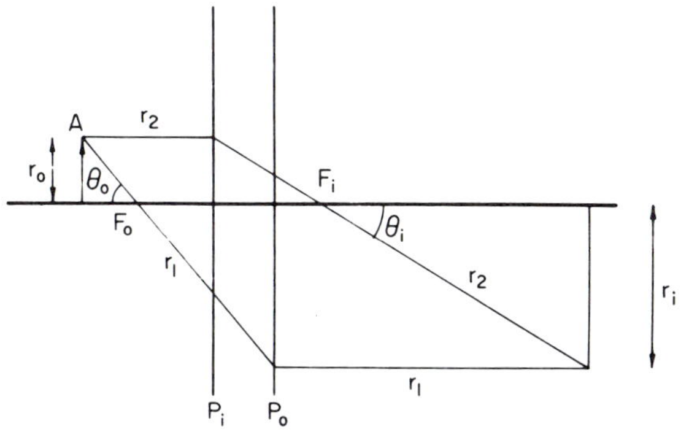

Fig. 4.3. Lens relations in electron optics. Ray r_2 leaves from A parallel to the axis and passes through the focal point F_i. Ray r_1 passes through F_0 and emerges parallel to the axis.

If we now substitute r_1 in Eq. (4.15) and multiply by r_2, and then substitute r_2, multiply by r_1, and subtract the two resulting equations, we obtain

$$r_2 \frac{d}{dz}\{[\phi(z)]^{1/2} r_1'\} - r_1 \frac{d}{dz}\{[\phi(z)]^{1/2} r_2'\} = 0 \tag{4.16}$$

and by adding and subtracting $[\phi(z)]^{1/2} r_1' r_2'$, we can write

$$d/dz\{r_2[\phi(z)]^{1/2} r_1' - r_1[\phi(z)]^{1/2} r_2')\} = 0 \tag{4.17}$$

We now integrate between $-\infty$ and $+\infty$ or between two axial points z_o and z_i where we are sure the rays obey the initial conditions. We obtain

$$r_1(z_i)[\phi(z_i)]^{1/2} r_2'(z_i) + r_2(z_0)[\phi(z_o)]^{1/2} r_1'(z_o) = 0 \tag{4.18}$$

In passing through the lens, the trajectory is bent toward the axis. As soon as the electron is a short distance beyond the lens, it is in a field-free region, and its path is subsequently a straight line. From simple geometry, we can see that the image focal length f_i is given by

$$1/f_i = -r_2'(z_i)/r_2(z_o) \tag{4.19}$$

and similarly

$$1/f_o = -r_1'(z_o)/r_1(z_i) \tag{4.20}$$

Hence

$$f_i/f_o = -[\phi(z_i)]^{1/2}/[\phi(z_o)]^{1/2} \tag{4.21}$$

which is identical to Eq. (4.1) if we follow our sign convention. By definition

$$r_1(z_i)/r_2(z_o) = r_i/r_o = M \tag{4.22}$$

where M is the magnification.

In addition, we have

$$r_2'(z_i)/r_1'(z_o) = \tan\theta_i/\tan\theta_o \tag{4.23}$$

If we substitute Eqs. (4.22) and (4.23) in Eq. (4.18), we obtain

$$r_o[\phi(z_o)]^{1/2} \tan\theta_o = -r_i[\phi(z_i)]^{1/2} \tan\theta_i \tag{4.24}$$

Since r_o and r_i are above and below the axis, respectively, Eq. (4.24) reduces to Eq. (4.13) for small angles.

4.8 The Action of an Electrostatic Lens

Let us consider the simple case of two equidiameter cylinders placed coaxially as shown in Fig. 4.4. If a ray proceeds from left to right it experiences an inward force because the gradient of the potential has a radial

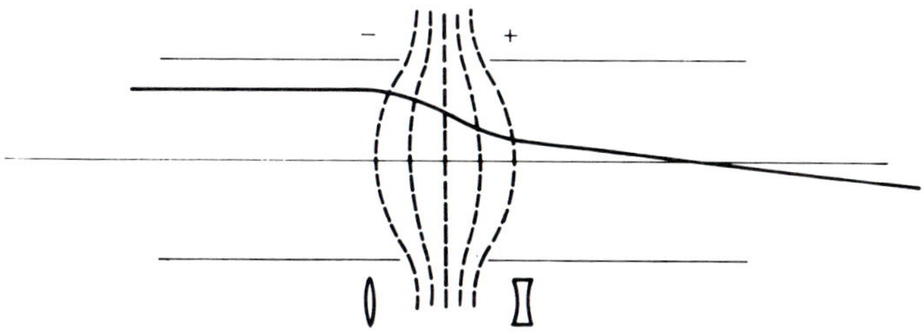

Fig. 4.4. The lens action in an electrostatic immersion lens. If the − and + signs are interchanged, then the convergent and divergent lenses should also be interchanged.

component directed toward the axis. When the ray proceeds to the right, the action is divergent because there the gradient is directed outward from the axis. In general, when a ray proceeds in the direction of increasing potential, a concave equipotential has a divergent effect and a convex equipotential a convergent effect. The opposite is true if the ray proceeds in the direction of decreasing potential. If the radial field were acting alone the net effect would be nil. However, if the ray proceeds in the direction of increasing potential, the longitudinal component accelerates the electrons as they cross the field, and therefore the electrons spend a longer time in the region where the equipotentials are convex, i.e., where there is a convergent effect, than they do in the region where there is a divergent effect. Thus, the net result is a convergent action. It can readily be seen that if the ray proceeds in a direction of decreasing potential, the net result is still convergent.

A similar effect occurs in a three-electrode assembly as shown in Fig. 4.5. Here we have a convergent region separating two divergent regions. Again, the longitudinal component accelerating and decelerating the electrons plays

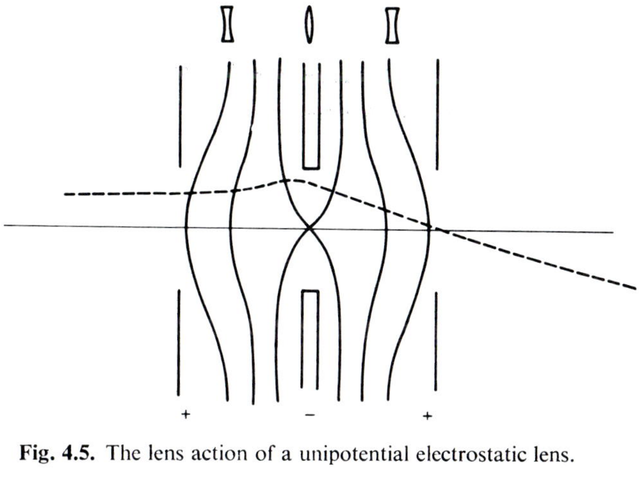

Fig. 4.5. The lens action of a unipotential electrostatic lens.

a very important role. In this case, the electrons spend a longer time in the convergent region than in the divergent region. The net result is again convergent. The same applies if the outer electrodes are negative and the central electrode is positive.

From the above observations, one can immediately conclude that any of the two geometries shown in Figs. 4.4 and 4.5 act as electron lenses no matter what the order of the potentials may be. In fact, all electron lenses are convergent because in any lens a region of divergence is always either followed or preceded by a stronger convergent region. There is one exception to this rule, as we shall see later on. In the special case of a plane cathode near an accelerating diaphragm, the net action is divergent because the convergent region does not exist.

An important property of electrostatic lenses, which is primarily due to the above lens action, is the fact that it is always necessary to use potentials at the various electrodes of the lens comparable to the beam accelerating potential in order to obtain a sufficient differential effect for the lens action. The use of much lower potentials will result in a negligible convergence.

4.9 The Types of Electrostatic Lenses

(a) *Immersion Lenses.* Figure 4.6 shows the various kinds of immersion lenses. Their common characteristic is that the potentials in the object and image spaces are constant and unequal. Their name derives from their analogy to the oil immersion objectives of light optics in which the object and image lie in media of constant, but different, indices of refraction, namely, oil and air.

Most immersion lenses consist of two cylinders, but some could have more than two electrodes. Thus, in accordance with our definition, lens *d* of Fig. 4.6 is an immersion lens.

(b) *Unipotential Lenses.* These are electrostatic electron lenses characterized by equal constant potentials in image and object space. Figure 4.7 shows examples of such lenses. Because the potentials of the outer electrodes are the same and the potential of the central electrode is sometimes zero, one voltage is used. This type of lens is referred to as an "*einzel* lens" from the German word *einzel* meaning single. Most often, the geometry of the lens is symmetrical as in Fig. 4.7a–c. Such lenses are often referred to as "symmetrical electrostatic lenses."

The *einzel* lens of Fig. 4.7d is not symmetrical. Notice that the unipotential lens shown in Fig. 4.7c consists of only two electrodes.

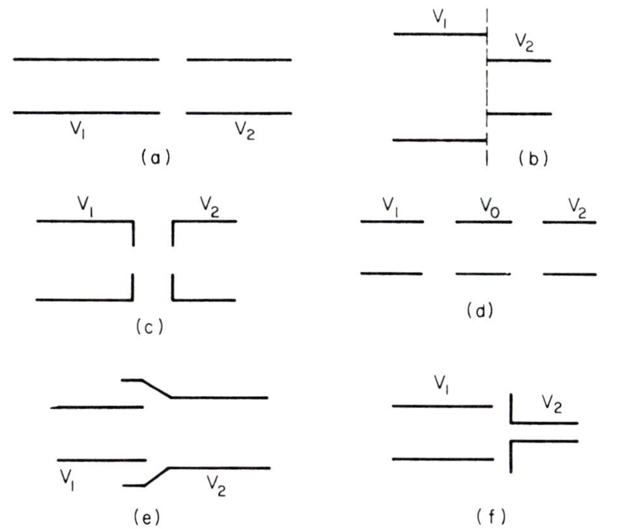

Fig. 4.6a–f. Various types of electrostatic immersion lenses.

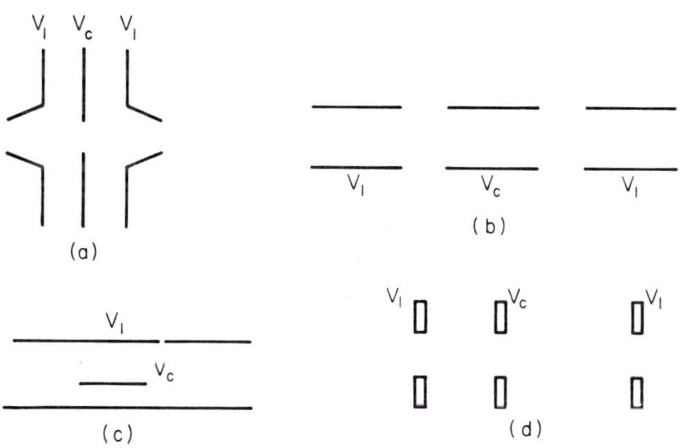

Fig. 4.7a–d. Various types of electrostatic unipotential lenses.

(c) *Aperture Lenses.* In these lenses, an aperture separates two regions of different but constant fields. Figure 4.8 shows an example of such a lens.

(d) *Cathode Lenses.* If the field of a lens is abruptly terminated on one side by an electrode with a surface normal to the axis (generally at zero potential), this type of lens is referred to as "cathode lens." This zero potential surface is generally the cathode, which is at the same time the object surface. Figure 4.9 shows examples of these. Since the object is immersed in the lens, some authors have referred to this arrangement as "immersion lenses." This is a misnomer.

(e) *Screen Lenses.* In order to benefit from all the enormous experience gained in light optics, researchers have attempted to obtain the most faithful analog of the light optical glass lens. With this in mind, the two wire screens as shown in Fig. 4.10 are shaped like two glass lens surfaces and are connected to two different potentials. Unfortunately, the image quality is impaired because the mesh becomes a source of secondary electrons which produces hazy images. In addition, the mesh is the cause of some blurring of the image. Figure 4.11 shows the distortion of the field caused by the mesh. Each line of force ends at the grid with a pronounced bend.

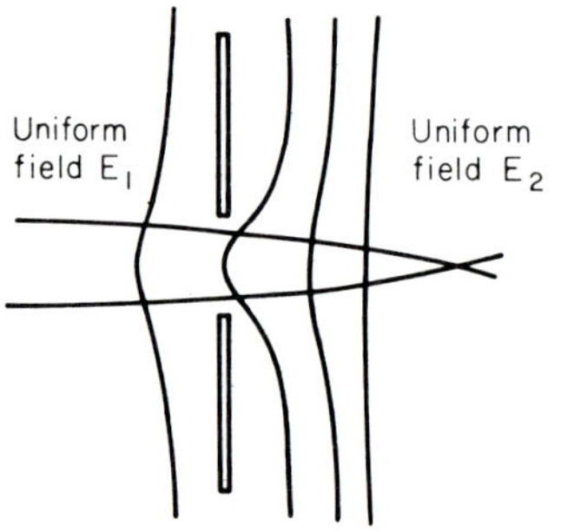

Fig. 4.8. Example of an aperture lens. The aperture separates two uniform fields.

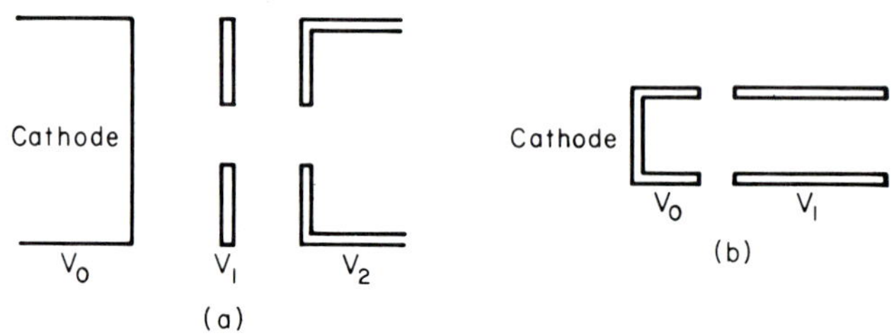

Fig. 4.9. Some examples of cathode lenses.

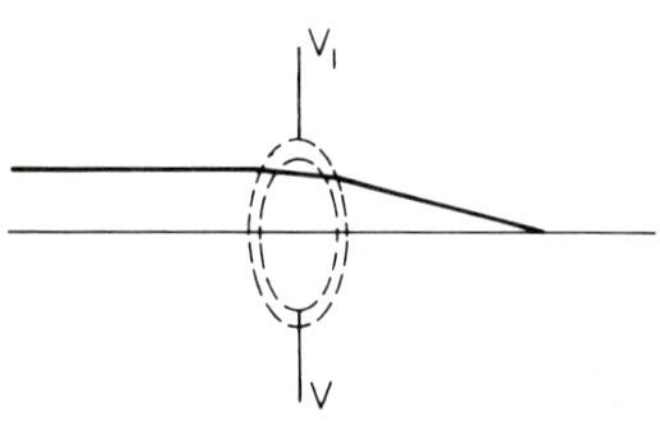

Fig. 4.10. Geometry of the screen lens.

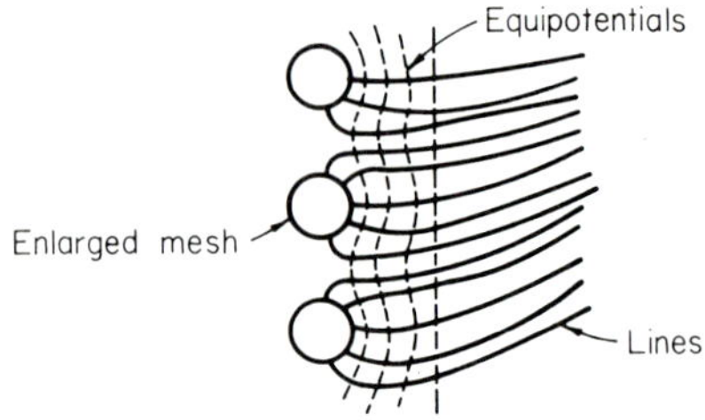

Fig. 4.11. Distortion of the field caused by the mesh in Fig. 4.10.

In order to overcome the above-mentioned difficulties, it would seem that the best approach would be to replace the meshes by thin, metallic membranes transparent to electrons. So far, no serious attempt has been made in this direction. Naturally, all the relations obtained for uniformly varying fields, as with all other types of lenses, are no longer applicable here. This kind of lens will not be discussed any further in this text.

(f) *Gas Lenses.* Historically speaking, these were the first types of lenses used with cathode-ray tubes. It is well known that an electron beam traversing a medium of poor vacuum will ionize the gas in its path. These ions will confine the beam to a narrow restriction and keep it in focus. This method of focusing is too difficult to control and is therefore no longer used.

(g) *Grid Lenses.* Figure 4.12 shows an example of a grid lens. The only difference between this lens and the other kinds discussed earlier is the addition of a mesh, or grid, inside the lens. In the example shown in Fig. 4.12 the mesh replaces the central electrode of the symmetrical *einzel* lens. Such lenses seem to have better optical qualities than those described above. However, the problems encountered in the screen lenses also exist here.

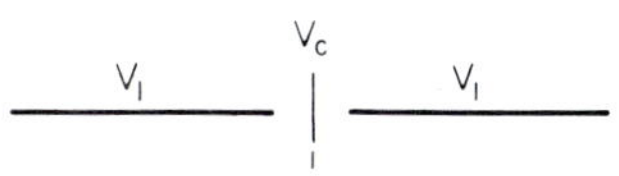

Fig. 4.12. Example of a grid lens.

4.10 Focal Length of the Weak Electrostatic Lens

We use the paraxial-ray equation (4.15)

$$[\phi(z)]^{1/2}\frac{d}{dz}\left\{[\phi(z)]^{1/2}\frac{dr}{dz}\right\}=\frac{-r\phi''(z)}{4} \tag{4.25}$$

We shall now integrate between z_o and z_i. Here, z_o and z_i are the object and image positions, respectively.

$$\left\{[\phi(z)]^{1/2}\frac{dr}{dz}\right\}_{z_i}-\left\{[\phi(z)]^{1/2}\frac{dr}{dz}\right\}_{z_o}=-\frac{1}{4}\int_{z_o}^{z_i}\frac{r\,\phi''(z)}{[\phi(z)]^{1/2}}\,dz \tag{4.26}$$

For an incident trajectory parallel to the axis, $r_o'=0$, and

$$r_i'=-\frac{1}{4[\phi(z_i)]^{1/2}}\int_{z_o}^{z_i}\frac{r\,\phi''(z)}{[\phi(z)]^{1/2}}\,dz \tag{4.27}$$

The image focal length f_i is given by

$$f_i^{-1} = -r_i'/r_o \tag{4.28}$$

hence

$$\frac{r_o}{f_i} = \frac{1}{4(\phi_i)^{1/2}} \int_{z_o}^{z_i} \frac{r\,\phi''(z)}{[\phi(z)]^{1/2}}\, dz \tag{4.29}$$

Since we are dealing with a weak lens, we can safely assume that r varies only slightly while the beam passes through the lens. Of course, the slope of the trajectory varies considerably, but its actual distance from the axis while inside the lens varies very little. Thus,

$$r \cong r_o \tag{4.30}$$

The image focal length is then equal to

$$\frac{1}{f_i} = \frac{1}{4(\phi_i)^{1/2}} \int_{z_o}^{z_i} \frac{\phi''(z)}{[\phi(z)]^{1/2}}\, dz \tag{4.31}$$

A similar analysis for the case of an electron entering the lens parallel to the axis from the image side yields

$$\frac{1}{f_o} = \frac{-1}{4(\phi_o)^{1/2}} \int_{z_o}^{z_i} \frac{\phi''(z)}{[\phi(z)]^{1/2}}\, dz \tag{4.32}$$

As a check, if we take the ratio of Eqs. (4.31) and (4.32), we obtain

$$\frac{f_o}{f_i} = \frac{-(\phi_o)^{1/2}}{(\phi_i)^{1/2}} \tag{4.33}$$

which is exactly the relation given in Eq. (4.21).

Figure 4.13 shows the distribution of the potential and its derivatives for the case of a two-cylinder, equidiameter immersion lens. We must remember that the above relations are valid for all types of electrostatic lenses as long as the radius inside the lens region does not vary very much.

The use of the second derivative $\phi''(z)$ in obtaining the focal length of a weak lens is very undesirable. Equation (4.31) can be integrated by parts. Set

$$u = [\phi(z)]^{-1/2} \qquad \text{and} \qquad dv = \phi''(z)\, dz \tag{4.34}$$

Then

$$du = -\tfrac{1}{2}[\phi(z)]^{-3/2}\, \phi'(z)\, dz \qquad \text{and} \qquad v = \phi'(z) \tag{4.35}$$

Integrating, we obtain

$$\frac{1}{f_i} = \frac{1}{4\phi(z_i)} \left\{ \frac{\phi'(z_i)}{[\phi(z_i)]^{1/2}} - \frac{\phi'(z_o)}{[\phi(z_o)]^{1/2}} \right\} + \frac{1}{8[\phi(z_i)]^{1/2}} \int_{z_o}^{z_i} \frac{[\phi'(z)]^2}{[\phi(z)]^{3/2}}\, dz \tag{4.36}$$

For the case in which $\phi'(z)$ is zero on both sides of the lens, which is very often the case, we have

$$\frac{1}{f_i} = \frac{1}{8[\phi(z_i)]^{1/2}} \int_{z_o}^{z_i} \frac{[\phi'(z)]^2}{[\phi(z)]^{3/2}}\, dz \tag{4.37}$$

and similarly for the object-side focal length

$$\frac{1}{f_o} = \frac{-1}{8[\phi(z_o)]^{1/2}} \int_{z_o}^{z_i} \frac{[\phi'(z)]^2}{[\phi(z)]^{3/2}}\, dz \tag{4.38}$$

The second derivative has been eliminated in this case. Notice that the derivative of the axial potential, even though it may be negative, is squared inside the integral, and therefore the quantity on the right-hand side of Eq. (4.37) or Eq. (4.38) is always positive. This is in agreement with the statement made earlier that electrostatic electron lenses are always convergent.

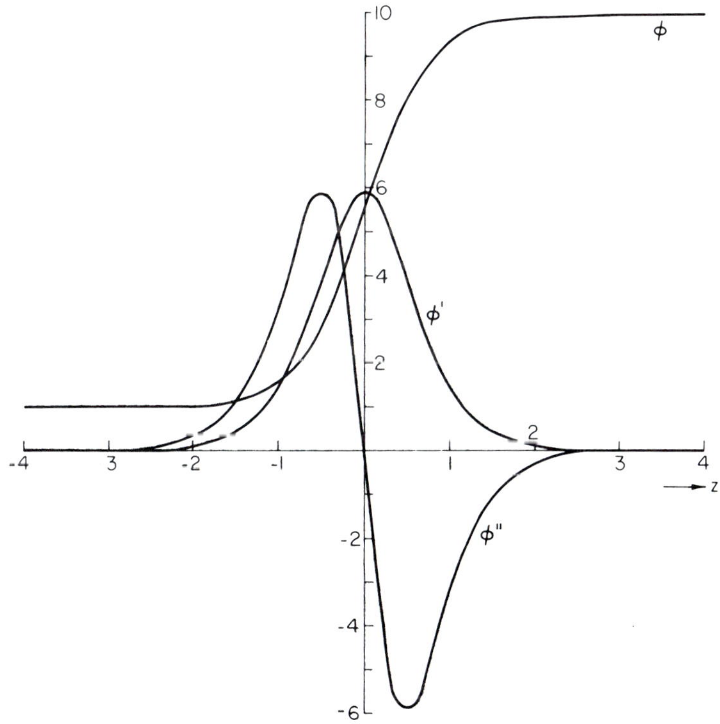

Fig. 4.13. Axial potential and its derivatives with respect to z of a two-cylinder equi-diameter immersion lens. All values are normalized with respect to the radius of the cylinders and the potential V_1 on the left electrode (ϕ in units of V_1; z in units of R, the radius).

4.11 Focal Length Starting with the Reduced Formula

Rather than derive an expression for the focal-length starting from the general second-order differential equation for paraxial rays, we shall start with the reduced equation (2.38)

$$\frac{d^2R}{dz^2} = -\frac{3}{16}\left[\frac{\phi'(z)}{\phi(z)}\right]^2 R \tag{4.39}$$

in which $R = r\phi^{1/4}$

If we assume R inside the lens to be constant, and $R_o' = 0$, since $1/\bar{F}_i = -R_i'/R_o$, we have

$$\frac{1}{\bar{F}_i} = \frac{3}{16}\int_{z_o}^{z_i}\left(\frac{\phi'(z)}{\phi(z)}\right)^2 dz \tag{4.40}$$

Here $\bar{F}$ is the "reduced" focal length. It must be multiplied by the reducing factor to obtain the true focal length f, as shown below.

If we consider the case in which the potential is constant on both sides of the lens, $\phi'(z) = 0$ there, and

$$R_i' = r_i'\phi_i^{1/4} \tag{4.41}$$

and

$$\frac{1}{f_i} = -\frac{r_i'}{r_o} = \left(\frac{\phi_o}{\phi_i}\right)^{1/4}\frac{1}{\bar{F}_i} \tag{4.42}$$

hence

$$\frac{1}{f_i} = \frac{3}{16}\left(\frac{\phi_o}{\phi_i}\right)^{1/4}\int_{z_o}^{z_i}\left(\frac{\phi'(z)}{\phi(z)}\right)^2 dz \tag{4.43}$$

and similarly

$$\frac{1}{f_o} = -\frac{3}{16}\left(\frac{\phi_i}{\phi_o}\right)^{1/4}\int_{z_o}^{z_i}\left(\frac{\phi'(z)}{\phi(z)}\right)^2 dz \tag{4.44}$$

For the case of the unipotential lens, where $\phi_o = \phi_i$, we have

$$\frac{1}{f_i} = \frac{-1}{f_o} = \frac{1}{f} = \frac{3}{16}\int_{z_o}^{z_i}\left[\frac{\phi'(z)}{\phi(z)}\right]^2 dz \tag{4.45}$$

Equations (4.37), and (4.38) and Eqs. (4.43) and (4.44) represent two sets of formulas for the focal lengths of weak electrostatic electron lenses. In the first relations we assumed r to be constant inside the lens. In the second we assumed $R = r\phi^{1/4}$ to be constant inside the lens. We find, in practice, that the approximation used in Eqs. (4.43) and (4.44) yields values which are very close to those obtained by integrating the equations of motion numerically. The two sets of equations were obtained from first principles using initial relations which were entirely valid. Similar assumptions were made in both cases. One would therefore expect that the results obtained from the two relations would not differ appreciably. In what follows, we shall show that this is not the case.

Let us take an extreme case and consider an infinitely weak lens. We can then assume ϕ as constant and consider the ratio $\phi(z_i)/\phi(z_o)$ as approximately equal to one. Equation (4.37) reduces to

$$f_i^{-1} = \{8[\phi(z)]^2\}^{-1} \int_{z_o}^{z_i} [\phi'(z)]^2 \, dz \tag{4.46}$$

while Eq. (4.43) reduces to

$$\frac{1}{f_i} = \frac{3}{16\phi^2(z)} \int_{z_o}^{z_i} [\phi'(z)]^2 \, dz \tag{4.47}$$

The two equations differ by a factor $\frac{3}{2}$! Thus, we come to the conclusion that the two equations are not identical. The next question is to determine which of the two is closer to the correct answer. It seemed originally that similar assumptions were made in the derivation of the two sets of equations. However, a closer look will reveal that this is not quite true. In Eq. (4.43), the ratio $[\phi'(z)/\phi(z)]^2$ is always positive. Any error we might have generated by assuming $R =$ constant within the lens results in a relatively small error in the overall computations. On the other hand, if we consider the relation given by Eq. (4.37), we must remember that it is a rearrangement of Eq. (4.31) where $\phi''(z)$ appears inside the integral. In any electrostatic lens where the potential is constant on both sides of the lens, which is very often the case, in other words where the field is zero on both sides of the lens, $\phi''(z)$ changes sign at least once. Therefore, any error caused by assuming r to be constant within the lens will yield a larger error in the overall result. One must therefore conclude that Eqs. (4.37) and (4.38) can be used without appreciable error only where $\phi''(z)$ does not change sign. If one attempts to analyze an electrostatic electron lens where $\phi''(z)$ changes sign, the lens must be divided into separate sections where in each section $\phi''(z)$ does not change sign. Then Eqs. (4.37) and (4.38) can be used, and the series of lenses can be combined using a method described later on. In Fig. 4.14, the ratios $\phi''/\sqrt{\phi}$ and $(\phi'/\phi)^2$ are plotted as functions of z. On the same graph, the values of r, R, and R'' are shown exactly as they were calculated numerically with a computer. It is clear that r changes appreciably more than R. In addition, the slope of R changes monotonically. This is not the case with r.

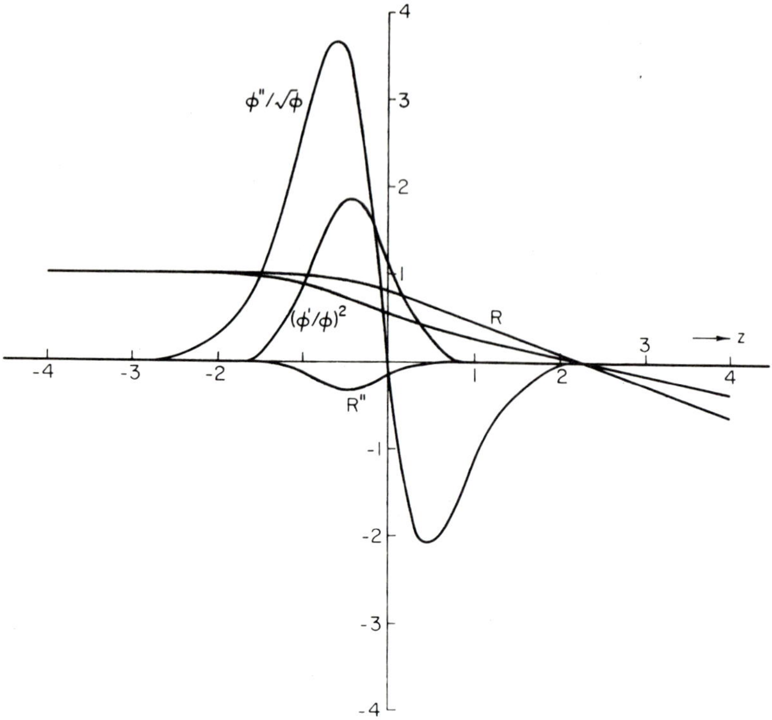

Fig. 4.14. Plot of $\phi''/\sqrt{\phi}$, $(\phi'/\phi)^2$, R, R'', and r of a two-cylinder equidiameter electrostatic lens with negligible spacing. The voltage ratio is $V_2/V_1 = 10$. All values normalized with respect to the radius R of the cylinders and the voltage V_1 on the left electrode. Here $\phi_i = V_2$; $\phi_o = V_1$.

4.12 Position of Principal Planes

We shall use Eqs. (4.43) and (4.44) since, as we have seen, they yield final results with the fewest errors. We have already seen how the "reduced" focal length F can be related to the actual focal length f of a lens. Accordingly, we shall define as "reduced" principal plane, the principal plane which corresponds to the reduced focal length F.

We have

$$f_o = \bar{F}_o(\phi_o/\phi_i)^{1/4} \tag{4.48}$$

and

$$f_i = \bar{F}_i(\phi_i/\phi_o)^{1/4} \tag{4.49}$$

hence, the ratio

$$f_o/f_i = (\bar{F}_o/\bar{F}_i)(\phi_o/\phi_i)^{1/2} \tag{4.50}$$

But, we already know that

$$f_o/f_i = -(\phi_o/\phi_i)^{1/2} \tag{4.51}$$

Therefore, we conclude that in any electron lens where the potentials on the image and object sides are constant, but not necessarily equal,

$$-\bar{F}_o = \bar{F}_i = \bar{F} \tag{4.52}$$

We remind the reader that the actual focal lengths f_o and f_i are equal in magnitude only if the potentials on both sides of the lens are equal as in a symmetrical lens.

The focal lengths obtained from the reduced equation are equal in magnitude for all values of ϕ_o and ϕ_i. The corresponding reduced principal planes are confused and coincide with the center of gravity G. The latter is defined as the center of gravity of the area limited by the curve

$$Q(z) = [\phi'(z)/\phi(z)]^2 \tag{4.53}$$

We shall adopt the following sign convention: a motion along the line to the right is positive, and to the left is negative. Let F_o and F_i be the focal points. Then $GF_i = -GF_o = -\bar{F}_o = \bar{F}_i$. The position of the focal points is the same for the reduced and unreduced equation because the factor $[\phi(z)]^{1/4}$ does not affect the point where the ray crosses the axis.

We will assume that the lens is accelerating, i.e., $\phi_o < \phi_i$. If P_i is the image-side principal point, we have

$$P_i F_i - f_i = \bar{F}_i(\phi_i/\phi_o)^{1/4} > \bar{F}_i \tag{4.54}$$

$$P_i F_i > GF_i \tag{4.55}$$

Therefore, P_i is on the left of G.

If P_o is the object-side principal point, we have

$$P_o F_o = f_o = \bar{F}_o(\phi_o/\phi_i)^{1/4} \tag{4.56}$$

or

$$F_o P_o = -\bar{F}_o(\phi_o/\phi_i)^{1/4} = \bar{F}_i(\phi_o/\phi_i)^{1/4} < \bar{F}_i \tag{4.57}$$

Therefore,

$$F_o P_o < F_o G \tag{4.58}$$

and P_o is on the left of G. Thus, both principal points are on the side of G where the lens has the lowest potential. Furthermore, we have

$$P_i G = P_i F_i - GF_i = f_i - \bar{F}_i = \bar{F}_i[(\phi_i/\phi_o)^{1/4} - 1] > 0 \tag{4.59}$$

and

$$P_o G = F_o G - F_o P_o = -\bar{F}_o - (-f_o) = \bar{F}_i - \bar{F}_i(\phi_o/\phi_i)^{1/4}$$
$$= \bar{F}_i[1 - (\phi_o/\phi_i)^{1/4}] > 0 \tag{4.60}$$

Therefore,

$$\frac{P_i G}{P_o G} = \frac{\bar{F}_i\{(\phi_i/\phi_o)^{1/4} - 1\}}{\bar{F}_i\{1 - (\phi_i/\phi_o)^{1/4}\}} = \frac{\phi_i^{1/4} - \phi_o^{1/4}}{\phi_i^{1/4} - \phi_o^{1/4}} \frac{\phi_i^{1/4}}{\phi_o^{1/4}} = \left(\frac{\phi_i}{\phi_o}\right)^{1/4} = \left(\frac{n_i}{n_o}\right)^{1/4} \tag{4.61}$$

Thus, $0 > (P_i G/P_o G) > 1$, and P_i must be on the left of P_o which means that the principal planes are crossed. Their locations are shown in Fig. 4.15. From Eq. (4.59) we can see that $P_i G$ becomes very large for large values of ϕ_i/ϕ_o and that $P_i G \to 0$ as $\phi_i \to \phi_o$. In addition, P_i coincides with F_o when $\phi_i/\phi_o = 2^4 = 16$. We then have

$$P_o G = \bar{F}_i(1 - \tfrac{1}{2}) = \tfrac{1}{2}\bar{F}_i \tag{4.62}$$

Here, $P_o G$ tends to $\bar{F}_i$ for very large values of ϕ_i, but it can never exceed $\bar{F}_i$. For $\phi_i \to \phi_o$ it tends to zero. Therefore, P_o stays between F_o and G (see Fig. 4.16).

Note that all the above considerations do not apply if intermediate images are formed within the lens. In that case, the lens must be sectionized before a detailed analysis is made.

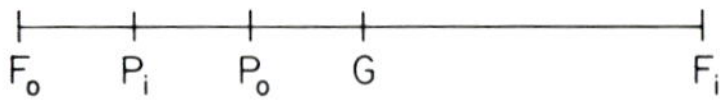

Fig. 4.15. The location of the center of gravity, principal points, and focal points; $GF_i = \bar{F}$, the reduced focal length.

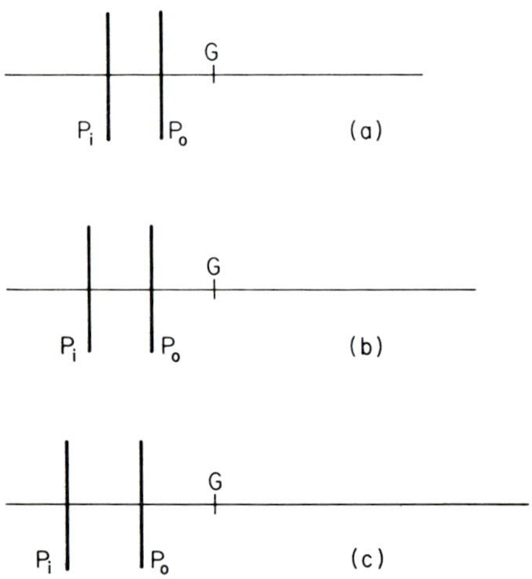

Fig. 4.16a–c. The relative position of the principal planes as the lens gets stronger. The lens gets stronger from (a) to (b) to (c).

4.13 Combination of Thin Lenses

Earlier we saw how an electrostatic lens consists of a convergent and a divergent region. The convergent region corresponds to the region where the second derivative of the axial potential is positive, and the divergent corresponds to where it is negative. The errors can be reduced greatly if we consider the overall lens to be made up of two thin lenses, one convergent and one divergent. If the cardinal points of these thin lenses are known, the corresponding cardinal points of the overall lens can easily be determined.

In order to derive a general relation, we shall not confine our analysis to only two lenses. Let us consider a series of several thin lenses. Figure 4.17

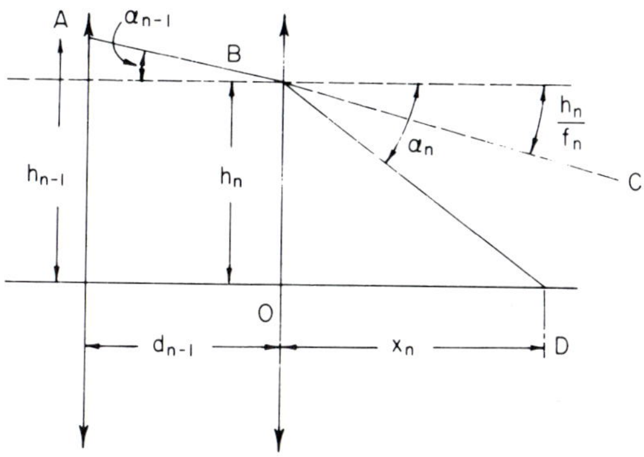

Fig. 4.17. Combination of thin lenses. Line *AB* is the trajectory as it enters lens *n*. Line *BD* is the trajectory as it leaves lens *n*. Line *BC* is the trajectory of the ray for an incoming parallel ray, in other words, the action of lens *n* alone with $\alpha_{n-1} = 0$.

shows the beam at the *n*th lens. If h_n is the height of the particle from the axis, α_{n-1} the angle that the trajectory of the particle makes with the axis before entering lens *n*, α_n the angle after it leaves the lens, and d_{n-1} the distance between lens $n-1$ and lens *n*, then

$$\alpha_n = \alpha_{n-1} + (h_n/f_n) \tag{4.63}$$

$$h_n = h_{n-1} - \alpha_{n-1} d_{n-1} \tag{4.64}$$

or

$$\alpha_n = \alpha_{n-1}\left(1 - \frac{d_{n-1}}{f_n}\right) + \frac{h_{n-1}}{f_n} \tag{4.65}$$

All these relations are, of course, for paraxial rays. Therefore, we can say that the *n*th lens operates on α and *h* with the matrix

$$\begin{bmatrix} 1 - \dfrac{d_{n-1}}{f_n} & \dfrac{1}{f_n} \\ -d_{n-1} & 1 \end{bmatrix} \tag{4.66}$$

If we assume that the incident ray on the first lens is parallel to the axis, the matrix operator is

$$\begin{bmatrix} 0 & \dfrac{1}{f_1} \\ 0 & 1 \end{bmatrix} \tag{4.67}$$

We can obtain the overall focal length by multiplying the matrices. Thus, for example, if we consider a combination of three lenses, we have

$$\begin{bmatrix} 1-\dfrac{d_2}{f_3} & \dfrac{1}{f_3} \\ -d_2 & 1 \end{bmatrix} \times \begin{bmatrix} 1-\dfrac{d_1}{f_2} & \dfrac{1}{f_2} \\ -d_1 & 1 \end{bmatrix} \times \begin{bmatrix} 0 & \dfrac{1}{f_1} \\ 0 & 1 \end{bmatrix}$$

$$= \begin{bmatrix} 0 & \dfrac{1}{f_1}+\dfrac{1}{f_2}+\dfrac{1}{f_3}-\dfrac{d_1}{f_1 f_2}-\dfrac{d_2}{f_2 f_3}+\dfrac{d_1 d_2}{f_1 f_2 f_3}-\dfrac{d_1+d_2}{f_1 f_3} \\ 0 & 1-\dfrac{d_1+d_2}{f_1}-\dfrac{d_2}{f_2}+\dfrac{d_1 d_2}{f_1 f_2} \end{bmatrix} \tag{4.68}$$

The combined focal length is then f for which

$$\frac{1}{f}=\frac{1}{f_1}+\frac{1}{f_2}+\frac{1}{f_3}-\frac{d_1}{f_1 f_2}-\frac{d_2}{f_2 f_3}-\frac{d_1+d_2}{f_3 f_1}+\frac{d_1 d_2}{f_1 f_2 f_3} \tag{4.69}$$

If the lens consists of n components, the overall focal length of the combination is given by

$$f = h_0/\alpha_n \tag{4.70}$$

Hence, the distance of the focal point from the last lens is, in general,

$$x_n = h_n/\alpha_n = f h_n/h_0 \tag{4.71}$$

and

$$h_n = (x_n/f)\, h_0 \tag{4.72}$$

Therefore, the 4th element of the matrix is equal to x_n/f.

For the case of three lenses

$$x_3 = f\left[1-\frac{d_1+d_2}{f_1}-\frac{d_2}{f_2}+\frac{d_1 d_2}{f_1 f_2}\right] \tag{4.73}$$

4.14 The Cardinal Points of a Strong Lens

In strong lenses, we can no longer assume that the radius remains constant inside the lens. If we do, we will obtain a focal length somewhat smaller than the exact value. This is because the analysis will be based on a height from the axis larger than the actual height thereby leading to the incorrect conclusion that the converging power of the lens is stronger than it really is. In all unipotential lenses the radius of the ray changes appreciably inside the lens, and therefore the analysis of a weak lens does not apply. The following sections give two methods that can be used for unipotential lenses and any other strong electrostatic lens.

4.15 Method of Sectionizing the Lens

For a given voltage distribution of the lens system, the parameter $[\phi'(z)/\phi(z)]^2$ is plotted as a function of z. Figure 4.18 shows an example of such a curve for a two-cylinder equidiameter lens. First, the center of gravity G on the z axis is determined by finding a line perpendicular to the axis which divides the area under the curve into two equal parts. Next, G_1 and G_2 are

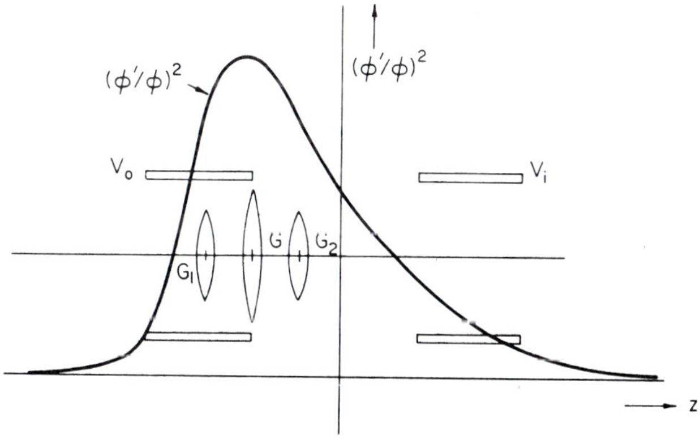

Fig. 4.18. Method of sectionizing the lens: G is the center of gravity of the complete curve; G_1 and G_2 are the centers of gravity of each half separated by G.

obtained as points on the z axis through which lines perpendicular to the axis divide each half area of the curve into two equal areas. Instead of one lens, we place two lenses, one at G_1 and one at G_2, having focal lengths given by

$$(\bar{F}_1^{-1})_{\text{at } G_1} = \tfrac{3}{16} \int_{z_0}^{G} [\phi'(z)/\phi(z)]^2 \, dz \qquad (4.74)$$

and

$$(\bar{F}_2^{-1})_{\text{at } G_2} = \tfrac{3}{16} \int_G^{z_i} [\phi'(z)/\phi(z)]^2 \, dz \tag{4.75}$$

respectively. Since the areas are equal, the integrals of Eqs. (4.74) and Eq. (4.75) are equal and $\bar{F}_1 = \bar{F}_2 = 2\bar{F}$. Combining the two lenses as in Section 4.13, we obtain for the corrected reduced focal length $\bar{F}_c$

$$\frac{1}{\bar{F}_c} = \frac{1}{\bar{F}}\left(1 - \frac{d}{4\bar{F}}\right) \tag{4.76}$$

where d is the distance between G_1 and G_2. It is therefore not necessary to determine the location of G.

This method, while very attractive because of its simplicity, is not very accurate. In the next chapter a comparison will be made between this method and the one described in the following section.

4.16 Method of Successive Approximation

Equation (4.43) or Eq. (4.44) represents only the first term of an infinite series for the focal length. If R changes inside the lens, as it does in strong lenses and in most unipotential lenses, additional terms of the series must be used. The procedure is to replace R by better and better approximations in the exact integrated paraxial-ray equation. This successive approximation is realized as follows:

$$R' = R_0' + R_1' + R_2' + \cdots = \sum_{n=0}^{\infty} R_n' \tag{4.77}$$

The initial conditions are $R = R_0$ and $R_0' = 0$. Then

$$R_1' = -\tfrac{3}{16} \int_{z_o}^{z_i} (\phi'/\phi)^2 R_0 \, dz = -\tfrac{3}{16} R_0 \int_{z_o}^{z_i} (\phi'/\phi)^2 \, dz \tag{4.78}$$

Next

$$R_1 = -\tfrac{3}{16} R_0 \int_{z_o}^{z_i} \int_{z_o}^{z} (\phi'/\phi)^2 \, dz \, dz \tag{4.79}$$

and

$$R_2' = -\tfrac{3}{16} \int_{z_o}^{z_i} (\phi'/\phi)^2 R_1 \, dz$$

$$= +\tfrac{9}{256} R_0 \int_{z_o}^{z_i} (\phi'/\phi)^2 \int_{z_o}^{z} \int_{z_o}^{z} (\phi'/\phi)^2 \, dz \, dz \, dz \tag{4.80}$$

and

$$R_2 = \int_{z_o}^{z_i} R_2' \, dz \quad \text{etc.} \tag{4.81}$$

and since $1/\bar{F}_i = -R_i'/R_0$, we have

$$\bar{F}_i^{-1} = +(3/16)\int_{z_o}^{z_i} (\phi'/\phi)^2 \, dz$$
$$- (9/256)\int_{z_o}^{z_i} (\phi'/\phi)^2 \int_{z_o}^{z} \int_{z_o}^{z} (\phi'/\phi)^2 \, dz \, dz \, dz + \cdots \tag{4.82}$$

In this notation, the last integral on the right is evaluated first to obtain a certain function $f(z)$. This function is then included in the integrand of the next integral and a new function $f(z)$ is obtained. This procedure is continued until all the integrals have been evaluated.

The series converges very rapidly. Generally speaking, it is enough to include only two terms of the series. This will become evident when we analyze various lenses in the next chapter.

4.17 Development up to Second Approximation

In general, we have

$$R(z) = R(z_o) + \left.\frac{dR}{dz}\right|_{z=z_o} (z - z_o)$$
$$- \frac{3}{16}\int_{z_o}^{z}\int_{z_o}^{z}\left[\frac{\phi'(z)}{\phi(z)}\right]^2 \left[R(z_o) + \left.\frac{dR}{dz}\right|_{z=z_o} (z - z_o)\right] dz \, dz$$
$$+ \frac{9}{256}\int_{z_o}^{z}\left\{\int_{z_o}^{z}\left[\frac{\phi'(z)}{\phi(z)}\right]^2 \int_{z_o}^{z}\left[\int_{z_o}^{z}\left(\frac{\phi'(z)}{\phi(z)}\right)^2\right.\right.$$
$$\left.\left.\times \left(R(z_o) + \left.\frac{dR}{dz}\right|_{z=z_o} (z - z_o)\right) dz\right] dz \, dz\right\} dz \tag{4.83}$$

Let

$$Q = [\phi'(z)/\phi(z)]^2 \tag{4.84}$$

For rays parallel to the axis on the object side, we have

$$\left.\frac{dR}{dz}\right|_{z=z_o} = 0$$

and Eq. (4.83) reduces to

$$R(z) = R(z_o) - (3/16)R(z_o) \int_{z_o}^{z} \int_{z_o}^{z} Q \, dz \, dz$$

$$+ (9/256)R(z_o) \int_{z_o}^{z} \int_{z_o}^{z} Q \int_{z_o}^{z} \int_{z_o}^{z} Q \, dz \, dz \, dz \, dz \tag{4.85}$$

We already know that

$$R(z_o) = \phi(z_o)^{1/4} r(z_o), \tag{4.86}$$

and

$$r'(z) = \frac{R'(z)}{\phi(z)^{1/4}} - \frac{1}{4} \frac{R(z)\phi'(z)}{\phi(z)^{5/4}} \tag{4.87}$$

Therefore,

$$R'(z) = -(3/16)R(z_o) \int_{z_o}^{z} Q \, dz + (9/256)R(z_o) \int_{z_o}^{z} Q \int_{z_o}^{z} \left(\int_{z_o}^{z} Q \, dz \right) dz \, dz \tag{4.88}$$

and

$$R(z) = [\phi(z_o)]^{1/4} r(z_o) \left\{ 1 - \frac{3}{16} \int_{z_o}^{z} \int_{z_o}^{z} Q \, dz \, dz \right.$$

$$\left. + \frac{9}{256} \int_{z_o}^{z} \left[\int_{z_o}^{z} Q \int_{z_o}^{z} \left(\int_{z_o}^{z} Q \, dz \right) dz \, dz \right] dz \right\} \tag{4.89}$$

A similar relation for $r(z)$ can be obtained by dividing by $[\phi(z)]^{1/4}$. Next we can easily obtain $R'(z)$ and $r'(z)$.

$$R'(z) = -(3/16)[\phi(z_o)]^{1/4} r(z_o) \left[\int_{z_o}^{z} Q \, dz - (3/16) \int_{z_o}^{z} Q \int_{z_o}^{z} \left(\int_{z_o}^{z} Q \, dz \right) dz \, dz \right] \tag{4.90}$$

and

$$r'(z) = -\frac{3}{16} \left[\frac{\phi(z_o)}{\phi(z)} \right]^{1/4} r(z_o) \left[\int_{z_o}^{z} Q \, dz - \frac{3}{16} \int_{z_o}^{z} Q \int_{z_o}^{z} \left(\int_{z_o}^{z} Q \, dz \right) dz \, dz \right]$$

$$- \frac{1}{4} \frac{\phi(z_o)^{1/4} \phi'(z)}{\phi(z)^{5/4}} r(z_o) \left\{ 1 - \frac{3}{16} \int_{z_o}^{z} \int_{z_o}^{z} Q \, dz \, dz \right.$$

$$\left. + \frac{9}{256} \int_{z_o}^{z} \left[\int_{z_o}^{z} Q \int_{z_o}^{z} \left(\int_{z_o}^{z} Q \, dz \right) dz \, dz \right] dz \right\} \tag{4.91}$$

The ray coming in parallel at distance $r(z_o)$ from the axis has the following relation in the image space:

$$r(z) = r_i + (z - z_i)r'(z_i) \tag{4.92}$$

Here, r_i is the distance of the ray from the axis at an arbitrary position z_i where we are sure that $\phi(z_i)$ is constant and $\phi'(z_i) = 0$. We define the midfocal length as the distance between the focal point and the reference plane of the lens (generally taken as the geometrical center in a symmetrical lens).

When the ray intersects the axis, $r(z) = 0$, and we then have

$$z_{m_i} = z_i - (r_i/r_i') \tag{4.93}$$

Here z_{m_i} is the image-side midfocal length. We thus obtain

$$\begin{aligned} z_{m_i} = z_i - \Bigg\{&[\phi(z_o)/\phi(z_i)]^{1/4} r(z_o)\bigg\{1 - (3/16)\int_{z_o}^{z_i}\int_{z_o}^{z} Q\,dz\,dz \\ &+ (9/256)\int_{z_o}^{z_i}\left[\int_{z_o}^{z} Q\int_{z_o}^{z}\left(\int_{z_o}^{z} Q\,dz\right)dz\,dz\right]dz\bigg\}\Bigg\} \\ &\times\Bigg\{-(3/16)[\phi(z_o)/\phi(z_i)]^{1/4} r(z_o)\bigg[\int_{z_o}^{z_i} Q\,dz \\ &- (3/16)\int_{z_o}^{z_i} Q\int_{z_o}^{z}\left(\int_{z_o}^{z} Q\,dz\right)dz\,dz\bigg]\Bigg\}^{-1} \end{aligned} \tag{4.94}$$

or

$$\begin{aligned} z_{m_i} = z_i + (16/3)\bigg\{&1 - (3/16)\int_{z_o}^{z_i}\int_{z_o}^{z} Q\,dz\,dz \\ &+ (9/256)\int_{z_o}^{z_i}\left[\int_{z_o}^{z} Q\int_{z_o}^{z}\left(\int_{z_o}^{z} Q\,dz\right)dz\,dz\right]dz\bigg\} \\ &\times\left\{\int_{z_o}^{z_i} Q\,dz - (3/16)\int_{z_o}^{z_i} Q\int_{z_o}^{z}\left(\int_{z_o}^{z} Q\,dz\right)dz\,dz\right\}^{-1} \end{aligned} \tag{4.95}$$

Now, for incoming parallel rays, the image-side focal length is $f_i = -[r(z_o)/r'(z_i)]$ with $r'(z_o) = 0$, and we write

$$\begin{aligned} f_i = -r(z_o)\bigg\{&-(3/16)[\phi(z_o)/\phi(z_i)]^{1/4}\, r(z_o)\bigg[\int_{z_o}^{z_i} Q\,dz \\ &- (3/16)\int_{z_o}^{z_i} Q\int_{z_o}^{z}\left(\int_{z_o}^{z} Q\,dz\right)dz\,dz\bigg]\bigg\}^{-1} \end{aligned} \tag{4.96}$$

$$f_i = +(16/3)[\phi(z_i)/\phi(z_o)]^{1/4}\left[\int_{z_o}^{z_i} Q\,dz - (3/16)\int_{z_o}^{z_i} Q\int_{z_o}^{z}\left(\int_{z_o}^{z} Q\,dz\right)dz\,dz\right]^{-1} \tag{4.97}$$

The principal point for the image side is then

$$
\begin{aligned}
z_{p_i} &= z_{m_i} - f_i \\
&= z_i + (16/3)\left\{1 - (3/16)\int_{z_o}^{z_i}\int_{z_o}^{z} Q\,dz\,dz \right. \\
&\quad \left. + (9/256)\int_{z_o}^{z}\left[\int_{z_o}^{z} Q\int_{z_o}^{z}\left(\int_{z_o}^{z} Q\,dz\right)dz\,dz\right]dz - [\phi(z_i)/\phi(z_o)]^{1/4}\right\} \\
&\quad \times\left[\int_{z_o}^{z_i} Q\,dz - (3/16)\int_{z_o}^{z_i} Q\int_{z_o}^{z}\left(\int_{z_o}^{z} Q\,dz\right)dz\,dz\right]^{-1}
\end{aligned}
\tag{4.98}
$$

and, if we limit ourselves to the first approximation, we obtain

$$
\begin{aligned}
z_{p_i} &= z_{m_i} - f_i \\
&= z_i + (16/3)\left[1 - (3/16)\int_{z_o}^{z_i}\int_{z_o}^{z} Q\,dz\,dz - [\phi(z_i)/\phi(z_o)]^{1/4}\right]\left(\int_{z_o}^{z_i} Q\,dz\right)^{-1}
\end{aligned}
\tag{4.99}
$$

V

The Electrostatic Immersion Lens

In this chapter we shall study two kinds of immersion lenses. The symmetrical two-cylinder lens as well as the immersion lens consisting of concentric cylinders of different radii will be treated in detail. As we have seen in the previous chapter, these lenses will perform both in the accelerating and in the decelerating mode. Their cardinal points will obviously not be the same in each case.

5.1 Symmetrical Two-Cylinder Lens with Negligible Gap

We shall first consider the lens with negligible gap width as shown in Fig. 5.1 and obtain all the cardinal points as a function of the voltage ratio

$$v = \phi_i/\phi_o = V_2/V_1 \tag{5.1}$$

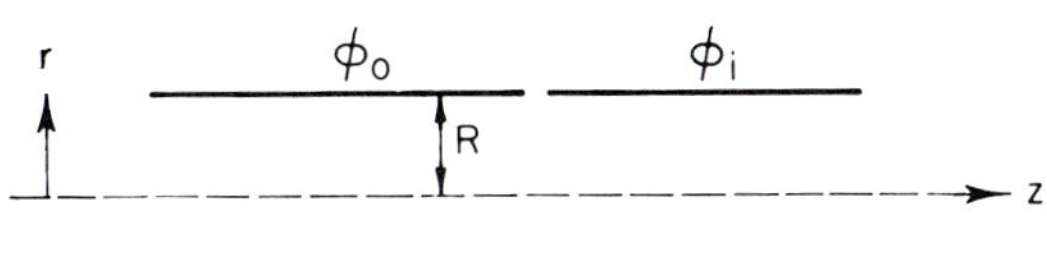

Fig. 5.1. Equidiameter two-cylinder electrostatic lens.

The image- and object-side focal lengths are related to the potentials. A similar relationship holds for the principal planes as given by Eq. (4.61). However, we need to know the center of gravity of the lens to apply Eq. (4.61).

Using Eq. (3.98) for the axial potential distribution, we shall determine analytically the cardinal points of this lens.

From Chapter IV, we know that the focal length of a weak lens is given by

$$\frac{1}{f_i} = \frac{3}{16}\left(\frac{\phi_o}{\phi_i}\right)^{1/4} \int_{z_o}^{z_i} \left[\frac{\phi'(z)}{\phi(z)}\right]^2 dz \tag{5.2}$$

From Chapter III, we have the axial potential distribution of a two-electrode lens. If we let

$$\chi = (\sigma - 1)/(\sigma + 1) \tag{5.3}$$

and use Eq. (3.98), we obtain

$$\frac{\phi'(z)}{\phi(z)} = \omega\chi \frac{\operatorname{sech}^2 \omega z}{1 + \chi \tanh \omega z} \tag{5.4}$$

where $\omega = 1.31845$.

The focal length is then given by

$$\frac{1}{f_i} = \frac{3}{16}\sigma^{-1/4}\chi^2\omega^2 \int_{-\infty}^{+\infty} \frac{\operatorname{sech}^4 \omega z}{(1 + \chi \tanh \omega z)^2} dz \tag{5.5}$$

We make the following substitution:

$$\xi = \tanh \omega z \tag{5.6}$$

then

$$\frac{1}{f_i} = \frac{3}{16}\sigma^{-1/4}\omega\chi^2 \int_{-1}^{+1} \frac{1 - \xi^2}{(1 + \chi\xi)^2} d\xi = \frac{3}{8}\sigma^{-1/4}\omega\left(\frac{1}{\chi}\log\sigma - 2\right) \tag{5.7}$$

Substituting for the value of ω, we finally obtain

$$f_i = \frac{2.02\sigma^{+1/4}}{[(\sigma + 1)/(\sigma - 1)]\log\sigma - 2} \tag{5.8}$$

This formula is dependable for weak lenses, i.e., for small values of σ. We shall see later on that σ can be as high as 4 before serious errors develop.

Next, let us obtain the focal length for strong lenses using the method of successive approximation. We use Eq. (4.82), and write

$$J = \int_{zo}^{z_i} \left[\frac{\phi'(z)}{\phi(z)}\right]^2 \int_{z_o}^{z} \int_{z_o}^{z_2} \left[\frac{\phi'(z_1)}{\phi(z_1)}\right]^2 dz_1\, dz_2\, dz \tag{5.9}$$

In general, we know that

$$\int_a^b dy \int_a^y f(x, y)\, dx = \int_a^b dx \int_x^b f(x, y)\, dy \tag{5.10}$$

We will therefore interchange the order of integration in Eq. (5.9), and write

$$J = \int_{z_o}^{z_i} \left[\frac{\phi'(z)}{\phi(z)}\right]^2 dz \int_{z_o}^{z} \left[\frac{\phi'(z_1)}{\phi(z_1)}\right]^2 dz_1 \int_{z_1}^{z} dz_2 \tag{5.11}$$

or

$$J = \int_{z_o}^{z_i} \left[\frac{\phi'(z)}{\phi(z)}\right]^2 dz \int_{z_o}^{z} \left[\frac{\phi'(z_1)}{\phi(z_1)}\right]^2 dz_1 (z - z_1) \tag{5.12}$$

$$J = \int_{z_o}^{z_i} \left[\frac{\phi'(z)}{\phi(z)}\right]^2 z\, dz \int_{z_o}^{z} \left[\frac{\phi'(z_1)}{\phi(z_1)}\right]^2 dz_1$$
$$- \int_{z_o}^{z_i} \left[\frac{\phi'(z)}{\phi(z)}\right]^2 dz \int_{z_o}^{z} \left[\frac{\phi'(z_1)}{\phi(z_1)}\right] z_1\, dz_1 \tag{5.13}$$

or

$$J = J_1 - J_2 \tag{5.14}$$

In J_2 we again change the order of integration according to the following relation:

$$\int_a^b g(y)\, dy \int_a^y f(x)\, dx = \int_a^b f(x)\, dx \int_x^b g(y)\, dy \tag{5.15}$$

Then

$$J_2 = \int_{z_o}^{z_i} z_1 \left[\frac{\phi'(z_1)}{\phi(z_1)}\right]^2 dz_1 \int_{z_1}^{z_i} \left[\frac{\phi'(z)}{\phi(z)}\right]^2 dz \tag{5.16}$$

Interchanging z and z_1, we get

$$J_2 = \int_{z_o}^{z_i} z \left[\frac{\phi'(z)}{\phi(z)}\right]^2 dz \int_{z}^{z_i} \left[\frac{\phi'(z_1)}{\phi(z_1)}\right]^2 dz_1 \tag{5.17}$$

and

$$J = J_1 - J_2 = \int_{z_o}^{z_i} z \left[\frac{\phi'(z)}{\phi(z)}\right]^2 dz \left\{\int_{z_o}^{z} \left[\frac{\phi'(z_1)}{\phi(z_1)}\right]^2 dz_1 - \int_{z}^{z_i} \left[\frac{\phi'(z_1)}{\phi(z_1)}\right]^2 dz_1\right\} \tag{5.18}$$

We have seen earlier that

$$\frac{\phi'(z)}{\phi(z)} = \omega\chi \frac{1 - \xi^2}{1 + \chi\xi} \tag{5.19}$$

with $\xi = \tanh \omega z$ and

$$z = \frac{1}{\omega} \operatorname{arctanh} \xi = \frac{1}{\omega} \log\left(\frac{1+\xi}{1-\xi}\right)^{1/2}$$

We can easily show that $dz = [\omega(1 - \xi^2)]^{-1}\, d\xi$, and

$$J = \int_{-1}^{+1} \frac{1}{\omega} \log\left(\frac{1+\xi}{1-\xi}\right)^{1/2} \omega^2 \chi^2 \frac{(1-\xi^2)^2}{(1+\chi\xi)^2} \frac{d\xi}{\omega(1-\xi^2)}$$
$$\times \left[\int_{-1}^{\xi} \frac{\omega^2\chi^2(1-\xi_1{}^2)^2}{(1+\chi\xi_1)^2} \frac{d\xi_1}{\omega(1-\xi_1{}^2)} - \int_{\xi}^{+1} \frac{\omega^2\chi^2(1-\xi_1{}^2)^2}{(1+\chi\xi_1)^2} \frac{d\xi_1}{\omega(1-\xi_1{}^2)}\right] \tag{5.20}$$

$$J = \frac{\omega}{2} \chi^4 \int_{-1}^{+1} \log\left(\frac{1+\xi}{1-\xi}\right) \frac{1-\xi^2}{(1+\chi\xi)^2}\, d\xi$$
$$\times \left[\int_{-1}^{\xi} \frac{1-\xi_1{}^2}{(1+\chi\xi_1)^2}\, d\xi_1 - \int_{\xi}^{+1} \frac{1-\xi_1{}^2}{(1+\chi\xi_1)^2}\, d\xi_1\right] \tag{5.21}$$

Now

$$\int \frac{1-\xi_1{}^2}{(1+\chi\xi_1)^2}\, d\xi_1 = \frac{1}{\chi^3}\left[\frac{1-\chi^2}{1+\chi\xi_1} + 2\log(1+\chi\xi_1) - \chi\xi_1\right] \tag{5.22}$$

Using Eq. (5.22), Eq. (5.21) can be written after some reduction, as

$$J = \omega\chi \int_{-1}^{+1} \frac{1-\xi^2}{(1+\chi\xi)^2}$$
$$\times \left[\frac{1-\chi^2}{1+\chi\xi} - (1+\chi\xi) + 2\log(1+\chi\xi) - \log(1-\chi^2)\right] \log\frac{1+\xi}{1-\xi}\, d\xi \tag{5.23}$$

Let

$$I(p, q) = \chi^p \int_{-1}^{+1} \frac{\xi^p}{(1+\chi\xi)^q} \log\left(\frac{1+\xi}{1-\xi}\right) d\xi \tag{5.24}$$

Then

$$J = 2\omega\chi \int_{-1}^{+1} \frac{1-\xi^2}{(1+\chi\xi)^2} \log(1+\chi\xi) \log\left(\frac{1+\xi}{1-\xi}\right) d\xi - \frac{\omega}{\chi}(1-\chi^2)^2 I(0, 3)$$
$$+ \frac{\omega}{\chi}(1-\chi^2)[2 + \log(1-\chi^2)]I(0, 2) - \frac{2\omega}{\chi}\log(1-\chi^2)I(0, 1)$$
$$+ \frac{\omega}{\chi} I(1, 0) - \frac{\omega}{\chi}[1 - \log(1-\chi^2)] \int_{-1}^{+1} \log\left(\frac{1+\xi}{1-\xi}\right) d\xi \tag{5.25}$$

We can easily show that

$$\int_{-1}^{+1} \log\left(\frac{1+\xi}{1-\xi}\right) d\xi = 0 \tag{5.26}$$

We thus obtain

$$\begin{aligned} J = 2\omega\chi \int_{-1}^{+1} \frac{1-\xi^2}{(1+\chi\xi)^2} \log(1+\chi\xi) \log\left(\frac{1+\xi}{1-\xi}\right) d\xi \\ - \frac{\omega}{\chi}[(1-\chi^2)^2 I(0,3) - (1-\chi^2)[2+\log(1-\chi^2)]I(0,2) \\ + 2\log(1-\chi^2)I(0,1) - I(1,0)] \end{aligned} \tag{5.27}$$

Now

$$I(1,0) = \chi \int_{-1}^{+1} \xi \log\left(\frac{1+\xi}{1-\xi}\right) d\xi = 2\chi \tag{5.28}$$

and we will show in the appendix to this chapter that

$$I(0,1) = \int_{-1}^{+1} \log\left[\left(\frac{1+\xi}{1-\xi}\right)\left(\frac{1}{1+\chi\xi}\right)\right] d\xi = -\frac{1}{2\chi}\left(\log\frac{1+\chi}{1-\chi}\right)^2 \tag{5.29}$$

$$\begin{aligned} I(p+1, q+1) &= \chi^{p+1} \int_{-1}^{+1} \frac{\xi^{p+1}}{(1+\chi\xi)^{q+1}} \log\left(\frac{1+\xi}{1-\xi}\right) d\xi \\ &= \chi^p \int_{-1}^{+1} \frac{(1+\chi\xi-1)\xi^p}{(1+\chi\xi)^{q+1}} \log\left(\frac{1+\xi}{1-\xi}\right) d\xi \end{aligned} \tag{5.30}$$

or

$$\begin{aligned} I(p+1, q+1) = \chi^p \int_{-1}^{+1} \frac{\xi^p}{(1+\chi\xi)^q} \log\left(\frac{1+\xi}{1-\xi}\right) d\xi \\ - \chi^p \int_{-1}^{+1} \frac{\xi^p}{(1+\chi\xi)^{q+1}} \log\left(\frac{1+\xi}{1-\xi}\right) d\xi \end{aligned} \tag{5.31}$$

In general, therefore

$$I(p+1, q+1) = I(p,q) - I(p, q+1) \tag{5.32}$$

Furthermore,

$$\frac{d}{d\chi} I(0,q) = \frac{d}{d\chi} \int_{-1}^{+1} \frac{1}{(1+\chi\xi)^q} \log\left(\frac{1+\xi}{1-\xi}\right) d\xi = -\frac{q}{\chi} I(1, q+1) \tag{5.33}$$

Therefore,

$$I(0, q+1) = I(0, q) + \frac{\chi}{q}\frac{d}{d\chi} I(0, q) \tag{5.34}$$

and

$$I(0, 2) = I(0, 1) + \frac{\chi}{1}\frac{d}{d\chi} I(0, 1) \tag{5.35}$$

or

$$I(0, 2) = -\frac{2}{1-\chi^2}\log\frac{1+\chi}{1-\chi} \tag{5.36}$$

Similarly, we have

$$I(0, 3) = I(0, 2) + \frac{\chi}{2}\frac{d}{d\chi} I(0, 2) \tag{5.37}$$

and

$$I(0, 3) = -\frac{2}{(1-\chi^2)^2}\left[\log\left(\frac{1+\chi}{1-\chi}\right) + \chi\right] \tag{5.38}$$

Using all these relations, Eq. (5.27) simplifies to

$$J = \frac{\omega}{\chi}\left\{4\chi - 2[1 + \log(1-\chi^2)]\log\sigma + \frac{1}{\chi}\log(1-\chi^2)(\log\sigma)^2\right\} + 2\omega\chi\beta \tag{5.39}$$

where β is defined by

$$\beta = \int_{-1}^{+1} \frac{1-\xi^2}{(1+\chi\xi)^2}\log(1+\chi\xi)\log\left(\frac{1+\xi}{1-\xi}\right) d\xi \tag{5.40}$$

It is possible to solve Eq. (5.40) by expanding log $(1 + \chi\xi)$ into a series and obtaining a series expression in terms of $I(p, q)$. However, it was found much simpler to solve it numerically with the help of a computer. Using Eq. (4.97) from the previous chapter, we obtain

$$f_i^{-1} = \sigma^{-1/4}(A - B) \tag{5.41}$$

in which

$$A = \frac{3\omega}{8}\left(\frac{1}{\chi}\log\sigma - 2\right) \tag{5.42}$$

and

$$B = \left(\frac{3}{16}\right)^2 J$$

$$= \left(\frac{3}{16}\right)^2 \frac{\omega}{\chi}\left\{4\chi - 2[1 + \log(1-\chi^2)]\log\sigma + \frac{1}{\chi}\log(1-\chi^2)(\log\sigma)^2 + 2\chi^2\beta\right\} \tag{5.43}$$

We can see immediately that $\sigma^{-1/4}A$ of Eq. (5.41) represents the thin-lens formula derived in Eq. (5.7). Part B is the second term of the infinite series. Table 5.1[1] shows the values of f_i, A, B, β, using the above equations, and f_i obtained with the thin-lens formula given by Eq. (5.8) for various ratios of V_2/V_1. We can see that even for strong lenses, B is appreciably smaller than A. This, therefore, definitely justifies our previous decision to limit ourselves up to the second term of the series.

One can pursue the same kind of analysis and obtain the position of the principal planes. However, we found it much simpler to compute these values by numerical integration using a digital computer. The focal length, the position of the principal plane, and the midfocal length are tabulated in Table 5.2 for various values of V_2/V_1 for the case of an accelerating lens. A comparison of the values obtained this way with those listed in Table 5.1 clearly shows the agreement of the two results.

Similar values were obtained with the help of a digital computer for the decelerating lens. We can see that in all cases the relationship between f_o and f_i holds.

5.2 Method of Sectionizing the Lens

An attempt was made to determine the cardinal points by sectionizing the lens as discussed in the previous chapter. The following is an example of the calculation:

$$V_2/V_1 = 10$$

The focal lengths according to the formulas given by Eqs. (4.40) and (4.42) are:

$$\frac{1}{F} = \frac{3}{16}\int_{z_o}^{z_i}\left(\frac{\phi'}{\phi}\right)^2 dz = \frac{3}{16}\cdot 2.14967$$

and

$$\frac{1}{f} = \left(\frac{1}{10}\right)^{1/4}\frac{3}{16}\cdot 2.14967 = 0.2265$$

[1]All tables for Chapter V will be found at the end of the chapter.

or

$$f = 4.42$$

The exact value obtained by numerical integration as discussed in Section 5.3 is $f = 4.95$. There is, therefore, an appreciable error. The lens is then sectionized:

Half area = 2.14967/2 = 1.07483;
Position of 1/4 point is at $z = -0.74505$;
Position of 3/4 point is at $z = -0.12715$;
Distance d between the two points = 0.61790.

Corrected focal length

$$\frac{1}{F_c} = \frac{1}{F}\left(1 - \frac{d}{4F}\right) = 0.378$$

and

$$f = 10^{1/4}/0.378 = 4.7$$

The error is still about 5%.

Similar errors were obtained for other values of V_2/V_1. It is obvious, therefore, that this method is not suitable for the analysis of this type of lens if a sufficient accuracy is required.

5.3 Analysis of the Symmetrical Two-Cylinder Lens Using a Digital Computer

The reduced paraxial-ray formula as given by Eq. (2.38) is of the form

$$\frac{d^2R}{dz^2} = -K^2R \tag{5.44}$$

There are different formulas available in the literature to solve this equation numerically. The following formula was found to be very accurate and convenient in this case:

$$R_{n+1} = R_n - R_{n-4} - R_{n-5} + \frac{\omega^2}{48}(67R''_n - 8R''_{n-1} + 122R''_{n-2} - 8R''_{n+2} + 67R''_{n-4}) \tag{5.45}$$

The analysis was started from a point sufficiently outside the field to ascertain that the derivatives along the axis be zero. The numerical analysis was then performed step by step until a point was reached after passing through the lens where the field was negligible.

In all the cases, we entered parallel to the axis at a normalized height of 1.

The slope of the trajectory after passing through the lens was determined outside the lens in a field-free region, and the values of f, z_m, and z_p were then easily computed.

The same analysis was done for the lens with spacing s. Here, Eq. (3.97) was used to determine the axial potential and its derivatives. The results for spacings from 0.1 to 2 in intervals of 0.1 are listed in Tables 5.3–5.22. Figure 5.2 shows the variation of the focal length and midfocal length as a function of the voltage ratio. In Fig. 5.3 the separation of the principal planes is plotted as a function of V_2/V_1.

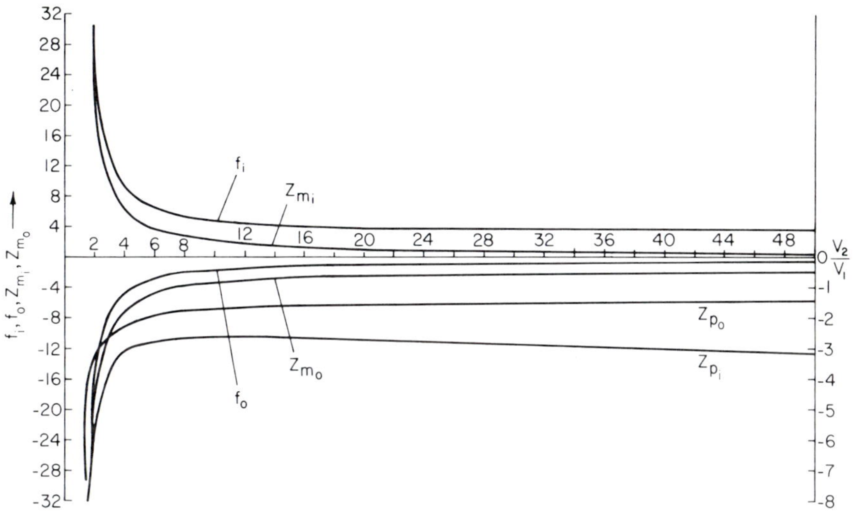

Fig. 5.2. Focal length and midfocal length as functions of voltage ratio for the symmetrical electrostatic lens.

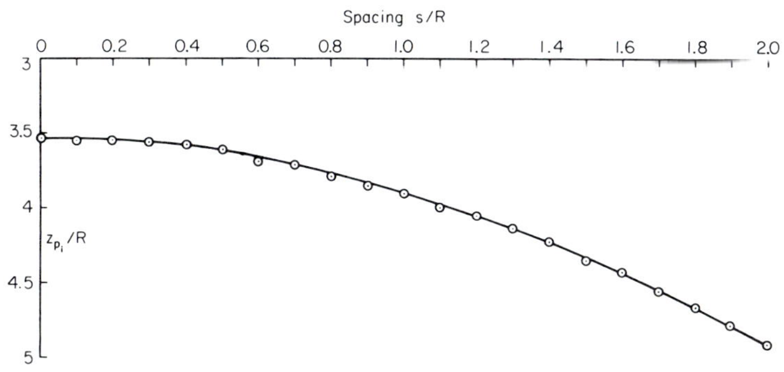

Fig. 5.3. Variation of the position of the principal plane with spacing s. The voltage ratio is 3.

5.4 Analysis of the Results

Upon examining the results obtained above, the following observations can be made:

(a) The focal length in the direction of increasing potential is always greater than the focal length in the direction of decreasing potential.

(b) The focal length of the symmetrical two-tube lens increases as s increases; that is, the lens gets weaker as the axial spacing of the cylinders increases. The change is small for small spacings, but it increases as the spacing is increased. At high voltage ratios, the effect is quite small.

(c) The position of the principal planes does not change much with voltage ratio except at very low values.

(d) As the spacing s is increased, a slight movement of the principal planes toward the low-voltage cylinder occurs. At high-voltage ratios, this effect is negligible.

(e) The lens thickness defined by the separation of the principal planes increases monotonically with voltage ratio, with decreasing rate as the voltage ratio increases.

5.5 The Asymmetrical Two-Tube Lens

In the case in which two coaxial cylinders of unequal diameters are placed end-to-end as shown in Fig. 5.4, the determination of the axial potential distribution and its derivatives is very complex. An analytical solution has been published. We shall not attempt to reproduce, here, the complete details of the analysis since this is outside the scope of this book. However, we will use the results to obtain the characteristics of the lens.

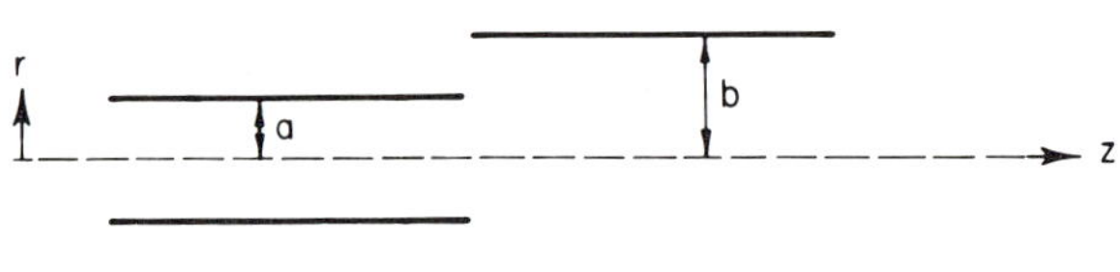

Fig. 5.4. Asymmetrical two-cylinder electrostatic lens.

Here, a cylinder of radius b is assumed to be of infinite length and is maintained at zero potential. A coaxial cylinder of radius a and of semi-infinite extent with a thin wall (actually negligible in our case) is at potential V. This is shown in Fig. 5.5. While this is not an exact model of the standard geometry shown in Fig. 5.4, the potential distribution along the axis and its derivatives are practically the same in both cases.

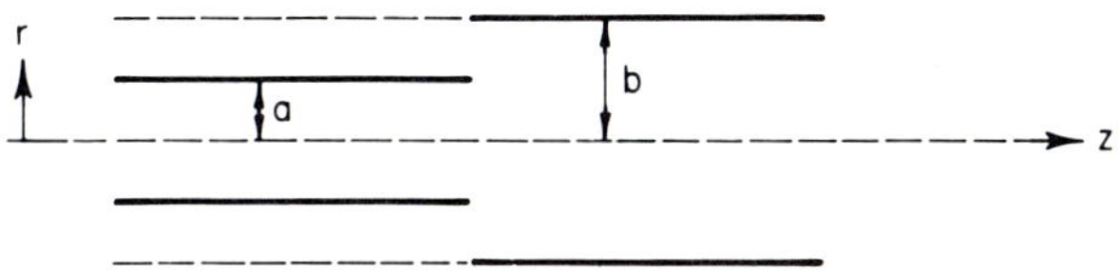

Fig. 5.5. Assumed geometry of the lens shown in Fig. 5.4.

An exact solution was obtained by Lebedev and Skal'skaya[2] using a method based on a solution of paired integral equations by means of the theory of functions of complex variables.

The following is the potential distribution:

$$\frac{u}{V} = 1 - \sum_{m=1}^{\infty} \frac{\exp(-\gamma_m z/a) J_0(\gamma_m r/a)}{\gamma_m J_1(\gamma_m) K_+(j\gamma_m/a)} \tag{5.46}$$

in the region $0 \leq r \leq a$ and $z > 0$, and

$$\frac{u}{V} = \frac{1}{\log(b/a)} \sum_{m=1}^{\infty} \frac{K_+(j\gamma_m/b)\exp(\gamma_m z/b)}{\gamma_m^2 J_1^2(\gamma_m)} J_0\left(\frac{\gamma_m a}{b}\right) J_0\left(\frac{\gamma_m r}{b}\right) \tag{5.47}$$

in the region $0 \leq r \leq a$ and $z < 0$.

In the two equations above γ_m are the roots of

$$J_0(\gamma) = 0 \tag{5.48}$$

where $J_0(\gamma)$ is the Bessel function of the first kind, order zero, and $J_1(\gamma)$ is of order 1, and

$$K_+\left(\frac{j\gamma_m}{a}\right) = f\left(\frac{b-a}{a}\gamma_m\right), \quad K_+\left(\frac{j\gamma_m}{b}\right) = f\left(\frac{b-a}{b}\gamma_m\right) \tag{5.49}$$

The $f(y)$ functions are real functions defined by

$$f(y) = \frac{P[by/(b-a)]\exp\{-y[c/(b-a) + \pi S]/\pi\}}{P[ay/(b-a)]\,Q(y)} \tag{5.50}$$

where

$$c = (b-a)\log(b-a) + a\log a - b\log b \tag{5.51}$$

and the functions $P(\alpha)$, S, and $Q(\alpha)$ are defined by

$$P(\alpha) = \prod_{n=1}^{\infty}\left(1 + \frac{\alpha}{\gamma_n}\right)\exp(-\alpha/\gamma_n), \qquad Q(\alpha) = \prod_{n=1}^{\infty}\left(1 + \frac{\alpha}{\delta_n}\right)\exp(-\alpha/\delta_n)$$

[2] See Lebedev and Skal'skaya (1960) in list of further reading at the end of this chapter.

and

$$S = \sum_{n=1}^{\infty} \left(\frac{1}{\delta_n} - \frac{1}{\gamma_n} \right) \tag{5.52}$$

In the above relations, δ_n are the roots of the cross product

$$J_0\left(\frac{a\delta}{b-a}\right) Y_0\left(\frac{b\delta}{b-a}\right) - J_0\left(\frac{b\delta}{b-a}\right) Y_0\left(\frac{a\delta}{b-a}\right) = 0 \tag{5.53}$$

where Y_0 is Bessel's function of the second kind. The series for $P(\alpha)$, $Q(\alpha)$, and S were found to converge not very rapidly. We have chosen instead the following series:

$$\begin{aligned} \log P(\alpha) = \log \frac{\Gamma(3/4)}{\Gamma[(3/4) + (\alpha/\pi)]} - \frac{\alpha}{\pi}\left(\gamma - \frac{\pi}{2} + 3 \log 2\right) \\ + \sum_{n=1}^{\infty} \left[\log\left(1 + \frac{\alpha}{\gamma_n}\right) - \frac{\alpha}{\gamma_n} - \log\left(1 + \frac{\alpha}{\gamma_n'}\right) + \frac{\alpha}{\gamma_n'} \right] \end{aligned} \tag{5.54}$$

$$\begin{aligned} \log Q(x) = -\frac{\gamma\alpha}{\pi} - \log \Gamma\left(1 + \frac{x}{\pi}\right) \\ + \sum_{n=1}^{\infty} \left[\log\left(1 + \frac{\alpha}{\delta_n}\right) - \frac{\alpha}{\delta_n} - \log\left(1 + \frac{\alpha}{n\pi}\right) + \frac{\alpha}{n\pi} \right] \end{aligned} \tag{5.55}$$

$$S = \frac{1}{2} - \frac{3}{\pi} \log 2 + \sum_{n=1}^{\infty} \left(\frac{1}{\delta_n} - \frac{1}{n\pi} - \frac{1}{\gamma_n} + \frac{1}{\gamma_n'} \right) \tag{5.56}$$

In the three series above, we define $\gamma = 0.5772\ldots$ which is Euler's constant and $\gamma' = n\pi - \pi/4$.

Using the relations above, a digital computer program was written to determine the axial voltage distribution and its derivatives. The cardinal points were obtained using exactly the same procedures outlined in Section 5.3. The results are listed in Tables 5.23–5.38 for various ratios b/a. Since this is an asymmetrical lens, the cardinal points are obtained for two cases: (a) when the voltage of the smaller electrode is higher than that of the larger electrode, and (b) when the voltage of the smaller electrode is lower than that of the larger electrode. In both cases, the accelerating and decelerating conditions are considered.

5.6 Analysis of the Results

The following conclusions can be drawn in this case:

(a) For any given voltage ratio V_2/V_1, the focal length increases monotonically with increasing diameter ratio b/a, with decreasing slope as b/a increases.

(b) For high-voltage ratios and large-diameter ratios, the focal length varies by a negligible amount.

Appendix A5

We shall prove Eq. (5.29), namely that

$$I = \int_{-1}^{+1} \frac{1}{1+\chi\xi} \log\left(\frac{1+\xi}{1-\xi}\right) d\xi = -\frac{1}{2\chi}\left(\log\frac{1-\chi}{1+\chi}\right)^2 \tag{A5.1}$$

Let

$$u = \frac{1+\xi}{1-\xi} \tag{A5.2}$$

Then

$$\xi = \frac{u-1}{u+1}, \qquad d\xi = \frac{2}{(u+1)^2}\,du \tag{A5.3}$$

Let

$$\alpha = \frac{1-\chi}{1+\chi} \tag{A5.4}$$

Then

$$I = 2\int_0^{\infty} \frac{\log u}{(u+1)[u+1+\chi(u-1)]}\,du \tag{A5.5}$$

$$I = 2\int_0^{\infty} \frac{\log u}{(u+1)(u+\alpha)(\chi+1)}\,du$$

$$= \frac{2}{(\chi+1)(\alpha-1)}\int_0^{\infty}\left(\frac{\log u}{u+1} - \frac{\log u}{u+\alpha}\right) du \tag{A5.6}$$

$$I = -\frac{1}{\chi}\lim_{\gamma\to\infty}\int_0^{\gamma}\left(\frac{\log u}{u+1} - \frac{\log u}{u+\alpha}\right) du \tag{A5.7}$$

From tables of integrals, we know that

$$I = -\frac{1}{\chi}\lim_{\gamma\to\infty}\left[\int_1^{\gamma+1}\frac{\log(u-1)}{u}\,du - \int_{\alpha}^{\gamma+\alpha}\frac{\log(u-\alpha)}{u}\,du\right]$$

$$= -\frac{1}{\chi}\lim_{\gamma\to\infty}\left\{\left[\frac{(\log u)^2}{2} + \frac{1}{u} + \frac{1}{2^2u^2} + \frac{1}{3^2u^3} + \cdots\right]_1^{\gamma+1}\right.$$

$$\left. - \left[\frac{(\log u)^2}{2} + \frac{\alpha}{u} + \frac{\alpha^2}{2^2u^2} + \frac{\alpha^3}{3^2u^3} + \cdots\right]_{\alpha}^{\gamma+\alpha}\right\} \tag{A5.8}$$

At the lower limit the powers in $1/u$ all cancel, and at the upper limit the powers in $1/u$ tend to zero. Also

$$\lim_{\gamma\to\infty} \{[\log(\gamma + 1)]^2 - [\log(\gamma + \alpha)]^2\}$$

$$= \lim_{\gamma\to\infty} \left\{\left[\log \gamma + \log\left(1 + \frac{1}{\gamma}\right)\right]^2 - \left[\log \gamma + \log\left(1 + \frac{\alpha}{\gamma}\right)\right]^2\right\}$$

$$= \lim_{\gamma\to\infty} \left\{2 \log \gamma\left(\frac{1-\alpha}{\gamma}\right) + \frac{1-\alpha^2}{\gamma^2}\right\} = 0 \qquad (A5.9)$$

Since

$$\lim_{\gamma\to\infty} \frac{\log \gamma}{\gamma} = -\lim_{\gamma\to\infty} \frac{1}{\gamma} \log \frac{1}{\gamma} = 0 \qquad (A5.10)$$

then

$$I = -\frac{1}{2\chi}(\log \alpha)^2 = -\frac{1}{2\chi}\left(\log \frac{1-\chi}{1+\chi}\right)^2 \qquad (A5.11)$$

Further Reading

Goddard, L. S. (1944). *Proc. Phys. Soc.* (*London*) **56**, 372.
Goddard, L. S. (1946). *Proc. Cambridge Phil. Soc.* **42**, 106.
Lebedev, N. N., and Skal'skaya, I. P. (1960). *Zh. Tekhn. Fiz.* **30**, 472.

TABLE 5.1[a]

V_2/V_1	β	A	B	f_i [Eq. (5.8)]	f_i [Eq. (5.41)]
1.50	0.144591	0.013510	0.000043	81.92	82.18
2.00	0.279533	0.039276	0.000372	30.28	30.57
2.50	0.420418	0.068233	0.001133	18.43	18.74
3.00	0.569478	0.097508	0.002334	13.50	13.83
3.50	0.726283	0.126059	0.003931	10.85	11.20
4.00	0.889845	0.153507	0.005871	9.21	9.58
4.50	1.05914	0.179739	0.008103	8.10	8.49
5.00	1.23325	0.204759	0.010583	7.30	7.70
5.50	1.41138	0.228617	0.013272	6.70	7.11
6.00	1.59286	0.251384	0.016139	6.23	6.65
6.50	1.77711	0.273135	0.019157	5.85	6.29
7.00	1.96365	0.293944	0.022302	5.53	5.99
7.50	2.15208	0.313881	0.025557	5.27	5.74
8.00	2.34205	0.333012	0.028904	5.05	5.53
8.50	2.53326	0.351395	0.032331	4.86	5.35
9.00	2.72546	0.369085	0.035825	4.69	5.20
9.50	2.91843	0.386130	0.039376	4.55	5.06
10.00	3.11199	0.402575	0.042977	4.42	4.95
10.50	3.30596	0.418461	0.046619	4.30	4.84
11.00	3.50020	0.433823	0.050296	4.20	4.75
11.50	3.69460	0.448696	0.054002	4.10	4.67
12.00	3.88905	0.463108	0.057733	4.02	4.59
12.50	4.08345	0.477089	0.061484	3.94	4.52
13.00	4.27772	0.490663	0.065251	3.87	4.46
13.50	4.47179	0.503853	0.069031	3.80	4.41
14.00	4.66559	0.516681	0.072821	3.74	4.36
14.50	4.85907	0.529166	0.076619	3.69	4.31
15.00	5.05218	0.541326	0.080422	3.64	4.27
15.50	5.24488	0.553178	0.084228	3.59	4.23
16.00	5.43714	0.564737	0.088036	3.54	4.20
16.50	5.62890	0.576018	0.091843	3.50	4.16
17.00	5.82016	0.587033	0.095649	3.46	4.13
17.50	6.01088	0.597796	0.099452	3.42	4.10
18.00	6.20104	0.608317	0.103251	3.39	4.08
18.50	6.39063	0.618608	0.107044	3.35	4.05
19.00	6.57961	0.628678	0.110832	3.32	4.03
19.50	6.76799	0.638538	0.114613	3.29	4.01
20.00	6.95575	0.648194	0.118386	3.26	3.99

[a] All values are normalized with respect to R, the radius of the cylinders.

TABLE 5.2

CARDINAL POINTS OF TWO EQUIDIAMETER CYLINDERS WITH NEGLIGIBLE SPACING[a]

V_2/V_1	f_i/R	z_{m_i}/R	z_{p_i}/R	f_o/R	z_{m_o}/R	z_{p_o}/R
1.50	82.18	74.18	−8.00	−67.10	−74.33	−7.23
2.00	30.57	25.57	−5.00	−21.62	−25.83	−4.21
2.50	18.74	14.72	−4.02	−11.85	−15.07	−3.20
3.00	13.83	10.29	−3.54	−7.98	−10.70	−2.72
3.50	11.20	7.94	−3.26	−5.98	−8.41	−2.43
4.00	9.57	6.49	−3.08	−4.79	−7.02	−2.23
4.50	8.48	5.52	−2.96	−4.00	−6.09	−2.09
5.00	7.69	4.82	−2.88	−3.44	−5.43	−1.99
5.50	7.10	4.29	−2.81	−3.03	−4.94	−1.91
6.00	6.64	3.88	−2.77	−2.71	−4.56	−1.85
6.50	6.28	3.55	−2.73	−2.46	−4.26	−1.79
7.00	5.98	3.27	−2.70	−2.26	−4.01	−1.75
7.50	5.73	3.04	−2.68	−2.09	−3.81	−1.72
8.00	5.52	2.85	−2.67	−1.95	−3.64	−1.69
8.50	5.33	2.68	−2.66	−1.83	−3.49	−1.66
9.00	5.18	2.53	−2.65	−1.73	−3.36	−1.64
9.50	5.04	2.40	−2.64	−1.64	−3.25	−1.62
10.00	4.92	2.29	−2.64	−1.56	−3.16	−1.60
10.50	4.82	2.18	−2.64	−1.49	−3.07	−1.59
11.00	4.72	2.09	−2.64	−1.42	−3.00	−1.57
11.50	4.64	2.00	−2.64	−1.37	−2.93	−1.56
12.00	4.56	1.92	−2.64	−1.32	−2.87	−1.55
12.50	4.50	1.85	−2.64	−1.27	−2.81	−1.54
13.00	4.43	1.79	−2.65	−1.23	−2.76	−1.53
13.50	4.38	1.73	−2.66	−1.19	−2.71	−1.52
14.00	4.32	1.67	−2.65	−1.16	−2.67	−1.51
14.50	4.28	1.62	−2.67	−1.12	−2.63	−1.51
15.00	4.23	1.57	−2.66	−1.09	−2.59	−1.50
15.50	4.19	1.52	−2.67	−1.07	−2.56	−1.50
16.00	4.16	1.48	−2.68	−1.04	−2.53	−1.49
16.50	4.12	1.44	−2.68	−1.01	−2.50	−1.49
17.00	4.09	1.40	−2.69	−0.99	−2.47	−1.48
17.50	4.06	1.36	−2.70	−0.97	−2.45	−1.48
18.00	4.03	1.33	−2.70	−0.95	−2.42	−1.47
18.50	4.01	1.29	−2.71	−0.93	−2.40	−1.47
19.00	3.98	1.26	−2.72	−0.91	−2.38	−1.47
19.50	3.96	1.23	−2.73	−0.90	−2.36	−1.46
20.00	3.94	1.20	−2.73	−0.88	−2.34	−1.46

[a] All values are normalized with respect to R, the radius of the cylinders.

TABLE 5.3

CARDINAL POINTS OF TWO EQUIDIAMETER CYLINDERS SEPARATED BY A DISTANCE $s = 0.1$[a]

V_2/V_1	f_i/R	z_{m_i}/R	z_{p_i}/R	f_o/R	z_{m_o}/R	z_{p_o}/R
1.50	82.18	74.18	−8.00	−67.10	−74.33	−7.23
2.00	30.57	25.57	−5.00	−21.61	−25.83	−4.22
2.50	18.74	14.72	−4.02	−11.85	−15.07	−3.22
3.00	13.82	10.29	−3.54	−7.98	−10.70	−2.72
3.50	11.19	7.93	−3.26	−5.98	−8.41	−2.43
4.00	9.57	6.49	−3.08	−4.79	−7.02	−2.23
4.50	8.48	5.52	−2.96	−4.00	−6.09	−2.09
5.00	7.69	4.82	−2.87	−3.44	−5.43	−1.99
5.50	7.10	4.29	−2.81	−3.03	−4.94	−1.91
6.00	6.64	3.88	−2.77	−2.71	−4.56	−1.85
6.50	6.27	3.54	−2.73	−2.46	−4.25	−1.79
7.00	5.97	3.27	−2.70	−2.26	−4.01	−1.75
7.50	5.72	3.04	−2.68	−2.09	−3.81	−1.72
8.00	5.51	2.85	−2.67	−1.95	−3.64	−1.69
8.50	5.33	2.68	−2.66	−1.83	−3.49	−1.66
9.00	5.18	2.53	−2.65	−1.73	−3.36	−1.64
9.50	5.04	2.40	−2.64	−1.64	−3.25	−1.62
10.00	4.92	2.28	−2.64	−1.56	−3.16	−1.60
10.50	4.82	2.18	−2.64	−1.49	−3.07	−1.59
11.00	4.72	2.09	−2.64	−1.42	−3.00	−1.57
11.50	4.64	2.00	−2.64	−1.37	−2.93	−1.56
12.00	4.56	1.92	−2.64	−1.32	−2.87	−1.55
12.50	4.49	1.85	−2.64	−1.27	−2.81	−1.54
13.00	4.43	1.79	−2.65	−1.23	−2.76	−1.53
13.50	4.38	1.73	−2.65	−1.19	−2.71	−1.52
14.00	4.32	1.67	−2.65	−1.16	−2.67	−1.51
14.50	4.28	1.62	−2.66	−1.12	−2.63	−1.51
15.00	4.23	1.57	−2.67	1.09	−2.59	−1.50
15.50	4.19	1.52	−2.67	−1.06	−2.56	−1.50
16.00	4.15	1.48	−2.68	−1.04	−2.53	−1.49
16.50	4.12	1.44	−2.68	−1.01	2.50	−1 49
17.00	4.09	1.40	2.69	−0.99	−2.47	−1.48
17.50	4.06	1.36	−2.70	−0.97	−2.45	−1.48
18.00	4.03	1.33	−2.70	−0.95	−2.42	−1.47
18.50	4.01	1.29	−2.71	−0.93	−2.40	−1.47
19.00	3.98	1.26	−2.72	−0.91	−2.38	−1.47
19.50	3.96	1.23	−2.73	−0.90	−2.36	−1.46
20.00	3.94	1.20	−2.73	−0.88	−2.34	−1.46
20.50	3.92	1.18	−2.74	−0.87	−2.32	−1.46
21.00	3.90	1.15	−2.75	−0.85	−2.31	−1.45
21.50	3.88	1.13	−2.76	−0.84	−2.29	−1.45
22.00	3.87	1.10	−2.76	−0.82	−2.27	−1.45

[a] All values are normalized with respect to R, the radius of the cylinders.

TABLE 5.3 (continued)

V_2/V_1	f_i/R	z_{m_i}/R	z_{p_i}/R	f_o/R	z_{m_o}/R	z_{p_o}/R
22.50	3.85	1.08	−2.77	−0.81	−2.26	−1.45
23.00	3.84	1.06	−2.78	−0.80	−2.25	−1.45
23.50	3.82	1.03	−2.79	−0.79	−2.23	−1.44
24.00	3.81	1.01	−2.80	−0.78	−2.22	−1.44
24.50	3.80	0.99	−2.80	−0.77	−2.21	−1.44
25.00	3.79	0.97	−2.81	−0.76	−2.20	−1.44
25.50	3.78	0.96	−2.82	−0.75	−2.18	−1.44
26.00	3.77	0.94	−2.83	−0.74	−2.17	−1.44
26.50	3.76	0.92	−2.84	−0.73	−2.16	−1.43
27.00	3.75	0.90	−2.84	−0.72	−2.15	−1.43
27.50	3.74	0.89	−2.85	−0.71	−2.14	−1.43
28.00	3.73	0.87	−2.86	−0.71	−2.14	−1.43
28.50	3.72	0.86	−2.87	−0.70	−2.13	−1.43
29.00	3.72	0.84	−2.88	−0.69	−2.12	−1.43
29.50	3.71	0.83	−2.88	−0.68	−2.11	−1.43
30.00	3.70	0.81	−2.89	−0.68	−2.10	−1.43
30.50	3.70	0.80	−2.90	−0.67	−2.09	−1.43
31.00	3.69	0.79	−2.91	−0.66	−2.09	−1.42
31.50	3.69	0.77	−2.91	−0.66	−2.08	−1.42
32.00	3.68	0.76	−2.92	−0.65	−2.07	−1.42
32.50	3.68	0.75	−2.93	−0.64	−2.07	−1.42
33.00	3.67	0.73	−2.94	−0.64	−2.06	−1.42
33.50	3.67	0.72	−2.95	−0.63	−2.05	−1.42
34.00	3.66	0.71	−2.95	−0.63	−2.05	−1.42
34.50	3.66	0.70	−2.96	−0.62	−2.04	−1.42
35.00	3.66	0.69	−2.97	−0.62	−2.04	−1.42
35.50	3.65	0.68	−2.98	−0.61	−2.03	−1.42
36.00	3.65	0.67	−2.98	−0.61	−2.03	−1.42
36.50	3.65	0.66	−2.99	−0.60	−2.02	−1.42
37.00	3.65	0.65	−3.00	−0.60	−2.02	−1.42
37.50	3.64	0.64	−3.01	−0.60	−2.01	−1.42
38.00	3.64	0.63	−3.01	−0.59	−2.01	−1.42
38.50	3.64	0.62	−3.02	−0.59	−2.00	−1.42
39.00	3.64	0.61	−3.03	−0.58	−2.00	−1.42
39.50	3.63	0.60	−3.03	−0.58	−1.99	−1.41
40.00	3.63	0.59	−3.04	−0.57	−1.98	−1.41
40.50	3.63	0.58	−3.05	−0.57	−1.98	−1.41
41.00	3.63	0.57	−3.06	−0.57	−1.97	−1.41
41.50	3.63	0.56	−3.06	−0.56	−1.97	−1.41
42.00	3.63	0.55	−3.07	−0.56	−1.97	−1.41
42.50	3.63	0.55	−3.08	−0.56	−1.96	−1.41
43.00	3.62	0.54	−3.09	−0.55	−1.96	−1.41
43.50	3.62	0.53	−3.09	−0.55	−1.96	−1.41
44.00	3.62	0.52	−3.10	−0.55	−1.95	−1.41
44.50	3.62	0.52	−3.11	−0.54	−1.95	−1.41

TABLE 5.3 (continued)

V_2/V_1	f_i/R	z_{m_i}/R	z_{p_i}/R	f_o/R	z_{m_o}/R	z_{p_o}/R
45.00	3.62	0.51	−3.12	−0.54	−1.95	−1.41
45.50	3.62	0.50	−3.12	−0.54	−1.94	−1.41
46.00	3.62	0.49	−3.13	−0.53	−1.94	−1.40
46.50	3.62	0.49	−3.14	−0.53	−1.94	−1.40
47.00	3.62	0.48	−3.14	−0.53	−1.93	−1.40
47.50	3.62	0.47	−3.15	−0.53	−1.93	−1.40
48.00	3.62	0.46	−3.16	−0.52	−1.93	−1.40
48.50	3.62	0.46	−3.16	−0.52	−1.92	−1.40
49.00	3.62	0.45	−3.17	−0.52	−1.92	−1.40
49.50	3.62	0.45	−3.17	−0.52	−1.92	−1.40
50.00	3.62	0.44	−3.18	−0.51	−1.92	−1.40

TABLE 5.4

CARDINAL POINTS OF TWO EQUIDIAMETER CYLINDERS SEPARATED BY A DISTANCE $s = 0.2$[a]

V_2/V_1	f_i/R	z_{m_i}/R	z_{p_i}/R	f_o/R	z_{m_o}/R	z_{p_o}/R
1.50	82.46	74.43	−8.03	−67.33	−74.59	−7.26
2.00	30.67	25.65	−5.02	−21.69	−25.92	−4.23
2.50	18.80	14.77	−4.03	−11.89	−15.12	−3.23
3.00	13.87	10.32	−3.55	−8.01	−10.74	−2.73
3.50	11.23	7.96	−3.27	−6.00	−8.44	−2.43
4.00	9.60	6.51	−3.09	−4.80	−7.04	−2.24
4.50	8.51	5.53	−2.97	−4.01	−6.11	−2.10
5.00	7.72	4.83	−2.88	−3.45	−5.45	−1.99
5.50	7.12	4.30	−2.82	−3.04	−4.95	−1.91
6.00	6.66	3.89	−2.77	−2.72	−4.57	−1.85
6.50	6.29	3.56	−2.74	−2.47	−4.27	−1.80
7.00	5.99	3.28	−2.71	−2.26	−4.02	−1.76
7.50	5.74	3.05	−2.69	−2.10	−3.82	−1.72
8.00	5.53	2.85	−2.68	−1.96	−3.65	−1.69
8.50	5.35	2.68	−2.66	−1.83	−3.50	−1.67
9.00	5.19	2.54	−2.66	−1.73	−3.37	−1.64
9.50	5.06	2.41	−2.65	−1.64	−3.26	−1.62
10.00	4.94	2.29	−2.65	−1.56	−3.17	−1.61
10.50	4.83	2.19	−2.65	−1.49	−3.08	−1.59
11.00	4.74	2.09	−2.65	−1.43	−3.01	−1.58
11.50	4.65	2.01	−2.65	−1.37	−2.94	−1.56
12.00	4.58	1.93	−2.65	−1.32	−2.87	−1.55
12.50	4.51	1.86	−2.65	−1.27	−2.82	−1.54
13.00	4.44	1.79	−2.65	−1.23	−2.77	−1.53

[a] All values are normalized with respect to R, the radius of the cylinders.

TABLE 5.4 (continued)

V_2/V_1	f_i/R	z_{m_i}/R	z_{p_i}/R	f_o/R	z_{m_o}/R	z_{p_o}/R
13.50	4.39	1.73	−2.66	−1.19	−2.72	−1.53
14.00	4.34	1.67	−2.66	−1.16	−2.68	−1.52
14.50	4.29	1.62	−2.67	−1.13	−2.64	−1.51
15.00	4.24	1.57	−2.67	−1.10	−2.60	−1.51
15.50	4.20	1.52	−2.68	−1.07	−2.57	−1.50
16.00	4.17	1.48	−2.69	−1.04	−2.54	−1.50
16.50	4.13	1.44	−2.69	−1.02	−2.51	−1.49
17.00	4.10	1.40	−2.70	−0.99	−2.48	−1.49
17.50	4.07	1.36	−2.71	−0.97	−2.45	−1.48
18.00	4.04	1.33	−2.71	−0.95	−2.43	−1.48
18.50	4.02	1.30	−2.72	−0.93	−2.41	−1.47
19.00	3.99	1.26	−2.73	−0.92	−2.39	−1.47
19.50	3.97	1.23	−2.73	−0.90	−2.37	−1.47
20.00	3.95	1.21	−2.74	−0.88	−2.35	−1.46
20.50	3.93	1.18	−2.75	−0.87	−2.33	−1.46
21.00	3.91	1.15	−2.76	−0.85	−2.31	−1.46
21.50	3.89	1.13	−2.76	−0.84	−2.30	−1.46
22.00	3.88	1.10	−2.77	−0.83	−2.28	−1.45
22.50	3.86	1.08	−2.78	−0.81	−2.27	−1.45
23.00	3.85	1.06	−2.79	−0.80	−2.25	−1.45
23.50	3.83	1.04	−2.80	−0.79	−2.24	−1.45
24.00	3.82	1.02	−2.80	−0.78	−2.23	−1.45
24.50	3.81	1.00	−2.81	−0.77	−2.21	−1.44
25.00	3.80	0.98	−2.82	−0.76	−2.20	−1.44
25.50	3.78	0.96	−2.83	−0.75	−2.19	−1.44
26.00	3.77	0.94	−2.84	−0.74	−2.18	−1.44
26.50	3.77	0.92	−2.84	−0.73	−2.17	−1.44
27.00	3.76	0.90	−2.85	−0.72	−2.16	−1.44
27.50	3.75	0.89	−2.86	−0.71	−2.15	−1.44
28.00	3.74	0.87	−2.87	−0.71	−2.14	−1.43
28.50	3.73	0.86	−2.88	−0.70	−2.13	−1.43
29.00	3.73	0.84	−2.88	−0.69	−2.12	−1.43
29.50	3.72	0.83	−2.89	−0.68	−2.12	−1.43
30.00	3.71	0.81	−2.90	−0.68	−2.11	−1.43
30.50	3.71	0.80	−2.91	−0.67	−2.10	−1.43
31.00	3.70	0.79	−2.91	−0.66	−2.09	−1.43
31.50	3.69	0.77	−2.92	−0.66	−2.09	−1.43
32.00	3.69	0.76	−2.93	−0.65	−2.08	−1.43
32.50	3.69	0.75	−2.94	−0.65	−2.07	−1.43
33.00	3.68	0.74	−2.95	−0.64	−2.07	−1.43
33.50	3.68	0.72	−2.95	−0.64	−2.06	−1.43
34.00	3.67	0.71	−2.96	−0.63	−2.05	−1.42
34.50	3.67	0.70	−2.97	−0.62	−2.05	−1.42
35.00	3.67	0.69	−2.98	−0.62	−2.04	−1.42
35.50	3.66	0.68	−2.98	−0.61	−2.04	−1.42

TABLE 5.4 (continued)

V_2/V_1	f_i/R	z_{m_i}/R	z_{p_i}/R	f_o/R	z_{m_o}/R	z_{p_o}/R
36.00	3.66	0.67	−2.99	−0.61	−2.03	−1.42
36.50	3.66	0.66	−3.00	−0.61	−2.03	−1.42
37.00	3.65	0.65	−3.01	−0.60	−2.02	−1.42
37.50	3.65	0.64	−3.02	−0.60	−2.02	−1.42
38.00	3.65	0.63	−3.02	−0.59	−2.01	−1.42
38.50	3.65	0.62	−3.03	−0.59	−2.01	−1.42
39.00	3.65	0.61	−3.04	−0.58	−2.00	−1.42
39.50	3.64	0.60	−3.05	−0.58	−2.00	−1.42
40.00	3.64	0.59	−3.05	−0.58	−1.99	−1.42
40.50	3.64	0.58	−3.06	−0.57	−1.99	−1.42
41.00	3.64	0.57	−3.07	−0.57	−1.99	−1.42
41.50	3.64	0.56	−3.08	−0.56	−1.98	−1.42
42.00	3.64	0.56	−3.08	−0.56	−1.98	−1.42
42.50	3.64	0.55	−3.09	−0.56	−1.97	−1.42
43.00	3.64	0.54	−3.10	−0.55	−1.97	−1.42
43.50	3.64	0.53	−3.10	−0.55	−1.97	−1.42
44.00	3.63	0.52	−3.11	−0.55	−1.96	−1.42
44.50	3.63	0.52	−3.12	−0.54	−1.96	−1.42
45.00	3.63	0.51	−3.13	−0.54	−1.96	−1.42
45.50	3.63	0.50	−3.13	−0.54	−1.95	−1.42
46.00	3.63	0.49	−3.14	−0.54	−1.95	−1.42
46.50	3.63	0.49	−3.15	−0.53	−1.95	−1.41
47.00	3.63	0.48	−3.16	−0.53	−1.94	−1.41
47.50	3.63	0.47	−3.16	−0.53	−1.94	−1.41
48.00	3.63	0.46	−3.17	−0.52	−1.94	−1.41
48.50	3.63	0.46	−3.18	−0.52	−1.94	−1.41
49.00	3.63	0.45	−3.18	−0.52	−1.93	−1.41
49.50	3.63	0.44	−3.19	−0.52	−1.93	−1.41
50.00	3.64	0.44	3.20	−0.51	−1.93	−1.41

TABLE 5.5

CARDINAL POINTS OF TWO EQUIDIAMETER CYLINDERS SEPARATED BY A DISTANCE $s = 0.3$[a]

V_2/V_1	f_i/R	z_{m_i}/R	z_{p_i}/R	f_o/R	z_{m_o}/R	z_{p_o}/R
1.50	82.93	74.86	−8.07	−67.71	−75.02	−7.30
2.00	30.84	25.80	−5.04	−21.81	−26.07	−4.26
2.50	18.90	14.85	−4.05	−11.96	−15.20	−3.25
3.00	13.95	10.38	−3.57	−8.05	−10.80	−2.75
3.50	11.29	8.00	−3.29	−6.04	−8.48	−2.45
4.00	9.66	6.55	−3.11	−4.83	−7.08	−2.25

[a] All values are normalized with respect to R, the radius of the cylinders.

TABLE 5.5 (continued)

V_2/V_1	f_i/R	z_{m_i}/R	z_{p_i}/R	f_o/R	z_{m_o}/R	z_{p_o}/R
4.50	8.55	5.56	−2.99	−4.03	−6.14	−2.11
5.00	7.76	4.86	−2.90	−3.47	−5.47	−2.01
5.50	7.16	4.32	−2.84	−3.05	−4.98	−1.93
6.00	6.70	3.91	−2.79	−2.73	−4.60	−1.86
6.50	6.33	3.57	−2.75	−2.48	−4.29	−1.81
7.00	6.02	3.30	−2.73	−2.28	−4.04	−1.77
7.50	5.77	3.06	−2.71	−2.11	−3.84	−1.73
8.00	5.56	2.87	−2.69	−1.96	−3.67	−1.70
8.50	5.38	2.70	−2.68	−1.84	−3.52	−1.67
9.00	5.22	2.55	−2.67	−1.74	−3.39	−1.65
9.50	5.08	2.42	−2.66	−1.65	−3.28	−1.63
10.00	4.96	2.30	−2.66	−1.57	−3.18	−1.61
10.50	4.85	2.19	−2.66	−1.50	−3.10	−1.60
11.00	4.76	2.10	−2.66	−1.43	−3.02	−1.59
11.50	4.67	2.01	−2.66	−1.38	−2.95	−1.57
12.00	4.60	1.94	−2.66	−1.33	−2.89	−1.56
12.50	4.53	1.86	−2.66	−1.28	−2.83	−1.55
13.00	4.46	1.80	−2.67	−1.24	−2.78	−1.54
13.50	4.41	1.74	−2.67	−1.20	−2.73	−1.53
14.00	4.35	1.68	−2.68	−1.16	−2.69	−1.53
14.50	4.31	1.63	−2.68	−1.13	−2.65	−1.52
15.00	4.26	1.58	−2.69	−1.10	−2.61	−1.51
15.50	4.22	1.53	−2.69	−1.07	−2.58	−1.51
16.00	4.18	1.49	−2.70	−1.05	−2.55	−1.50
16.50	4.15	1.44	−2.71	−1.02	−2.52	−1.50
17.00	4.12	1.41	−2.71	−1.00	−2.49	−1.49
17.50	4.09	1.37	−2.72	−0.98	−2.47	−1.49
18.00	4.06	1.33	−2.73	−0.96	−2.44	−1.49
18.50	4.03	1.30	−2.73	−0.94	−2.42	−1.48
19.00	4.01	1.27	−2.74	−0.92	−2.40	−1.48
19.50	3.99	1.24	−2.75	−0.90	−2.38	−1.47
20.00	3.96	1.21	−2.75	−0.89	−2.36	−1.47
20.50	3.94	1.18	−2.76	−0.87	−2.34	−1.47
21.00	3.93	1.16	−2.77	−0.86	−2.32	−1.47
21.50	3.91	1.13	−2.78	−0.84	−2.31	−1.46
22.00	3.89	1.11	−2.79	−0.83	−2.29	−1.46
22.50	3.88	1.08	−2.79	−0.82	−2.28	−1.46
23.00	3.86	1.06	−2.80	−0.81	−2.26	−1.46
23.50	3.85	1.04	−2.81	−0.79	−2.25	−1.46
24.00	3.83	1.02	−2.82	−0.78	−2.24	−1.45
24.50	3.82	1.00	−2.82	−0.77	−2.22	−1.45
25.00	3.81	0.98	−2.83	−0.76	−2.21	−1.45
25.50	3.80	0.96	−2.84	−0.75	−2.20	−1.45
26.00	3.79	0.94	−2.85	−0.74	−2.19	−1.45
26.50	3.78	0.92	−2.86	−0.73	−2.18	−1.45
27.00	3.77	0.91	−2.86	−0.73	−2.17	−1.44

TABLE 5.5 (continued)

V_2/V_1	f_i/R	z_{mi}/R	z_{pi}/R	f_o/R	z_{mo}/R	z_{po}/R
27.50	3.76	0.89	−2.87	−0.72	−2.16	−1.44
28.00	3.75	0.87	−2.88	−0.71	−2.15	−1.44
28.50	3.75	0.86	−2.89	−0.70	−2.14	−1.44
29.00	3.74	0.84	−2.90	−0.69	−2.13	−1.44
29.50	3.73	0.83	−2.90	−0.69	−2.13	−1.44
30.00	3.73	0.81	−2.91	−0.68	−2.12	−1.44
30.50	3.72	0.80	−2.92	−0.67	−2.11	−1.44
31.00	3.71	0.79	−2.93	−0.67	−2.10	−1.44
31.50	3.71	0.77	−2.94	−0.66	−2.10	−1.44
32.00	3.70	0.76	−2.94	−0.65	−2.09	−1.43
32.50	3.70	0.75	−2.95	−0.65	−2.08	−1.43
33.00	3.70	0.74	−2.96	−0.64	−2.08	−1.43
33.50	3.69	0.72	−2.97	−0.64	−2.07	−1.43
34.00	3.69	0.71	−2.97	−0.63	−2.06	−1.43
34.50	3.68	0.70	−2.98	−0.63	−2.06	−1.43
35.00	3.68	0.69	−2.99	−0.62	−2.05	−1.43
35.50	3.68	0.68	−3.00	−0.62	−2.05	−1.43
36.00	3.67	0.67	−3.01	−0.61	−2.04	−1.43
36.50	3.67	0.66	−3.01	−0.61	−2.04	−1.43
37.00	3.67	0.65	−3.02	−0.60	−2.03	−1.43
37.50	3.67	0.64	−3.03	−0.60	−2.03	−1.43
38.00	3.66	0.63	−3.04	−0.59	−2.02	−1.43
38.50	3.66	0.62	−3.04	−0.59	−2.02	−1.43
39.00	3.66	0.61	−3.05	−0.59	−2.01	−1.43
39.50	3.66	0.60	−3.06	−0.58	−2.01	−1.43
40.00	3.66	0.59	−3.07	−0.58	−2.00	−1.43
40.50	3.65	0.58	−3.07	−0.57	−2.00	−1.43
41.00	3.65	0.57	−3.08	−0.57	−2.00	−1.43
41.50	3.65	0.56	−3.09	−0.57	−1.99	−1.42
42.00	3.65	0.55	3.10	−0.56	−1.99	−1.42
42.50	3.65	0.55	−3.10	−0.56	−1.98	−1.42
43.00	3.65	0.54	−3.11	−0.56	−1.98	−1.42
43.50	3.65	0.53	−3.12	−0.55	−1.98	−1.42
44.00	3.65	0.52	−3.13	−0.55	−1.97	−1.42
44.50	3.65	0.51	−3.13	−0.55	−1.97	−1.42
45.00	3.65	0.51	−3.14	−0.54	−1.97	−1.42
45.50	3.65	0.50	−3.15	−0.54	−1.96	−1.42
46.00	3.65	0.49	−3.15	−0.54	−1.96	−1.42
46.50	3.65	0.48	−3.16	−0.53	−1.96	−1.42
47.00	3.65	0.48	−3.17	−0.53	−1.95	−1.42
47.50	3.65	0.47	−3.18	−0.53	−1.95	−1.42
48.00	3.65	0.46	−3.18	−0.53	−1.95	−1.42
48.50	3.65	0.46	−3.19	−0.52	−1.94	−1.42
49.00	3.65	0.45	−3.20	−0.52	−1.94	−1.42
49.50	3.65	0.44	−3.20	−0.52	−1.94	−1.42
50.00	3.65	0.44	−3.21	−0.52	−1.94	−1.42

TABLE 5.6
CARDINAL POINTS OF TWO EQUIDIAMETER CYLINDERS SEPARATED BY A DISTANCE $s = 0.4$[a]

V_2/V_1	f_i/R	z_{m_i}/R	z_{p_i}/R	f_o/R	z_{m_o}/R	z_{p_o}/R
1.50	83.59	75.45	−8.14	−68.25	−75.61	−7.36
2.00	31.08	26.00	−5.08	−21.98	−26.27	−4.29
2.50	19.05	14.97	−4.08	−12.05	−15.32	−3.27
3.00	14.05	10.45	−3.60	−8.11	−10.88	−2.77
3.50	11.38	8.06	−3.31	−6.08	−8.55	−2.47
4.00	9.73	6.59	−3.13	−4.86	−7.13	−2.27
4.50	8.61	5.60	−3.01	−4.06	−6.19	−2.13
5.00	7.81	4.89	−2.92	−3.49	−5.52	−2.02
5.50	7.21	4.35	−2.86	−3.08	−5.01	−1.94
6.00	6.74	3.93	−2.81	−2.75	−4.63	−1.88
6.50	6.37	3.60	−2.77	−2.50	−4.32	−1.82
7.00	6.06	3.32	−2.75	−2.29	−4.07	−1.78
7.50	5.81	3.08	−2.73	−2.12	−3.87	−1.74
8.00	5.59	2.89	−2.71	−1.98	−3.69	−1.71
8.50	5.41	2.71	−2.70	−1.86	−3.54	−1.69
9.00	5.25	2.56	−2.69	−1.75	−3.42	−1.66
9.50	5.11	2.43	−2.68	−1.66	−3.30	−1.64
10.00	4.99	2.31	−2.68	−1.58	−3.21	−1.63
10.50	4.89	2.21	−2.68	−1.51	−3.12	−1.61
11.00	4.79	2.11	−2.68	−1.44	−3.04	−1.60
11.50	4.70	2.03	−2.68	−1.39	−2.97	−1.58
12.00	4.63	1.95	−2.68	−1.34	−2.91	−1.57
12.50	4.56	1.87	−2.68	−1.29	−2.85	−1.56
13.00	4.49	1.81	−2.69	−1.25	−2.80	−1.55
13.50	4.43	1.74	−2.69	−1.21	−2.75	−1.55
14.00	4.38	1.69	−2.69	−1.17	−2.71	−1.54
14.50	4.33	1.63	−2.70	−1.14	−2.67	−1.53
15.00	4.29	1.58	−2.71	−1.11	−2.63	−1.53
15.50	4.25	1.54	−2.71	−1.08	−2.60	−1.52
16.00	4.21	1.49	−2.72	−1.05	−2.57	−1.51
16.50	4.17	1.45	−2.72	−1.03	−2.54	−1.51
17.00	4.14	1.41	−2.73	−1.00	−2.51	−1.50
17.50	4.11	1.37	−2.74	−0.98	−2.48	−1.50
18.00	4.08	1.34	−2.74	−0.96	−2.46	−1.50
18.50	4.06	1.31	−2.75	−0.94	−2.44	−1.49
19.00	4.03	1.27	−2.76	−0.92	−2.41	−1.49
19.50	4.01	1.24	−2.77	−0.91	−2.39	−1.49
20.00	3.99	1.21	−2.77	−0.89	−2.37	−1.48
20.50	3.97	1.19	−2.78	−0.88	−2.36	−1.48
21.00	3.95	1.16	−2.79	−0.86	−2.34	−1.48
21.50	3.93	1.13	−2.80	−0.85	−2.32	−1.47
22.00	3.91	1.11	−2.80	−0.83	−2.31	−1.47

[a] All values are normalized with respect to R, the radius of the cylinders.

TABLE 5.6 (continued)

V_2/V_1	f_i/R	z_{m_i}/R	z_{p_i}/R	f_o/R	z_{m_o}/R	z_{p_o}/R
22.50	3.90	1.09	−2.81	−0.82	−2.29	−1.47
23.00	3.88	1.06	−2.82	−0.81	−2.28	−1.47
23.50	3.87	1.04	−2.83	−0.80	−2.26	−1.47
24.00	3.86	1.02	−2.84	−0.79	−2.25	−1.46
24.50	3.84	1.00	−2.84	−0.78	−2.24	−1.46
25.00	3.83	0.98	−2.85	−0.77	−2.23	−1.46
25.50	3.82	0.96	−2.86	−0.76	−2.22	−1.46
26.00	3.81	0.94	−2.87	−0.75	−2.21	−1.46
26.50	3.80	0.93	−2.87	−0.74	−2.19	−1.46
27.00	3.79	0.91	−2.88	−0.73	−2.18	−1.46
27.50	3.78	0.89	−2.89	−0.72	−2.18	−1.45
28.00	3.77	0.88	−2.90	−0.71	−2.17	−1.45
28.50	3.77	0.86	−2.91	−0.71	−2.16	−1.45
29.00	3.76	0.85	−2.91	−0.70	−2.15	−1.45
29.50	3.75	0.83	−2.92	−0.69	−2.14	−1.45
30.00	3.75	0.82	−2.93	−0.68	−2.13	−1.45
30.50	3.74	0.80	−2.94	−0.68	−2.12	−1.45
31.00	3.73	0.79	−2.95	−0.67	−2.12	−1.45
31.50	3.73	0.77	−2.95	−0.66	−2.11	−1.45
32.00	3.72	0.76	−2.96	−0.66	−2.10	−1.44
32.50	3.72	0.75	−2.97	−0.65	−2.10	−1.44
33.00	3.71	0.74	−2.98	−0.65	−2.09	−1.44
33.50	3.71	0.72	−2.99	−0.64	−2.08	−1.44
34.00	3.71	0.71	−2.99	−0.64	−2.08	−1.44
34.50	3.70	0.70	−3.00	−0.63	−2.07	−1.44
35.00	3.70	0.69	−3.01	−0.63	−2.07	−1.44
35.50	3.70	0.68	−3.02	−0.62	−2.06	−1.44
36.00	3.69	0.67	−3.02	−0.62	−2.06	−1.44
36.50	3.69	0.66	−3.03	−0.61	−2.05	−1.44
37.00	3.69	0.65	−3.04	−0.61	−2.04	−1.44
37.50	3.68	0.64	−3.05	−0.60	−2.04	−1.44
38.00	3.68	0.63	−3.05	−0.60	−2.03	−1.44
38.50	3.68	0.62	−3.06	−0.59	−2.03	−1.44
39.00	3.68	0.61	−3.07	−0.59	−2.03	−1.44
39.50	3.68	0.60	−3.08	−0.58	−2.02	−1.44
40.00	3.67	0.59	−3.08	−0.58	−2.02	−1.44
40.50	3.67	0.58	−3.09	−0.58	−2.01	−1.44
41.00	3.67	0.57	−3.10	−0.57	−2.01	−1.44
41.50	3.67	0.56	−3.11	−0.57	−2.00	−1.43
42.00	3.67	0.55	−3.11	−0.57	−2.00	−1.43
42.50	3.67	0.55	−3.12	−0.56	−2.00	−1.43
43.00	3.67	0.54	−3.13	−0.56	−1.99	−1.43
43.50	3.67	0.53	−3.14	−0.56	−1.99	−1.43
44.00	3.67	0.52	−3.14	−0.55	−1.99	−1.43
44.50	3.67	0.51	−3.15	−0.55	−1.98	−1.43

TABLE 5.6 (continued)

V_2/V_1	f_i/R	z_{m_i}/R	z_{p_i}/R	f_o/R	z_{m_o}/R	z_{p_o}/R
45.00	3.66	0.51	−3.16	−0.55	−1.98	−1.43
45.50	3.66	0.50	−3.17	−0.54	−1.98	−1.43
46.00	3.66	0.49	−3.17	−0.54	−1.97	−1.43
46.50	3.66	0.48	−3.18	−0.54	−1.97	−1.43
47.00	3.66	0.48	−3.19	−0.53	−1.97	−1.43
47.50	3.66	0.47	−3.19	−0.53	−1.96	−1.43
48.00	3.66	0.46	−3.20	−0.53	−1.96	−1.43
48.50	3.66	0.46	−3.21	−0.53	−1.96	−1.43
49.00	3.66	0.45	−3.22	−0.52	−1.95	−1.43
49.50	3.66	0.44	−3.22	−0.52	−1.95	−1.43
50.00	3.67	0.44	−3.23	−0.52	−1.95	−1.43

TABLE 5.7

CARDINAL POINTS OF TWO EQUIDIAMETER CYLINDERS SEPARATED BY A DISTANCE $s = 0.5$[a]

V_2/V_1	f_i/R	z_{m_i}/R	z_{p_i}/R	f_o/R	z_{m_o}/R	z_{p_o}/R
1.50	84.42	76.20	−8.22	−68.93	−76.36	−7.43
2.00	31.39	26.26	−5.13	−22.20	−26.53	−4.33
2.50	19.24	15.11	−4.13	−12.17	−15.47	−3.31
3.00	14.19	10.55	−3.63	−8.19	−10.99	−2.79
3.50	11.48	8.14	−3.35	−6.14	−8.63	−2.49
4.00	9.82	6.65	−3.16	−4.91	−7.20	−2.29
4.50	8.69	5.65	−3.04	−4.10	−6.24	−2.15
5.00	7.88	4.93	−2.95	−3.53	−5.57	−2.04
5.50	7.28	4.39	−2.88	−3.10	−5.06	−1.96
6.00	6.80	3.97	−2.84	−2.78	−4.67	−1.89
6.50	6.43	3.63	−2.80	−2.52	−4.36	−1.84
7.00	6.12	3.34	−2.77	−2.31	−4.11	−1.80
7.50	5.86	3.11	−2.75	−2.14	−3.90	−1.76
8.00	5.64	2.91	−2.73	−2.00	−3.72	−1.73
8.50	5.46	2.73	−2.72	−1.87	−3.57	−1.70
9.00	5.30	2.58	−2.71	−1.77	−3.45	−1.68
9.50	5.16	2.45	−2.71	−1.67	−3.33	−1.66
10.00	5.03	2.33	−2.70	−1.59	−3.23	−1.64
10.50	4.93	2.22	−2.70	−1.52	−3.15	−1.63
11.00	4.83	2.13	−2.70	−1.46	−3.07	−1.61
11.50	4.74	2.04	−2.70	−1.40	−3.00	−1.60
12.00	4.66	1.96	−2.70	−1.35	−2.93	−1.59
12.50	4.59	1.89	−2.71	−1.30	−2.88	−1.58
13.00	4.53	1.82	−2.71	−1.26	−2.82	−1.57

[a] All values are normalized with respect to R, the radius of the cylinders.

TABLE 5.7 (continued)

V_2/V_1	f_i/R	z_{m_i}/R	z_{p_i}/R	f_o/R	z_{m_o}/R	z_{p_o}/R
13.50	4.47	1.76	−2.71	−1.22	−2.78	−1.56
14.00	4.42	1.70	−2.72	−1.18	−2.73	−1.55
14.50	4.37	1.64	−2.72	−1.15	−2.69	−1.55
15.00	4.32	1.59	−2.73	−1.12	−2.66	−1.54
15.50	4.28	1.55	−2.73	−1.09	−2.62	−1.53
16.00	4.24	1.50	−2.74	−1.06	−2.59	−1.53
16.50	4.21	1.46	−2.75	−1.04	−2.56	−1.52
17.00	4.17	1.42	−2.75	−1.01	−2.53	−1.52
17.50	4.14	1.38	−2.76	−0.99	−2.50	−1.51
18.00	4.11	1.35	−2.77	−0.97	−2.48	−1.51
18.50	4.09	1.31	−2.77	−0.95	−2.46	−1.51
19.00	4.06	1.28	−2.78	−0.93	−2.43	−1.50
19.50	4.04	1.25	−2.79	−0.91	−2.41	−1.50
20.00	4.02	1.22	−2.80	−0.90	−2.39	−1.50
20.50	4.00	1.19	−2.80	−0.88	−2.38	−1.49
21.00	3.98	1.17	−2.81	−0.87	−2.36	−1.49
21.50	3.96	1.14	−2.82	−0.85	−2.34	−1.49
22.00	3.94	1.11	−2.83	−0.84	−2.33	−1.49
22.50	3.93	1.09	−2.83	−0.83	−2.31	−1.48
23.00	3.91	1.07	−2.84	−0.82	−2.30	−1.48
23.50	3.90	1.05	−2.85	−0.80	−2.28	−1.48
24.00	3.88	1.02	−2.86	−0.79	−2.27	−1.48
24.50	3.87	1.00	−2.87	−0.78	−2.26	−1.48
25.00	3.86	0.98	−2.87	−0.77	−2.25	−1.47
25.50	3.85	0.97	−2.88	−0.76	−2.23	−1.47
26.00	3.84	0.95	−2.89	−0.75	−2.22	−1.47
26.50	3.83	0.93	−2.90	−0.74	−2.21	−1.47
27.00	3.82	0.91	−2.91	−0.73	−2.20	−1.47
27.50	3.81	0.89	−2.91	−0.73	−2.19	−1.47
28.00	3.80	0.88	−2.92	−0.72	−2.18	−1.47
28.50	3.79	0.86	−2.93	−0.71	−2.18	−1.46
29.00	3.79	0.85	−2.94	−0.70	−2.17	−1.46
29.50	3.78	0.83	−2.95	−0.70	−2.16	−1.46
30.00	3.77	0.82	−2.95	−0.69	−2.15	−1.46
30.50	3.77	0.80	−2.96	−0.68	−2.14	−1.46
31.00	3.76	0.79	−2.97	−0.68	−2.13	−1.46
31.50	3.75	0.78	−2.98	−0.67	−2.13	−1.46
32.00	3.75	0.76	−2.99	−0.66	−2.12	−1.46
32.50	3.74	0.75	−2.99	−0.66	−2.11	−1.46
33.00	3.74	0.74	−3.00	−0.65	−2.11	−1.46
33.50	3.73	0.73	−3.01	−0.65	−2.10	−1.46
34.00	3.73	0.71	−3.02	−0.64	−2.09	−1.46
34.50	3.73	0.70	−3.02	−0.63	−2.09	−1.45
35.00	3.72	0.69	−3.03	−0.63	−2.08	−1.45
35.50	3.72	0.68	−3.04	−0.62	−2.08	−1.45

TABLE 5.7 (continued)

V_2/V_1	f_i/R	z_{m_i}/R	z_{p_i}/R	f_o/R	z_{m_o}/R	z_{p_o}/R
36.00	3.72	0.67	−3.05	−0.62	−2.07	−1.45
36.50	3.71	0.66	−3.06	−0.61	−2.07	−1.45
37.00	3.71	0.65	−3.06	−0.61	−2.06	−1.45
37.50	3.71	0.64	−3.07	−0.61	−2.06	−1.45
38.00	3.71	0.63	−3.08	−0.60	−2.05	−1.45
38.50	3.70	0.62	−3.09	−0.60	−2.05	−1.45
39.00	3.70	0.61	−3.09	−0.59	−2.04	−1.45
39.50	3.70	0.60	−3.10	−0.59	−2.04	−1.45
40.00	3.70	0.59	−3.11	−0.58	−2.03	−1.45
40.50	3.70	0.58	−3.12	−0.58	−2.03	−1.45
41.00	3.69	0.57	−3.12	−0.58	−2.03	−1.45
41.50	3.69	0.56	−3.13	−0.57	−2.02	−1.45
42.00	3.69	0.55	−3.14	−0.57	−2.02	−1.45
42.50	3.69	0.55	−3.15	−0.57	−2.01	−1.45
43.00	3.69	0.54	−3.15	−0.56	−2.01	−1.45
43.50	3.69	0.53	−3.16	−0.56	−2.01	−1.45
44.00	3.69	0.52	−3.17	−0.56	−2.00	−1.45
44.50	3.69	0.51	−3.17	−0.55	−2.00	−1.45
45.00	3.69	0.51	−3.18	−0.55	−2.00	−1.45
45.50	3.69	0.50	−3.19	−0.55	−1.99	−1.45
46.00	3.69	0.49	−3.20	−0.54	−1.99	−1.45
46.50	3.69	0.48	−3.20	−0.54	−1.99	−1.44
47.00	3.69	0.48	−3.21	−0.54	−1.98	−1.44
47.50	3.69	0.47	−3.22	−0.53	−1.98	−1.44
48.00	3.69	0.46	−3.23	−0.53	−1.98	−1.44
48.50	3.69	0.45	−3.23	−0.53	−1.97	−1.44
49.00	3.69	0.45	−3.24	−0.53	−1.97	−1.44
49.50	3.69	0.44	−3.25	−0.52	−1.97	−1.44
50.00	3.69	0.43	−3.25	−0.52	−1.96	−1.44

TABLE 5.8
CARDINAL POINTS OF TWO EQUIDIAMETER CYLINDERS SEPARATED BY A DISTANCE $s = 0.6$[a]

V_2/V_1	f_i/R	z_{m_i}/R	z_{p_i}/R	f_o/R	z_{m_o}/R	z_{p_o}/R
1.50	85.43	77.12	−8.32	−69.76	−77.28	−7.52
2.00	31.76	26.57	−5.20	−22.46	−26.84	−4.38
2.50	19.46	15.29	−4.17	−12.31	−15.65	−3.34
3.00	14.35	10.67	−3.68	−8.29	−11.11	−2.83
3.50	11.61	8.23	−3.39	−6.21	−8.73	−2.52
4.00	9.93	6.73	−3.20	−4.96	−7.28	−2.32

[a] All values are normalized with respect to R, the radius of the cylinders.

TABLE 5.8 (continued)

V_2/V_1	f_i/R	z_{m_i}/R	z_{p_i}/R	f_o/R	z_{m_o}/R	z_{p_o}/R
4.50	8.79	5.71	−3.07	−4.14	−6.31	−2.17
5.00	7.97	4.99	−2.98	−3.56	−5.63	−2.06
5.50	7.36	4.44	−2.92	−3.14	−5.12	−1.98
6.00	6.88	4.01	−2.87	−2.81	−4.72	−1.92
6.50	6.49	3.66	−2.83	−2.55	−4.41	−1.86
7.00	6.18	3.38	−2.80	−2.34	−4.15	−1.82
7.50	5.92	3.14	−2.78	−2.16	−3.94	−1.78
8.00	5.70	2.94	−2.76	−2.02	−3.76	−1.75
8.50	5.51	2.76	−2.75	−1.89	−3.61	−1.72
9.00	5.35	2.61	−2.74	−1.78	−3.48	−1.70
9.50	5.21	2.47	−2.74	−1.69	−3.37	−1.68
10.00	5.08	2.35	−2.73	−1.61	−3.27	−1.66
10.50	4.97	2.24	−2.73	−1.53	−3.18	−1.64
11.00	4.88	2.15	−2.73	−1.47	−3.10	−1.63
11.50	4.79	2.06	−2.73	−1.41	−3.03	−1.62
12.00	4.71	1.98	−2.73	−1.36	−2.97	−1.61
12.50	4.64	1.90	−2.74	−1.31	−2.91	−1.60
13.00	4.57	1.83	−2.74	−1.27	−2.85	−1.59
13.50	4.51	1.77	−2.74	−1.23	−2.81	−1.58
14.00	4.46	1.71	−2.75	−1.19	−2.76	−1.57
14.50	4.41	1.66	−2.75	−1.16	−2.72	−1.56
15.00	4.36	1.60	−2.76	−1.13	−2.68	−1.56
15.50	4.32	1.56	−2.76	−1.10	−2.65	−1.55
16.00	4.28	1.51	−2.77	−1.07	−2.61	−1.54
16.50	4.24	1.47	−2.77	−1.04	−2.58	−1.54
17.00	4.21	1.43	−2.78	−1.02	−2.56	−1.53
17.50	4.18	1.39	−2.79	−1.00	−2.53	−1.53
18.00	4.15	1.35	−2.80	−0.98	−2.50	−1.53
18.50	4.12	1.32	−2.80	−0.96	−2.48	−1.52
19.00	4.10	1.29	−2.81	−0.94	−2.46	−1.52
19.50	4.07	1.26	−2.82	−0.92	−2.44	−1.52
20.00	4.05	1.23	−2.82	−0.91	−2.42	−1.51
20.50	4.03	1.20	−2.83	−0.89	−2.40	−1.51
21.00	4.01	1.17	−2.84	−0.88	−2.38	−1.51
21.50	3.99	1.15	−2.85	−0.86	−2.37	−1.50
22.00	3.98	1.12	−2.86	−0.85	−2.35	−1.50
22.50	3.96	1.10	−2.86	−0.83	−2.33	−1.50
23.00	3.94	1.07	−2.87	−0.82	−2.32	−1.50
23.50	3.93	1.05	−2.88	−0.81	−2.31	−1.50
24.00	3.92	1.03	−2.89	−0.80	−2.29	−1.49
24.50	3.90	1.01	−2.89	−0.79	−2.28	−1.49
25.00	3.89	0.99	−2.90	−0.78	−2.27	−1.49
25.50	3.88	0.97	−2.91	−0.77	−2.26	−1.49
26.00	3.87	0.95	−2.92	−0.76	−2.25	−1.49
26.50	3.86	0.93	−2.93	−0.75	−2.24	−1.49
27.00	3.85	0.92	−2.93	−0.74	−2.22	−1.48

TABLE 5.8 (continued)

V_1/V_2	f_i/R	z_{mi}/R	z_{pi}/R	f_o/R	z_{mo}/R	z_{po}/R
27.50	3.84	0.90	−2.94	−0.73	−2.22	−1.48
28.00	3.83	0.88	−2.95	−0.72	−2.21	−1.48
28.50	3.82	0.87	−2.96	−0.72	−2.20	−1.48
29.00	3.82	0.85	−2.97	−0.71	−2.19	−1.48
29.50	3.81	0.83	−2.97	−0.70	−2.18	−1.48
30.00	3.80	0.82	−2.98	−0.69	−2.17	−1.48
30.50	3.80	0.81	−2.99	−0.69	−2.16	−1.48
31.00	3.79	0.79	−3.00	−0.68	−2.16	−1.48
31.50	3.78	0.78	−3.01	−0.67	−2.15	−1.47
32.00	3.78	0.76	−3.01	−0.67	−2.14	−1.47
32.50	3.77	0.75	−3.02	−0.66	−2.13	−1.47
33.00	3.77	0.74	−3.03	−0.66	−2.13	−1.47
33.50	3.76	0.73	−3.04	−0.65	−2.12	−1.47
34.00	3.76	0.71	−3.04	−0.64	−2.12	−1.47
34.50	3.76	0.70	−3.05	−0.64	−2.11	−1.47
35.00	3.75	0.69	−3.06	−0.63	−2.10	−1.47
35.50	3.75	0.68	−3.07	−0.63	−2.10	−1.47
36.00	3.74	0.67	−3.08	−0.62	−2.09	−1.47
36.50	3.74	0.66	−3.08	−0.62	−2.09	−1.47
37.00	3.74	0.65	−3.09	−0.61	−2.08	−1.47
37.50	3.74	0.64	−3.10	−0.61	−2.08	−1.47
38.00	3.73	0.63	−3.11	−0.61	−2.07	−1.47
38.50	3.73	0.62	−3.11	−0.60	−2.07	−1.47
39.00	3.73	0.61	−3.12	−0.60	−2.06	−1.47
39.50	3.73	0.60	−3.13	−0.59	−2.06	−1.46
40.00	3.73	0.59	−3.14	−0.59	−2.05	−1.46
40.50	3.72	0.58	−3.14	−0.59	−2.05	−1.46
41.00	3.72	0.57	−3.15	−0.58	−2.04	−1.46
41.50	3.72	0.56	−3.16	−0.58	−2.04	−1.46
42.00	3.72	0.55	−3.17	−0.57	−2.04	−1.46
42.50	3.72	0.54	−3.17	−0.57	−2.03	−1.46
43.00	3.72	0.54	−3.18	−0.57	−2.03	−1.46
43.50	3.72	0.53	−3.19	−0.56	−2.03	−1.46
44.00	3.72	0.52	−3.20	−0.56	−2.02	−1.46
44.50	3.71	0.51	−3.20	−0.56	−2.02	−1.46
45.00	3.71	0.50	−3.21	−0.55	−2.01	−1.46
45.50	3.71	0.50	−3.22	−0.55	−2.01	−1.46
46.00	3.71	0.49	−3.22	−0.55	−2.01	−1.46
46.50	3.71	0.48	−3.23	−0.54	−2.00	−1.46
47.00	3.71	0.47	−3.24	−0.54	−2.00	−1.46
47.50	3.71	0.47	−3.25	−0.54	−2.00	−1.46
48.00	3.71	0.46	−3.25	−0.54	−2.00	−1.46
48.50	3.71	0.45	−3.26	−0.53	−1.99	−1.46
49.00	3.71	0.45	−3.27	−0.53	−1.99	−1.46
49.50	3.71	0.44	−3.27	−0.53	−1.99	−1.46
50.00	3.71	0.43	−3.28	−0.53	−1.98	−1.46

TABLE 5.9

CARDINAL POINTS OF TWO EQUIDIAMETER CYLINDERS SEPARATED BY A DISTANCE $s = 0.7$[a]

V_2/V_1	f_i/R	z_{m_i}/R	z_{p_i}/R	f_o/R	z_{m_o}/R	z_{p_o}/R
1.50	86.62	78.18	−8.43	−70.72	−78.35	−7.63
2.00	32.20	26.93	−5.27	−22.77	−27.21	−4.44
2.50	19.72	15.49	−4.23	−12.47	−15.86	−3.39
3.00	14.54	10.82	−3.73	−8.40	−11.26	−2.87
3.50	11.77	8.33	−3.43	−6.29	−8.84	−2.55
4.00	10.06	6.81	−3.24	−5.03	−7.38	−2.35
4.50	8.90	5.79	−3.11	−4.20	−6.40	−2.20
5.00	8.07	5.05	−3.02	−3.61	−5.70	−2.09
5.50	7.45	4.49	−2.96	−3.18	−5.18	−2.01
6.00	6.96	4.06	−2.91	−2.84	−4.78	−1.94
6.50	6.57	3.70	−2.87	−2.58	−4.46	−1.89
7.00	6.26	3.42	−2.84	−2.36	−4.21	−1.84
7.50	5.99	3.17	−2.82	−2.19	−3.99	−1.80
8.00	5.77	2.97	−2.80	−2.04	−3.81	−1.77
8.50	5.58	2.79	−2.79	−1.91	−3.66	−1.74
9.00	5.41	2.63	−2.78	−1.80	−3.53	−1.72
9.50	5.27	2.50	−2.77	−1.71	−3.41	−1.70
10.00	5.14	2.37	−2.77	−1.63	−3.31	−1.68
10.50	5.03	2.27	−2.77	−1.55	−3.22	−1.67
11.00	4.93	2.17	−2.76	−1.49	−3.14	−1.65
11.50	4.84	2.08	−2.76	−1.43	−3.07	−1.64
12.00	4.76	1.99	−2.77	−1.37	−3.00	−1.63
12.50	4.69	1.92	−2.77	−1.33	−2.94	−1.62
13.00	4.62	1.85	−2.77	−1.28	−2.89	−1.61
13.50	4.56	1.79	−2.78	−1.24	−2.84	−1.60
14.00	4.51	1.73	−2.78	−1.20	−2.79	−1.59
14.50	4.45	1.67	−2.78	−1.17	−2.75	−1.58
15.00	4.41	1.62	−2.79	−1.14	−2.71	−1.58
15.50	4.37	1.57	−2.80	−1.11	−2.68	−1.57
16.00	4.33	1.52	−2.80	−1.08	−2.65	−1.56
16.50	4.29	1.48	−2.81	−1.06	−2.62	−1.56
17.00	4.25	1.44	−2.81	−1.03	−2.59	−1.55
17.50	4.22	1.40	−2.82	−1.01	−2.56	−1.55
18.00	4.19	1.36	−2.83	−0.99	−2.53	−1.55
18.50	4.16	1.33	−2.84	−0.97	−2.51	−1.54
19.00	4.14	1.30	−2.84	−0.95	−2.49	−1.54
19.50	4.11	1.27	−2.85	−0.93	−2.47	−1.53
20.00	4.09	1.24	−2.86	−0.92	−2.45	−1.53
20.50	4.07	1.21	−2.86	−0.90	−2.43	−1.53
21.00	4.05	1.18	−2.87	−0.88	−2.41	−1.53
21.50	4.03	1.15	−2.88	−0.87	−2.39	−1.52
22.00	4.01	1.13	−2.89	−0.86	−2.38	−1.52

[a] All values are normalized with respect to R, the radius of the cylinders.

TABLE 5.9 (continued)

V_2/V_1	f_i/R	z_{m_i}/R	z_{p_i}/R	f_o/R	z_{m_o}/R	z_{p_o}/R
22.50	4.00	1.10	−2.90	−0.84	−2.36	−1.52
23.00	3.98	1.08	−2.90	−0.83	−2.35	−1.52
23.50	3.97	1.06	−2.91	−0.82	−2.33	−1.51
24.00	3.95	1.03	−2.92	−0.81	−2.32	−1.51
24.50	3.94	1.01	−2.93	−0.80	−2.31	−1.51
25.00	3.93	0.99	−2.93	−0.79	−2.29	−1.51
25.50	3.92	0.97	−2.94	−0.78	−2.28	−1.51
26.00	3.91	0.96	−2.95	−0.77	−2.27	−1.51
26.50	3.90	0.94	−2.96	−0.76	−2.26	−1.50
27.00	3.89	0.92	−2.97	−0.75	−2.25	−1.50
27.50	3.88	0.90	−2.97	−0.74	−2.24	−1.50
28.00	3.87	0.88	−2.98	−0.73	−2.23	−1.50
28.50	3.86	0.87	−2.99	−0.72	−2.22	−1.50
29.00	3.85	0.85	−3.00	−0.72	−2.21	−1.50
29.50	3.84	0.84	−3.01	−0.71	−2.20	−1.50
30.00	3.84	0.82	−3.01	−0.70	−2.20	−1.50
30.50	3.83	0.81	−3.02	−0.69	−2.19	−1.49
31.00	3.82	0.79	−3.03	−0.69	−2.18	−1.49
31.50	3.82	0.78	−3.04	−0.68	−2.17	−1.49
32.00	3.81	0.77	−3.05	−0.67	−2.17	−1.49
32.50	3.81	0.75	−3.05	−0.67	−2.16	−1.49
33.00	3.80	0.74	−3.06	−0.66	−2.15	−1.49
33.50	3.80	0.73	−3.07	−0.66	−2.15	−1.49
34.00	3.79	0.72	−3.08	−0.65	−2.14	−1.49
34.50	3.79	0.70	−3.09	−0.65	−2.13	−1.49
35.00	3.79	0.69	−3.09	−0.64	−2.13	−1.49
35.50	3.78	0.68	−3.10	−0.63	−2.12	−1.49
36.00	3.78	0.67	−3.11	−0.63	−2.12	−1.49
36.50	3.77	0.66	−3.12	−0.62	−2.11	−1.49
37.00	3.77	0.65	−3.12	−0.62	−2.11	−1.49
37.50	3.77	0.64	−3.13	−0.62	−2.10	−1.48
38.00	3.77	0.63	−3.14	−0.61	−2.10	−1.48
38.50	3.76	0.62	−3.15	−0.61	−2.09	−1.48
39.00	3.76	0.61	−3.15	−0.60	−2.09	−1.48
39.50	3.76	0.60	−3.16	−0.60	−2.08	−1.48
40.00	3.76	0.59	−3.17	−0.59	−2.08	−1.48
40.50	3.76	0.58	−3.18	−0.59	−2.07	−1.48
41.00	3.75	0.57	−3.18	−0.59	−2.07	−1.48
41.50	3.75	0.56	−3.19	−0.58	−2.06	−1.48
42.00	3.75	0.55	−3.20	−0.58	−2.06	−1.48
42.50	3.75	0.54	−3.21	−0.58	−2.06	−1.48
43.00	3.75	0.53	−3.21	−0.57	−2.05	−1.48
43.50	3.75	0.53	−3.22	−0.57	−2.05	−1.48
44.00	3.75	0.52	−3.23	−0.56	−2.04	−1.48
44.50	3.75	0.51	−3.24	−0.56	−2.04	−1.48

TABLE 5.9 (continued)

V_2/V_1	f_i/R	z_{m_i}/R	z_{p_i}/R	f_o/R	z_{m_o}/R	z_{p_o}/R
45.00	3.75	0.50	−3.24	−0.56	−2.04	−1.48
45.50	3.74	0.49	−3.25	−0.56	−2.03	−1.48
46.00	3.74	0.49	−3.26	−0.55	−2.03	−1.48
46.50	3.74	0.48	−3.27	−0.55	−2.03	−1.48
47.00	3.74	0.47	−3.27	−0.55	−2.02	−1.48
47.50	3.74	0.46	−3.28	−0.54	−2.02	−1.48
48.00	3.74	0.46	−3.29	−0.54	−2.02	−1.48
48.50	3.74	0.45	−3.29	−0.54	−2.02	−1.48
49.00	3.74	0.44	−3.30	−0.53	−2.01	−1.48
49.50	3.74	0.44	−3.31	−0.53	−2.01	−1.48
50.00	3.74	0.43	−3.32	−0.53	−2.01	−1.48

TABLE 5.10

CARDINAL POINTS OF TWO EQUIDIAMETER CYLINDERS SEPARATED BY A DISTANCE $s = 0.8$[a]

V_2/V_1	f_i/R	z_{m_i}/R	z_{p_i}/R	f_o/R	z_{m_o}/R	z_{p_o}/R
1.50	87.96	79.40	−8.56	−71.82	−79.57	−7.74
2.00	32.69	27.34	−5.35	−23.12	−27.63	−4.51
2.50	20.02	15.72	−4.30	−12.66	−16.10	−3.44
3.00	14.76	10.98	−3.78	−8.52	−11.43	−2.91
3.50	11.94	8.46	−3.48	−6.38	−8.98	−2.59
4.00	10.20	6.91	−3.29	−5.10	−7.48	−2.38
4.50	9.03	5.87	−3.16	−4.26	−6.49	−2.23
5.00	8.19	5.12	−3.07	−3.66	−5.78	−2.12
5.50	7.55	4.55	−3.00	−3.22	−5.26	−2.04
6.00	7.06	4.11	−2.95	−2.88	−4.85	−1.97
6.50	6.66	3.75	−2.91	−2.61	−4.53	−1.91
7.00	6.34	3.46	−2.88	−2.40	−4.27	−1.87
7.50	6.07	3.21	−2.86	−2.22	−4.05	−1.83
8.00	5.84	3.00	−2.84	−2.07	−3.86	−1.80
8.50	5.65	2.82	−2.83	−1.94	−3.71	1.77
9.00	5.48	2.67	−2.82	−1.83	−3.57	−1.75
9.50	5.34	2.53	−2.81	−1.73	−3.46	−1.73
10.00	5.21	2.40	−2.81	−1.65	−3.35	−1.71
10.50	5.09	2.29	−2.80	−1.57	−3.26	−1.69
11.00	4.99	2.19	−2.80	−1.51	−3.18	−1.68
11.50	4.90	2.10	−2.80	−1.45	−3.11	−1.66
12.00	4.82	2.02	−2.80	−1.39	−3.04	−1.65
12.50	4.75	1.94	−2.81	−1.34	−2.98	−1.64
13.00	4.68	1.87	−2.81	−1.30	−2.93	−1.63

[a] All values are normalized with respect to R, the radius of the cylinders.

TABLE 5.10 (continued)

V_2/V_1	f_i/R	z_{m_i}/R	z_{p_i}/R	f_o/R	z_{m_o}/R	z_{p_o}/R
13.50	4.62	1.80	−2.81	−1.26	−2.88	−1.62
14.00	4.56	1.74	−2.82	−1.22	−2.83	−1.61
14.50	4.51	1.69	−2.82	−1.18	−2.79	−1.61
15.00	4.46	1.63	−2.83	−1.15	−2.75	−1.60
15.50	4.42	1.58	−2.83	−1.12	−2.71	−1.59
16.00	4.38	1.54	−2.84	−1.09	−2.68	−1.59
16.50	4.34	1.49	−2.85	−1.07	−2.65	−1.58
17.00	4.30	1.45	−2.85	−1.04	−2.62	−1.58
17.50	4.27	1.41	−2.86	−1.02	−2.59	−1.57
18.00	4.24	1.38	−2.87	−1.00	−2.57	−1.57
18.50	4.21	1.34	−2.87	−0.98	−2.54	−1.56
19.00	4.19	1.31	−2.88	−0.96	−2.52	−1.56
19.50	4.16	1.27	−2.89	−0.94	−2.50	−1.56
20.00	4.14	1.24	−2.89	−0.93	−2.48	−1.55
20.50	4.12	1.22	−2.90	−0.91	−2.46	−1.55
21.00	4.10	1.19	−2.91	−0.89	−2.44	−1.55
21.50	4.08	1.16	−2.92	−0.88	−2.42	−1.54
22.00	4.06	1.13	−2.92	−0.87	−2.41	−1.54
22.50	4.04	1.11	−2.93	−0.85	−2.39	−1.54
23.00	4.03	1.09	−2.94	−0.84	−2.38	−1.54
23.50	4.01	1.06	−2.95	−0.83	−2.36	−1.54
24.00	4.00	1.04	−2.96	−0.82	−2.35	−1.53
24.50	3.98	1.02	−2.96	−0.80	−2.34	−1.53
25.00	3.97	1.00	−2.97	−0.79	−2.32	−1.53
25.50	3.96	0.98	−2.98	−0.78	−2.31	−1.53
26.00	3.95	0.96	−2.99	−0.77	−2.30	−1.53
26.50	3.94	0.94	−3.00	−0.76	−2.29	−1.53
27.00	3.93	0.92	−3.00	−0.76	−2.28	−1.52
27.50	3.92	0.91	−3.01	−0.75	−2.27	−1.52
28.00	3.91	0.89	−3.02	−0.74	−2.26	−1.52
28.50	3.90	0.87	−3.03	−0.73	−2.25	−1.52
29.00	3.89	0.86	−3.04	−0.72	−2.24	−1.52
29.50	3.88	0.84	−3.04	−0.72	−2.23	−1.52
30.00	3.88	0.83	−3.05	−0.71	−2.22	−1.52
30.50	3.87	0.81	−3.06	−0.70	−2.22	−1.52
31.00	3.86	0.80	−3.07	−0.69	−2.21	−1.51
31.50	3.86	0.78	−3.08	−0.69	−2.20	−1.51
32.00	3.85	0.77	−3.08	−0.68	−2.19	−1.51
32.50	3.85	0.76	−3.09	−0.67	−2.19	−1.51
33.00	3.84	0.74	−3.10	−0.67	−2.18	−1.51
33.50	3.84	0.73	−3.11	−0.66	−2.17	−1.51
34.00	3.83	0.72	−3.11	−0.66	−2.17	−1.51
34.50	3.83	0.70	−3.12	−0.65	−2.16	−1.51
35.00	3.82	0.69	−3.13	−0.65	−2.15	−1.51
35.50	3.82	0.68	−3.14	−0.64	−2.15	−1.51

TABLE 5.10 (continued)

V_2/V_1	f_i/R	z_{m_i}/R	z_{p_i}/R	f_o/R	z_{m_o}/R	z_{p_o}/R
36.00	3.82	0.67	−3.15	−0.64	−2.14	−1.51
36.50	3.81	0.66	−3.15	−0.63	−2.14	−1.51
37.00	3.81	0.65	−3.16	−0.63	−2.13	−1.51
37.50	3.81	0.64	−3.17	−0.62	−2.13	−1.51
38.00	3.80	0.63	−3.18	−0.62	−2.12	−1.51
38.50	3.80	0.62	−3.18	−0.61	−2.12	−1.50
39.00	3.80	0.61	−3.19	−0.61	−2.11	−1.50
39.50	3.80	0.60	−3.20	−0.60	−2.11	−1.50
40.00	3.79	0.59	−3.21	−0.60	−2.10	−1.50
40.50	3.79	0.58	−3.21	−0.60	−2.10	−1.50
41.00	3.79	0.57	−3.22	−0.59	−2.09	−1.50
41.50	3.79	0.56	−3.23	−0.59	−2.09	−1.50
42.00	3.79	0.55	−3.24	−0.58	−2.09	−1.50
42.50	3.79	0.54	−3.24	−0.58	−2.08	−1.50
43.00	3.78	0.53	−3.25	−0.58	−2.08	−1.50
43.50	3.78	0.52	−3.26	−0.57	−2.07	−1.50
44.00	3.78	0.52	−3.27	−0.57	−2.07	−1.50
44.50	3.78	0.51	−3.27	−0.57	−2.07	−1.50
45.00	3.78	0.50	−3.28	−0.56	−2.06	−1.50
45.50	3.78	0.49	−3.29	−0.56	−2.06	−1.50
46.00	3.78	0.48	−3.30	−0.56	−2.06	−1.50
46.50	3.78	0.48	−3.30	−0.55	−2.05	−1.50
47.00	3.78	0.47	−3.31	−0.55	−2.05	−1.50
47.50	3.78	0.46	−3.32	−0.55	−2.05	−1.50
48.00	3.78	0.45	−3.32	−0.55	−2.04	−1.50
48.50	3.78	0.45	−3.33	−0.54	−2.04	−1.50
49.00	3.78	0.44	−3.34	−0.54	−2.04	−1.50
49.50	3.78	0.43	−3.35	−0.54	−2.03	−1.50
50.00	3.78	0.43	−3.35	−0.53	−2.03	−1.50

TABLE 5.11

CARDINAL POINTS OF TWO EQUIDIAMETER CYLINDERS SEPARATED BY A DISTANCE $s = 0.9$[a]

V_2/V_1	f_i/R	z_{m_i}/R	z_{p_i}/R	f_o/R	z_{m_o}/R	z_{p_o}/R
1.50	89.46	80.75	−8.71	−73.05	−80.93	−7.88
2.00	33.24	27.80	−5.44	−23.51	−28.10	−4.59
2.50	20.35	15.98	−4.37	−12.87	−16.37	−3.50
3.00	15.00	11.15	−3.85	−8.66	−11.62	−2.96
3.50	12.13	8.59	−3.54	−6.49	−9.12	−2.64
4.00	10.36	7.02	−3.35	−5.18	−7.61	−2.42

[a] All values are normalized with respect to R, the radius of the cylinders.

TABLE 5.11 (continued)

V_2/V_1	f_i/R	z_{m_i}/R	z_{p_i}/R	f_o/R	z_{m_o}/R	z_{p_o}/R
4.50	9.17	5.96	−3.21	−4.32	−6.59	−2.27
5.00	8.31	5.19	−3.12	−3.72	−5.88	−2.16
5.50	7.67	4.62	−3.05	−3.27	−5.34	−2.07
6.00	7.16	4.17	−3.00	−2.92	−4.93	−2.00
6.50	6.76	3.81	−2.96	−2.65	−4.60	−1.95
7.00	6.43	3.51	−2.93	−2.43	−4.33	−1.90
7.50	6.16	3.26	−2.90	−2.25	−4.11	−1.86
8.00	5.93	3.04	−2.89	−2.10	−3.92	−1.83
8.50	5.73	2.86	−2.87	−1.97	−3.77	−1.80
9.00	5.56	2.70	−2.86	−1.85	−3.63	−1.77
9.50	5.41	2.56	−2.85	−1.76	−3.51	−1.75
10.00	5.28	2.43	−2.85	−1.67	−3.40	−1.73
10.50	5.17	2.32	−2.85	−1.59	−3.31	−1.72
11.00	5.06	2.22	−2.85	−1.53	−3.23	−1.70
11.50	4.97	2.12	−2.85	−1.47	−3.15	−1.69
12.00	4.89	2.04	−2.85	−1.41	−3.09	−1.68
12.50	4.81	1.96	−2.85	−1.36	−3.03	−1.67
13.00	4.74	1.89	−2.85	−1.31	−2.97	−1.66
13.50	4.68	1.82	−2.86	−1.27	−2.92	−1.65
14.00	4.62	1.76	−2.86	−1.23	−2.87	−1.64
14.50	4.57	1.70	−2.86	−1.20	−2.83	−1.63
15.00	4.52	1.65	−2.87	−1.17	−2.79	−1.62
15.50	4.47	1.60	−2.87	−1.14	−2.75	−1.62
16.00	4.43	1.55	−2.88	−1.11	−2.72	−1.61
16.50	4.39	1.51	−2.89	−1.08	−2.69	−1.61
17.00	4.36	1.47	−2.89	−1.06	−2.66	−1.60
17.50	4.33	1.43	−2.90	−1.03	−2.63	−1.60
18.00	4.29	1.39	−2.91	−1.01	−2.60	−1.59
18.50	4.27	1.35	−2.91	−0.99	−2.58	−1.59
19.00	4.24	1.32	−2.92	−0.97	−2.56	−1.58
19.50	4.21	1.29	−2.93	−0.95	−2.54	−1.58
20.00	4.19	1.25	−2.94	−0.94	−2.51	−1.58
20.50	4.17	1.22	−2.94	−0.92	−2.50	−1.57
21.00	4.15	1.20	−2.95	−0.90	−2.48	−1.57
21.50	4.13	1.17	−2.96	−0.89	−2.46	−1.57
22.00	4.11	1.14	−2.97	−0.88	−2.44	−1.57
22.50	4.09	1.12	−2.97	−0.86	−2.43	−1.56
23.00	4.07	1.09	−2.98	−0.85	−2.41	−1.56
23.50	4.06	1.07	−2.99	−0.84	−2.40	−1.56
24.00	4.04	1.05	−3.00	−0.83	−2.38	−1.56
24.50	4.03	1.03	−3.01	−0.81	−2.37	−1.56
25.00	4.02	1.00	−3.01	−0.80	−2.36	−1.55
25.50	4.01	0.98	−3.02	−0.79	−2.35	−1.55
26.00	3.99	0.96	−3.03	−0.78	−2.33	−1.55
26.50	3.98	0.95	−3.04	−0.77	−2.32	−1.55
27.00	3.97	0.93	−3.05	−0.76	−2.31	−1.55

TABLE 5.11 (continued)

V_2/V_1	f_i/R	z_{m_i}/R	z_{p_i}/R	f_o/R	z_{m_o}/R	z_{p_o}/R
27.50	3.96	0.91	−3.05	−0.76	−2.30	−1.55
28.00	3.95	0.89	−3.06	−0.75	−2.29	−1.54
28.50	3.94	0.88	−3.07	−0.74	−2.28	−1.54
29.00	3.94	0.86	−3.08	−0.73	−2.27	−1.54
29.50	3.93	0.84	−3.08	−0.72	−2.26	−1.54
30.00	3.92	0.83	−3.09	−0.72	−2.26	−1.54
30.50	3.91	0.81	−3.10	−0.71	−2.25	−1.54
31.00	3.91	0.80	−3.11	−0.70	−2.24	−1.54
31.50	3.90	0.78	−3.12	−0.70	−2.23	−1.54
32.00	3.89	0.77	−3.12	−0.69	−2.22	−1.54
32.50	3.89	0.76	−3.13	−0.68	−2.22	−1.54
33.00	3.88	0.74	−3.14	−0.68	−2.21	−1.53
33.50	3.88	0.73	−3.15	−0.67	−2.20	−1.53
34.00	3.87	0.72	−3.16	−0.66	−2.20	−1.53
34.50	3.87	0.71	−3.16	−0.66	−2.19	−1.53
35.00	3.87	0.69	−3.17	−0.65	−2.18	−1.53
35.50	3.86	0.68	−3.18	−0.65	−2.18	−1.53
36.00	3.86	0.67	−3.19	−0.64	−2.17	−1.53
36.50	3.85	0.66	−3.19	−0.64	−2.17	−1.53
37.00	3.85	0.65	−3.20	−0.63	−2.16	−1.53
37.50	3.85	0.64	−3.21	−0.63	−2.16	−1.53
38.00	3.84	0.63	−3.22	−0.62	−2.15	−1.53
38.50	3.84	0.62	−3.23	−0.62	−2.15	−1.53
39.00	3.84	0.61	−3.23	−0.61	−2.14	−1.53
39.50	3.84	0.60	−3.24	−0.61	−2.14	−1.53
40.00	3.83	0.59	−3.25	−0.61	−2.13	−1.53
40.50	3.83	0.58	−3.26	−0.60	−2.13	−1.53
41.00	3.83	0.57	−3.26	−0.60	−2.12	−1.53
41.50	3.83	0.56	−3.27	−0.59	−2.12	−1.52
42.00	3.83	0.55	−3.28	−0.59	−2.12	−1.52
42.50	3.83	0.54	−3.29	−0.59	−2.11	−1.52
43.00	3.82	0.53	−3.29	−0.58	−2.11	−1.52
43.50	3.82	0.52	−3.30	−0.58	−2.10	−1.52
44.00	3.82	0.51	−3.31	−0.58	−2.10	−1.52
44.50	3.82	0.51	−3.32	−0.57	−2.10	−1.52
45.00	3.82	0.50	−3.32	−0.57	−2.09	−1.52
45.50	3.82	0.49	−3.33	−0.57	−2.09	−1.52
46.00	3.82	0.48	−3.34	−0.56	−2.09	−1.52
46.50	3.82	0.47	−3.34	−0.56	−2.08	−1.52
47.00	3.82	0.47	−3.35	−0.56	−2.08	−1.52
47.50	3.82	0.46	−3.36	−0.55	−2.08	−1.52
48.00	3.82	0.45	−3.37	−0.55	−2.07	−1.52
48.50	3.82	0.44	−3.37	−0.55	−2.07	−1.52
49.00	3.82	0.44	−3.38	−0.55	−2.07	−1.52
49.50	3.82	0.43	−3.39	−0.54	−2.06	−1.52
50.00	3.82	0.42	−3.39	−0.54	−2.06	−1.52

TABLE 5.12

CARDINAL POINTS OF TWO EQUIDIAMETER CYLINDERS SEPARATED BY A DISTANCE $s = 1.0$[a]

V_2/V_1	f_i/R	z_{m_i}/R	z_{p_i}/R	f_o/R	z_{m_o}/R	z_{p_o}/R
1.50	91.12	82.24	−8.87	−74.40	−82.42	−8.02
2.00	33.85	28.31	−5.54	−23.93	−28.61	−4.68
2.50	20.72	16.27	−4.45	−13.10	−16.67	−3.57
3.00	15.27	11.35	−3.92	−8.81	−11.83	−3.01
3.50	12.34	8.74	−3.61	−6.60	−9.28	−2.68
4.00	10.54	7.14	−3.41	−5.27	−7.74	−2.47
4.50	9.33	6.06	−3.27	−4.40	−6.71	−2.31
5.00	8.45	5.28	−3.17	−3.78	−5.98	−2.20
5.50	7.79	4.69	−3.10	−3.32	−5.43	−2.11
6.00	7.28	4.23	−3.05	−2.97	−5.01	−2.04
6.50	6.87	3.86	−3.01	−2.70	−4.68	−1.98
7.00	6.54	3.56	−2.98	−2.47	−4.40	−1.93
7.50	6.26	3.31	−2.95	−2.29	−4.18	−1.89
8.00	6.02	3.09	−2.93	−2.13	−3.99	−1.86
8.50	5.82	2.90	−2.92	−2.00	−3.83	−1.83
9.00	5.65	2.74	−2.91	−1.88	−3.69	−1.81
9.50	5.50	2.59	−2.90	−1.78	−3.57	−1.78
10.00	5.36	2.46	−2.90	−1.70	−3.46	−1.76
10.50	5.24	2.35	−2.89	−1.62	−3.37	−1.75
11.00	5.14	2.24	−2.89	−1.55	−3.28	−1.73
11.50	5.04	2.15	−2.89	−1.49	−3.21	−1.72
12.00	4.96	2.06	−2.89	−1.43	−3.14	−1.71
12.50	4.88	1.98	−2.90	−1.38	−3.07	−1.69
13.00	4.81	1.91	−2.90	−1.33	−3.02	−1.68
13.50	4.75	1.84	−2.90	−1.29	−2.97	−1.68
14.00	4.69	1.78	−2.91	−1.25	−2.92	−1.67
14.50	4.63	1.72	−2.91	−1.22	−2.88	−1.66
15.00	4.58	1.67	−2.92	−1.18	−2.84	−1.65
15.50	4.54	1.62	−2.92	−1.15	−2.80	−1.65
16.00	4.49	1.57	−2.93	−1.12	−2.76	−1.64
16.50	4.46	1.52	−2.93	−1.10	−2.73	−1.63
17.00	4.42	1.48	−2.94	−1.07	−2.70	−1.63
17.50	4.39	1.44	−2.95	−1.05	−2.67	−1.62
18.00	4.35	1.40	−2.95	−1.03	−2.65	−1.62
18.50	4.32	1.36	−2.96	−1.01	−2.62	−1.62
19.00	4.30	1.33	−2.97	−0.99	−2.60	−1.61
19.50	4.27	1.30	−2.97	−0.97	−2.57	−1.61
20.00	4.25	1.26	−2.98	−0.95	−2.55	−1.60
20.50	4.22	1.23	−2.99	−0.93	−2.53	−1.60
21.00	4.20	1.21	−3.00	−0.92	−2.52	−1.60
21.50	4.18	1.18	−3.00	−0.90	−2.50	−1.60
22.00	4.16	1.15	−3.01	−0.89	−2.48	−1.59

[a] All values are normalized with respect to R, the radius of the cylinders.

TABLE 5.12 (continued)

V_2/V_1	f_i/R	z_{m_i}/R	z_{p_i}/R	f_o/R	z_{m_o}/R	z_{p_o}/R
22.50	4.14	1.13	−3.02	−0.87	−2.46	−1.59
23.00	4.13	1.10	−3.03	−0.86	−2.45	−1.59
23.50	4.11	1.08	−3.03	−0.85	−2.43	−1.59
24.00	4.10	1.05	−3.04	−0.84	−2.42	−1.58
24.50	4.08	1.03	−3.05	−0.82	−2.41	−1.58
25.00	4.07	1.01	−3.06	−0.81	−2.39	−1.58
25.50	4.06	0.99	−3.07	−0.80	−2.38	−1.58
26.00	4.04	0.97	−3.07	−0.79	−2.37	−1.58
26.50	4.03	0.95	−3.08	−0.78	−2.36	−1.57
27.00	4.02	0.93	−3.09	−0.77	−2.35	−1.57
27.50	4.01	0.91	−3.10	−0.77	−2.34	−1.57
28.00	4.00	0.90	−3.11	−0.76	−2.33	−1.57
28.50	3.99	0.88	−3.11	−0.75	−2.32	−1.57
29.00	3.99	0.86	−3.12	−0.74	−2.31	−1.57
29.50	3.98	0.85	−3.13	−0.73	−2.30	−1.57
30.00	3.97	0.83	−3.14	−0.72	−2.29	−1.57
30.50	3.96	0.82	−3.15	−0.72	−2.28	−1.56
31.00	3.95	0.80	−3.15	−0.71	−2.27	−1.56
31.50	3.95	0.79	−3.16	−0.70	−2.27	−1.56
32.00	3.94	0.77	−3.17	−0.70	−2.26	−1.56
32.50	3.94	0.76	−3.18	−0.69	−2.25	−1.56
33.00	3.93	0.74	−3.19	−0.68	−2.24	−1.56
33.50	3.93	0.73	−3.19	−0.68	−2.24	−1.56
34.00	3.92	0.72	−3.20	−0.67	−2.23	−1.56
34.50	3.92	0.71	−3.21	−0.67	−2.22	−1.56
35.00	3.91	0.69	−3.22	−0.66	−2.22	−1.56
35.50	3.91	0.68	−3.22	−0.66	−2.21	−1.56
36.00	3.90	0.67	−3.23	−0.65	−2.21	−1.56
36.50	3.90	0.66	−3.24	−0.65	−2.20	−1.56
37.00	3.90	0.65	−3.25	−0.64	−2.19	−1.55
37.50	3.89	0.64	−3.26	−0.64	−2.19	−1.55
38.00	3.89	0.63	3.26	−0.63	−2.18	−1.55
38.50	3.89	0.62	−3.27	−0.63	−2.18	−1.55
39.00	3.88	0.60	−3.28	−0.62	−2.17	−1.55
39.50	3.88	0.59	−3.29	−0.62	−2.17	−1.55
40.00	3.88	0.58	−3.29	−0.61	−2.16	−1.55
40.50	3.88	0.58	−3.30	−0.61	−2.16	−1.55
41.00	3.87	0.57	−3.31	−0.61	−2.16	−1.55
41.50	3.87	0.56	−3.32	−0.60	−2.15	−1.55
42.00	3.87	0.55	−3.32	−0.60	−2.15	−1.55
42.50	3.87	0.54	−3.33	−0.59	−2.14	−1.55
43.00	3.87	0.53	−3.34	−0.59	−2.14	−1.55
43.50	3.87	0.52	−3.35	−0.59	−2.13	−1.55
44.00	3.86	0.51	−3.35	−0.58	−2.13	−1.55
44.50	3.86	0.50	−3.36	−0.58	−2.13	−1.55

TABLE 5.12 (continued)

V_2/V_1	f_i/R	z_{m_i}/R	z_{p_i}/R	f_o/R	z_{m_o}/R	z_{p_o}/R
45.00	3.86	0.49	−3.37	−0.58	−2.12	−1.55
45.50	3.86	0.49	−3.38	−0.57	−2.12	−1.55
46.00	3.86	0.48	−3.38	−0.57	−2.12	−1.55
46.50	3.86	0.47	−3.39	−0.57	−2.11	−1.55
47.00	3.86	0.46	−3.40	−0.56	−2.11	−1.55
47.50	3.86	0.45	−3.40	−0.56	−2.11	−1.55
48.00	3.86	0.45	−3.41	−0.56	−2.10	−1.55
48.50	3.86	0.44	−3.42	−0.55	−2.10	−1.55
49.00	3.86	0.43	−3.43	−0.55	−2.10	−1.55
49.50	3.86	0.42	−3.43	−0.55	−2.09	−1.55
50.00	3.86	0.42	−3.44	−0.55	−2.09	−1.55

TABLE 5.13

CARDINAL POINTS OF TWO EQUIDIAMETER CYLINDERS SEPARATED BY A DISTANCE $s = 1.1$[a]

V_2/V_1	f_i/R	z_{m_i}/R	z_{p_i}/R	f_o/R	z_{m_o}/R	z_{p_o}/R
1.50	92.91	83.86	−9.05	−75.86	−84.04	−8.18
2.00	34.51	28.86	−5.65	−24.40	−29.17	−4.77
2.50	21.12	16.58	−4.54	−13.36	−16.99	−3.64
3.00	15.55	11.56	−3.99	−8.98	−12.05	−3.07
3.50	12.57	8.90	−3.68	−6.72	−9.46	−2.74
4.00	10.74	7.26	−3.47	−5.37	−7.88	−2.51
4.50	9.50	6.16	−3.33	−4.48	−6.83	−2.36
5.00	8.60	5.37	−3.23	−3.85	−6.09	−2.24
5.50	7.93	4.77	−3.16	−3.38	−5.53	−2.15
6.00	7.41	4.30	−3.10	−3.02	−5.10	−2.08
6.50	6.99	3.93	−3.06	−2.74	−4.76	−2.02
7.00	6.65	3.62	−3.03	−2.51	−4.48	−1.97
7.50	6.36	3.36	−3.01	−2.32	−4.25	−1.93
8.00	6.12	3.14	−2.99	−2.17	−4.06	−1.89
8.50	5.92	2.95	−2.97	−2.03	−3.89	−1.86
9.00	5.74	2.78	−2.96	−1.91	−3.75	−1.84
9.50	5.59	2.63	−2.95	−1.81	−3.63	−1.82
10.00	5.45	2.50	−2.95	−1.72	−3.52	−1.80
10.50	5.33	2.38	−2.95	−1.64	−3.42	−1.78
11.00	5.22	2.28	−2.94	−1.57	−3.34	−1.76
11.50	5.12	2.18	−2.94	−1.51	−3.26	−1.75
12.00	5.03	2.09	−2.94	−1.45	−3.19	−1.74
12.50	4.96	2.01	−2.95	−1.40	−3.13	−1.73
13.00	4.88	1.94	−2.95	−1.35	−3.07	−1.72

[a] All values are normalized with respect to R, the radius of the cylinders.

TABLE 5.13 (continued)

V_2/V_1	f_i/R	z_{m_i}/R	z_{p_i}/R	f_o/R	z_{m_o}/R	z_{p_o}/R
13.50	4.82	1.87	−2.95	−1.31	−3.02	−1.71
14.00	4.76	1.80	−2.96	−1.27	−2.97	−1.70
14.50	4.70	1.74	−2.96	−1.24	−2.92	−1.69
15.00	4.65	1.69	−2.96	−1.20	−2.88	−1.68
15.50	4.61	1.63	−2.97	−1.17	−2.85	−1.68
16.00	4.56	1.59	−2.98	−1.14	−2.81	−1.67
16.50	4.52	1.54	−2.98	−1.11	−2.78	−1.66
17.00	4.48	1.50	−2.99	−1.09	−2.75	−1.66
17.50	4.45	1.45	−2.99	−1.06	−2.72	−1.65
18.00	4.42	1.42	−3.00	−1.04	−2.69	−1.65
18.50	4.39	1.38	−3.01	−1.02	−2.66	−1.64
19.00	4.36	1.34	−3.02	−1.00	−2.64	−1.64
19.50	4.33	1.31	−3.02	−0.98	−2.62	−1.64
20.00	4.31	1.28	−3.03	−0.96	−2.60	−1.63
20.50	4.28	1.25	−3.04	−0.95	−2.58	−1.63
21.00	4.26	1.22	−3.04	−0.93	−2.56	−1.63
21.50	4.24	1.19	−3.05	−0.91	−2.54	−1.62
22.00	4.22	1.16	−3.06	−0.90	−2.52	−1.62
22.50	4.20	1.13	−3.07	−0.89	−2.50	−1.62
23.00	4.18	1.11	−3.08	−0.87	−2.49	−1.62
23.50	4.17	1.08	−3.08	−0.86	−2.47	−1.61
24.00	4.15	1.06	−3.09	−0.85	−2.46	−1.61
24.50	4.14	1.04	−3.10	−0.84	−2.45	−1.61
25.00	4.12	1.02	−3.11	−0.82	−2.43	−1.61
25.50	4.11	1.00	−3.12	−0.81	−2.42	−1.61
26.00	4.10	0.98	−3.12	−0.80	−2.41	−1.60
26.50	4.09	0.96	−3.13	−0.79	−2.40	−1.60
27.00	4.08	0.94	−3.14	−0.78	−2.39	−1.60
27.50	4.07	0.92	−3.15	−0.78	−2.38	−1.60
28.00	4.06	0.90	−3.16	−0.77	−2.36	−1.60
28.50	4.05	0.88	−3.16	−0.76	−2.36	−1.60
29.00	4.04	0.87	−3.17	−0.75	−2.35	−1.60
29.50	4.03	0.85	−3.18	−0.74	−2.34	−1.59
30.00	4.02	0.83	−3.19	−0.73	−2.33	−1.59
30.50	4.01	0.82	−3.19	−0.73	−2.32	−1.59
31.00	4.01	0.80	−3.20	−0.72	−2.31	−1.59
31.50	4.00	0.79	−3.21	−0.71	−2.30	−1.59
32.00	3.99	0.77	−3.22	−0.71	−2.30	−1.59
32.50	3.99	0.76	−3.23	−0.70	−2.29	−1.59
33.00	3.98	0.75	−3.23	−0.69	−2.28	−1.59
33.50	3.98	0.73	−3.24	−0.69	−2.27	−1.59
34.00	3.97	0.72	−3.25	−0.68	−2.27	−1.59
34.50	3.96	0.71	−3.26	−0.68	−2.26	−1.59
35.00	3.96	0.69	−3.27	−0.67	−2.25	−1.58
35.50	3.96	0.68	−3.27	−0.66	−2.25	−1.58

TABLE 5.13 (continued)

V_2/V_1	f_i/R	z_{m_i}/R	z_{p_i}/R	f_o/R	z_{m_o}/R	z_{p_o}/R
36.00	3.95	0.67	−3.28	−0.66	−2.24	−1.58
36.50	3.95	0.66	−3.29	−0.65	−2.24	−1.58
37.00	3.94	0.65	−3.30	−0.65	−2.23	−1.58
37.50	3.94	0.64	−3.30	−0.64	−2.22	−1.58
38.00	3.94	0.62	−3.31	−0.64	−2.22	−1.58
38.50	3.93	0.61	−3.32	−0.63	−2.21	−1.58
39.00	3.93	0.60	−3.33	−0.63	−2.21	−1.58
39.50	3.93	0.59	−3.34	−0.63	−2.20	−1.58
40.00	3.93	0.58	−3.34	−0.62	−2.20	−1.58
40.50	3.92	0.57	−3.35	−0.62	−2.19	−1.58
41.00	3.92	0.56	−3.36	−0.61	−2.19	−1.58
41.50	3.92	0.55	−3.37	−0.61	−2.19	−1.58
42.00	3.92	0.54	−3.37	−0.60	−2.18	−1.58
42.50	3.92	0.53	−3.38	−0.60	−2.18	−1.58
43.00	3.91	0.53	−3.39	−0.60	−2.17	−1.58
43.50	3.91	0.52	−3.40	−0.59	−2.17	−1.58
44.00	3.91	0.51	−3.40	−0.59	−2.17	−1.58
44.50	3.91	0.50	−3.41	−0.59	−2.16	−1.58
45.00	3.91	0.49	−3.42	−0.58	−2.16	−1.57
45.50	3.91	0.48	−3.42	−0.58	−2.15	−1.57
46.00	3.91	0.47	−3.43	−0.58	−2.15	−1.57
46.50	3.91	0.47	−3.44	−0.57	−2.15	−1.57
47.00	3.90	0.46	−3.45	−0.57	−2.14	−1.57
47.50	3.90	0.45	−3.45	−0.57	−2.14	−1.57
48.00	3.90	0.44	−3.46	−0.56	−2.14	−1.57
48.50	3.90	0.43	−3.47	−0.56	−2.13	−1.57
49.00	3.90	0.43	−3.48	−0.56	−2.13	−1.57
49.50	3.90	0.42	−3.48	−0.55	−2.13	−1.57
50.00	3.90	0.41	−3.49	−0.55	−2.12	−1.57

TABLE 5.14

CARDINAL POINTS OF TWO EQUIDIAMETER CYLINDERS SEPARATED BY A DISTANCE $s = 1.2$[a]

V_2/V_1	f_i/R	z_{m_i}/R	z_{p_i}/R	f_o/R	z_{m_o}/R	z_{p_o}/R
1.50	94.84	85.60	−9.24	−77.44	−85.79	−8.35
2.00	35.21	29.45	−5.77	−24.90	−29.76	−4.87
2.50	21.54	16.91	−4.63	−13.62	−17.33	−3.71
3.00	15.86	11.79	−4.07	−9.16	−12.29	−3.14
3.50	12.82	9.07	−3.75	−6.85	−9.65	−2.79
4.00	10.94	7.40	−3.54	−5.47	−8.04	−2.57

[a] All values are normalized with respect to R, the radius of the cylinders.

TABLE 5.14 (continued)

V_2/V_1	f_i/R	z_{m_i}/R	z_{p_i}/R	f_o/R	z_{m_o}/R	z_{p_o}/R
4.50	9.68	6.28	−3.40	−4.56	−6.97	−2.40
5.00	8.77	5.47	−3.30	−3.92	−6.20	−2.28
5.50	8.08	4.86	−3.22	−3.45	−5.64	−2.19
6.00	7.55	4.38	−3.17	−3.08	−5.20	−2.12
6.50	7.12	4.00	−3.12	−2.79	−4.85	−2.06
7.00	6.77	3.68	−3.09	−2.56	−4.57	−2.01
7.50	6.48	3.41	−3.06	−2.37	−4.33	−1.97
8.00	6.23	3.19	−3.04	−2.20	−4.14	−1.93
8.50	6.02	2.99	−3.03	−2.07	−3.97	−1.90
9.00	5.84	2.82	−3.02	−1.95	−3.82	−1.88
9.50	5.68	2.67	−3.01	−1.84	−3.70	−1.85
10.00	5.54	2.54	−3.00	−1.75	−3.58	−1.83
10.50	5.42	2.42	−3.00	−1.67	−3.49	−1.81
11.00	5.31	2.31	−3.00	−1.60	−3.40	−1.80
11.50	5.21	2.21	−3.00	−1.54	−3.32	−1.78
12.00	5.12	2.12	−3.00	−1.48	−3.25	−1.77
12.50	5.04	2.04	−3.00	−1.42	−3.18	−1.76
13.00	4.96	1.96	−3.00	−1.38	−3.12	−1.75
13.50	4.90	1.89	−3.01	−1.33	−3.07	−1.74
14.00	4.83	1.83	−3.01	−1.29	−3.02	−1.73
14.50	4.78	1.76	−3.01	−1.25	−2.98	−1.72
15.00	4.73	1.71	−3.02	−1.22	−2.93	−1.71
15.50	4.68	1.65	−3.02	−1.19	−2.90	−1.71
16.00	4.63	1.60	−3.03	−1.16	−2.86	−1.70
16.50	4.59	1.56	−3.03	−1.13	−2.83	−1.70
17.00	4.55	1.51	−3.04	−1.10	−2.79	−1.69
17.50	4.52	1.47	−3.05	−1.08	−2.76	−1.68
18.00	4.48	1.43	−3.05	−1.06	−2.74	−1.68
18.50	4.45	1.39	−3.06	−1.04	−2.71	−1.68
19.00	4.42	1.36	−3.07	−1.01	−2.69	−1.67
19.50	4.40	1.32	−3.08	−1.00	−2.66	−1.67
20.00	4.37	1.29	−3.08	−0.98	−2.64	1.66
20.50	4.35	1.26	−3.09	−0.96	−2.62	−1.66
21.00	4.32	1.23	−3.10	−0.94	−2.60	−1.66
21.50	4.30	1.20	−3.11	−0.93	−2.58	−1.65
22.00	4.28	1.17	−3.11	−0.91	−2.57	−1.65
22.50	4.26	1.14	−3.12	−0.90	−2.55	−1.65
23.00	4.25	1.12	−3.13	−0.89	−2.53	−1.65
23.50	4.23	1.09	−3.14	−0.87	−2.52	−1.64
24.00	4.21	1.07	−3.14	−0.86	−2.50	−1.64
24.50	4.20	1.05	−3.15	−0.85	−2.49	−1.64
25.00	4.18	1.02	−3.16	−0.84	−2.47	−1.64
25.50	4.17	1.00	−3.17	−0.83	−2.46	−1.64
26.00	4.16	0.98	−3.18	−0.82	−2.45	−1.63
26.50	4.14	0.96	−3.18	−0.81	−2.44	−1.63

TABLE 5.14 (continued)

V_2/V_1	f_i/R	z_{m_i}/R	z_{p_i}/R	f_o/R	z_{m_o}/R	z_{p_o}/R
27.00	4.13	0.94	−3.19	−0.80	−2.43	−1.63
27.50	4.12	0.92	−3.20	−0.79	−2.42	−1.63
28.00	4.11	0.90	−3.21	−0.78	−2.41	−1.63
28.50	4.10	0.89	−3.22	−0.77	−2.40	−1.63
29.00	4.09	0.87	−3.22	−0.76	−2.39	−1.63
29.50	4.08	0.85	−3.23	−0.75	−2.38	−1.62
30.00	4.08	0.84	−3.24	−0.74	−2.37	−1.62
30.50	4.07	0.82	−3.25	−0.74	−2.36	−1.62
31.00	4.06	0.81	−3.26	−0.73	−2.35	−1.62
31.50	4.05	0.79	−3.26	−0.72	−2.34	−1.62
32.00	4.05	0.78	−3.27	−0.72	−2.33	−1.62
32.50	4.04	0.76	−3.28	−0.71	−2.33	−1.62
33.00	4.03	0.75	−3.29	−0.70	−2.32	−1.62
33.50	4.03	0.73	−3.29	−0.70	−2.31	−1.62
34.00	4.02	0.72	−3.30	−0.69	−2.31	−1.62
34.50	4.02	0.71	−3.31	−0.68	−2.30	−1.61
35.00	4.01	0.69	−3.32	−0.68	−2.29	−1.61
35.50	4.01	0.68	−3.33	−0.67	−2.29	−1.61
36.00	4.00	0.67	−3.33	−0.67	−2.28	−1.61
36.50	4.00	0.66	−3.34	−0.66	−2.27	−1.61
37.00	4.00	0.65	−3.35	−0.66	−2.27	−1.61
37.50	3.99	0.63	−3.36	−0.65	−2.26	−1.61
38.00	3.99	0.62	−3.36	−0.65	−2.26	−1.61
38.50	3.98	0.61	−3.37	−0.64	−2.25	−1.61
39.00	3.98	0.60	−3.38	−0.64	−2.25	−1.61
39.50	3.98	0.59	−3.39	−0.63	−2.24	−1.61
40.00	3.98	0.58	−3.40	−0.63	−2.24	−1.61
40.50	3.97	0.57	−3.40	−0.62	−2.23	−1.61
41.00	3.97	0.56	−3.41	−0.62	−2.23	−1.61
41.50	3.97	0.55	−3.42	−0.62	−2.22	−1.61
42.00	3.97	0.54	−3.43	−0.61	−2.22	−1.61
42.50	3.96	0.53	−3.43	−0.61	−2.21	−1.61
43.00	3.96	0.52	−3.44	−0.60	−2.21	−1.61
43.50	3.96	0.51	−3.45	−0.60	−2.21	−1.60
44.00	3.96	0.50	−3.46	−0.60	−2.20	−1.60
44.50	3.96	0.50	−3.46	−0.59	−2.20	−1.60
45.00	3.96	0.49	−3.47	−0.59	−2.19	−1.60
45.50	3.96	0.48	−3.48	−0.59	−2.19	−1.60
46.00	3.95	0.47	−3.48	−0.58	−2.19	−1.60
46.50	3.95	0.46	−3.49	−0.58	−2.18	−1.60
47.00	3.95	0.45	−3.50	−0.58	−2.18	−1.60
47.50	3.95	0.45	−3.51	−0.57	−2.18	−1.60
48.00	3.95	0.44	−3.51	−0.57	−2.17	−1.60
48.50	3.95	0.43	−3.52	−0.57	−2.17	−1.60
49.00	3.95	0.42	−3.53	−0.56	−2.17	−1.60
49.50	3.95	0.41	−3.54	−0.56	−2.16	−1.60
50.00	3.95	0.41	−3.54	−0.56	−2.16	−1.60

TABLE 5.15

CARDINAL POINTS OF TWO EQUIDIAMETER CYLINDERS SEPARATED BY A DISTANCE $s = 1.3$[a]

V_2/V_1	f_i/R	z_{m_i}/R	z_{p_i}/R	f_o/R	z_{m_o}/R	z_{p_o}/R
1.50	96.89	87.45	−9.44	−79.11	−87.65	−8.53
2.00	35.96	30.07	−5.89	−25.43	−30.40	−4.97
2.50	22.00	17.27	−4.73	−13.91	−17.70	−3.79
3.00	16.19	12.03	−4.16	−9.35	−12.55	−3.20
3.50	13.08	9.25	−3.83	−6.99	−9.85	−2.85
4.00	11.16	7.55	−3.62	−5.58	−8.20	−2.62
4.50	9.87	6.40	−3.47	−4.65	−7.11	−2.46
5.00	8.94	5.57	−3.37	−4.00	−6.33	−2.33
5.50	8.24	4.95	−3.29	−3.51	−5.75	−2.24
6.00	7.69	4.46	−3.23	−3.14	−5.30	−2.16
6.50	7.25	4.07	−3.19	−2.85	−4.95	−2.10
7.00	6.90	3.74	−3.15	−2.61	−4.66	−2.05
7.50	6.60	3.47	−3.13	−2.41	−4.42	−2.01
8.00	6.35	3.24	−3.11	−2.24	−4.22	−1.97
8.50	6.13	3.04	−3.09	−2.10	−4.04	−1.94
9.00	5.95	2.87	−3.08	−1.98	−3.90	−1.91
9.50	5.78	2.71	−3.07	−1.88	−3.77	−1.89
10.00	5.64	2.58	−3.06	−1.78	−3.65	−1.87
10.50	5.51	2.45	−3.06	−1.70	−3.55	−1.85
11.00	5.40	2.34	−3.06	−1.63	−3.46	−1.83
11.50	5.30	2.24	−3.06	−1.56	−3.38	−1.82
12.00	5.21	2.15	−3.06	−1.50	−3.31	−1.81
12.50	5.12	2.07	−3.06	−1.45	−3.24	−1.79
13.00	5.05	1.99	−3.06	−1.40	−3.18	−1.78
13.50	4.98	1.92	−3.06	−1.35	−3.13	−1.77
14.00	4.92	1.85	−3.07	−1.31	−3.08	−1.76
14.50	4.86	1.79	−3.07	−1.28	−3.03	−1.76
15.00	4.80	1.73	−3.07	−1.24	−2.99	−1.75
15.50	4.75	1.67	−3.08	−1.21	−2.95	−1.74
16.00	4.71	1.62	−3.09	−1.18	−2.91	−1.74
16.50	4.67	1.57	−3.09	−1.15	−2.88	−1.73
17.00	4.63	1.53	−3.10	−1.12	−2.85	−1.72
17.50	4.59	1.49	−3.10	−1.10	−2.82	−1.72
18.00	4.56	1.45	−3.11	−1.07	−2.79	−1.71
18.50	4.52	1.41	−3.12	−1.05	−2.76	−1.71
19.00	4.49	1.37	−3.12	−1.03	−2.74	−1.70
19.50	4.46	1.33	−3.13	−1.01	−2.71	−1.70
20.00	4.44	1.30	−3.14	−0.99	−2.69	−1.70
20.50	4.41	1.27	−3.15	−0.97	−2.67	−1.69
21.00	4.39	1.24	−3.15	−0.96	−2.65	−1.69
21.50	4.37	1.21	−3.16	−0.94	−2.63	−1.69
22.00	4.35	1.18	−3.17	−0.93	−2.61	−1.68
22.50	4.33	1.15	−3.18	−0.91	−2.59	−1.68

[a] All values are normalized with respect to R, the radius of the cylinders.

TABLE 5.15 (continued)

V_2/V_1	f_i/R	z_{m_i}/R	z_{p_i}/R	f_o/R	z_{m_o}/R	z_{p_o}/R
23.00	4.31	1.13	−3.18	−0.90	−2.58	−1.68
23.50	4.29	1.10	−3.19	−0.89	−2.56	−1.68
24.00	4.28	1.08	−3.20	−0.87	−2.55	−1.67
24.50	4.26	1.05	−3.21	−0.86	−2.53	−1.67
25.00	4.25	1.03	−3.22	−0.85	−2.52	−1.67
25.50	4.23	1.01	−3.22	−0.84	−2.51	−1.67
26.00	4.22	0.99	−3.23	−0.83	−2.49	−1.67
26.50	4.21	0.97	−3.24	−0.82	−2.48	−1.66
27.00	4.19	0.95	−3.25	−0.81	−2.47	−1.66
27.50	4.18	0.93	−3.25	−0.80	−2.46	−1.66
28.00	4.17	0.91	−3.26	−0.79	−2.45	−1.66
28.50	4.16	0.89	−3.27	−0.78	−2.44	−1.66
29.00	4.15	0.87	−3.28	−0.77	−2.43	−1.66
29.50	4.14	0.86	−3.29	−0.76	−2.42	−1.66
30.00	4.13	0.84	−3.29	−0.75	−2.41	−1.65
30.50	4.13	0.82	−3.30	−0.75	−2.40	−1.65
31.00	4.12	0.81	−3.31	−0.74	−2.39	−1.65
31.50	4.11	0.79	−3.32	−0.73	−2.38	−1.65
32.00	4.10	0.78	−3.33	−0.73	−2.38	−1.65
32.50	4.10	0.76	−3.33	−0.72	−2.37	−1.65
33.00	4.09	0.75	−3.34	−0.71	−2.36	−1.65
33.50	4.08	0.73	−3.35	−0.71	−2.35	−1.65
34.00	4.08	0.72	−3.36	−0.70	−2.35	−1.65
34.50	4.07	0.71	−3.37	−0.69	−2.34	−1.65
35.00	4.07	0.69	−3.37	−0.69	−2.33	−1.65
35.50	4.06	0.68	−3.38	−0.68	−2.33	−1.64
36.00	4.06	0.67	−3.39	−0.68	−2.32	−1.64
36.50	4.05	0.66	−3.40	−0.67	−2.31	−1.64
37.00	4.05	0.64	−3.40	−0.67	−2.31	−1.64
37.50	4.05	0.63	−3.41	−0.66	−2.30	−1.64
38.00	4.04	0.62	−3.42	−0.66	−2.30	−1.64
38.50	4.04	0.61	−3.43	−0.65	−2.29	−1.64
39.00	4.03	0.60	−3.44	−0.65	−2.29	−1.64
39.50	4.03	0.59	−3.44	−0.64	−2.28	−1.64
40.00	4.03	0.58	−3.45	−0.64	−2.28	−1.64
40.50	4.03	0.57	−3.46	−0.63	−2.27	−1.64
41.00	4.02	0.56	−3.47	−0.63	−2.27	−1.64
41.50	4.02	0.55	−3.47	−0.62	−2.26	−1.64
42.00	4.02	0.54	−3.48	−0.62	−2.26	−1.64
42.50	4.02	0.53	−3.49	−0.62	−2.25	−1.64
43.00	4.01	0.52	−3.50	−0.61	−2.25	−1.64
43.50	4.01	0.51	−3.50	−0.61	−2.24	−1.64
44.00	4.01	0.50	−3.51	−0.60	−2.24	−1.64
44.50	4.01	0.49	−3.52	−0.60	−2.24	−1.64
45.00	4.01	0.48	−3.53	−0.60	−2.23	−1.63

TABLE 5.15 (continued)

V_2/V_1	f_i/R	z_{m_i}/R	z_{p_i}/R	f_o/R	z_{m_o}/R	z_{p_o}/R
45.50	4.01	0.47	−3.53	−0.59	−2.23	−1.63
46.00	4.01	0.46	−3.54	−0.59	−2.22	−1.63
46.50	4.00	0.46	−3.55	−0.59	−2.22	−1.63
47.00	4.00	0.45	−3.55	−0.58	−2.22	−1.63
47.50	4.00	0.44	−3.56	−0.58	−2.21	−1.63
48.00	4.00	0.43	−3.57	−0.58	−2.21	−1.63
48.50	4.00	0.42	−3.58	−0.57	−2.21	−1.63
49.00	4.00	0.42	−3.58	−0.57	−2.20	−1.63
49.50	4.00	0.41	−3.59	−0.57	−2.20	−1.63
50.00	4.00	0.40	−3.60	−0.57	−2.20	−1.63

TABLE 5.16

CARDINAL POINTS OF TWO EQUIDIAMETER CYLINDERS SEPARATED BY A DISTANCE $s = 1.4$[a]

V_2/V_1	f_i/R	z_{m_i}/R	z_{p_i}/R	f_o/R	z_{m_o}/R	z_{p_o}/R
1.50	99.06	89.41	−9.65	−80.89	−89.61	−8.73
2.00	36.76	30.74	−6.02	−25.99	−31.08	−5.08
2.50	22.48	17.64	−4.83	−14.21	−18.09	−3.87
3.00	16.54	12.29	−4.25	−9.55	−12.82	−3.27
3.50	13.36	9.45	−3.91	−7.14	−10.06	−2.91
4.00	11.40	7.70	−3.69	−5.70	−8.38	−2.68
4.50	10.07	6.53	−3.54	−4.75	−7.26	−2.51
5.00	9.12	5.68	−3.44	−4.08	−6.46	−2.38
5.50	8.40	5.04	−3.36	−3.58	−5.87	−2.29
6.00	7.84	4.54	−3.30	−3.20	−5.41	−2.21
6.50	7.40	4.14	−3.25	−2.90	−5.05	−2.15
7.00	7.03	3.81	−3.22	−2.66	−4.75	−2.09
7.50	6.73	3.53	−3.19	−2.46	−4.51	−2.05
8.00	6.47	3.30	−3.17	−2.29	−4.30	−2.01
8.50	6.25	3.09	−3.15	−2.14	−4.12	−1.98
9.00	6.06	2.92	−3.14	−2.02	−3.97	−1.95
9.50	5.89	2.76	−3.13	−1.91	−3.84	−1.93
10.00	5.74	2.62	−3.12	−1.82	−3.72	−1.91
10.50	5.61	2.49	−3.12	−1.73	−3.62	−1.89
11.00	5.50	2.38	−3.12	−1.66	−3.53	−1.87
11.50	5.39	2.28	−3.12	−1.59	−3.45	−1.86
12.00	5.30	2.18	−3.12	−1.53	−3.37	−1.84
12.50	5.21	2.10	−3.12	−1.47	−3.31	−1.83
13.00	5.14	2.02	−3.12	−1.42	−3.25	−1.82
13.50	5.06	1.94	−3.12	−1.38	−3.19	−1.81

[a] All values are normalized with respect to R, the radius of the cylinders.

TABLE 5.16 (continued)

V_2/V_1	f_i/R	z_{m_i}/R	z_{p_i}/R	f_o/R	z_{m_o}/R	z_{p_o}/R
14.00	5.00	1.87	−3.13	−1.34	−3.14	−1.80
14.50	4.94	1.81	−3.13	−1.30	−3.09	−1.79
15.00	4.89	1.75	−3.13	−1.26	−3.05	−1.78
15.50	4.83	1.70	−3.14	−1.23	−3.01	−1.78
16.00	4.79	1.64	−3.14	−1.20	−2.97	−1.77
16.50	4.74	1.59	−3.15	−1.17	−2.93	−1.76
17.00	4.70	1.55	−3.16	−1.14	−2.90	−1.76
17.50	4.67	1.50	−3.16	−1.12	−2.87	−1.75
18.00	4.63	1.46	−3.17	−1.09	−2.84	−1.75
18.50	4.60	1.42	−3.18	−1.07	−2.81	−1.74
19.00	4.57	1.38	−3.18	−1.05	−2.79	−1.74
19.50	4.54	1.35	−3.19	−1.03	−2.76	−1.74
20.00	4.51	1.31	−3.20	−1.01	−2.74	−1.73
20.50	4.48	1.28	−3.20	−0.99	−2.72	−1.73
21.00	4.46	1.25	−3.21	−0.97	−2.70	−1.73
21.50	4.44	1.22	−3.22	−0.96	−2.68	−1.72
22.00	4.42	1.19	−3.23	−0.94	−2.66	−1.72
22.50	4.40	1.16	−3.23	−0.93	−2.64	−1.72
23.00	4.38	1.13	−3.24	−0.91	−2.63	−1.71
23.50	4.36	1.11	−3.25	−0.90	−2.61	−1.71
24.00	4.34	1.08	−3.26	−0.89	−2.59	−1.71
24.50	4.33	1.06	−3.27	−0.87	−2.58	−1.71
25.00	4.31	1.04	−3.27	−0.86	−2.57	−1.70
25.50	4.30	1.01	−3.28	−0.85	−2.55	−1.70
26.00	4.28	0.99	−3.29	−0.84	−2.54	−1.70
26.50	4.27	0.97	−3.30	−0.83	−2.53	−1.70
27.00	4.26	0.95	−3.31	−0.82	−2.52	−1.70
27.50	4.24	0.93	−3.31	−0.81	−2.50	−1.70
28.00	4.23	0.91	−3.32	−0.80	−2.49	−1.69
28.50	4.22	0.89	−3.33	−0.79	−2.48	−1.69
29.00	4.21	0.88	−3.34	−0.78	−2.47	−1.69
29.50	4.20	0.86	−3.34	−0.77	−2.46	−1.69
30.00	4.19	0.84	−3.35	−0.77	−2.45	−1.69
30.50	4.19	0.82	−3.36	−0.76	−2.44	−1.69
31.00	4.18	0.81	−3.37	−0.75	−2.44	−1.69
31.50	4.17	0.79	−3.38	−0.74	−2.43	−1.68
32.00	4.16	0.78	−3.38	−0.74	−2.42	−1.68
32.50	4.16	0.76	−3.39	−0.73	−2.41	−1.68
33.00	4.15	0.75	−3.40	−0.72	−2.40	−1.68
33.50	4.14	0.73	−3.41	−0.72	−2.40	−1.68
34.00	4.14	0.72	−3.42	−0.71	−2.39	−1.68
34.50	4.13	0.71	−3.42	−0.70	−2.38	−1.68
35.00	4.12	0.69	−3.43	−0.70	−2.38	−1.68
35.50	4.12	0.68	−3.44	−0.69	−2.37	−1.68
36.00	4.11	0.67	−3.45	−0.69	−2.36	−1.68

TABLE 5.16 (continued)

V_2/V_1	f_i/R	z_{m_i}/R	z_{p_i}/R	f_o/R	z_{m_o}/R	z_{p_o}/R
36.50	4.11	0.66	−3.46	−0.68	−2.36	−1.68
37.00	4.11	0.64	−3.46	−0.67	−2.35	−1.68
37.50	4.10	0.63	−3.47	−0.67	−2.34	−1.67
38.00	4.10	0.62	−3.48	−0.66	−2.34	−1.67
38.50	4.09	0.61	−3.49	−0.66	−2.33	−1.67
39.00	4.09	0.60	−3.49	−0.65	−2.33	−1.67
39.50	4.09	0.59	−3.50	−0.65	−2.32	−1.67
40.00	4.08	0.57	−3.51	−0.65	−2.32	−1.67
40.50	4.08	0.56	−3.52	−0.64	−2.31	−1.67
41.00	4.08	0.55	−3.52	−0.64	−2.31	−1.67
41.50	4.08	0.54	−3.53	−0.63	−2.30	−1.67
42.00	4.07	0.53	−3.54	−0.63	−2.30	−1.67
42.50	4.07	0.52	−3.55	−0.62	−2.29	−1.67
43.00	4.07	0.51	−3.55	−0.62	−2.29	−1.67
43.50	4.07	0.50	−3.56	−0.62	−2.29	−1.67
44.00	4.06	0.50	−3.57	−0.61	−2.28	−1.67
44.50	4.06	0.49	−3.58	−0.61	−2.28	−1.67
45.00	4.06	0.48	−3.58	−0.61	−2.27	−1.67
45.50	4.06	0.47	−3.59	−0.60	−2.27	−1.67
46.00	4.06	0.46	−3.60	−0.60	−2.27	−1.67
46.50	4.06	0.45	−3.61	−0.59	−2.26	−1.67
47.00	4.06	0.44	−3.61	−0.59	−2.26	−1.67
47.50	4.05	0.43	−3.62	−0.59	−2.25	−1.67
48.00	4.05	0.43	−3.63	−0.59	−2.25	−1.67
48.50	4.05	0.42	−3.64	−0.58	−2.25	−1.67
49.00	4.05	0.41	−3.64	−0.58	−2.24	−1.66
49.50	4.05	0.40	−3.65	−0.58	−2.24	−1.66
50.00	4.05	0.39	−3.66	−0.57	−2.24	−1.66

TABLE 5.17

CARDINAL POINTS OF TWO EQUIDIAMETER CYLINDERS SEPARATED BY A DISTANCE $s = 1.5$[a]

V_2/V_1	f_i/R	z_{m_i}/R	z_{p_i}/R	f_o/R	z_{m_o}/R	z_{p_o}/R
1.50	101.35	91.47	−9.87	−82.75	−91.68	−8.93
2.00	37.60	31.43	−6.16	−26.58	−31.78	−5.20
2.50	22.98	18.04	−4.94	−14.53	−18.50	−3.96
3.00	16.90	12.56	−4.35	−9.76	−13.11	−3.35
3.50	13.65	9.65	−4.00	−7.30	−10.28	−2.98
4.00	11.64	7.86	−3.78	−5.82	−8.56	−2.74
4.50	10.28	6.66	−3.62	−4.85	−7.41	−2.56

[a] All values are normalized with respect to R, the radius of the cylinders.

TABLE 5.17 (continued)

V_2/V_1	f_i/R	z_{m_i}/R	z_{p_i}/R	f_o/R	z_{m_o}/R	z_{p_o}/R
5.00	9.31	5.80	−3.51	−4.16	−6.60	−2.44
5.50	8.57	5.14	−3.43	−3.66	−5.99	−2.34
6.00	8.00	4.63	−3.37	−3.27	−5.52	−2.26
6.50	7.54	4.22	−3.32	−2.96	−5.15	−2.19
7.00	7.17	3.88	−3.29	−2.71	−4.85	−2.14
7.50	6.86	3.60	−3.26	−2.50	−4.60	−2.10
8.00	6.59	3.36	−3.24	−2.33	−4.39	−2.06
8.50	6.37	3.15	−3.22	−2.18	−4.21	−2.03
9.00	6.17	2.97	−3.21	−2.06	−4.05	−2.00
9.50	6.00	2.80	−3.20	−1.95	−3.92	−1.97
10.00	5.85	2.66	−3.19	−1.85	−3.80	−1.95
10.50	5.72	2.53	−3.18	−1.76	−3.70	−1.93
11.00	5.60	2.42	−3.18	−1.69	−3.60	−1.91
11.50	5.49	2.31	−3.18	−1.62	−3.52	−1.90
12.00	5.39	2.21	−3.18	−1.56	−3.44	−1.88
12.50	5.31	2.13	−3.18	−1.50	−3.37	−1.87
13.00	5.23	2.05	−3.18	−1.45	−3.31	−1.86
13.50	5.15	1.97	−3.18	−1.40	−3.25	−1.85
14.00	5.09	1.90	−3.19	−1.36	−3.20	−1.84
14.50	5.03	1.83	−3.19	−1.32	−3.15	−1.83
15.00	4.97	1.77	−3.20	−1.28	−3.11	−1.82
15.50	4.92	1.72	−3.20	−1.25	−3.06	−1.82
16.00	4.87	1.66	−3.21	−1.22	−3.03	−1.81
16.50	4.82	1.61	−3.21	−1.19	−2.99	−1.80
17.00	4.78	1.56	−3.22	−1.16	−2.96	−1.80
17.50	4.74	1.52	−3.22	−1.13	−2.92	−1.79
18.00	4.71	1.48	−3.23	−1.11	−2.90	−1.79
18.50	4.67	1.44	−3.24	−1.09	−2.87	−1.78
19.00	4.64	1.40	−3.24	−1.06	−2.84	−1.78
19.50	4.61	1.36	−3.25	−1.04	−2.82	−1.77
20.00	4.58	1.33	−3.26	−1.02	−2.79	−1.77
20.50	4.56	1.29	−3.27	−1.01	−2.77	−1.76
21.00	4.53	1.26	−3.27	−0.99	−2.75	−1.76
21.50	4.51	1.23	−3.28	−0.97	−2.73	−1.76
22.00	4.49	1.20	−3.29	−0.96	−2.71	−1.76
22.50	4.47	1.17	−3.30	−0.94	−2.69	−1.75
23.00	4.45	1.14	−3.30	−0.93	−2.68	−1.75
23.50	4.43	1.12	−3.31	−0.91	−2.66	−1.75
24.00	4.41	1.09	−3.32	−0.90	−2.64	−1.74
24.50	4.39	1.07	−3.33	−0.89	−2.63	−1.74
25.00	4.38	1.04	−3.33	−0.88	−2.62	−1.74
25.50	4.36	1.02	−3.34	−0.86	−2.60	−1.74
26.00	4.35	1.00	−3.35	−0.85	−2.59	−1.74
26.50	4.33	0.98	−3.36	−0.84	−2.58	−1.73
27.00	4.32	0.96	−3.37	−0.83	−2.56	−1.73

TABLE 5.17 (continued)

V_2/V_1	f_i/R	z_{m_i}/R	z_{p_i}/R	f_o/R	z_{m_o}/R	z_{p_o}/R
27.50	4.31	0.94	−3.37	−0.82	−2.55	−1.73
28.00	4.30	0.92	−3.38	−0.81	−2.54	−1.73
28.50	4.29	0.90	−3.39	−0.80	−2.53	−1.73
29.00	4.28	0.88	−3.40	−0.79	−2.52	−1.73
29.50	4.27	0.86	−3.41	−0.79	−2.51	−1.72
30.00	4.26	0.84	−3.41	−0.78	−2.50	−1.72
30.50	4.25	0.83	−3.42	−0.77	−2.49	−1.72
31.00	4.24	0.81	−3.43	−0.76	−2.48	−1.72
31.50	4.23	0.79	−3.44	−0.75	−2.47	−1.72
32.00	4.22	0.78	−3.45	−0.75	−2.47	−1.72
32.50	4.22	0.76	−3.45	−0.74	−2.46	−1.72
33.00	4.21	0.75	−3.46	−0.73	−2.45	−1.72
33.50	4.20	0.73	−3.47	−0.73	−2.44	−1.72
34.00	4.20	0.72	−3.48	−0.72	−2.43	−1.71
34.50	4.19	0.71	−3.48	−0.71	−2.43	−1.71
35.00	4.18	0.69	−3.49	−0.71	−2.42	−1.71
35.50	4.18	0.68	−3.50	−0.70	−2.41	−1.71
36.00	4.17	0.67	−3.51	−0.70	−2.41	−1.71
36.50	4.17	0.65	−3.52	−0.69	−2.40	−1.71
37.00	4.16	0.64	−3.52	−0.68	−2.39	−1.71
37.50	4.16	0.63	−3.53	−0.68	−2.39	−1.71
38.00	4.16	0.62	−3.54	−0.67	−2.38	−1.71
38.50	4.15	0.60	−3.55	−0.67	−2.38	−1.71
39.00	4.15	0.59	−3.55	−0.66	−2.37	−1.71
39.50	4.14	0.58	−3.56	−0.66	−2.37	−1.71
40.00	4.14	0.57	−3.57	−0.65	−2.36	−1.71
40.50	4.14	0.56	−3.58	−0.65	−2.36	−1.71
41.00	4.13	0.55	−3.59	−0.65	−2.35	−1.71
41.50	4.13	0.54	−3.59	−0.64	−2.35	−1.70
42.00	4.13	0.53	−3.60	−0.64	−2.34	−1.70
42.50	4.13	0.52	−3.61	−0.63	−2.34	−1.70
43.00	4.12	0.51	−3.62	−0.63	−2.33	−1.70
43.50	4.12	0.50	3.62	0.62	2.33	−1.70
44.00	4.12	0.49	−3.63	−0.62	−2.32	−1.70
44.50	4.12	0.48	−3.64	−0.62	−2.32	−1.70
45.00	4.12	0.47	−3.64	−0.61	−2.32	−1.70
45.50	4.11	0.46	−3.65	−0.61	−2.31	−1.70
46.00	4.11	0.45	−3.66	−0.61	−2.31	−1.70
46.50	4.11	0.44	−3.67	−0.60	−2.30	−1.70
47.00	4.11	0.44	−3.67	−0.60	−2.30	−1.70
47.50	4.11	0.43	−3.68	−0.60	−2.30	−1.70
48.00	4.11	0.42	−3.69	−0.59	−2.29	−1.70
48.50	4.11	0.41	−3.70	−0.59	−2.29	−1.70
49.00	4.11	0.40	−3.70	−0.59	−2.29	−1.70
49.50	4.10	0.39	−3.71	−0.58	−2.28	−1.70
50.00	4.10	0.39	−3.72	−0.58	−2.28	−1.70

TABLE 5.18
CARDINAL POINTS OF TWO EQUIDIAMETER CYLINDERS SEPARATED BY A DISTANCE $s = 1.6$[a]

V_2/V_1	f_i/R	z_{m_i}/R	z_{p_i}/R	f_o/R	z_{m_o}/R	z_{p_o}/R
1.50	103.73	93.62	−10.11	−84.69	−93.83	−9.14
2.00	38.47	32.16	−6.31	−27.20	−32.52	−5.32
2.50	23.50	18.45	−5.06	−14.87	−18.92	−4.06
3.00	17.28	12.84	−4.45	−9.98	−13.41	−3.43
3.50	13.95	9.86	−4.09	−7.46	−10.51	−3.05
4.00	11.90	8.03	−3.86	−5.95	−8.75	−2.80
4.50	10.51	6.80	−3.70	−4.95	−7.58	−2.62
5.00	9.51	5.92	−3.59	−4.25	−6.74	−2.49
5.50	8.75	5.25	−3.51	−3.73	−6.12	−2.39
6.00	8.17	4.72	−3.44	−3.33	−5.64	−2.31
6.50	7.70	4.30	−3.40	−3.02	−5.26	−2.24
7.00	7.32	3.96	−3.36	−2.76	−4.95	−2.19
7.50	7.00	3.67	−3.33	−2.55	−4.70	−2.14
8.00	6.72	3.42	−3.31	−2.38	−4.48	−2.10
8.50	6.49	3.20	−3.29	−2.23	−4.30	−2.07
9.00	6.29	3.02	−3.28	−2.10	−4.14	−2.04
9.50	6.12	2.85	−3.26	−1.98	−4.00	−2.02
10.00	5.96	2.71	−3.26	−1.89	−3.88	−1.99
10.50	5.83	2.57	−3.25	−1.80	−3.77	−1.97
11.00	5.70	2.45	−3.25	−1.72	−3.68	−1.96
11.50	5.59	2.35	−3.25	−1.65	−3.59	−1.94
12.00	5.49	2.25	−3.25	−1.59	−3.51	−1.93
12.50	5.40	2.16	−3.25	−1.53	−3.44	−1.91
13.00	5.32	2.07	−3.25	−1.48	−3.38	−1.90
13.50	5.25	2.00	−3.25	−1.43	−3.32	−1.89
14.00	5.18	1.93	−3.25	−1.38	−3.26	−1.88
14.50	5.12	1.86	−3.26	−1.34	−3.21	−1.87
15.00	5.06	1.80	−3.26	−1.31	−3.17	−1.86
15.50	5.00	1.74	−3.27	−1.27	−3.13	−1.86
16.00	4.96	1.68	−3.27	−1.24	−3.09	−1.85
16.50	4.91	1.63	−3.28	−1.21	−3.05	−1.84
17.00	4.87	1.58	−3.28	−1.18	−3.02	−1.84
17.50	4.83	1.54	−3.29	−1.15	−2.98	−1.83
18.00	4.79	1.49	−3.30	−1.13	−2.95	−1.82
18.50	4.75	1.45	−3.30	−1.11	−2.92	−1.82
19.00	4.72	1.41	−3.31	−1.08	−2.90	−1.81
19.50	4.69	1.37	−3.32	−1.06	−2.87	−1.81
20.00	4.66	1.34	−3.32	−1.04	−2.85	−1.81
20.50	4.63	1.30	−3.33	−1.02	−2.83	−1.80
21.00	4.61	1.27	−3.34	−1.01	−2.80	−1.80
21.50	4.58	1.24	−3.34	−0.99	−2.78	−1.80
22.00	4.56	1.21	−3.35	−0.97	−2.76	−1.79

[a] All values are normalized with respect to R, the radius of the cylinders.

TABLE 5.18 (continued)

V_2/V_1	f_i/R	z_{m_i}/R	z_{p_i}/R	f_o/R	z_{m_o}/R	z_{p_o}/R
22.50	4.54	1.18	−3.36	−0.96	−2.75	−1.79
23.00	4.52	1.15	−3.37	−0.94	−2.73	−1.79
23.50	4.50	1.12	−3.37	−0.93	−2.71	−1.78
24.00	4.48	1.10	−3.38	−0.91	−2.70	−1.78
24.50	4.46	1.07	−3.39	−0.90	−2.68	−1.78
25.00	4.45	1.05	−3.40	−0.89	−2.67	−1.78
25.50	4.43	1.03	−3.41	−0.88	−2.65	−1.77
26.00	4.42	1.00	−3.41	−0.87	−2.64	−1.77
26.50	4.40	0.98	−3.42	−0.86	−2.63	−1.77
27.00	4.39	0.96	−3.43	−0.84	−2.61	−1.77
27.50	4.38	0.94	−3.44	−0.83	−2.60	−1.77
28.00	4.37	0.92	−3.45	−0.82	−2.59	−1.77
28.50	4.35	0.90	−3.45	−0.82	−2.58	−1.76
29.00	4.34	0.88	−3.46	−0.81	−2.57	−1.76
29.50	4.33	0.86	−3.47	−0.80	−2.56	−1.76
30.00	4.32	0.85	−3.48	−0.79	−2.55	−1.76
30.50	4.31	0.83	−3.49	−0.78	−2.54	−1.76
31.00	4.30	0.81	−3.49	−0.77	−2.53	−1.76
31.50	4.30	0.79	−3.50	−0.77	−2.52	−1.76
32.00	4.29	0.78	−3.51	−0.76	−2.51	−1.75
32.50	4.28	0.76	−3.52	−0.75	−2.50	−1.75
33.00	4.27	0.75	−3.52	−0.74	−2.50	−1.75
33.50	4.27	0.73	−3.53	−0.74	−2.49	−1.75
34.00	4.26	0.72	−3.54	−0.73	−2.48	−1.75
34.50	4.25	0.70	−3.55	−0.72	−2.47	−1.75
35.00	4.25	0.69	−3.56	−0.72	−2.47	−1.75
35.50	4.24	0.68	−3.56	−0.71	−2.46	−1.75
36.00	4.24	0.66	−3.57	−0.71	−2.45	−1.75
36.50	4.23	0.65	−3.58	−0.70	−2.45	−1.75
37.00	4.22	0.64	−3.59	−0.69	−2.44	−1.75
37.50	4.22	0.63	−3.59	−0.69	−2.43	−1.75
38.00	4.22	0.61	−3.60	−0.68	2.43	−1.74
38.50	4.21	0.60	−3.61	0.68	−2.42	−1.74
39.00	4.21	0.59	−3.62	−0.67	−2.42	−1.74
39.50	4.20	0.58	−3.63	−0.67	−2.41	−1.74
40.00	4.20	0.57	−3.63	−0.66	−2.41	−1.74
40.50	4.20	0.56	−3.64	−0.66	−2.40	−1.74
41.00	4.19	0.54	−3.65	−0.65	−2.40	−1.74
41.50	4.19	0.53	−3.66	−0.65	−2.39	−1.74
42.00	4.19	0.52	−3.66	−0.65	−2.39	−1.74
42.50	4.18	0.51	−3.67	−0.64	−2.38	−1.74
43.00	4.18	0.50	−3.68	−0.64	−2.38	−1.74
43.50	4.18	0.49	−3.69	−0.63	−2.37	−1.74
44.00	4.18	0.48	−3.69	−0.63	−2.37	−1.74
44.50	4.17	0.47	−3.70	−0.63	−2.36	−1.74

TABLE 5.18 (continued)

V_2/V_1	f_i/R	z_{m_i}/R	z_{p_i}/R	f_o/R	z_{m_o}/R	z_{p_o}/R
45.00	4.17	0.46	−3.71	−0.62	−2.36	−1.74
45.50	4.17	0.46	−3.72	−0.62	−2.36	−1.74
46.00	4.17	0.45	−3.72	−0.61	−2.35	−1.74
46.50	4.17	0.44	−3.73	−0.61	−2.35	−1.74
47.00	4.17	0.43	−3.74	−0.61	−2.34	−1.74
47.50	4.16	0.42	−3.75	−0.60	−2.34	−1.74
48.00	4.16	0.41	−3.75	−0.60	−2.34	−1.74
48.50	4.16	0.40	−3.76	−0.60	−2.33	−1.73
49.00	4.16	0.39	−3.77	−0.59	−2.33	−1.73
49.50	4.16	0.39	−3.77	−0.59	−2.33	−1.73
50.00	4.16	0.38	−3.78	−0.59	−2.32	−1.73

TABLE 5.19

CARDINAL POINTS OF TWO EQUIDIAMETER CYLINDERS SEPARATED BY A DISTANCE $s = 1.7$[a]

V_2/V_1	f_i/R	z_{m_i}/R	z_{p_i}/R	f_o/R	z_{m_o}/R	z_{p_o}/R
1.50	106.21	95.86	−10.35	−86.72	−96.08	−9.36
2.00	39.37	32.92	−6.46	−27.84	−33.29	−5.45
2.50	24.05	18.87	−5.18	−15.21	−19.36	−4.15
3.00	17.68	13.13	−4.55	−10.21	−13.71	−3.51
3.50	14.27	10.08	−4.19	−7.63	−10.75	−3.12
4.00	12.16	8.21	−3.95	−6.08	−8.95	−2.87
4.50	10.74	6.95	−3.79	−5.06	−7.75	−2.68
5.00	9.71	6.04	−3.67	−4.34	−6.89	−2.55
5.50	8.94	5.35	−3.59	−3.81	−6.26	−2.45
6.00	8.34	4.82	−3.52	−3.41	−5.77	−2.36
6.50	7.86	4.39	−3.47	−3.08	−5.38	−2.29
7.00	7.47	4.03	−3.43	−2.82	−5.06	−2.24
7.50	7.14	3.73	−3.40	−2.61	−4.80	−2.19
8.00	6.86	3.48	−3.38	−2.43	−4.58	−2.15
8.50	6.62	3.26	−3.36	−2.27	−4.39	−2.12
9.00	6.42	3.07	−3.35	−2.14	−4.23	−2.09
9.50	6.24	2.90	−3.34	−2.02	−4.08	−2.06
10.00	6.08	2.75	−3.33	−1.92	−3.96	−2.04
10.50	5.94	2.62	−3.32	−1.83	−3.85	−2.02
11.00	5.81	2.49	−3.32	−1.75	−3.75	−2.00
11.50	5.70	2.38	−3.32	−1.68	−3.66	−1.98
12.00	5.60	2.28	−3.31	−1.62	−3.58	−1.97
12.50	5.51	2.19	−3.31	−1.56	−3.51	−1.96

[a] All values are normalized with respect to R, the radius of the cylinders.

TABLE 5.19 (continued)

V_2/V_1	f_i/R	z_{m_i}/R	z_{p_i}/R	f_o/R	z_{m_o}/R	z_{p_o}/R
13.00	5.42	2.11	−3.32	−1.50	−3.45	−1.94
13.50	5.34	2.03	−3.32	−1.45	−3.39	−1.93
14.00	5.27	1.95	−3.32	−1.41	−3.33	−1.92
14.50	5.21	1.88	−3.32	−1.37	−3.28	−1.91
15.00	5.15	1.82	−3.33	−1.33	−3.23	−1.90
15.50	5.09	1.76	−3.33	−1.29	−3.19	−1.90
16.00	5.04	1.71	−3.34	−1.26	−3.15	−1.89
16.50	5.00	1.65	−3.34	−1.23	−3.11	−1.88
17.00	4.95	1.60	−3.35	−1.20	−3.08	−1.88
17.50	4.91	1.55	−3.36	−1.17	−3.04	−1.87
18.00	4.87	1.51	−3.36	−1.15	−3.01	−1.86
18.50	4.84	1.47	−3.37	−1.12	−2.98	−1.86
19.00	4.80	1.43	−3.37	−1.10	−2.96	−1.85
19.50	4.77	1.39	−3.38	−1.08	−2.93	−1.85
20.00	4.74	1.35	−3.39	−1.06	−2.91	−1.85
20.50	4.71	1.32	−3.40	−1.04	−2.88	−1.84
21.00	4.69	1.28	−3.40	−1.02	−2.86	−1.84
21.50	4.66	1.25	−3.41	−1.01	−2.84	−1.83
22.00	4.64	1.22	−3.42	−0.99	−2.82	−1.83
22.50	4.61	1.19	−3.43	−0.97	−2.80	−1.83
23.00	4.59	1.16	−3.43	−0.96	−2.78	−1.83
23.50	4.57	1.13	−3.44	−0.94	−2.77	−1.82
24.00	4.55	1.11	−3.45	−0.93	−2.75	−1.82
24.50	4.54	1.08	−3.46	−0.92	−2.73	−1.82
25.00	4.52	1.06	−3.46	−0.90	−2.72	−1.82
25.50	4.50	1.03	−3.47	−0.89	−2.70	−1.81
26.00	4.49	1.01	−3.48	−0.88	−2.69	−1.81
26.50	4.47	0.99	−3.49	−0.87	−2.68	−1.81
27.00	4.46	0.96	−3.50	−0.86	−2.67	−1.81
27.50	4.45	0.94	−3.50	−0.85	−2.65	−1.81
28.00	4.43	0.92	−3.51	−0.84	−2.64	−1.80
28.50	4.42	0.90	−3.52	−0.83	−2.63	−1.80
29.00	4.41	0.88	−3.53	−0.82	−2.62	−1.80
29.50	4.40	0.86	−3.53	−0.81	−2.61	−1.80
30.00	4.39	0.85	−3.54	−0.80	−2.60	−1.80
30.50	4.38	0.83	−3.55	−0.79	−2.59	−1.80
31.00	4.37	0.81	−3.56	−0.78	−2.58	−1.79
31.50	4.36	0.79	−3.57	−0.78	−2.57	−1.79
32.00	4.35	0.78	−3.57	−0.77	−2.56	−1.79
32.50	4.34	0.76	−3.58	−0.76	−2.55	−1.79
33.00	4.34	0.75	−3.59	−0.75	−2.55	−1.79
33.50	4.33	0.73	−3.60	−0.75	−2.54	−1.79
34.00	4.32	0.72	−3.61	−0.74	−2.53	−1.79
34.50	4.32	0.70	−3.61	−0.73	−2.52	−1.79
35.00	4.31	0.69	−3.62	−0.73	−2.51	−1.79

TABLE 5.19 (continued)

V_2/V_1	f_i/R	z_{m_i}/R	z_{p_i}/R	f_o/R	z_{m_o}/R	z_{p_o}/R
35.50	4.30	0.67	−3.63	−0.72	−2.51	−1.79
36.00	4.30	0.66	−3.64	−0.72	−2.50	−1.78
36.50	4.29	0.66	−3.64	−0.71	−2.49	−1.78
37.00	4.29	0.63	−3.65	−0.70	−2.49	−1.78
37.50	4.28	0.62	−3.66	−0.70	−2.48	−1.78
38.00	4.28	0.61	−3.67	−0.69	−2.48	−1.78
38.50	4.27	0.60	−3.68	−0.69	−2.47	−1.78
39.00	4.27	0.59	−3.68	−0.68	−2.46	−1.78
39.50	4.26	0.57	−3.69	−0.68	−2.46	−1.78
40.00	4.26	0.56	−3.70	−0.67	−2.45	−1.78
40.50	4.26	0.55	−3.71	−0.67	−2.45	−1.78
41.00	4.25	0.54	−3.71	−0.66	−2.44	−1.78
41.50	4.25	0.53	−3.72	−0.66	−2.44	−1.78
42.00	4.25	0.52	−3.73	−0.66	−2.43	−1.78
42.50	4.24	0.51	−3.74	−0.65	−2.43	−1.78
43.00	4.24	0.50	−3.74	−0.65	−2.42	−1.78
43.50	4.24	0.49	−3.75	−0.64	−2.42	−1.78
44.00	4.24	0.48	−3.76	−0.64	−2.41	−1.77
44.50	4.23	0.47	−3.77	−0.63	−2.41	−1.77
45.00	4.23	0.46	−3.77	−0.63	−2.40	−1.77
45.50	4.23	0.45	−3.78	−0.63	−2.40	−1.77
46.00	4.23	0.44	−3.79	−0.62	−2.40	−1.77
46.50	4.23	0.43	−3.80	−0.62	−2.39	−1.77
47.00	4.22	0.42	−3.80	−0.62	−2.39	−1.77
47.50	4.22	0.41	−3.81	−0.61	−2.38	−1.77
48.00	4.22	0.40	−3.82	−0.61	−2.38	−1.77
48.50	4.22	0.39	−3.83	−0.61	−2.38	−1.77
49.00	4.22	0.39	−3.83	−0.60	−2.37	−1.77
49.50	4.22	0.38	−3.84	−0.60	−2.37	−1.77
50.00	4.22	0.37	−3.85	−0.60	−2.37	−1.77

TABLE 5.20

CARDINAL POINTS OF TWO EQUIDIAMETER CYLINDERS SEPARATED BY A DISTANCE $s = 1.8$[a]

V_2/V_1	f_i/R	z_{m_i}/R	z_{p_i}/R	f_o/R	z_{m_o}/R	z_{p_o}/R
1.50	108.77	98.17	−10.60	−88.81	−98.40	−9.59
2.00	40.31	33.70	−6.61	−28.51	−34.09	−5.58
2.50	24.61	19.31	−5.30	−15.57	−19.82	−4.25
3.00	18.09	13.43	−4.66	−10.44	−14.03	−3.59
3.50	14.59	10.31	−4.29	−7.80	−11.00	−3.20
4.00	12.43	8.39	−4.04	−6.22	−9.15	−2.93
4.50	10.97	7.10	−3.88	−5.17	−7.92	−2.75
5.00	9.92	6.17	−3.76	−4.44	−7.05	−2.61
5.50	9.13	5.46	−3.67	−3.89	−6.40	−2.50
6.00	8.52	4.92	−3.60	−3.48	−5.89	−2.42
6.50	8.02	4.47	−3.55	−3.15	−5.50	−2.35
7.00	7.62	4.11	−3.51	−2.88	−5.17	−2.29
7.50	7.28	3.81	−3.48	−2.66	−4.90	−2.24
8.00	7.00	3.55	−3.45	−2.47	−4.68	−2.20
8.50	6.76	3.32	−3.44	−2.32	−4.48	−2.17
9.00	6.55	3.12	−3.42	−2.18	−4.32	−2.14
9.50	6.36	2.95	−3.41	−2.06	−4.17	−2.11
10.00	6.20	2.80	−3.40	−1.96	−4.04	−2.08
10.50	6.05	2.66	−3.39	−1.87	−3.93	−2.06
11.00	5.92	2.54	−3.39	−1.79	−3.83	−2.04
11.50	5.81	2.42	−3.39	−1.71	−3.74	−2.03
12.00	5.70	2.32	−3.39	−1.65	−3.66	−2.01
12.50	5.61	2.22	−3.39	−1.59	−3.59	−2.00
13.00	5.52	2.14	−3.39	−1.53	−3.52	−1.99
13.50	5.44	2.06	−3.39	−1.48	−3.46	−1.98
14.00	5.37	1.98	−3.39	−1.44	−3.40	−1.96
14.50	5.30	1.91	−3.39	−1.39	−3.35	−1.96
15.00	5.24	1.85	−3.40	−1.35	−3.30	−1.95
15.50	5.19	1.78	−3.40	−1.32	−3.26	−1.94
16.00	5.13	1.73	−3.41	−1.28	−3.21	−1.93
16.50	5.08	1.67	−3.41	−1.25	3.18	−1.92
17.00	5.04	1.62	−3.42	−1.22	−3.14	−1.92
17.50	5.00	1.57	−3.42	−1.19	−3.11	−1.91
18.00	4.96	1.53	−3.43	−1.17	−3.07	−1.91
18.50	4.92	1.48	−3.44	−1.14	−3.04	−1.90
19.00	4.88	1.44	−3.44	−1.12	−3.02	−1.90
19.50	4.85	1.40	−3.45	−1.10	−2.99	−1.89
20.00	4.82	1.36	−3.46	−1.08	−2.96	−1.89
20.50	4.79	1.33	−3.46	−1.06	−2.94	−1.88
21.00	4.76	1.29	−3.47	−1.04	−2.92	−1.88
21.50	4.74	1.26	−3.48	−1.02	−2.90	−1.88
22.00	4.71	1.23	−3.49	−1.01	−2.88	−1.87

[a] All values are normalized with respect to R, the radius of the cylinders.

TABLE 5.20 (continued)

V_2/V_1	f_i/R	z_{m_i}/R	z_{p_i}/R	f_o/R	z_{m_o}/R	z_{p_o}/R
22.50	4.69	1.20	−3.49	−0.99	−2.86	−1.87
23.00	4.67	1.17	−3.50	−0.97	−2.84	−1.87
23.50	4.65	1.14	−3.51	−0.96	−2.82	−1.86
24.00	4.63	1.11	−3.52	−0.94	−2.81	−1.86
24.50	4.61	1.09	−3.52	−0.93	−2.79	−1.86
25.00	4.59	1.06	−3.53	−0.92	−2.77	−1.86
25.50	4.58	1.04	−3.54	−0.91	−2.76	−1.85
26.00	4.56	1.01	−3.55	−0.89	−2.75	−1.85
26.50	4.55	0.99	−3.55	−0.88	−2.73	−1.85
27.00	4.53	0.97	−3.56	−0.87	−2.72	−1.85
27.50	4.52	0.95	−3.57	−0.86	−2.71	−1.84
28.00	4.50	0.93	−3.58	−0.85	−2.69	−1.84
28.50	4.49	0.91	−3.59	−0.84	−2.68	−1.84
29.00	4.48	0.89	−3.59	−0.83	−2.67	−1.84
29.50	4.47	0.87	−3.60	−0.82	−2.66	−1.84
30.00	4.46	0.85	−3.61	−0.81	−2.65	−1.84
30.50	4.45	0.83	−3.62	−0.81	−2.64	−1.84
31.00	4.44	0.81	−3.63	−0.80	−2.63	−1.83
31.50	4.43	0.79	−3.63	−0.79	−2.62	−1.83
32.00	4.42	0.78	−3.64	−0.78	−2.61	−1.83
32.50	4.41	0.76	−3.65	−0.77	−2.60	−1.83
33.00	4.40	0.75	−3.66	−0.77	−2.60	−1.83
33.50	4.40	0.73	−3.67	−0.76	−2.59	−1.83
34.00	4.39	0.71	−3.67	−0.75	−2.58	−1.83
34.50	4.38	0.70	−3.68	−0.75	−2.57	−1.83
35.00	4.37	0.69	−3.69	−0.74	−2.56	−1.83
35.50	4.37	0.67	−3.70	−0.73	−2.56	−1.82
36.00	4.36	0.66	−3.70	−0.73	−2.55	−1.82
36.50	4.36	0.64	−3.71	−0.72	−2.54	−1.82
37.00	4.35	0.63	−3.72	−0.72	−2.54	−1.82
37.50	4.35	0.62	−3.73	−0.71	−2.53	−1.82
38.00	4.34	0.60	−3.74	−0.70	−2.52	−1.82
38.50	4.34	0.59	−3.74	−0.70	−2.52	−1.82
39.00	4.33	0.58	−3.75	−0.69	−2.51	−1.82
39.50	4.33	0.57	−3.76	−0.69	−2.51	−1.82
40.00	4.32	0.56	−3.77	−0.68	−2.50	−1.82
40.50	4.32	0.54	−3.77	−0.68	−2.50	−1.82
41.00	4.31	0.53	−3.78	−0.67	−2.49	−1.82
41.50	4.31	0.52	−3.79	−0.67	−2.48	−1.82
42.00	4.31	0.51	−3.80	−0.66	−2.48	−1.82
42.50	4.30	0.50	−3.80	−0.66	−2.47	−1.81
43.00	4.30	0.49	−3.81	−0.66	−2.47	−1.81
43.50	4.30	0.48	−3.82	−0.65	−2.47	−1.81
44.00	4.30	0.47	−3.83	−0.65	−2.46	−1.81
44.50	4.29	0.46	−3.83	−0.64	−2.46	−1.81

TABLE 5.20 (continued)

V_2/V_1	f_i/R	z_{m_i}/R	z_{p_i}/R	f_o/R	z_{m_o}/R	z_{p_o}/R
45.00	4.29	0.45	−3.84	−0.64	−2.45	−1.81
45.50	4.29	0.44	−3.85	−0.64	−2.45	−1.81
46.00	4.29	0.43	−3.86	−0.63	−2.44	−1.81
46.50	4.28	0.42	−3.86	−0.63	−2.44	−1.81
47.00	4.28	0.41	−3.87	−0.62	−2.44	−1.81
47.50	4.28	0.40	−3.88	−0.62	−2.43	−1.81
48.00	4.28	0.39	−3.89	−0.62	−2.43	−1.81
48.50	4.28	0.39	−3.89	−0.61	−2.42	−1.81
49.00	4.28	0.38	−3.90	−0.61	−2.42	−1.81
49.50	4.27	0.37	−3.91	−0.61	−2.42	−1.81
50.00	4.27	0.36	−3.91	−0.60	−2.41	−1.81

TABLE 5.21

CARDINAL POINTS OF TWO EQUIDIAMETER CYLINDERS SEPARATED BY A DISTANCE $s = 1.9$[a]

V_2/V_1	f_i/R	z_{m_i}/R	z_{p_i}/R	f_o/R	z_{m_o}/R	z_{p_o}/R
1.50	111.42	100.56	−10.86	−90.97	−100.79	−9.82
2.00	41.28	34.51	−6.77	−29.19	−34.91	−5.72
2.50	25.20	19.77	−5.43	−15.94	−20.29	−4.35
3.00	18.51	13.74	−4.77	−10.69	−14.36	−3.68
3.50	14.93	10.54	−4.39	−7.98	−11.25	−3.27
4.00	12.71	8.57	−4.14	−6.36	−9.36	−3.00
4.50	11.22	7.25	−3.97	−5.29	−8.10	−2.81
5.00	10.14	6.30	−3.85	−4.54	−7.21	−2.67
5.50	9.33	5.58	−3.75	−3.98	−6.54	−2.56
6.00	8.70	5.02	−3.69	−3.55	−6.03	−2.47
6.50	8.19	4.56	−3.63	−3.21	−5.62	−2.40
7.00	7.78	4.19	−3.59	−2.94	−5.28	−2.34
7.50	7.44	3.88	−3.56	−2.72	−5.01	−2.29
8.00	7.14	3.61	−3.53	−2.53	−4.78	−2.25
8.50	6.89	3.38	−3.51	−2.36	−4.58	−2.22
9.00	6.68	3.18	−3.50	−2.23	−4.41	−2.18
9.50	6.49	3.00	−3.48	−2.10	−4.26	−2.16
10.00	6.32	2.85	−3.47	−2.00	−4.13	−2.13
10.50	6.17	2.70	−3.47	−1.90	−4.02	−2.11
11.00	6.04	2.58	−3.46	−1.82	−3.91	−2.09
11.50	5.92	2.46	−3.46	−1.75	−3.82	−2.07
12.00	5.81	2.35	−3.46	−1.68	−3.74	−2.06
12.50	5.72	2.26	−3.46	−1.62	−3.66	−2.04

[a] All values are normalized with respect to R, the radius of the cylinders.

TABLE 5.21 (continued)

V_2/V_1	f_i/R	z_{m_i}/R	z_{p_i}/R	f_o/R	z_{m_o}/R	z_{p_o}/R
13.00	5.63	2.17	−3.46	−1.56	−3.59	−2.03
13.50	5.55	2.08	−3.46	−1.51	−3.53	−2.02
14.00	5.47	2.01	−3.46	−1.46	−3.47	−2.01
14.50	5.40	1.94	−3.47	−1.42	−3.42	−2.00
15.00	5.34	1.87	−3.47	−1.38	−3.37	−1.99
15.50	5.28	1.81	−3.47	−1.34	−3.32	−1.98
16.00	5.23	1.75	−3.48	−1.31	−3.28	−1.97
16.50	5.18	1.69	−3.48	−1.27	−3.24	−1.97
17.00	5.13	1.64	−3.49	−1.24	−3.20	−1.96
17.50	5.08	1.59	−3.49	−1.22	−3.17	−1.95
18.00	5.04	1.54	−3.50	−1.19	−3.14	−1.95
18.50	5.01	1.50	−3.51	−1.16	−3.11	−1.94
19.00	4.97	1.46	−3.51	−1.14	−3.08	−1.94
19.50	4.94	1.42	−3.52	−1.12	−3.05	−1.93
20.00	4.90	1.38	−3.53	−1.10	−3.03	−1.93
20.50	4.87	1.34	−3.53	−1.08	−3.00	−1.92
21.00	4.85	1.30	−3.54	−1.06	−2.98	−1.92
21.50	4.82	1.27	−3.55	−1.04	−2.96	−1.92
22.00	4.79	1.24	−3.56	−1.02	−2.94	−1.91
22.50	4.77	1.21	−3.56	−1.01	−2.92	−1.91
23.00	4.75	1.18	−3.57	−0.99	−2.90	−1.91
23.50	4.73	1.15	−3.58	−0.97	−2.88	−1.90
24.00	4.71	1.12	−3.59	−0.96	−2.86	−1.90
24.50	4.69	1.09	−3.59	−0.95	−2.85	−1.90
25.00	4.67	1.07	−3.60	−0.93	−2.83	−1.90
25.50	4.65	1.04	−3.61	−0.92	−2.81	−1.89
26.00	4.63	1.02	−3.62	−0.91	−2.80	−1.89
26.50	4.62	0.99	−3.62	−0.90	−2.79	−1.89
27.00	4.60	0.97	−3.63	−0.89	−2.77	−1.89
27.50	4.59	0.95	−3.64	−0.88	−2.76	−1.89
28.00	4.58	0.93	−3.65	−0.86	−2.75	−1.88
28.50	4.56	0.91	−3.66	−0.85	−2.74	−1.88
29.00	4.55	0.89	−3.66	−0.85	−2.73	−1.88
29.50	4.54	0.87	−3.67	−0.84	−2.71	−1.88
30.00	4.53	0.85	−3.68	−0.83	−2.70	−1.88
30.50	4.52	0.83	−3.69	−0.82	−2.69	−1.88
31.00	4.51	0.81	−3.70	−0.81	−2.68	−1.87
31.50	4.50	0.79	−3.70	−0.80	−2.67	−1.87
32.00	4.49	0.78	−3.71	−0.79	−2.66	−1.87
32.50	4.48	0.76	−3.72	−0.79	−2.66	−1.87
33.00	4.47	0.74	−3.73	−0.78	−2.65	−1.87
33.50	4.46	0.73	−3.73	−0.77	−2.64	−1.87
34.00	4.45	0.71	−3.74	−0.76	−2.63	−1.87
34.50	4.45	0.70	−3.75	−0.76	−2.62	−1.87
35.00	4.44	0.68	−3.76	−0.75	−2.62	−1.86

TABLE 5.21 (continued)

V_2/V_1	f_i/R	z_{m_i}/R	z_{p_i}/R	f_o/R	z_{m_o}/R	z_{p_o}/R
35.50	4.43	0.67	−3.77	−0.74	−2.61	−1.86
36.00	4.43	0.65	−3.77	−0.74	−2.60	−1.86
36.50	4.42	0.64	−3.78	−0.73	−2.59	−1.86
37.00	4.42	0.63	−3.79	−0.73	−2.59	−1.86
37.50	4.41	0.61	−3.80	−0.72	−2.58	−1.86
38.00	4.40	0.60	−3.80	−0.71	−2.57	−1.86
38.50	4.40	0.59	−3.81	−0.71	−2.57	−1.86
39.00	4.39	0.57	−3.82	−0.70	−2.56	−1.86
39.50	4.39	0.56	−3.83	−0.70	−2.56	−1.86
40.00	4.39	0.55	−3.84	−0.69	−2.55	−1.86
40.50	4.38	0.54	−3.84	−0.69	−2.54	−1.86
41.00	4.38	0.53	−3.85	−0.68	−2.54	−1.86
41.50	4.37	0.52	−3.86	−0.68	−2.53	−1.85
42.00	4.37	0.50	−3.87	−0.67	−2.53	−1.85
42.50	4.37	0.49	−3.87	−0.67	−2.52	−1.85
43.00	4.36	0.48	−3.88	−0.67	−2.52	−1.85
43.50	4.36	0.47	−3.89	−0.66	−2.51	−1.85
44.00	4.36	0.46	−3.90	−0.66	−2.51	−1.85
44.50	4.35	0.45	−3.90	−0.65	−2.50	−1.85
45.00	4.35	0.44	−3.91	−0.65	−2.50	−1.85
45.50	4.35	0.43	−3.92	−0.64	−2.50	−1.85
46.00	4.35	0.42	−3.93	−0.64	−2.49	−1.85
46.50	4.34	0.41	−3.93	−0.64	−2.49	−1.85
47.00	4.34	0.40	−3.94	−0.63	−2.48	−1.85
47.50	4.34	0.39	−3.95	−0.63	−2.48	−1.85
48.00	4.34	0.38	−3.95	−0.63	−2.48	−1.85
48.50	4.34	0.38	−3.96	−0.62	−2.47	−1.85
49.00	4.34	0.37	−3.97	−0.62	−2.47	−1.85
49.50	4.33	0.36	3.98	−0.62	−2.46	−1.85
50.00	4.33	0.35	−3.98	−0.61	−2.46	−1.85

TABLE 5.22
CARDINAL POINTS OF TWO EQUIDIAMETER CYLINDERS SEPARATED BY A DISTANCE $s = 2$ [a]

V_2/V_1	f_i/R	z_{m_i}/R	z_{p_i}/R	f_o/R	z_{m_o}/R	z_{p_o}/R
1.50	114.14	103.01	−11.12	−93.19	−103.25	−10.06
2.00	42.27	35.34	−6.94	−29.89	−35.75	−5.85
2.50	25.79	20.23	−5.56	−16.31	−20.77	−4.46
3.00	18.94	14.05	−4.89	−10.94	−14.70	−3.77
3.50	15.27	10.78	−4.49	−8.16	−11.51	−3.35
4.00	13.00	8.76	−4.24	−6.50	−9.58	−3.08
4.50	11.47	7.41	−4.06	−5.41	−8.29	−2.88
5.00	10.37	6.43	−3.94	−4.64	−7.37	−2.73
5.50	9.54	5.69	−3.84	−4.07	−6.69	−2.62
6.00	8.89	5.12	−3.77	−3.63	−6.16	−2.53
6.50	8.37	4.65	−3.71	−3.28	−5.74	−2.46
7.00	7.94	4.27	−3.67	−3.00	−5.40	−2.40
7.50	7.59	3.95	−3.64	−2.77	−5.12	−2.35
8.00	7.29	3.68	−3.61	−2.58	−4.88	−2.30
8.50	7.03	3.44	−3.59	−2.41	−4.68	−2.27
9.00	6.81	3.24	−3.57	−2.27	−4.51	−2.23
9.50	6.62	3.06	−3.56	−2.15	−4.35	−2.21
10.00	6.45	2.89	−3.55	−2.04	−4.22	−2.18
10.50	6.29	2.75	−3.54	−1.94	−4.10	−2.16
11.00	6.16	2.62	−3.54	−1.86	−4.00	−2.14
11.50	6.03	2.50	−3.54	−1.78	−3.90	−2.12
12.00	5.92	2.39	−3.53	−1.71	−3.82	−2.11
12.50	5.82	2.29	−3.53	−1.65	−3.74	−2.09
13.00	5.73	2.20	−3.53	−1.59	−3.67	−2.08
13.50	5.65	2.11	−3.53	−1.54	−3.60	−2.07
14.00	5.57	2.04	−3.54	−1.49	−3.54	−2.05
14.50	5.50	1.96	−3.54	−1.44	−3.49	−2.04
15.00	5.44	1.89	−3.54	−1.40	−3.44	−2.04
15.50	5.38	1.83	−3.55	−1.37	−3.39	−2.03
16.00	5.32	1.77	−3.55	−1.33	−3.35	−2.02
16.50	5.27	1.71	−3.56	−1.30	−3.31	−2.01
17.00	5.22	1.66	−3.56	−1.27	−3.27	−2.00
17.50	5.18	1.61	−3.57	−1.24	−3.24	−2.00
18.00	5.13	1.56	−3.57	−1.21	−3.20	−1.99
18.50	5.09	1.51	−3.58	−1.18	−3.17	−1.99
19.00	5.06	1.47	−3.59	−1.16	−3.14	−1.98
19.50	5.02	1.43	−3.59	−1.14	−3.11	−1.98
20.00	4.99	1.39	−3.60	−1.12	−3.09	−1.97
20.50	4.96	1.35	−3.61	−1.09	−3.06	−1.97
21.00	4.93	1.32	−3.61	−1.08	−3.04	−1.96
21.50	4.90	1.28	−3.62	−1.06	−3.02	−1.96
22.00	4.87	1.25	−3.63	−1.04	−3.00	−1.96

[a] All values are normalized with respect to R, the radius of the cylinders.

TABLE 5.22 (continued)

V_2/V_1	f_i/R	z_{m_i}/R	z_{p_i}/R	f_o/R	z_{m_o}/R	z_{p_o}/R
22.50	4.85	1.22	−3.63	−1.02	−2.98	−1.95
23.00	4.83	1.18	−3.64	−1.01	−2.96	−1.95
23.50	4.80	1.15	−3.65	−0.99	−2.94	−1.95
24.00	4.78	1.13	−3.66	−0.98	−2.92	−1.94
24.50	4.76	1.10	−3.66	−0.96	−2.90	−1.94
25.00	4.74	1.07	−3.67	−0.95	−2.89	−1.94
25.50	4.73	1.05	−3.68	−0.94	−2.87	−1.94
26.00	4.71	1.02	−3.69	−0.92	−2.86	−1.93
26.50	4.69	1.00	−3.70	−0.91	−2.84	−1.93
27.00	4.68	0.97	−3.70	−0.90	−2.83	−1.93
27.50	4.66	0.95	−3.71	−0.89	−2.82	−1.93
28.00	4.65	0.93	−3.72	−0.88	−2.80	−1.93
28.50	4.64	0.91	−3.73	−0.87	−2.79	−1.92
29.00	4.62	0.89	−3.73	−0.86	−2.78	−1.92
29.50	4.61	0.87	−3.74	−0.85	−2.77	−1.92
30.00	4.60	0.85	−3.75	−0.84	−2.76	−1.92
30.50	4.59	0.83	−3.76	−0.83	−2.75	−1.92
31.00	4.58	0.81	−3.77	−0.82	−2.74	−1.92
31.50	4.57	0.79	−3.77	−0.81	−2.73	−1.91
32.00	4.56	0.78	−3.78	−0.81	−2.72	−1.91
32.50	4.55	0.76	−3.79	−0.80	−2.71	−1.91
33.00	4.54	0.74	−3.80	−0.79	−2.70	−1.91
33.50	4.53	0.73	−3.81	−0.78	−2.69	−1.91
34.00	4.52	0.71	−3.81	−0.78	−2.68	−1.91
34.50	4.51	0.69	−3.82	−0.77	−2.68	−1.91
35.00	4.51	0.68	−3.83	−0.76	−2.67	−1.91
35.50	4.50	0.66	−3.84	−0.76	−2.66	−1.90
36.00	4.49	0.65	−3.84	−0.75	−2.65	−1.90
36.50	4.49	0.64	−3.85	−0.74	−2.65	−1.90
37.00	4.48	0.62	−3.86	−0.74	−2.64	−1.90
37.50	4.48	0.61	−3.87	−0.73	−2.63	−1.90
38.00	4.47	0.59	−3.88	−0.73	−2.63	−1.90
38.50	4.46	0.58	−3.88	−0.72	−2.62	−1.90
39.00	4.46	0.57	−3.89	−0.71	−2.61	−1.90
39.50	4.45	0.56	−3.90	−0.71	−2.61	−1.90
40.00	4.45	0.54	−3.91	−0.70	−2.60	−1.90
40.50	4.44	0.53	−3.91	−0.70	−2.60	−1.90
41.00	4.44	0.52	−3.92	−0.69	−2.59	−1.90
41.50	4.44	0.51	−3.93	−0.69	−2.58	−1.90
42.00	4.43	0.50	−3.94	−0.68	−2.58	−1.89
42.50	4.43	0.49	−3.94	−0.68	−2.57	−1.89
43.00	4.43	0.47	−3.95	−0.67	−2.57	−1.89
43.50	4.42	0.46	−3.96	−0.67	−2.56	−1.89
44.00	4.42	0.45	−3.97	−0.67	−2.56	−1.89
44.50	4.42	0.44	−3.97	−0.66	−2.55	−1.89

TABLE 5.22 (continued)

V_2/V_1	f_i/R	z_{m_i}/R	z_{p_i}/R	f_o/R	z_{m_o}/R	z_{p_o}/R
45.00	4.41	0.43	−3.98	−0.66	−2.55	−1.89
45.50	4.41	0.42	−3.99	−0.65	−2.55	−1.89
46.00	4.41	0.41	−4.00	−0.65	−2.54	−1.89
46.50	4.41	0.40	−4.00	−0.65	−2.54	−1.89
47.00	4.40	0.39	−4.01	−0.64	−2.53	−1.89
47.50	4.40	0.38	−4.02	−0.64	−2.53	−1.89
48.00	4.40	0.37	−4.03	−0.63	−2.52	−1.89
48.50	4.40	0.36	−4.03	−0.63	−2.52	−1.89
49.00	4.40	0.36	−4.04	−0.63	−2.52	−1.89
49.50	4.39	0.35	−4.05	−0.62	−2.51	−1.89
50.00	4.39	0.34	−4.05	−0.62	−2.51	−1.89

TABLE 5.23

CARDINAL POINTS OF ASYMMETRICAL TWO-CYLINDER LENS OF RADII a AND b AND POTENTIALS V_2 AND V_1, RESPECTIVELY[a]

$b/a = 1.5$

V_2/V_1	f_i/a	z_{m_i}/a	z_{p_i}/a	f_o/a	z_{m_o}/a	z_{p_o}/a
1.50	103.14	92.82	−10.33	−84.23	−93.60	−9.37
2.00	39.03	32.36	−6.67	−27.60	−33.27	−5.67
2.50	24.18	18.72	−5.47	−15.30	−19.73	−4.44
3.00	17.99	13.11	−4.88	−10.39	−14.21	−3.82
3.50	14.67	10.12	−4.55	−7.84	−11.30	−3.46
4.00	12.62	8.28	−4.34	−6.31	−9.53	−3.21
4.50	11.24	7.04	−4.20	−5.30	−8.34	−3.04
5.00	10.24	6.15	−4.10	−4.58	−7.50	−2.92
5.50	9.50	5.47	−4.03	−4.05	−6.87	−2.82
6.00	8.92	4.94	−3.98	−3.64	−6.38	−2.74
6.50	8.45	4.51	−3.94	−3.32	−5.99	−2.68
7.00	8.07	4.16	−3.91	−3.05	−5.68	−2.63
7.50	7.76	3.86	−3.89	−2.83	−5.42	−2.58
8.00	7.49	3.61	−3.88	−2.65	−5.20	−2.55
8.50	7.27	3.39	−3.87	−2.49	−5.01	−2.52
9.00	7.07	3.20	−3.87	−2.36	−4.85	−2.49
9.50	6.90	3.03	−3.87	−2.24	−4.71	−2.46
10.00	6.75	2.88	−3.87	−2.14	−4.58	−2.44
10.50	6.62	2.75	−3.87	−2.04	−4.47	−2.43
11.00	6.51	2.63	−3.88	−1.96	−4.37	−2.41
11.50	6.40	2.52	−3.89	−1.89	−4.28	−2.40

[a] Here, $a < b$.

TABLE 5.23 (continued)

$b/a = 1.5$

V_2/V_1	f_i/a	z_{m_i}/a	z_{p_i}/a	f_o/a	z_{m_o}/a	z_{p_o}/a
12.00	6.31	2.41	−3.89	−1.82	−4.20	−2.38
12.50	6.22	2.32	−3.90	−1.76	−4.13	−2.37
13.00	6.15	2.24	−3.91	−1.71	−4.07	−2.36
13.50	6.08	2.16	−3.92	−1.65	−4.01	−2.35
14.00	6.02	2.08	−3.93	−1.61	−3.95	−2.34
14.50	5.96	2.01	−3.94	−1.57	−3.90	−2.33
15.00	5.91	1.95	−3.96	−1.53	−3.85	−2.33
15.50	5.86	1.89	−3.97	−1.49	−3.81	−2.32
16.00	5.81	1.83	−3.98	−1.45	−3.77	−2 32
16.50	5.77	1.78	−3.99	−1.42	−3.73	−2.31
17.00	5.73	1.73	−4.01	−1.39	−3.70	−2.30
17.50	5.70	1.68	−4.02	−1.36	−3.66	−2.30
18.00	5.67	1.63	−4.03	−1.34	−3.63	−2.30
18.50	5.64	1.59	−4.05	−1.31	−3.60	−2.29
19.00	5.61	1.55	−4.06	−1.29	−3.57	−2.29
19.50	5.58	1.51	−4.07	−1.26	−3.55	−2.28
20.00	5.56	1.47	−4.09	−1.24	−3.52	−2.28
20.50	5.54	1.44	−4.10	−1.22	−3.50	−2.28
21.00	5.52	1.40	−4.11	−1.20	−3.48	−2.28
21.50	5.50	1.37	−4.13	−1.19	−3.46	−2.27
22.00	5.48	1.34	−4.14	−1.17	−3.44	−2.27
22.50	5.46	1.31	−4.15	−1.15	−3.42	−2.27
23.00	5.45	1.28	−4.17	−1.14	−3.40	−2.27
23.50	5.43	1.25	−4.18	−1.12	−3.38	−2.26
24.00	5.42	1.22	−4.19	−1.11	−3.37	−2.26
24.50	5.40	1.20	−4.21	−1.09	−3.35	−2.26
25.00	5.39	1.17	−4.22	−1.08	−3.34	−2.26
25.50	5.38	1.15	−4.24	−1.07	−3.32	−2.26
26.00	5.37	1.12	−4.25	−1.05	−3.31	−2.26
26.50	5.36	1.10	−4.26	−1.04	−3.30	−2.25
27.00	5.35	1.08	−4.28	−1.03	−3.28	−2.25
27.50	5.34	1.05	−4.29	−1.02	−3.27	−2.25
28.00	5.34	1.03	−4.30	−1.01	−3.26	−2.25
28.50	5.33	1.01	−4.32	−1.00	−3.25	−2.25
29.00	5.32	0.99	−4.33	−0.99	−3.24	−2.25
29.50	5.32	0.97	−4.34	−0.98	−3.23	−2.25
30.00	5.31	0.95	−4.36	−0.97	−3.22	−2.25
30.50	5.31	0.94	−4.37	−0.96	−3.21	−2.25
31.00	5.30	0.92	−4.38	−0.95	−3.20	−2.25
31.50	5.30	0.90	−4.40	−0.94	−3.19	−2.24
32.00	5.29	0.88	−4.41	−0.94	−3.18	−2.24
32.50	5.29	0.87	−4.42	−0.93	−3.17	−2.24
33.00	5.29	0.85	−4.43	−0.92	−3.16	−2.24

TABLE 5.23 (continued)

$b/a = 1.5$

V_2/V_1	f_i/a	z_{m_i}/a	z_{p_i}/a	f_o/a	z_{m_o}/a	z_{p_o}/a
33.50	5.28	0.84	−4.45	−0.91	−3.15	−2.24
34.00	5.28	0.82	−4.46	−0.91	−3.15	−2.24
34.50	5.28	0.80	−4.47	−0.90	−3.14	−2.24
35.00	5.28	0.79	−4.49	−0.89	−3.13	−2.24
35.50	5.27	0.78	−4.50	−0.88	−3.13	−2.24
36.00	5.27	0.76	−4.51	−0.88	−3.12	−2.24
36.50	5.27	0.75	−4.52	−0.87	−3.11	−2.24
37.00	5.27	0.73	−4.54	−0.87	−3.11	−2.24
37.50	5.27	0.72	−4.55	−0.86	−3.10	−2.24
38.00	5.27	0.71	−4.56	−0.85	−3.09	−2.24
38.50	5.27	0.69	−4.57	−0.85	−3.09	−2.24
39.00	5.27	0.68	−4.59	−0.84	−3.08	−2.24
39.50	5.27	0.67	−4.60	−0.84	−3.08	−2.24
40.00	5.27	0.66	−4.61	−0.83	−3.07	−2.24
40.50	5.27	0.65	−4.62	−0.83	−3.06	−2.24
41.00	5.27	0.63	−4.64	−0.82	−3.06	−2.24
41.50	5.27	0.62	−4.65	−0.82	−3.05	−2.24
42.00	5.27	0.61	−4.66	−0.81	−3.05	−2.24
42.50	5.27	0.60	−4.67	−0.81	−3.04	−2.24
43.00	5.27	0.59	−4.68	−0.80	−3.04	−2.24
43.50	5.27	0.58	−4.70	−0.80	−3.04	−2.24
44.00	5.28	0.57	−4.71	−0.79	−3.03	−2.24
44.50	5.28	0.56	−4.72	−0.79	−3.03	−2.24
45.00	5.28	0.55	−4.73	−0.79	−3.02	−2.24
45.50	5.28	0.54	−4.74	−0.78	−3.02	−2.24
46.00	5.28	0.53	−4.75	−0.78	−3.01	−2.24
46.50	5.28	0.52	−4.77	−0.77	−3.01	−2.24
47.00	5.29	0.51	−4.78	−0.77	−3.01	−2.24
47.50	5.29	0.50	−4.79	−0.77	−3.00	−2.24
48.00	5.29	0.49	−4.80	−0.76	−3.00	−2.24
48.50	5.29	0.48	−4.81	−0.76	−2.99	−2.24
49.00	5.30	0.47	−4.82	−0.76	−2.99	−2.24
49.50	5.30	0.46	−4.84	−0.75	−2.99	−2.24
50.00	5.30	0.45	−4.85	−0.75	−2.98	−2.24

TABLE 5.24

CARDINAL POINTS OF ASYMMETRICAL TWO-CYLINDER LENS OF RADII a AND b AND POTENTIALS V_2 AND V_1, RESPECTIVELY[a]

$b/a = 2$

V_2/V_1	f_i/a	z_{m_i}/a	z_{p_i}/a	f_o/a	z_{m_o}/a	z_{p_o}/a
1.50	118.92	106.93	−11.99	−97.11	−107.99	−10.88
2.00	45.41	37.58	−7.83	−32.12	−38.80	−6.67
2.50	28.33	21.86	−6.48	−17.93	−23.20	−5.27
3.00	21.20	15.38	−5.82	−12.25	−16.82	−4.57
3.50	17.37	11.92	−5.45	−9.29	−13.44	−4.15
4.00	15.00	9.78	−5.22	−7.51	−11.39	−3.88
4.50	13.40	8.34	−5.06	−6.32	−10.01	−3.69
5.00	12.25	7.30	−4.96	−5.48	−9.03	−3.54
5.50	11.39	6.51	−4.88	−4.86	−8.29	−3.43
6.00	10.72	5.89	−4.83	−4.38	−7.73	−3.34
6.50	10.19	5.39	−4.80	−4.00	−7.27	−3.27
7.00	9.75	4.98	−4.77	−3.69	−6.90	−3.22
7.50	9.39	4.63	−4.76	−3.43	−6.60	−3.17
8.00	9.09	4.34	−4.75	−3.22	−6.34	−3.13
8.50	8.83	4.08	−4.74	−3.03	−6.12	−3.09
9.00	8.60	3.86	−4.75	−2.87	−5.93	−3.06
9.50	8.41	3.66	−4.75	−2.73	−5.77	−3.03
10.00	8.24	3.48	−4.76	−2.61	−5.62	−3.01
10.50	8.09	3.33	−4.76	−2.50	−5.49	−2.99
11.00	7.96	3.18	−4.77	−2.40	−5.38	−2.97
11.50	7.84	3.05	−4.79	−2.32	−5.27	−2.96
12.00	7.73	2.93	−4.80	−2.24	−5.18	−2.94
12.50	7.64	2.82	−4.81	−2.16	−5.10	−2.93
13.00	7.55	2.72	−4.83	−2.10	−5.02	−2.92
13.50	7.47	2.63	−4.84	−2.04	−4.95	−2.91
14.00	7.40	2.54	−4.86	1.98	−4.88	−2.90
14.50	7.34	2.46	−4.88	−1.93	−4.82	−2.89
15.00	7.28	2.39	−4.89	−1.88	−4.77	−2.88
15.50	7.23	2.32	−4.91	−1.84	4.72	−2.88
16.00	7.18	2.25	−4.93	−1.80	−4.67	−2.87
16.50	7.13	2.19	−4.94	−1.76	−4.62	−2.87
17.00	7.09	2.13	−4.96	−1.72	−4.58	−2.86
17.50	7.05	2.07	−4.98	−1.69	−4.54	−2.86
18.00	7.01	2.02	−5.00	−1.66	−4.51	−2.85
18.50	6.98	1.96	−5.02	−1.63	−4.47	−2.85
19.00	6.95	1.92	−5.03	−1.60	−4.44	−2.84
19.50	6.92	1.87	−5.05	−1.57	−4.41	−2.84
20.00	6.90	1.83	−5.07	−1.55	−4.38	−2.84
20.50	6.87	1.78	−5.09	−1.52	−4.35	−2.83

[a] Here, $a < b$.

TABLE 5.24 (continued)

$b/a = 2$

V_2/V_1	f_i/a	z_{m_i}/a	z_{p_i}/a	f_o/a	z_{m_o}/a	z_{p_o}/a
21.00	6.85	1.74	−5.11	−1.50	−4.33	−2.83
21.50	6.83	1.70	−5.13	−1.48	−4.30	−2.83
22.00	6.81	1.67	−5.14	−1.46	−4.28	−2.83
22.50	6.79	1.63	−5.16	−1.44	−4.26	−2.82
23.00	6.78	1.60	−5.18	−1.42	−4.24	−2.82
23.50	6.76	1.56	−5.20	−1.40	−4.22	−2.82
24.00	6.75	1.53	−5.22	−1.38	−4.20	−2.82
24.50	6.73	1.50	−5.23	−1.36	−4.18	−2.82
25.00	6.72	1.47	−5.25	−1.35	−4.16	−2.81
25.50	6.71	1.44	−5.27	−1.33	−4.14	−2.81
26.00	6.70	1.41	−5.29	−1.32	−4.13	−2.81
26.50	6.69	1.39	−5.30	−1.30	−4.11	−2.81
27.00	6.68	1.36	−5.32	−1.29	−4.10	−2.81
27.50	6.67	1.33	−5.34	−1.28	−4.08	−2.81
28.00	6.67	1.31	−5.36	−1.26	−4.07	−2.81
28.50	6.66	1.29	−5.37	−1.25	−4.06	−2.81
29.00	6.65	1.26	−5.39	−1.24	−4.04	−2.81
29.50	6.65	1.24	−5.41	−1.23	−4.03	−2.80
30.00	6.64	1.22	−5.42	−1.22	−4.02	−2.80
30.50	6.64	1.20	−5.44	−1.20	−4.01	−2.80
31.00	6.63	1.18	−5.46	−1.19	−4.00	−2.80
31.50	6.63	1.16	−5.47	−1.18	−3.99	−2.80
32.00	6.63	1.14	−5.49	−1.17	−3.98	−2.80
32.50	6.62	1.12	−5.51	−1.16	−3.97	−2.80
33.00	6.62	1.10	−5.52	−1.16	−3.96	−2.80
33.50	6.62	1.08	−5.54	−1.15	−3.95	−2.80
34.00	6.62	1.06	−5.56	−1.14	−3.94	−2.80
34.50	6.61	1.04	−5.57	−1.13	−3.93	−2.80
35.00	6.61	1.03	−5.59	−1.12	−3.92	−2.80
35.50	6.61	1.01	−5.60	−1.11	−3.91	−2.80
36.00	6.61	0.99	−5.62	−1.11	−3.90	−2.80
36.50	6.61	0.98	−5.63	−1.10	−3.90	−2.80
37.00	6.61	0.96	−5.65	−1.09	−3.89	−2.80
37.50	6.61	0.95	−5.67	−1.08	−3.88	−2.80
38.00	6.61	0.93	−5.68	−1.08	−3.87	−2.80
38.50	6.61	0.92	−5.70	−1.07	−3.87	−2.80
39.00	6.61	0.90	−5.71	−1.06	−3.86	−2.80
39.50	6.61	0.89	−5.73	−1.06	−3.85	−2.80
40.00	6.62	0.87	−5.74	−1.05	−3.85	−2.80
40.50	6.62	0.86	−5.76	−1.04	−3.84	−2.80
41.00	6.62	0.85	−5.77	−1.04	−3.84	−2.80
41.50	6.62	0.83	−5.79	−1.03	−3.83	−2.80
42.00	6.62	0.82	−5.80	−1.02	−3.82	−2.80

TABLE 5.24 (continued)

$b/a = 2$

V_2/V_1	f_i/a	z_{m_i}/a	z_{p_i}/a	f_o/a	z_{m_o}/a	z_{p_o}/a
42.50	6.62	0.81	−5.82	−1.02	−3.82	−2.80
43.00	6.63	0.80	−5.83	−1.01	−3.81	−2.80
43.50	6.63	0.78	−5.85	−1.01	−3.81	−2.80
44.00	6.63	0.77	−5.86	−1.00	−3.80	−2.80
44.50	6.63	0.76	−5.87	−1.00	−3.80	−2.80
45.00	6.64	0.75	−5.89	−0.99	−3.79	−2.80
45.50	6.64	0.74	−5.90	−0.99	−3.79	−2.80
46.00	6.64	0.72	−5.92	−0.98	−3.78	−2.80
46.50	6.65	0.71	−5.93	−0.98	−3.78	−2.80
47.00	6.65	0.70	−5.95	−0.97	−3.77	−2.80
47.50	6.65	0.69	−5.96	−0.97	−3.77	−2.80
48.00	6.65	0.68	−5.97	−0.96	−3.76	−2.80
48.50	6.66	0.67	−5.99	−0.96	−3.76	−2.80
49.00	6.66	0.66	−6.00	−0.95	−3.76	−2.80
49.50	6.67	0.65	−6.01	−0.95	−3.75	−2.80
50.00	6.67	0.64	−6.03	−0.95	−3.75	−2.80

TABLE 5.25

CARDINAL POINTS OF ASYMMETRICAL TWO-CYLINDER LENS OF RADII a AND b AND POTENTIALS V_2 AND V_1, RESPECTIVELY[a]

$b/a = 2.5$

V_2/V_1	f_i/a	z_{m_i}/a	z_{p_i}/a	f_o/a	z_{m_o}/a	z_{p_o}/a
1.50	132.38	119.00	−13.37	−108.16	−120.31	−12.14
2.00	50.93	42.12	−8.81	−36.05	−43.56	−7.51
2.50	31.96	24.63	−7.32	−20.24	−26.20	−5.96
3.00	24.02	17.40	−6.61	−13.89	−19.08	−5.19
3.50	19.75	13.54	−6.21	10.57	−15.31	−4.74
4.00	17.12	11.16	−5.96	−8.57	−13.01	−4.43
4.50	15.34	9.54	−5.80	−7.24	−11.46	−4.22
5.00	14.06	8.37	−5.69	−6.30	−10.36	−4.06
5.50	13.10	7.49	−5.61	−5.60	−9.54	−3.94
6.00	12.35	6.80	−5.56	−5.06	−8.90	−3.85
6.50	11.76	6.24	−5.52	−4.63	−8.39	−3.77
7.00	11.28	5.77	−5.50	−4.27	−7.98	−3.70
7.50	10.88	5.39	−5.49	−3.98	−7.64	−3.65
8.00	10.54	5.06	−5.48	−3.74	−7.35	−3.61

[a] Here, $a < b$.

TABLE 5.25 (continued)

$b/a = 2.5$

V_2/V_1	f_i/a	z_{m_i}/a	z_{p_i}/a	f_o/a	z_{m_o}/a	z_{p_o}/a
8.50	10.25	4.77	−5.48	−3.53	−7.10	−3.57
9.00	10.01	4.52	−5.49	−3.35	−6.88	−3.54
9.50	9.79	4.30	−5.50	−3.19	−6.70	−3.51
10.00	9.61	4.10	−5.51	−3.05	−6.53	−3.48
10.50	9.44	3.92	−5.52	−2.93	−6.39	−3.46
11.00	9.30	3.76	−5.54	−2.81	−6.26	−3.44
11.50	9.17	3.62	−5.55	−2.71	−6.14	−3.43
12.00	9.05	3.48	−5.57	−2.62	−6.03	−3.41
12.50	8.95	3.36	−5.59	−2.54	−5.94	−3.40
13.00	8.86	3.25	−5.61	−2.47	−5.85	−3.38
13.50	8.77	3.15	−5.63	−2.40	−5.77	−3.37
14.00	8.69	3.05	−5.65	−2.33	−5.70	−3.36
14.50	8.62	2.96	−5.67	−2.28	−5.63	−3.35
15.00	8.56	2.87	−5.69	−2.22	−5.57	−3.35
15.50	8.50	2.80	−5.71	−2.17	−5.51	−3.34
16.00	8.45	2.72	−5.73	−2.12	−5.46	−3.33
16.50	8.40	2.65	−5.75	−2.08	−5.40	−3.33
17.00	8.36	2.59	−5.77	−2.04	−5.36	−3.32
17.5C	8.32	2.52	−5.79	−2.00	−5.31	−3.31
18.00	8.28	2.46	−5.82	−1.96	−5.27	−3.31
18.50	8.24	2 41	−5.84	−1.93	−5.23	−3.31
19.00	8.21	2.35	−5.86	−1.90	−5.20	−3.30
19.50	8.18	2.30	−5.88	−1.86	−5.16	−3.30
20.00	8.16	2.26	−5.90	−1.83	−5.13	−3.29
20.50	8.13	2.21	−5.92	−1.81	−5.10	−3.29
21.00	8.11	2.17	−5.94	−1.78	−5.07	−3.29
21.50	8.09	2.12	−5.97	−1.76	−5.04	−3.29
22.00	8.07	2.08	−5.99	−1.73	−5.01	−3.28
22.50	8.05	2.04	−6.01	−1.71	−4.99	−3.28
23.00	8.03	2.01	−6.03	−1.69	−4.97	−3.28
23.50	8.02	1.97	−6.05	−1.67	−4.94	−3.28
24.00	8.00	1.94	−6.07	−1.64	−4.92	−3.28
24.50	7.99	1.90	−6.09	−1.63	−4.90	−3.27
25.00	7.98	1.87	−6.11	−1.61	−4.88	−3.27
25.50	7.97	1.84	−6.13	−1.59	−4.86	−3.27
26.00	7.96	1.81	−6.15	−1.57	−4.84	−3.27
26.50	7.95	1.78	−6.17	−1.56	−4.82	−3.27
27.00	7.94	1.75	−6.19	−1.54	−4.81	−3.27
27.50	7.93	1.72	−6.21	−1.52	−4.79	−3.27
28.00	7.93	1.70	−6.23	−1.51	−4.77	−3.27
28.50	7.92	1.67	−6.25	−1.49	−4.76	−3.26
29.00	7.92	1.65	−6.27	−1.48	−4.75	−3.26
29.50	7.91	1.62	−6.29	−1.47	−4.73	−3.26

TABLE 5.25 (continued)

$b/a = 2.5$

V_2/V_1	f_i/a	z_{m_i}/a	z_{p_i}/a	f_o/a	z_{m_o}/a	z_{p_o}/a
30.00	7.91	1.60	−6.31	−1.45	−4.72	−3.26
30.50	7.90	1.58	−6.33	−1.44	−4.70	−3.26
31.00	7.90	1.56	−6.35	−1.43	−4.69	−3.26
31.50	7.90	1.53	−6.36	−1.42	−4.68	−3.26
32.00	7.90	1.51	−6.38	−1.41	−4.67	−3.26
32.50	7.89	1.49	−6.40	−1.40	−4.66	−3.26
33.00	7.89	1.47	−6.42	−1.38	−4.65	−3.26
33.50	7.89	1.45	−6.44	−1.37	−4.63	−3.26
34.00	7.89	1.43	−6.46	−1.36	−4.62	−3.26
34.50	7.89	1.42	−6.47	−1.35	−4.61	−3.26
35.00	7.89	1.40	−6.49	−1.34	−4.60	−3.26
35.50	7.89	1.38	−6.51	−1.34	−4.60	−3.26
36.00	7.89	1.36	−6.53	−1.33	−4.59	−3.26
36.50	7.89	1.35	−6.54	−1.32	−4.58	−3.26
37.00	7.89	1.33	−6.56	−1.31	−4.57	−3.26
37.50	7.89	1.32	−6.58	−1.30	−4.56	−3.26
38.00	7.90	1.30	−6.60	−1.29	−4.55	−3.26
38.50	7.90	1.28	−6.61	−1.28	−4.54	−3.26
39.00	7.90	1.27	−6.63	−1.28	−4.54	−3.26
39.50	7.90	1.26	−6.65	−1.27	−4.53	−3.26
40.00	7.90	1.24	−6.66	−1.26	−4.52	−3.26
40.50	7.91	1.23	−6.68	−1.25	−4.51	−3.26
41.00	7.91	1.21	−6.70	−1.25	−4.51	−3.26
41.50	7.91	1.20	−6.71	−1.24	−4.50	−3.26
42.00	7.92	1.19	−6.73	−1.23	−4.49	−3.26
42.50	7.92	1.17	−6.74	−1.23	−4.49	−3.26
43.00	7.92	1.16	−6.76	−1.22	−4.48	−3.26
43.50	7.93	1.15	−6.78	−1.21	−4.47	−3.26
44.00	7.93	1.14	−6.79	−1.21	−4.47	−3.26
44.50	7.93	1.13	−6.81	−1.20	−4.46	−3.26
45.00	7.94	1.11	−6.82	−1.19	−4.46	−3.26
45.50	7.94	1.10	−6.84	−1.19	−4.45	−3.26
46.00	7.95	1.09	−6.85	−1.18	4.45	−3.26
46.50	7.95	1.08	−6.87	−1.18	−4.44	−3.26
47.00	7.95	1.07	−6.88	−1.17	−4.44	−3.26
47.50	7.96	1.06	−6.90	−1.17	−4.43	−3.26
48.00	7.96	1.05	−6.91	−1.16	−4.43	−3.26
48.50	7.97	1.04	−6.93	−1.16	−4.42	−3.27
49.00	7.97	1.03	−6.94	−1.15	−4.42	−3.27
49.50	7.98	1.02	−6.96	−1.15	−4.41	−3.27
50.00	7.98	1.01	−6.97	−1.14	−4.41	−3.27

TABLE 5.26
CARDINAL POINTS OF ASYMMETRICAL TWO-CYLINDER LENS OF RADII a AND b AND POTENTIALS V_2 AND V_1, RESPECTIVELY[a]

$b/a = 3$

V_2/V_1	f_i/a	z_{m_i}/a	z_{p_i}/a	f_o/a	z_{m_o}/a	z_{p_o}/a
1.50	145.57	130.86	−14.71	−118.99	−132.35	−13.36
2.00	56.35	46.60	−9.75	−39.91	−48.23	−8.31
2.50	35.51	27.38	−8.13	−22.51	−29.13	−6.62
3.00	26.79	19.43	−7.36	−15.50	−21.28	−5.78
3.50	22.09	15.17	−6.92	−11.84	−17.12	−5.28
4.00	19.20	12.54	−6.66	−9.63	−14.58	−4.95
4.50	17.24	10.76	−6.48	−8.15	−12.87	−4.72
5.00	15.84	9.47	−6.37	−7.11	−11.65	−4.54
5.50	14.78	8.50	−6.29	−6.33	−10.74	−4.41
6.00	13.97	7.73	−6.24	−5.73	−10.03	−4.31
6.50	13.32	7.12	−6.20	−5.25	−9.47	−4.22
7.00	12.79	6.61	−6.18	−4.86	−9.01	−4.15
7.50	12.35	6.18	−6.17	−4.53	−8.63	−4.10
8.00	11.98	5.82	−6.17	−4.26	−8.31	−4.05
8.50	11.67	5.50	−6.17	−4.03	−8.03	−4.01
9.00	11.41	5.23	−6.18	−3.82	−7.79	−3.97
9.50	11.17	4.98	−6.19	−3.65	−7.59	−3.94
10.00	10.97	4.77	−6.20	−3.49	−7.40	−3.91
10.50	10.79	4.57	−6.22	−3.35	−7.24	−3.89
11.00	10.64	4.40	−6.24	−3.23	−7.10	−3.87
11.50	10.50	4.24	−6.26	−3.12	−6.97	−3.85
12.00	10.38	4.10	−6.28	−3.02	−6.85	−3.83
12.50	10.27	3.96	−6.30	−2.92	−6.74	−3.82
13.00	10.17	3.84	−6.32	−2.84	−6.64	−3.80
13.50	10.08	3.73	−6.35	−2.76	−6.55	−3.79
14.00	10.00	3.63	−6.37	−2.69	−6.47	−3.78
14.50	9.92	3.53	−6.39	−2.63	−6.40	−3.77
15.00	9.86	3.44	−6.42	−2.56	−6.33	−3.76
15.50	9.80	3.35	−6.44	−2.51	−6.26	−3.76
16.00	9.74	3.27	−6.47	−2.45	−6.20	−3.75
16.50	9.69	3.20	−6.49	−2.40	−6.15	−3.74
17.00	9.64	3.13	−6.52	−2.36	−6.09	−3.74
17.50	9.60	3.06	−6.54	−2.31	−6.04	−3.73
18.00	9.56	3.00	−6.57	−2.27	−6.00	−3.72
18.50	9.53	2.94	−6.59	−2.23	−5.95	−3.72
19.00	9.50	2.88	−6.61	−2.20	−5.91	−3.72
19.50	9.47	2.83	−6.64	−2.16	−5.88	−3.71
20.00	9.44	2.78	−6.66	−2.13	−5.84	−3.71
20.50	9.42	2.73	−6.69	−2.10	−5.80	−3.71

[a] Here, $a < b$.

TABLE 5.26 (continued)

$b/a = 3$

V_2/V_1	f_i/a	z_{m_i}/a	z_{p_i}/a	f_o/a	z_{m_o}/a	z_{p_o}/a
21.00	9.39	2.68	−6.71	−2.07	−5.77	−3.70
21.50	9.37	2.64	−6.74	−2.04	−5.74	−3.70
22.00	9.36	2.60	−6.76	−2.01	−5.71	−3.70
22.50	9.34	2.56	−6.78	−1.99	−5.68	−3.69
23.00	9.32	2.52	−6.81	−1.96	−5.65	−3.69
23.50	9.31	2.48	−6.83	−1.94	−5.63	−3.69
24.00	9.30	2.44	−6.85	−1.92	−5.60	−3.69
24.50	9.28	2.41	−6.88	−1.89	−5.58	−3.69
25.00	9.27	2.37	−6.90	−1.87	−5.56	−3.69
25.50	9.26	2.34	−6.92	−1.85	−5.54	−3.68
26.00	9.26	2.31	−6.95	−1.83	−5.52	−3.68
26.50	9.25	2.28	−6.97	−1.81	−5.50	−3.68
27.00	9.24	2.25	−6.99	−1.80	−5.48	−3.68
27.50	9.24	2.22	−7.01	−1.78	−5.46	−3.68
28.00	9.23	2.20	−7.03	−1.76	−5.44	−3.68
28.50	9.23	2.17	−7.06	−1.75	−5.42	−3.68
29.00	9.22	2.15	−7.08	−1.73	−5.41	−3.68
29.50	9.22	2.12	−7.10	−1.72	−5.39	−3.68
30.00	9.22	2.10	−7.12	−1.70	−5.38	−3.68
30.50	9.22	2.07	−7.14	−1.69	−5.36	−3.68
31.00	9.21	2.05	−7.16	−1.67	−5.35	−3.68
31.50	9.21	2.03	−7.18	−1.66	−5.33	−3.68
32.00	9.21	2.01	−7.20	−1.65	−5.32	−3.67
32.50	9.21	1.99	−7.22	−1.63	−5.31	−3.67
33.00	9.21	1.97	−7.25	−1.62	−5.30	−3.67
33.50	9.22	1.95	−7.27	−1.61	−5.28	−3.67
34.00	9.22	1.93	−7.29	−1.60	−5.27	−3.67
34.50	9.22	1.91	−7.31	−1.59	−5.26	−3.67
35.00	9.22	1.89	−7.33	−1.58	−5.25	−3.67
35.50	9.22	1.88	−7.35	−1.57	−5.24	−3.67
36.00	9.22	1.86	−7.36	−1.56	−5.23	−3.67
36.50	9.23	1.84	−7.38	−1.55	−5.22	−3.67
37.00	9.23	1.83	−7.40	−1.54	−5.21	−3.67
37.50	9.23	1.81	−7.42	−1.53	−5.20	−3.67
38.00	9.24	1.80	−7.44	−1.52	−5.19	−3.67
38.50	9.24	1.78	−7.46	−1.51	−5.18	−3.68
39.00	9.25	1.77	−7.48	−1.50	−5.17	−3.68
39.50	9.25	1.75	−7.50	−1.49	−5.16	−3.68
40.00	9.25	1.74	−7.52	−1.48	−5.16	−3.68
40.50	9.26	1.72	−7.53	−1.47	−5.15	−3.68
41.00	9.26	1.71	−7.55	−1.46	−5.14	−3.68
41.50	9.27	1.70	−7.57	−1.46	−5.13	−3.68
42.00	9.27	1.69	−7.59	−1.45	−5.13	−3.68

TABLE 5.26 (continued)

$b/a = 3$

V_2/V_1	f_i/a	z_{m_i}/a	z_{p_i}/a	f_o/a	z_{m_o}/a	z_{p_o}/a
42.50	9.28	1.67	−7.61	−1.44	−5.12	−3.68
43.00	9.29	1.66	−7.62	−1.43	−5.11	−3.68
43.50	9.29	1.65	−7.64	−1.43	−5.10	−3.68
44.00	9.30	1.64	−7.66	−1.42	−5.10	−3.68
44.50	9.30	1.63	−7.68	−1.41	−5.09	−3.68
45.00	9.31	1.61	−7.69	−1.41	−5.08	−3.68
45.50	9.31	1.60	−7.71	−1.40	−5.08	−3.68
46.00	9.32	1.59	−7.73	−1.39	−5.07	−3.68
46.50	9.33	1.58	−7.75	−1.39	−5.07	−3.68
47.00	9.33	1.57	−7.76	−1.38	−5.06	−3.68
47.50	9.34	1.56	−7.78	−1.37	−5.05	−3.68
48.00	9.35	1.55	−7.80	−1.37	−5.05	−3.68
48.50	9.35	1.54	−7.81	−1.36	−5.04	−3.68
49.00	9.36	1.53	−7.83	−1.35	−5.04	−3.68
49.50	9.37	1.52	−7.84	−1.35	−5.03	−3.68
50.00	9.37	1.51	−7.86	−1.34	−5.03	−3.68

TABLE 5.27

CARDINAL POINTS OF ASYMMETRICAL TWO-CYLINDER LENS OF RADII a AND b AND POTENTIALS V_2 AND V_1, RESPECTIVELY[a]

$b/a = 3.5$

V_2/V_1	f_i/a	z_{m_i}/a	z_{p_i}/a	f_o/a	z_{m_o}/a	z_{p_o}/a
1.50	155.06	139.40	−15.66	−126.82	−141.04	−14.22
2.00	60.47	50.03	−10.44	−42.86	−51.76	−8.90
2.50	38.31	29.57	−8.74	−24.30	−31.42	−7.12
3.00	29.02	21.08	−7.94	−16.81	−23.04	−6.23
3.50	24.03	16.54	−7.49	−12.89	−18.59	−5.70
4.00	20.94	13.73	−7.21	−10.51	−15.87	−5.36
4.50	18.86	11.82	−7.04	−8.93	−14.04	−5.11
5.00	17.37	10.44	−6.92	−7.80	−12.73	−4.93
5.50	16.25	9.40	−6.85	−6.96	−11.75	−4.79
6.00	15.38	8.59	−6.80	−6.31	−10.99	−4.68
6.50	14.69	7.93	−6.77	−5.80	−10.39	−4.59
7.00	14.13	7.39	−6.75	−5.37	−9.89	−4.52
7.50	13.67	6.93	−6.74	−5.02	−9.48	−4.46
8.00	13.29	6.54	−6.74	−4.73	−9.14	−4.41

[a] Here, $a < b$.

TABLE 5.27 (continued)

$b/a = 3.5$

V_2/V_1	f_i/a	z_{m_i}/a	z_{p_i}/a	f_o/a	z_{m_o}/a	z_{p_o}/a
8.50	12.96	6.21	−6.75	−4.74	−8.84	−4.36
9.00	12.68	5.91	−6.76	−4.26	−8.58	−4.33
9.50	12.44	5.66	−6.78	−4.06	−8.36	−4.30
10.00	12.23	5.43	−6.80	−3.89	−8.16	−4.27
10.50	12.04	5.22	−6.82	−3.74	−7.99	−4.24
11.00	11.88	5.04	−6.84	−3.61	−7.83	−4.22
11.50	11.74	4.87	−6.87	−3.49	−7.69	−4.20
12.00	11.61	4.72	−6.89	−3.38	−7.56	−4.19
12.50	11.50	4.58	−6.92	−3.28	−7.45	−4.17
13.00	11.40	4.45	−6.95	−3.19	−7.34	−4.16
13.50	11.31	4.33	−6.97	−3.10	−7.25	−4.14
14.00	11.22	4.22	−7.00	−3.03	−7.16	−4.13
14.50	11.15	4.12	−7.03	−2.95	−7.08	−4.12
15.00	11.08	4.03	−7.06	−2.89	−7.00	−4.11
15.50	11.02	3.94	−7.09	−2.83	−6.93	−4.11
16.00	10.97	3.86	−7.12	−2.77	−6.87	−4.10
16.50	10.92	3.78	−7.14	−2.71	−6.81	−4.09
17.00	10.88	3.70	−7.17	−2.66	−6.75	−4.09
17.50	10.84	3.64	−7.20	−2.62	−6.70	−4.08
18.00	10.80	3.57	−7.23	−2.57	−6.65	−4.08
18.50	10.77	3.51	−7.26	−2.53	−6.60	−4.07
19.00	10.74	3.45	−7.29	−2.49	−6.55	−4.07
19.50	10.71	3.40	−7.31	−2.45	−6.51	−4.06
20.00	10.69	3.34	−7.34	−2.41	−6.47	−4.06
20.50	10.66	3.29	−7.37	−2.38	−6.44	−4.06
21.00	10.64	3.24	−7.40	−2.35	−6.40	−4.05
21.50	10.63	3.20	7.43	−2.32	−6.37	−4.05
22.00	10.61	3.16	−7.45	−2.29	−6.33	−4.05
22.50	10.59	3.11	−7.48	−2.26	−6.30	−4.05
23.00	10.58	3.07	−7.51	−2.23	−6.27	−4.05
23.50	10.57	3.04	−7.54	−2.20	−6.25	−4.04
24.00	10.56	3.00	−7.56	−2.18	−6.22	−4.04
24.50	10.55	2.96	−7.59	−2.15	−6.19	−4.04
25.00	10.54	2.93	−7.61	−2.13	−6.17	−4.04
25.50	10.54	2.90	−7.64	−2.11	−6.15	−4.04
26.00	10.53	2.87	−7.67	−2.09	−6.12	−4.04
26.50	10.53	2.84	−7.69	−2.07	−6.10	−4.04
27.00	10.52	2.81	−7.72	−2.05	−6.08	−4.03
27.50	10.52	2.78	−7.74	−2.03	−6.06	−4.03
28.00	10.52	2.75	−7.77	−2.01	−6.04	−4.03
28.50	10.52	2.72	−7.79	−1.99	−6.03	−4.03
29.00	10.52	2.70	−7.82	−1.98	−6.01	−4.03
29.50	10.52	2.67	−7.84	−1.96	−5.99	−4.03

TABLE 5.27 (continued)

$b/a = 3.5$

V_2/V_1	f_i/a	z_{m_i}/a	z_{p_i}/a	f_o/a	z_{m_o}/a	z_{p_o}/a
30.00	10.52	2.65	−7.87	−1.94	−5.97	−4.03
30.50	10.52	2.63	−7.89	−1.93	−5.96	−4.03
31.00	10.52	2.61	−7.92	−1.91	−5.94	−4.03
31.50	10.52	2.58	−7.94	−1.90	−5.93	−4.03
32.00	10.53	2.56	−7.96	−1.88	−5.91	−4.03
32.50	10.53	2.54	−7.99	−1.87	−5.90	−4.03
33.00	10.53	2.52	−8.01	−1.85	−5.89	−4.03
33.50	10.54	2.50	−8.03	−1.84	−5.87	−4.03
34.00	10.54	2.48	−8.06	−1.83	−5.86	−4.03
34.50	10.55	2.47	−8.08	−1.82	−5.85	−4.03
35.00	10.55	2.45	−8.10	−1.80	−5.84	−4.03
35.50	10.56	2.43	−8.12	−1.79	−5.82	−4.03
36.00	10.56	2.41	−8.15	−1.78	−5.81	−4.03
36.50	10.57	2.40	−8.17	−1.77	−5.80	−4.03
37.00	10.57	2.38	−8.19	−1.76	−5.79	−4.03
37.50	10.58	2.37	−8.21	−1.75	−5.78	−4.03
38.00	10.59	2.35	−8.23	−1.74	−5.77	−4.03
38.50	10.59	2.34	−8.26	−1.73	−5.76	−4.03
39.00	10.60	2.32	−8.28	−1.72	−5.75	−4.03
39.50	10.61	2.31	−8.30	−1.71	−5.74	−4.04
40.00	10.61	2.29	−8.32	−1.70	−5.73	−4.04
40.50	10.62	2.28	−8.34	−1.69	−5.73	−4.04
41.00	10.63	2.27	−8.36	−1.68	−5.72	−4.04
41.50	10.64	2.26	−8.38	−1.67	−5.71	−4.04
42.00	10.65	2.24	−8.40	−1.66	−5.70	−4.04
42.50	10.65	2.23	−8.42	−1.65	−5.69	−4.04
43.00	10.66	2.22	−8.44	−1.65	−5.69	−4.04
43.50	10.67	2.21	−8.46	−1.64	−5.68	−4.04
44.00	10.68	2.20	−8.48	−1.63	−5.67	−4.04
44.50	10.69	2.18	−8.50	−1.62	−5.66	−4.04
45.00	10.70	2.17	−8.52	−1.61	−5.66	−4.04
45.50	10.71	2.16	−8.54	−1.61	−5.65	−4.04
46.00	10.71	2.15	−8.56	−1.60	−5.64	−4.04
46.50	10.72	2.14	−8.58	−1.59	−5.64	−4.04
47.00	10.73	2.13	−8.60	−1.59	−5.63	−4.04
47.50	10.74	2.12	−8.62	−1.58	−5.62	−4.04
48.00	10.75	2.11	−8.64	−1.57	−5.62	−4.05
48.50	10.76	2.10	−8.66	−1.57	−5.61	−4.05
49.00	10.77	2.09	−8.68	−1.56	−4.61	−4.05
49.50	10.78	2.09	−8.70	−1.55	−5.60	−4.05
50.00	10.79	2.08	−8.71	−1.55	−5.59	−4.05

TABLE 5.28

CARDINAL POINTS OF ASYMMETRICAL TWO-CYLINDER LENS OF RADII a AND b AND POTENTIALS V_2 AND V_1, RESPECTIVELY[a]

$b/a = 4$

V_2/V_1	f_i/a	z_{m_i}/a	z_{p_i}/a	f_o/a	z_{m_o}/a	z_{p_o}/a
1.50	162.45	146.07	−16.38	−132.90	−147.78	−14.88
2.00	63.77	52.79	−10.98	−45.21	−54.57	−9.36
2.50	40.61	31.38	−9.23	−25.76	−33.27	−7.51
3.00	30.89	22.48	−8.40	−17.90	−24.48	−6.59
3.50	25.66	17.71	−7.94	−13.77	−19.81	−6.04
4.00	22.43	14.76	−7.67	−11.27	−16.95	−5.68
4.50	20.25	12.76	−7.49	−9.59	−15.02	−5.43
5.00	18.69	11.32	−7.38	−8.40	−13.64	−5.24
5.50	17.53	10.22	−7.31	−7.52	−12.61	−5.10
6.00	16.63	9.37	−7.26	−6.83	−11.81	−4.98
6.50	15.91	8.68	−7.24	−6.28	−11.17	−4.89
7.00	15.33	8.11	−7.23	−5.83	−10.65	−4.82
7.50	14.86	7.63	−7.22	−5.46	−10.22	−4.76
8.00	14.46	7.23	−7.23	−5.15	−9.85	−4.70
8.50	14.12	6.88	−7.24	−4.88	−9.54	−4.66
9.00	13.83	6.57	−7.26	−4.64	−9.27	−4.62
9.50	13.59	6.30	−7.28	−4.44	−9.03	−4.59
10.00	13.37	6.06	−7.31	−4.26	−8.82	−4.56
10.50	13.18	5.85	−7.33	−4.10	−8.64	−4.54
11.00	13.02	5.66	−7.36	−3.96	−8.47	−4.52
11.50	12.88	5.48	−7.39	−3.83	−8.32	−4.50
12.00	12.75	5.33	−7.42	−3.71	−8.19	−4.48
12.50	12.64	5.18	−7.45	−3.60	−8.07	−4.46
13.00	12.54	5.05	−7.49	−3.51	−7.96	−4.45
13.50	12.45	4.93	−7.52	−3.42	−7.86	−4.44
14.00	12.37	4.81	−7.55	−3.33	−7.76	−4.43
14.50	12.29	4.71	−7.58	−3.26	−7.68	−4.42
15.00	12.23	4.61	−7.62	−3.19	−7.60	−4.41
15.50	12.17	4.52	−7.65	−3.12	−7.52	−4.40
16.00	12.12	4.44	−7.68	−3.06	−7.45	−4.39
16.50	12.07	4.36	−7.72	−3.00	−7.39	−4.39
17.00	12.03	4.28	−7.75	−2.95	−7.33	−4.38
17.50	11.99	4.21	−7.78	−2.89	−7.27	−4.38
18.00	11.96	4.14	−7.82	−2.85	−7.22	−4.37
18.50	11.93	4.08	−7.85	−2.80	−7.17	−4.37
19.00	11.90	4.02	−7.88	−2.76	−7.12	−4.36
19.50	11.88	3.96	−7.92	−2.72	−7.08	−4.36
20.00	11.86	3.91	−7.95	−2.68	−7.04	−4.36
20.50	11.84	3.86	−7.98	−2.64	−7.00	−4.36

[a] Here, $a < b$.

TABLE 5.28 (continued)

$b/a = 4$

V_2/V_1	f_i/a	z_{m_i}/a	z_{p_i}/a	f_o/a	z_{m_o}/a	z_{p_o}/a
21.00	11.82	3.81	−8.01	−2.61	−6.96	−4.35
21.50	11.81	3.76	−8.04	−2.57	−6.92	−4.35
22.00	11.80	3.72	−8.08	−2.54	−6.89	−4.35
22.50	11.79	3.68	−8.11	−2.51	−6.86	−4.35
23.00	11.78	3.64	−8.14	−2.48	−6.83	−4.34
23.50	11.77	3.60	−8.17	−2.45	−6.80	−4.34
24.00	11.76	3.56	−8.20	−2.43	−6.77	−4.34
24.50	11.76	3.53	−8.23	−2.40	−6.74	−4.34
25.00	11.75	3.49	−8.26	−2.38	−6.72	−4.34
25.50	11.75	3.46	−8.29	−2.35	−6.69	−4.34
26.00	11.75	3.43	−8.32	−2.33	−6.67	−4.34
26.50	11.75	3.40	−8.35	−2.31	−6.64	−4.34
27.00	11.75	3.37	−8.38	−2.28	−6.62	−4.34
27.50	11.75	3.34	−8.41	−2.26	−6.60	−4.34
28.00	11.75	3.31	−8.44	−2.24	−6.58	−4.34
28.50	11.75	3.29	−8.46	−2.23	−6.56	−4.34
29.00	11.75	3.26	−8.49	−2.21	−6.54	−4.34
29.50	11.76	3.24	−8.52	−2.19	−6.53	−4.34
30.00	11.76	3.21	−8.55	−2.17	−6.51	−4.34
30.50	11.77	3.19	−8.58	−2.15	−6.49	−4.34
31.00	11.77	3.17	−8.60	−2.14	−6.47	−4.34
31.50	11.78	3.15	−8.63	−2.12	−6.46	−4.34
32.00	11.78	3.12	−8.66	−2.11	−6.44	−4.34
32.50	11.79	3.10	−8.69	−2.09	−6.43	−4.34
33.00	11.80	3.08	−8.71	−2.08	−6.41	−4.34
33.50	11.80	3.07	−8.74	−2.06	−6.40	−4.34
34.00	11.81	3.05	−8.76	−2.05	−6.39	−4.34
34.50	11.82	3.03	−8.79	−2.03	−6.37	−4.34
35.00	11.83	3.01	−8.82	−2.02	−6.36	−4.34
35.50	11.84	2.99	−8.84	−2.01	−6.35	−4.34
36.00	11.85	2.98	−8.87	−2.00	−6.34	−4.34
36.50	11.85	2.96	−8.89	−1.98	−6.33	−4.34
37.00	11.86	2.95	−8.92	−1.97	−6.31	−4.34
37.50	11.87	2.93	−8.94	−1.96	−6.30	−4.34
38.00	11.88	2.92	−8.97	−1.95	−6.29	−4.34
38.50	11.89	2.90	−8.99	−1.94	−6.28	−4.34
39.00	11.90	2.89	−9.02	−1.93	−6.27	−4.35
39.50	11.91	2.87	−9.04	−1.92	−6.26	−4.35
40.00	11.92	2.86	−9.06	−1.91	−6.25	−4.35
40.50	11.94	2.85	−9.09	−1.90	−6.24	−4.35
41.00	11.95	2.83	−9.11	−1.89	−6.24	−4.35
41.50	11.96	2.82	−9.14	−1.88	−6.23	−4.35
42.00	11.97	2.81	−9.16	−1.87	−6.22	−4.35

TABLE 5.28 (continued)

$b/a = 4$

V_2/V_1	f_i/a	z_{m_i}/a	z_{p_i}/a	f_o/a	z_{m_o}/a	z_{p_o}/a
42.50	11.98	2.80	−9.18	−1.86	−6.21	−4.35
43.00	11.99	2.79	−9.21	−1.85	−6.20	−4.35
43.50	12.00	2.77	−9.23	−1.84	−6.19	−4.35
44.00	12.01	2.76	−9.25	−1.83	−6.19	−4.35
44.50	12.03	2.75	−9.27	−1.82	−6.18	−4.35
45.00	12.04	2.74	−9.30	−1.82	−6.17	−4.36
45.50	12.05	2.73	−9.32	−1.81	−6.16	−4.36
46.00	12.06	2.72	−9.34	−1.80	−6.16	−4.36
46.50	12.07	2.71	−9.36	−1.79	−6.15	−4.36
47.00	12.09	2.70	−9.39	−1.78	−6.14	−4.36
47.50	12.10	2.69	−9.41	−1.78	−6.14	−4.36
48.00	12.11	2.68	−9.43	−1.77	−6.13	−4.36
48.50	12.12	2.67	−9.45	−1.76	−6.12	−4.36
49.00	12.14	2.66	−9.47	−1.75	−6.12	−4.36
49.50	12.15	2.65	−9.49	−1.75	−6.11	−4.36
50.00	12.16	2.65	−9.52	−1.74	−6.10	−4.36

TABLE 5.29

CARDINAL POINTS OF ASYMMETRICAL TWO-CYLINDER LENS OF RADII a AND b AND POTENTIALS V_2 AND V_1, RESPECTIVELY[a]

$b/a = 4.5$

V_2/V_1	f_i/a	z_{m_i}/a	z_{p_i}/a	f_o/a	z_{m_o}/a	z_{p_o}/a
1.50	178.29	160.36	−17.93	−145.87	−162.16	−16.29
2.00	69.91	57.93	−11.98	−49.58	−59.79	−10.21
2.50	44.49	34.44	−10.05	−28.24	−36.41	−8.17
3.00	33.84	24.70	−9.14	−19.62	−26.77	−7.15
3.50	28.12	19.48	−8.63	−15.10	−21.65	−6.55
4.00	24.58	16.26	−8.32	−12.35	−18.51	−6.15
4.50	22.21	14.08	−8.13	−10.52	−16.40	−5.87
5.00	20.51	12.51	−8.00	−9.22	−14.89	−5.67
5.50	19.24	11.32	−7.92	−8.25	−13.76	−5.51
6.00	18.26	10.39	−7.87	−7.50	−12.89	−5.38
6.50	17.48	9.64	−7.84	−6.90	−12.19	−5.28
7.00	16.86	9.03	−7.83	−6.41	−11.62	−5.20
7.50	16.34	8.51	−7.83	−6.01	−11.14	−5.13
8.00	15.91	8.08	−7.83	−5.67	−10.74	−5.08

[a] Here, $a < b$.

TABLE 5.29 (continued)

$b/a = 4.5$

V_2/V_1	f_i/a	z_{m_i}/a	z_{p_i}/a	f_o/a	z_{m_o}/a	z_{p_o}/a
8.50	15.55	7.70	−7.85	−5.37	−10.40	−5.03
9.00	15.24	7.37	−7.87	−5.12	−10.10	−4.99
9.50	14.97	7.08	−7.89	−4.90	−9.85	−4.95
10.00	14.74	6.83	−7.92	−4.70	−9.62	−4.92
10.50	14.54	6.60	−7.95	−4.52	−9.42	−4.89
11.00	14.37	6.39	−7.98	−4.37	−9.24	−4.87
11.50	14.22	6.21	−8.01	−4.23	−9.08	−4.85
12.00	14.08	6.04	−8.04	−4.10	−8.93	−4.83
12.50	13.96	5.89	−8.08	−3.98	−8.80	−4.81
13.00	13.86	5.74	−8.11	−3.88	−8.67	−4.80
13.50	13.76	5.61	−8.15	−3.78	−8.56	−4.78
14.00	13.68	5.49	−8.18	−3.69	−8.46	−4.77
14.50	13.61	5.38	−8.22	−3.60	−8.37	−4.76
15.00	13.54	5.28	−8.26	−3.53	−8.28	−4.75
15.50	13.48	5.18	−8.30	−3.45	−8.20	−4.74
16.00	13.43	5.09	−8.33	−3.39	−8.12	−4.74
16.50	13.38	5.01	−8.37	−3.32	−8.05	−4.73
17.00	13.34	4.93	−8.41	−3.26	−7.99	−4.72
17.50	13.30	4.86	−8.44	−3.21	−7.93	−4.72
18.00	13.27	4.79	−8.48	−3.16	−7.87	−4.71
18.50	13.24	4.72	−8.52	−3.11	−7.82	−4.71
19.00	13.21	4.66	−8.55	−3.06	−7.76	−4.70
19.50	13.19	4.60	−8.59	−3.02	−7.72	−4.70
20.00	13.17	4.54	−8.62	−2.97	−7.67	−4.70
20.50	13.15	4.49	−8.66	−2.93	−7.63	−4.69
21.00	13.13	4.44	−8.70	−2.89	−7.59	−4.69
21.50	13.12	4.39	−8.73	−2.86	−7.55	−4.69
22.00	13.11	4.34	−8.77	−2.82	−7.51	−4.69
22.50	13.10	4.30	−8.80	−2.79	−7.47	−4.69
23.00	13.09	4.26	−8.84	−2.76	−7.44	−4.68
23.50	13.09	4.22	−8.87	−2.73	−7.41	−4.68
24.00	13.08	4.18	−8.9	−2.70	−7.38	−4.68
24.50	13.08	4.14	−8.94	−2.67	−7.35	−4.68
25.00	13.08	4.11	−8.97	−2.64	−7.32	−4.68
25.50	13.08	4.07	−9.00	−2.62	−7.29	−4.68
26.00	13.08	4.04	−9.04	−2.59	−7.27	−4.68
26.50	13.08	4.01	−9.07	−2.57	−7.24	−4.68
27.00	13.08	3.98	−9.10	−2.54	−7.22	−4.68
27.50	13.08	3.95	−9.14	−2.52	−7.20	−4.68
28.00	13.09	3.92	−9.17	−2.50	−7.17	−4.68
28.50	13.09	3.89	−9.20	−2.48	−7.15	−4.68
29.00	13.10	3.87	−9.23	−2.46	−7.13	−4.68
29.50	13.10	3.84	−9.26	−2.44	−7.11	−4.68

TABLE 5.29 (continued)

$b/a = 4.5$

V_2/V_1	f_i/a	z_{m_i}/a	z_{p_i}/a	f_o/a	z_{m_o}/a	z_{p_o}/a
30.00	13.11	3.82	−9.29	−2.42	−7.09	−4.68
30.50	13.12	3.79	−9.32	−2.40	−7.08	−4.68
31.00	13.13	3.77	−9.36	−2.38	−7.06	−4.68
31.50	13.13	3.75	−9.39	−2.36	−7.04	−4.68
32.00	13.14	3.73	−9.42	−2.35	−7.02	−4.68
32.50	13.15	3.71	−9.45	−2.33	−7.01	−4.68
33.00	13.16	3.68	−9.48	−2.32	−6.99	−4.68
33.50	13.17	3.67	−9.51	−2.30	−6.98	−4.68
34.00	13.18	3.65	−9.53	−2.28	−6.96	−4.68
34.50	13.19	3.63	−9.56	−2.27	−6.95	−4.68
35.00	13.20	3.61	−9.59	−2.26	−6.94	−4.68
35.50	13.21	3.59	−9.62	−2.24	−6.92	−4.68
36.00	13.23	3.58	−9.65	−2.23	−6.91	−4.68
36.50	13.24	3.56	−9.68	−2.21	−6.90	−4.68
37.00	13.25	3.54	−9.71	−2.20	−6.88	−4.68
37.50	13.26	3.53	−9.73	−2.19	−6.87	−4.68
38.00	13.27	3.51	−9.76	−2.18	−6.86	−4.69
38.50	13.29	3.50	−9.79	−2.16	−6.85	−4.69
39.00	13.30	3.48	−9.82	−2.15	−6.84	−4.69
39.50	13.31	3.47	−9.84	−2.14	−6.83	−4.69
40.00	13.33	3.45	−9.87	−2.13	−6.82	−4.69
40.50	13.34	3.44	−9.90	−2.12	−6.81	−4.69
41.00	13.35	3.43	−9.93	−2.11	−6.80	−4.69
41.50	13.37	3.41	−9.95	−2.10	−6.79	−4.69
42.00	13.38	3.40	−9.98	−2.09	−6.78	−4.69
42.50	13.39	3.39	−10.00	−2.08	−6.77	−4.69
43.00	13.41	3.38	−10.03	−2.07	−6.76	−4.69
43.50	13.42	3.37	−10.06	2.06	−6.75	−4.70
44.00	13.44	3.36	−10.08	−2.05	−6.74	−4.70
44.50	13.45	3.34	−10.11	−2.04	−6.74	−4.70
45.00	13.47	3.33	−10.13	−2.03	6.73	−4.70
45.50	13.48	3.32	−10.16	−2.02	−6.72	−4.70
46.00	13.49	3.31	−10.18	−2.01	−6.71	−4.70
46.50	13.51	3.30	−10.21	−2.00	−6.70	−4.70
47.00	13.52	3.29	−10.23	−1.99	−6.70	−4.70
47.50	13.54	3.28	−10.26	−1.99	−6.69	−4.70
48.00	13.55	3.27	−10.28	−1.98	−6.68	−4.71
48.50	13.57	3.26	−10.31	−1.97	−6.68	−4.71
49.00	13.58	3.25	−10.33	−1.96	−6.67	−4.71
49.50	13.60	3.24	−10.35	−1.95	−6.66	−4.71
50.00	13.61	3.24	−10.38	−1.95	−6.66	−4.71

TABLE 5.30
CARDINAL POINTS OF ASYMMETRICAL TWO-CYLINDER LENS OF RADII a AND b AND POTENTIALS V_2 AND V_1, RESPECTIVELY[a]

$b/a = 5$

V_2/V_1	f_i/a	z_{m_i}/a	z_{p_i}/a	f_o/a	z_{m_o}/a	z_{p_o}/a
1.50	192.16	172.93	−19.23	−157.24	−174.70	−17.46
2.00	75.38	62.56	−12.82	−53.46	−64.37	−10.91
2.50	47.98	37.25	−10.73	−30.46	−39.16	−8.70
3.00	36.50	26.76	−9.74	−21.16	−28.77	−7.61
3.50	30.34	21.14	−9.19	−16.29	−23.24	−6.96
4.00	26.54	17.68	−8.86	−13.33	−19.86	−6.53
4.50	23.98	15.33	−8.65	−11.36	−17.59	−6.22
5.00	22.16	13.65	−8.51	−9.96	−15.96	−6.00
5.50	20.79	12.37	−8.42	−8.92	−14.75	−5.83
6.00	19.74	11.38	−8.37	−8.11	−13.80	−5.69
6.50	18.91	10.58	−8.34	−7.47	−13.05	−5.59
7.00	18.24	9.92	−8.32	−6.94	−12.44	−5.50
7.50	17.69	9.37	−8.32	−6.50	−11.93	−5.42
8.00	17.23	8.90	−8.33	−6.13	−11.50	−5.36
8.50	16.84	8.50	−8.34	−5.82	−11.13	−5.31
9.00	16.52	8.15	−8.37	−5.55	−10.81	−5.27
9.50	16.23	7.84	−8.39	−5.31	−10.53	−5.23
10.00	15.99	7.57	−8.42	−5.10	−10.29	−5.19
10.50	15.78	7.33	−8.45	−4.91	−10.07	−5.16
11.00	15.60	7.11	−8.49	−4.74	−9.88	−5.14
11.50	15.44	6.91	−8.52	−4.59	−9.70	−5.12
12.00	15.29	6.74	−8.56	−4.45	−9.55	−5.10
12.50	15.17	6.57	−8.60	−4.32	−9.40	−5.08
13.00	15.06	6.42	−8.63	−4.21	−9.27	−5.06
13.50	14.96	6.29	−8.67	−4.11	−9.15	−5.05
14.00	14.87	6.16	−8.71	−4.01	−9.04	−5.04
14.50	14.80	6.04	−8.75	−3.92	−8.94	−5.02
15.00	14.73	5.93	−8.79	−3.83	−8.85	−5.01
15.50	14.67	5.83	−8.83	−3.76	−8.76	−5.00
16.00	14.61	5.74	−8.87	−3.68	−8.68	−5.00
16.50	14.56	5.65	−8.91	−3.62	−8.61	−4.99
17.00	14.52	5.57	−8.95	−3.55	−8.54	−4.98
17.50	14.48	5.49	−8.99	−3.49	−8.47	−4.98
18.00	14.45	5.41	−9.04	−3.44	−8.41	−4.97
18.50	14.42	5.34	−9.08	−3.38	−8.35	−4.97
19.00	14.39	5.28	−9.12	−3.33	−8.29	−4.96
19.50	14.37	5.22	−9.15	−3.28	−8.24	−4.96
20.00	14.35	5.16	−9.19	−3.24	−8.19	−4.96
20.50	14.34	5.10	−9.23	−3.19	−8.15	−4.95

[a] Here, $a < b$.

TABLE 5.30 (continued)

$b/a = 5$

V_2/V_1	f_i/a	z_{m_i}/a	z_{p_i}/a	f_o/a	z_{m_o}/a	z_{p_o}/a
21.00	14.32	5.05	−9.27	−3.15	−8.10	−4.95
21.50	14.31	5.00	−9.31	−3.11	−8.06	−4.95
22.00	14.30	4.95	−9.35	−3.08	−8.02	−4.94
22.50	14.29	4.90	−9.39	−3.04	−7.98	−4.94
23.00	14.29	4.86	−9.43	−3.01	−7.95	−4.94
23.50	14.28	4.82	−9.46	−2.97	−7.91	−4.94
24.00	14.28	4.78	−9.50	−2.94	−7.88	−4.94
24.50	14.28	4.74	−9.54	−2.91	−7.85	−4.94
25.00	14.28	4.70	−9.58	−2.88	−7.82	−4.94
25.50	14.28	4.67	−9.61	−2.85	−7.79	−4.94
26.00	14.28	4.63	−9.65	−2.83	−7.76	−4.93
26.50	14.28	4.60	−9.68	−2.80	−7.73	−4.93
27.00	14.29	4.57	−9.72	−2.78	−7.71	−4.93
27.50	14.29	4.54	−9.76	−2.75	−7.68	−4.93
28.00	14.30	4.51	−9.79	−2.73	−7.66	−4.93
28.50	14.31	4.48	−9.83	−2.70	−7.64	−4.93
29.00	14.31	4.45	−9.86	−2.68	−7.62	−4.93
29.50	14.32	4.43	−9.90	−2.66	−7.60	−4.93
30.00	14.33	4.40	−9.93	−2.64	−7.58	−4.93
30.50	14.34	4.38	−9.96	−2.62	−7.56	−4.93
31.00	14.35	4.35	−10.00	−2.60	−7.54	−4.93
31.50	14.36	4.33	−10.03	−2.58	−7.52	−4.94
32.00	14.37	4.31	−10.06	−2.56	−7.50	−4.94
32.50	14.38	4.28	−10.10	−2.55	−7.48	−4.94
33.00	14.39	4.26	−10.13	−2.53	−7.47	−4.94
33.50	14.41	4.24	−10.16	−2.51	−7.45	−4.94
34.00	14.42	4.22	−10.19	−2.50	−7.43	−4.94
34.50	14.43	4.20	−10.23	−2.48	−7.42	−4.94
35.00	14.44	4.19	−10.26	−2.46	−7.40	−4.94
35.50	14.46	4.17	−10.29	−2.45	−7.39	−4.94
36.00	14.47	4.15	−10.32	−2.44	7.38	−4.94
36.50	14.49	4.13	−10.35	−2.42	−7.36	4.94
37.00	14.50	4.12	−10.38	−2.41	−7.35	−4.94
37.50	14.51	4.10	−10.41	−2.39	−7.34	−4.94
38.00	14.53	4.08	−10.45	−2.38	−7.33	−4.95
38.50	14.54	4.07	−10.48	−2.37	−7.31	−4.95
39.00	14.56	4.05	−10.51	−2.35	−7.30	−4.95
39.50	14.57	4.04	−10.54	−2.34	−7.29	−4.95
40.00	14.59	4.02	−10.57	−2.33	−7.28	−4.95
40.50	14.61	4.01	−10.59	−2.32	−7.27	−4.95
41.00	14.62	4.00	−10.62	−2.31	−7.26	−4.95
41.50	14.64	3.98	−10.65	−2.29	−7.25	−4.95
42.00	14.65	3.97	−10.68	−2.28	−7.24	−4.95

TABLE 5.30 (continued)

$b/a = 5$

V_2/V_1	f_i/a	z_{m_i}/a	z_{p_i}/a	f_o/a	z_{m_o}/a	z_{p_o}/a
42.50	14.67	3.96	−10.71	−2.27	−7.23	−4.96
43.00	14.69	3.95	−10.74	−2.26	−7.22	−4.96
43.50	14.70	3.93	−10.77	−2.25	−7.21	−4.96
44.00	14.72	3.92	−10.80	−2.24	−7.20	−4.96
44.50	14.74	3.91	−10.82	−2.23	−7.19	−4.96
45.00	14.75	3.90	−10.85	−2.22	−7.18	−4.96
45.50	14.77	3.89	−10.88	−2.21	−7.17	−4.96
46.00	14.79	3.88	−10.91	−2.20	−7.17	−4.96
46.50	14.80	3.87	−10.94	−2.19	−7.16	−4.97
47.00	14.82	3.86	−10.96	−2.18	−7.15	−4.97
47.50	14.84	3.85	−10.99	−2.17	−7.14	−4.97
48.00	14.85	3.84	−11.02	−2.16	−7.13	−4.97
48.50	14.87	3.83	−11.04	−2.16	−7.13	−4.97
49.00	14.89	3.82	−11.07	−2.15	−7.12	−4.97
49.50	14.90	3.81	−11.10	−2.14	−7.11	−4.97
50.00	14.92	3.80	−11.12	−2.13	−7.11	−4.97

TABLE 5.31

CARDINAL POINTS OF ASYMMETRICAL TWO-CYLINDER LENS OF RADII a AND b AND POTENTIALS V_2 AND V_1, RESPECTIVELY[a]

$a/b = 1.5$

V_2/V_1	f_i/b	z_{m_i}/b	z_{p_i}/b	f_o/b	z_{m_o}/b	z_{p_o}/b
1.50	99.82	90.39	−9.42	−81.50	−89.99	−8.49
2.00	36.89	31.16	−5.74	−26.09	−30.88	−4.79
2.50	22.45	17.93	−4.52	−14.20	−17.76	−3.56
3.00	16.46	12.54	−3.92	−9.50	−12.45	−2.94
3.50	13.26	9.69	−3.56	−7.09	−9.66	−2.58
4.00	11.29	7.95	−3.34	−5.64	−7.98	−2.33
4.50	9.96	6.77	−3.18	−4.69	−6.85	−2.16
5.00	9.00	5.93	−3.07	−4.03	−6.06	−2.03
5.50	8.28	5.30	−2.98	−3.53	−5.47	−1.93
6.00	7.73	4.80	−2.92	−3.15	−5.01	−1.85
6.50	7.28	4.41	−2.87	−2.86	−4.65	−1.79
7.00	6.92	4.08	−2.83	−2.61	−4.35	−1.74
7.50	6.61	3.81	−2.80	−2.41	−4.11	−1.69
8.00	6.36	3.57	−2.78	−2.25	−3.90	−1.65

[a] Here $a > b$.

TABLE 5.31 (continued)

$a/b = 1.5$

V_2/V_1	f_i/b	z_{m_i}/b	z_{p_i}/b	f_o/b	z_{m_o}/b	z_{p_o}/b
8.50	6.14	3.37	−2.76	−2.11	−3.73	−1.62
9.00	5.95	3.20	−2.75	−1.98	−3.58	−1.59
9.50	5.78	3.05	−2.74	−1.88	−3.44	−1.57
10.00	5.64	2.91	−2.73	−1.78	−3.33	−1.55
10.50	5.51	2.79	−2.72	−1.70	−3.23	−1.53
11.00	5.39	2.67	−2.72	−1.63	−3.14	−1.51
11.50	5.29	2.57	−2.72	−1.56	−3.05	−1.49
12.00	5.20	2.48	−2.72	−1.50	−2.98	−1.48
12.50	5.12	2.40	−2.72	−1.45	−2.91	−1.47
13.00	5.04	2.32	−2.72	−1.40	−2.85	−1.46
13.50	4.97	2.25	−2.72	−1.35	−2.80	−1.45
14.00	4.91	2.18	−2.72	−1.31	−2.75	−1.44
14.50	4.85	2.12	−2.72	−1.27	−2.70	−1.43
15.00	4.79	2.07	−2.73	−1.24	−2.66	−1.42
15.50	4.74	2.01	−2.73	−1.20	−2.62	−1.41
16.00	4.70	1.96	−2.74	−1.17	−2.58	−1.40
16.50	4.66	1.91	−2.74	−1.15	−2.54	−1.40
17.00	4.62	1.87	−2.75	−1.12	−2.51	−1.39
17.50	4.58	1.83	−2.75	−1.09	−2.48	−1.39
18.00	4.54	1.79	−2.76	−1.07	−2.45	−1.38
18.50	4.51	1.75	−2.77	−1.05	−2.43	−1.38
19.00	4.48	1.71	−2.77	−1.03	−2.40	−1.37
19.50	4.45	1.68	−2.78	−1.01	−2.38	−1.37
20.00	4.43	1.64	−2.78	−0.99	−2.35	−1.36
20.50	4.40	1.61	−2.79	−0.97	−2.33	−1.36
21.00	4.38	1.58	−2.80	−0.96	−2.31	−1.36
21.50	4.36	1.55	−2.81	−0.94	−2.29	−1.35
22.00	4.34	1.52	−2.81	−0.92	−2.28	−1.35
22.50	4.32	1.50	−2.82	−0.91	−2.26	−1.35
23.00	4.30	1.47	−2.83	−0.90	−2.24	−1.34
23.50	4.28	1.45	−2.83	−0.88	−2.23	−1.34
24.00	4.27	1.42	−2.84	−0.87	2.21	−1.34
24.50	4.25	1.40	−2.85	−0.86	−2.20	−1.34
25.00	4.24	1.38	−2.86	−0.85	−2.18	−1.34
25.50	4.22	1.36	−2.86	−0.84	−2.17	−1.33
26.00	4.21	1.34	−2.87	−0.83	−2.16	−1.33
26.50	4.20	1.32	−2.88	−0.82	−2.14	−1.33
27.00	4.19	1.30	−2.89	−0.81	−2.13	−1.33
27.50	4.17	1.28	−2.90	−0.80	−2.12	−1.33
28.00	4.16	1.26	−2.90	−0.79	−2.11	−1.32
28.50	4.15	1.24	−2.91	−0.78	−2.10	−1.32
29.00	4.14	1.23	−2.92	−0.77	−2.09	−1.32
29.50	4.14	1.21	−2.93	−0.76	−2.08	−1.32

TABLE 5.31 (continued)

$a/b = 1.5$

V_2/V_1	f_i/b	z_{m_i}/b	z_{p_i}/b	f_o/b	z_{m_o}/b	z_{p_o}/b
30.00	4.13	1.19	−2.93	−0.75	−2.07	−1.32
30.50	4.12	1.18	−2.94	−0.75	−2.06	−1.32
31.00	4.11	1.16	−2.95	−0.74	−2.05	−1.32
31.50	4.10	1.15	−2.96	−0.73	−2.05	−1.31
32.00	4.10	1.13	−2.96	−0.72	−2.04	−1.31
32.50	4.09	1.12	−2.97	−0.72	−2.03	−1.31
33.00	4.08	1.10	−2.98	−0.71	−2.02	−1.31
33.50	4.08	1.09	−2.99	−0.70	−2.02	−1.31
34.00	4.07	1.08	−3.00	−0.70	−2.01	−1.31
34.50	4.07	1.06	−3.00	−0.69	−2.00	−1.31
35.00	4.06	1.05	−3.01	−0.69	−1.99	−1.31
35.50	4.06	1.04	−3.02	−0.68	−1.99	−1.31
36.00	4.05	1.03	−3.03	−0.68	−1.98	−1.31
36.50	4.05	1.02	−3.03	−0.67	−1.98	−1.31
37.00	4.05	1.00	−3.04	−0.67	−1.97	−1.30
37.50	4.04	0.99	−3.05	−0.66	−1.96	−1.30
38.00	4.04	0.98	−3.06	−0.66	−1.96	−1.30
38.50	4.04	0.97	−3.06	−0.65	−1.95	−1.30
39.00	4.03	0.96	−3.07	−0.65	−1.95	−1.30
39.50	4.03	0.95	−3.08	−0.64	−1.94	−1.30
40.00	4.03	0.94	−3.09	−0.64	−1.94	−1.30
40.50	4.02	0.93	−3.09	−0.63	−1.93	−1.30
41.00	4.02	0.92	−3.10	−0.63	−1.93	−1.30
41.50	4.02	0.91	−3.11	−0.62	−1.92	−1.30
42.00	4.02	0.90	−3.12	−0.62	−1.92	−1.30
42.50	4.02	0.89	−3.12	−0.62	−1.91	−1.30
43.00	4.01	0.88	−3.13	−0.61	−1.91	−1.30
43.50	4.01	0.87	−3.14	−0.61	−1.91	−1.30
44.00	4.01	0.87	−3.15	−0.60	−1.90	−1.30
44.50	4.01	0.86	−3.15	−0.60	−1.90	−1.30
45.00	4.01	0.85	−3.16	−0.60	−1.89	−1.30
45.50	4.01	0.84	−3.17	−0.59	−1.89	−1.30
46.00	4.01	0.83	−3.17	−0.59	−1.89	−1.30
46.50	4.01	0.82	−3.18	−0.59	−1.88	−1.29
47.00	4.00	0.82	−3.19	−0.58	−1.88	−1.29
47.50	4.00	0.81	−3.20	−0.58	−1.87	−1.29
48.00	4.00	0.80	−3.20	−0.58	−1.87	−1.29
48.50	4.00	0.79	−3.21	−0.57	−1.87	−1.29
49.00	4.00	0.78	−3.22	−0.57	−1.86	−1.29
49.50	4.00	0.78	−3.22	−0.57	−1.86	−1.29
50.00	4.00	0.77	−3.23	−0.57	−1.86	−1.29

TABLE 5.32

CARDINAL POINTS OF ASYMMETRICAL TWO-CYLINDER LENS OF RADII a AND b AND POTENTIALS V_2 AND V_1, RESPECTIVELY[a]

$a/b = 2$

V_2/V_1	f_i/b	z_{m_i}/b	z_{p_i}/b	f_o/b	z_{m_o}/b	z_{p_o}/b
1.50	112.36	101.84	−10.52	−91.73	−101.20	−9.47
2.00	41.20	34.89	−6.32	−29.13	−34.39	−5.26
2.50	24.92	20.00	−4.92	−15.76	−19.62	−3.86
3.00	18.18	13.95	−4.23	−10.50	−13.65	−3.16
3.50	14.59	10.76	−3.83	−7.80	−10.54	−2.74
4.00	12.37	8.81	−3.56	−6.19	−8.65	−2.46
4.50	10.88	7.50	−3.38	−5.13	−7.40	−2.27
5.00	9.81	6.56	−3.25	−4.39	−6.51	−2.12
5.50	9.01	5.86	−3.15	−3.84	−5.85	−2.01
6.00	8.38	5.31	−3.07	−3.42	−5.34	−1.92
6.50	7.88	4.87	−3.02	−3.09	−4.94	−1.84
7.00	7.48	4.51	−2.97	−2.83	−4.61	−1.78
7.50	7.14	4.20	−2.93	−2.61	−4.34	−1.73
8.00	6.85	3.95	−2.90	−2.42	−4.11	−1.69
8.50	6.60	3.73	−2.88	−2.27	−3.92	−1.65
9.00	6.39	3.53	−2.86	−2.13	−3.75	−1.62
9.50	6.21	3.36	−2.84	−2.01	−3.60	−1.59
10.00	6.04	3.21	−2.83	−1.91	−3.48	−1.57
10.50	5.90	3.08	−2.82	−1.82	−3.36	−1.54
11.00	5.77	2.96	−2.81	−1.74	−3.26	−1.52
11.50	5.66	2.85	−2.81	−1.67	−3.17	−1.51
12.00	5.55	2.75	−2.81	−1.60	−3.09	−1.49
12.50	5.46	2.65	−2.80	−1.54	−3.02	−1.47
13.00	5.37	2.57	−2.80	−1.49	−2.95	−1.46
13.50	5.29	2.49	−2.80	−1.44	−2.89	−1.45
14.00	5.22	2.42	−2.80	−1.40	−2.83	−1.44
14.50	5.15	2.35	−2.80	−1.35	−2.78	−1.43
15.00	5.09	2.29	−2.81	−1.32	−2.73	−1.42
15.50	5.04	2.23	−2.81	−1.28	−2.69	−1.41
16.00	4.98	2.17	−2.81	−1.25	−2.65	−1.40
16.50	4.94	2.12	−2.81	−1.22	−2.61	−1.39
17.00	4.89	2.07	−2.82	−1.19	−2.57	−1.39
17.50	4.85	2.03	−2.82	−1.16	−2.54	−1.38
18.00	4.81	1.98	−2.83	−1.13	−2.51	−1.37
18.50	4.77	1.94	−2.83	−1.11	−2.48	−1.37
19.00	4.74	1.90	−2.84	−1.09	−2.45	−1.36
19.50	4.71	1.86	−2.84	−1.07	−2.42	−1.36
20.00	4.68	1.83	−2.85	−1.05	−2.4	−1.35
20.50	4.65	1.79	−2.85	−1.03	−2.38	−1.35

[a] Here, $a > b$.

TABLE 5.32 (continued)

$a/b = 2$

V_2/V_1	f_i/b	z_{m_i}/b	z_{p_i}/b	f_o/b	z_{m_o}/b	z_{p_o}/b
21.00	4.62	1.76	−2.86	−1.01	−2.35	−1.35
21.50	4.60	1.73	−2.87	−0.99	−2.33	−1.34
22.00	4.57	1.70	−2.87	−0.97	−2.31	−1.34
22.50	4.55	1.67	−2.88	−0.96	−2.29	−1.33
23.00	4.53	1.64	−2.89	−0.94	−2.27	−1.33
23.50	4.51	1.62	−2.89	−0.93	−2.26	−1.33
24.00	4.49	1.59	−2.90	−0.92	−2.24	−1.32
24.50	4.47	1.56	−2.91	−0.90	−2.23	−1.32
25.00	4.45	1.54	−2.91	−0.89	−2.21	−1.32
25.50	4.44	1.52	−2.92	−0.88	−2.20	−1.32
26.00	4.42	1.50	−2.93	−0.87	−2.18	−1.31
26.50	4.41	1.47	−2.93	−0.86	−2.17	−1.31
27.00	4.39	1.45	−2.94	−0.85	−2.16	−1.31
27.50	4.38	1.43	−2.95	−0.84	−2.14	−1.31
28.00	4.37	1.41	−2.95	−0.83	−2.13	−1.31
28.50	4.36	1.39	−2.96	−0.82	−2.12	−1.30
29.00	4.34	1.38	−2.97	−0.81	−2.11	−1.30
29.50	4.33	1.36	−2.98	−0.80	−2.10	−1.30
30.00	4.32	1.34	−2.98	−0.79	−2.09	−1.30
30.50	4.31	1.32	−2.99	−0.78	−2.08	−1.30
31.00	4.30	1.31	−3.00	−0.77	−2.07	−1.30
31.50	4.30	1.29	−3.01	−0.77	−2.06	−1.29
32.00	4.29	1.27	−3.01	−0.76	−2.05	−1.29
32.50	4.28	1.26	−3.02	−0.75	−2.04	−1.29
33.00	4.27	1.24	−3.03	−0.74	−2.03	−1.29
33.50	4.26	1.23	−3.03	−0.74	−2.03	−1.29
34.00	4.26	1.22	−3.04	−0.73	−2.02	−1.29
34.50	4.25	1.20	−3.05	−0.72	−2.01	−1.29
35.00	4.24	1.19	−3.06	−0.72	−2.00	−1.29
35.50	4.24	1.18	−3.06	−0.71	−2.00	−1.28
36.00	4.23	1.16	−3.07	−0.71	−1.99	−1.28
36.50	4.23	1.15	−3.08	−0.70	−1.98	−1.28
37.00	4.22	1.14	−3.08	−0.69	−1.98	−1.28
37.50	4.22	1.13	−3.09	−0.69	−1.97	−1.28
38.00	4.21	1.11	−3.10	−0.68	−1.96	−1.28
38.50	4.21	1.10	−3.11	−0.68	−1.96	−1.28
39.00	4.20	1.09	−3.11	−0.67	−1.95	−1.28
39.50	4.20	1.08	−3.12	−0.67	−1.95	−1.28
40.00	4.20	1.07	−3.13	−0.66	−1.94	−1.28
40.50	4.19	1.06	−3.13	−0.66	−1.93	−1.28
41.00	4.19	1.05	−3.14	−0.65	−1.93	−1.28
41.50	4.19	1.04	−3.15	−0.65	−1.92	−1.27
42.00	4.18	1.03	−3.16	−0.65	−1.92	−1.27

TABLE 5.32 (continued)

$a/b = 2$

V_2/V_1	f_i/b	z_{m_i}/b	z_{p_i}/b	f_o/b	z_{m_o}/b	z_{p_o}/b
42.50	4.18	1.02	−3.16	−0.64	−1.91	−1.27
43.00	4.18	1.01	−3.17	−0.64	−1.91	−1.27
43.50	4.17	1.00	−3.18	−0.63	−1.90	−1.27
44.00	4.17	0.99	−3.18	−0.63	−1.90	−1.27
44.50	4.17	0.98	−3.19	−0.63	−1.90	−1.27
45.00	4.17	0.97	−3.20	−0.62	−1.89	−1.27
45.50	4.17	0.96	−3.21	−0.62	−1.89	−1.27
46.00	4.16	0.95	−3.21	−0.61	−1.88	−1.27
46.50	4.16	0.94	−3.22	−0.61	−1.88	−1.27
47.00	4.16	0.93	−3.23	−0.61	−1.88	−1.27
47.50	4.16	0.93	−3.23	−0.60	−1.87	−1.27
48.00	4.16	0.92	−3.24	−0.60	−1.87	−1.27
48.50	4.16	0.91	−3.25	−0.60	−1.86	−1.27
49.00	4.16	0.90	−3.25	−0.59	−1.86	−1.27
49.50	4.15	0.89	−3.26	−0.59	−1.86	−1.27
50.00	4.15	0.89	−3.27	−0.59	−1.85	−1.27

TABLE 5.33

CARDINAL POINTS OF ASYMMETRICAL TWO-CYLINDER LENS OR RADII a AND b AND POTENTIALS V_2 AND V_1, RESPECTIVELY[a]

$a/b = 2.5$

V_2/V_1	f_i/b	z_{m_i}/b	z_{p_i}/b	f_o/b	z_{m_o}/b	z_{p_o}/b
1.50	122.79	111.34	−11.46	−100.24	−110.54	−10.31
2.00	44.76	37.95	−6.81	−31.64	−37.31	−5.67
2.50	26.94	21.67	−5.27	−17.04	−21.16	−4.12
3.00	19.58	15.08	−4.51	−11.30	−14.66	−3.35
3.50	15.66	11.60	−4.06	−8.37	−11.26	−2.89
4.00	13.25	9.48	−3.76	−6.62	−9.21	−2.59
4.50	11.63	8.07	−3.56	−5.48	−7.85	−2.37
5.00	10.46	7.05	−3.41	−4.68	−6.89	−2.21
5.50	9.59	6.29	−3.30	−4.09	−6.17	−2.08
6.00	8.91	5.69	−3.21	−3.64	−5.62	−1.98
6.50	8.36	5.22	−3.15	−3.28	−5.19	−1.90
7.00	7.92	4.83	−3.09	−2.99	−4.83	−1.84
7.50	7.55	4.50	−3.05	−2.76	−4.54	−1.78
8.00	7.24	4.23	−3.01	−2.56	−4.29	−1.73

[a] Here, $a > b$.

TABLE 5.33 (continued)

$a/b = 2.5$

V_2/V_1	f_i/b	z_{m_i}/b	z_{p_i}/b	f_o/b	z_{m_o}/b	z_{p_o}/b
8.50	6.97	3.99	−2.99	−2.39	−4.08	−1.69
9.00	6.74	3.78	−2.96	−2.25	−3.90	−1.66
9.50	6.54	3.60	−2.94	−2.12	−3.75	−1.62
10.00	6.36	3.44	−2.93	−2.01	−3.61	−1.60
10.50	6.21	3.29	−2.91	−1.92	−3.49	−1.57
11.00	6.07	3.16	−2.90	−1.83	−3.38	−1.55
11.50	5.94	3.04	−2.90	−1.75	−3.28	−1.53
12.00	5.83	2.94	−2.89	−1.68	−3.19	−1.51
12.50	5.72	2.84	−2.89	−1.62	−3.11	−1.49
13.00	5.63	2.75	−2.88	−1.56	−3.04	−1.48
13.50	5.54	2.66	−2.88	−1.51	−2.98	−1.47
14.00	5.46	2.59	−2.88	−1.46	−2.91	−1.45
14.50	5.39	2.51	−2.88	−1.42	−2.86	−1.44
15.00	5.33	2.45	−2.88	−1.38	−2.81	−1.43
15.50	5.26	2.38	−2.88	−1.34	−2.76	−1.42
16.00	5.21	2.32	−2.88	−1.30	−2.71	−1.41
16.50	5.15	2.27	−2.88	−1.27	−2.67	−1.40
17.00	5.10	2.22	−2.89	−1.24	−2.63	−1.40
17.50	5.06	2.17	−2.89	−1.21	−2.60	−1.39
18.00	5.01	2.12	−2.89	−1.18	−2.56	−1.38
18.50	4.97	2.08	−2.90	−1.16	−2.53	−1.38
19.00	4.94	2.03	−2.90	−1.13	−2.50	−1.37
19.50	4.90	1.99	−2.91	−1.11	−2.47	−1.36
20.00	4.87	1.96	−2.91	−1.09	−2.45	−1.36
20.50	4.84	1.92	−2.92	−1.07	−2.42	−1.35
21.00	4.81	1.88	−2.92	−1.05	−2.40	−1.35
21.50	4.78	1.85	−2.93	−1.03	−2.38	−1.35
22.00	4.75	1.82	−2.93	−1.01	−2.35	−1.34
22.50	4.73	1.79	−2.94	−1.00	−2.33	−1.34
23.00	4.70	1.76	−2.94	−0.98	−2.31	−1.33
23.50	4.68	1.73	−2.95	−0.97	−2.30	−1.33
24.00	4.66	1.70	−2.96	−0.95	−2.28	−1.33
24.50	4.64	1.68	−2.96	−0.94	−2.26	−1.32
25.00	4.62	1.65	−2.97	−0.92	−2.24	−1.32
25.50	4.60	1.63	−2.98	−0.91	−2.23	−1.32
26.00	4.58	1.60	−2.98	−0.90	−2.21	−1.31
26.50	4.57	1.58	−2.99	−0.89	−2.20	−1.31
27.00	4.55	1.56	−2.99	−0.88	−2.19	−1.31
27.50	4.54	1.54	−3.00	−0.87	−2.17	−1.31
28.00	4.52	1.52	−3.01	−0.85	−2.16	−1.31
28.50	4.51	1.50	−3.01	−0.84	−2.15	−1.30
29.00	4.50	1.48	−3.02	−0.84	−2.14	−1.30
29.50	4.48	1.46	−3.03	−0.83	−2.12	−1.30

TABLE 5.33 (continued)

$a/b = 2.5$

V_2/V_1	f_i/b	z_{m_i}/b	z_{p_i}/b	f_o/b	z_{m_o}/b	z_{p_o}/b
30.00	4.47	1.44	−3.03	−0.82	−2.11	−1.30
30.50	4.46	1.42	−3.04	−0.81	−2.10	−1.30
31.00	4.45	1.40	−3.05	−0.80	−2.09	−1.29
31.50	4.44	1.39	−3.05	−0.79	−2.08	−1.29
32.00	4.43	1.37	−3.06	−0.78	−2.07	−1.29
32.50	4.42	1.35	−3.07	−0.78	−2.06	−1.29
33.00	4.41	1.34	−3.08	−0.77	−2.06	−1.29
33.50	4.40	1.32	−3.08	−0.76	−2.05	−1.29
34.00	4.40	1.31	−3.09	−0.75	−2.04	−1.28
34.50	4.39	1.29	−3.10	−0.75	−2.03	−1.28
35.00	4.38	1.28	−3.10	−0.74	−2.02	−1.28
35.50	4.37	1.26	−3.11	−0.73	−2.01	−1.28
36.00	4.37	1.25	−3.12	−0.73	−2.01	−1.28
36.50	4.36	1.24	−3.12	−0.72	−2.00	−1.28
37.00	4.35	1.22	−3.13	−0.72	−1.99	−1.28
37.50	4.35	1.21	−3.14	−0.71	−1.99	−1.28
38.00	4.34	1.20	−3.14	−0.70	−1.98	−1.27
38.50	4.34	1.19	−3.15	−0.70	−1.97	−1.27
39.00	4.33	1.18	−3.16	−0.69	−1.97	−1.27
39.50	4.33	1.16	−3.16	−0.69	−1.96	−1.27
40.00	4.32	1.15	−3.17	−0.68	−1.95	−1.27
40.50	4.32	1.14	−3.18	−0.68	−1.95	−1.27
41.00	4.31	1.13	−3.18	−0.67	−1.94	−1.27
41.50	4.31	1.12	−3.19	−0.67	−1.94	−1.27
42.00	4.31	1.11	−3.20	−0.66	−1.93	−1.27
42.50	4.30	1.10	−3.21	−0.66	−1.93	−1.27
43.00	4.30	1.09	−3.21	−0.66	−1.92	−1.27
43.50	4.30	1.08	−3.22	−0.65	−1.92	−1.27
44.00	4.29	1.07	−3.23	−0.65	−1.91	−1.27
44.50	4.29	1.06	−3.23	−0.64	−1.91	−1.26
45.00	4.29	1.05	−3.24	−0.64	−1.90	−1.26
45.50	4.28	1.04	−3.25	−0.64	−1.90	−1.26
46.00	4.28	1.03	−3.25	−0.63	−1.89	−1.26
46.50	4.28	1.02	−3.26	−0.63	−1.89	−1.26
47.00	4.28	1.01	−3.27	−0.62	−1.89	−1.26
47.50	4.28	1.00	−3.27	−0.62	−1.88	−1.26
48.00	4.27	0.99	−3.28	−0.62	−1.88	−1.26
48.50	4.27	0.99	−3.29	−0.61	−1.87	−1.26
49.00	4.27	0.98	−3.29	−0.61	−1.87	−1.26
49.50	4.27	0.97	−3.30	−0.61	−1.87	−1.26
50.00	4.27	9.96	−3.31	−0.60	−1.86	−1.26

TABLE 5.34
CARDINAL POINTS OF ASYMMETRICAL TWO-CYLINDER LENS OF RADII a AND b AND POTENTIALS V_2 AND V_1, RESPECTIVELY[a]

$a/b = 3$

V_2/V_1	f_i/b	z_{m_i}/b	z_{p_i}/b	f_o/b	z_{m_o}/b	z_{p_o}/b
1.50	132.95	120.56	−12.39	−108.50	−119.64	−11.14
2.00	48.22	40.90	−7.32	−34.08	−40.16	−6.08
2.50	28.90	23.27	−5.63	−18.27	−22.67	−4.40
3.00	20.93	16.14	−4.79	−12.08	−15.64	−3.56
3.50	16.69	12.39	−4.30	−8.92	−11.97	−3.06
4.00	14.08	10.11	−3.97	−7.04	−9.76	−2.72
4.50	12.33	8.58	−3.75	−5.81	−8.30	−2.49
5.00	11.08	7.49	−3.58	−4.95	−7.26	−2.31
5.50	10.13	6.68	−3.46	−4.32	−6.49	−2.17
6.00	9.40	6.04	−3.36	−3.84	−5.90	−2.07
6.50	8.81	5.53	−3.28	−3.46	−5.43	−1.98
7.00	8.34	5.11	−3.22	−3.15	−5.05	−1.90
7.50	7.94	4.76	−3.17	−2.90	−4.74	−1.84
8.00	7.60	4.47	−3.13	−2.69	−4.48	−1.79
8.50	7.31	4.21	−3.10	−2.51	−4.25	−1.74
9.00	7.06	3.99	−3.07	−2.35	−4.06	−1.71
9.50	6.85	3.80	−3.05	−2.22	−3.89	−1.67
10.00	6.66	3.63	−3.03	−2.10	−3.74	−1.64
10.50	6.49	3.47	−3.01	−2.00	−3.61	−1.61
11.00	6.33	3.33	−3.00	−1.91	−3.50	−1.59
11.50	6.20	3.21	−2.99	−1.83	−3.39	−1.57
12.00	6.08	3.09	−2.98	−1.75	−3.30	−1.55
12.50	5.96	2.99	−2.97	−1.69	−3.22	−1.53
13.00	5.86	2.89	−2.97	−1.63	−3.14	−1.51
13.50	5.77	2.80	−2.96	−1.57	−3.07	−1.50
14.00	5.68	2.72	−2.96	−1.52	−3.00	−1.48
14.50	5.61	2.65	−2.96	−1.47	−2.94	−1.47
15.00	5.53	2.58	−2.96	−1.43	−2.89	−1.46
15.50	5.47	2.51	−2.96	−1.39	−2.84	−1.45
16.00	5.41	2.45	−2.96	−1.35	−2.79	−1.44
16.50	5.35	2.39	−2.96	−1.32	−2.75	−1.43
17.00	5.29	2.33	−2.96	−1.28	−2.70	−1.42
17.50	5.24	2.28	−2.96	−1.25	−2.67	−1.41
18.00	5.20	2.23	−2.97	−1.22	−2.63	−1.40
18.50	5.15	2.18	−2.97	−1.20	−2.60	−1.40
19.00	5.11	2.14	−2.97	−1.17	−2.56	−1.39
19.50	5.07	2.10	−2.98	−1.15	−2.53	−1.38
20.00	5.04	2.06	−2.98	−1.13	−2.50	−1.38
20.50	5.00	2.02	−2.98	−1.10	−2.48	−1.37

[a] Here, $a > b$.

TABLE 5.34 (continued)

$a/b = 3$

V_2/V_1	f_i/b	z_{m_i}/b	z_{p_i}/b	f_o/b	z_{m_o}/b	z_{p_o}/b
21.00	4.97	1.98	−2.99	−1.08	−2.45	−1.37
21.50	4.94	1.95	−2.99	−1.07	−2.43	−1.36
22.00	4.91	1.91	−3.00	−1.05	−2.41	−1.36
22.50	4.88	1.88	−3.00	−1.03	−2.38	−1.35
23.00	4.86	1.85	−3.01	−1.01	−2.36	−1.35
23.50	4.83	1.82	−3.01	−1.00	−2.34	−1.35
24.00	4.81	1.79	−3.02	−0.98	−2.32	−1.34
24.50	4.79	1.76	−3.02	−0.97	−2.31	−1.34
25.00	4.77	1.74	−3.03	−0.95	−2.29	−1.33
25.50	4.75	1.71	−3.03	−0.94	−2.27	−1.33
26.00	4.73	1.69	−3.04	−0.93	−2.26	−1.33
26.50	4.71	1.66	−3.05	−0.91	−2.24	−1.33
27.00	4.69	1.64	−3.05	−0.90	−2.23	−1.32
27.50	4.67	1.62	−3.06	−0.89	−2.21	−1.32
28.00	4.66	1.60	−3.06	−0.88	−2.20	−1.32
28.50	4.64	1.57	−3.07	−0.87	−2.18	−1.31
29.00	4.63	1.55	−3.08	−0.86	−2.17	−1.31
29.50	4.62	1.53	−3.08	−0.85	−2.16	−1.31
30.00	4.60	1.51	−3.09	−0.84	−2.15	−1.31
30.50	4.59	1.50	−3.09	−0.83	−2.14	−1.31
31.00	4.58	1.48	−3.10	−0.82	−2.13	−1.30
31.50	4.57	1.46	−3.11	−0.81	−2.12	−1.30
32.00	4.55	1.44	−3.11	−0.81	−2.11	−1.30
32.50	4.54	1.43	−3.12	−0.80	−2.10	−1.30
33.00	4.53	1.41	−3.13	−0.79	−2.09	−1.30
33.50	4.52	1.39	−3.13	−0.78	−2.08	−1.30
34.00	4.52	1.38	−3.14	−0.77	−2.07	−1.29
34.50	4.51	1.36	−3.14	−0.77	−2.06	−1.29
35.00	4.50	1.35	−3.15	−0.76	−2.05	−1.29
35.50	4.49	1.33	−3.16	−0.75	−2.04	−1.29
36.00	4.48	1.32	−3.16	−0.75	−2.03	−1.29
36.50	4.48	1.30	−3.17	−0.74	−2.03	−1.29
37.00	4.47	1.29	−3.18	−0.73	−2.02	−1.29
37.50	4.46	1.28	−3.18	−0.73	−2.01	−1.28
38.00	4.45	1.26	−3.19	−0.72	−2.01	−1.28
38.50	4.45	1.25	−3.20	−0.72	−2.00	−1.28
39.00	4.44	1.24	−3.20	−0.71	−1.99	−1.28
39.50	4.44	1.23	−3.21	−0.71	−1.99	−1.28
40.00	4.43	1.22	−3.22	−0.70	−1.98	−1.28
40.50	4.43	1.20	−3.22	−0.70	−1.97	−1.28
41.00	4.42	1.19	−3.23	−0.69	−1.97	−1.28
41.50	4.42	1.18	−3.24	−0.69	−1.96	−1.28
42.00	4.41	1.17	−3.24	−0.68	−1.96	−1.27

TABLE 5.34 (continued)

$a/b = 3$

V_2/V_1	f_i/b	z_{m_i}/b	z_{p_i}/b	f_o/b	z_{m_o}/b	z_{p_o}/b
42.50	4.41	1.16	−3.25	−0.68	−1.95	−1.27
43.00	4.40	1.15	−3.26	−0.67	−1.94	−1.27
43.50	4.40	1.14	−3.26	−0.67	−1.94	−1.27
44.00	4.40	1.13	−3.27	−0.66	−1.93	−1.27
44.50	4.39	1.12	−3.27	−0.66	−1.93	−1.27
45.00	4.39	1.11	−3.28	−0.65	−1.92	−1.27
45.50	4.38	1.10	−3.29	−0.65	−1.92	−1.27
46.00	4.38	1.09	−3.29	−0.65	−1.91	−1.27
46.50	4.38	1.08	−3.30	−0.64	−1.91	−1.27
47.00	4.38	1.07	−3.31	−0.64	−1.91	−1.27
47.50	4.37	1.06	−3.31	−0.63	−1.90	−1.27
48.00	4.37	1.05	−3.32	−0.63	−1.90	−1.27
48.50	4.37	1.04	−3.33	−0.63	−1.89	−1.27
49.00	4.37	1.03	−3.33	−0.62	−1.89	−1.26
49.50	4.36	1.02	−3.34	−0.62	−1.88	−1.26
50.00	4.36	1.02	−3.34	−0.62	−1.88	−1.26

TABLE 5.35

CARDINAL POINTS OF ASYMMETRICAL TWO-CYLINDER LENS OF RADII a AND b AND POTENTIALS V_2 AND V_1, RESPECTIVELY[a]

$a/b = 3.5$

V_2/V_1	f_i/b	z_{m_i}/b	z_{p_i}/b	f_o/b	z_{m_o}/b	z_{p_o}/b
1.50	139.03	126.08	−12.95	−113.40	−125.04	−11.64
2.00	50.09	42.49	−7.60	−35.40	−41.71	−6.31
2.50	29.88	24.07	−5.81	−18.89	−23.42	−4.54
3.00	21.56	16.64	−4.93	−12.44	−16.10	−3.65
3.50	17.14	12.74	−4.40	−9.16	−12.29	−3.13
4.00	14.43	10.37	−4.06	−7.21	−9.99	−2.78
4.50	12.61	8.79	−3.82	−5.94	−8.47	−2.53
5.00	11.31	7.67	−3.64	−5.06	−7.40	−2.34
5.50	10.33	6.82	−3.51	−4.41	−6.61	−2.20
6.00	9.57	6.17	−3.41	−3.91	−6.00	−2.09
6.50	8.97	5.64	−3.33	−3.52	−5.51	−2.00
7.00	8.47	5.21	−3.26	−3.20	−5.12	−1.92
7.50	8.06	4.86	−3.21	−2.94	−4.80	−1.85
8.00	7.72	4.55	−3.16	−2.73	−4.53	−1.80

[a] Here, $a > b$.

TABLE 5.35 (continued)

$a/b = 3.5$

V_2/V_1	f_i/b	z_{m_i}/b	z_{p_i}/b	f_o/b	z_{m_o}/b	z_{p_o}/b
8.50	7.42	4.29	−3.13	−2.54	−4.30	−1.75
9.00	7.16	4.07	−3.10	−2.39	−4.10	−1.71
9.50	6.94	3.87	−3.07	−2.25	−3.93	−1.68
10.00	6.74	3.69	−3.05	−2.13	−3.77	−1.64
10.50	6.57	3.53	−3.03	−2.03	−3.64	−1.61
11.00	6.41	3.39	−3.02	−1.93	−3.52	−1.59
11.50	6.27	3.26	−3.01	−1.85	−3.42	−1.57
12.00	6.14	3.15	−3.00	−1.77	−3.32	−1.55
12.50	6.03	3.04	−2.99	−1.71	−3.23	−1.53
13.00	5.92	2.94	−2.98	−1.64	−3.15	−1.51
13.50	5.83	2.85	−2.98	−1.59	−3.08	−1.49
14.00	5.74	2.77	−2.97	−1.53	−3.01	−1.48
14.50	5.66	2.69	−2.97	−1.49	−2.95	−1.47
15.00	5.59	2.62	−2.97	−1.44	−2.90	−1.45
15.50	5.52	2.55	−2.97	−1.40	−2.85	−1.44
16.00	5.46	2.49	−2.97	−1.36	−2.80	−1.43
16.50	5.40	2.43	−2.97	−1.33	−2.75	−1.42
17.00	5.34	2.37	−2.97	−1.30	−2.71	−1.41
17.50	5.29	2.32	−2.97	−1.26	−2.67	−1.41
18.00	5.24	2.27	−2.97	−1.24	−2.63	−1.40
18.50	5.20	2.22	−2.97	−1.21	−2.60	−1.39
19.00	5.15	2.18	−2.98	−1.18	−2.57	−1.38
19.50	5.11	2.13	−2.98	−1.16	−2.54	−1.38
20.00	5.08	2.09	−2.98	−1.14	−2.51	−1.37
20.50	5.04	2.05	−2.99	−1.11	−2.48	−1.37
21.00	5.01	2.02	−2.99	−1.09	−2.45	−1.36
21.50	4.98	1.98	−2.99	−1.07	−2.43	−1.35
22.00	4.95	1.95	−3.00	−1.05	−2.40	−1.35
22.50	4.92	1.91	−3.00	−1.04	−2.38	−1.35
23.00	4.89	1.88	3.01	−1.02	−2.36	−1.34
23.50	4.87	1.85	−3.01	−1.00	−2.34	−1.34
24.00	4.84	1.82	−3.02	−0.99	−2.32	−1.33
24.50	4.82	1.80	−3.02	−0.97	−2.30	−1.33
25.00	4.80	1.77	−3.03	−0.96	−2.28	−1.33
25.50	4.78	1.74	−3.03	−0.95	−2.27	−1.32
26.00	4.76	1.72	−3.04	−0.93	−2.25	−1.32
26.50	4.74	1.69	−3.05	−0.92	−2.24	−1.32
27.00	4.72	1.67	−3.05	−0.91	−2.22	−1.31
27.50	4.70	1.65	−3.06	−0.90	−2.21	−1.31
28.00	4.69	1.63	−3.06	−0.89	−2.19	−1.31
28.50	4.67	1.60	−3.07	−0.88	−2.18	−1.30
29.00	4.66	1.58	−3.07	−0.87	−2.17	−1.30
29.50	4.64	1.56	−3.08	−0.86	−2.15	−1.30

TABLE 5.35 (continued)

$a/b = 3.5$

V_2/V_1	f_i/b	z_{m_i}/b	z_{p_i}/b	f_o/b	z_{m_o}/b	z_{p_o}/b
30.00	4.63	1.54	−3.09	−0.85	−2.14	−1.30
30.50	4.62	1.53	−3.09	−0.84	−2.13	−1.30
31.00	4.61	1.51	−3.10	−0.83	−2.12	−1.29
31.50	4.59	1.49	−3.10	−0.82	−2.11	−1.29
32.00	4.58	1.47	−3.11	−0.81	−2.10	−1.29
32.50	4.57	1.45	−3.12	−0.80	−2.09	−1.29
33.00	4.56	1.44	−3.12	−0.79	−2.08	−1.29
33.50	4.55	1.42	−3.13	−0.79	−2.07	−1.28
34.00	4.54	1.41	−3.14	−0.78	−2.06	−1.28
34.50	4.53	1.39	−3.14	−0.77	−2.05	−1.28
35.00	4.52	1.38	−3.15	−0.76	−2.04	−1.28
35.50	4.52	1.36	−3.15	−0.76	−2.04	−1.28
36.00	4.51	1.35	−3.16	−0.75	−2.03	−1.28
36.50	4.50	1.33	−3.17	−0.74	−2.02	−1.27
37.00	4.49	1.32	−3.17	−0.74	−2.01	−1.27
37.50	4.49	1.31	−3.18	−0.73	−2.00	−1.27
38.00	4.48	1.29	−3.19	−0.73	−2.00	−1.27
38.50	4.47	1.28	−3.19	−0.72	−1.99	−1.27
39.00	4.47	1.27	−3.20	−0.72	−1.98	−1.27
39.50	4.46	1.25	−3.21	−0.71	−1.98	−1.27
40.00	4.45	1.24	−3.21	−0.70	−1.97	−1.27
40.50	4.45	1.23	−3.22	−0.70	−1.96	−1.27
41.00	4.44	1.22	−3.22	−0.69	−1.96	−1.26
41.50	4.44	1.21	−3.23	−0.69	−1.95	−1.26
42.00	4.43	1.20	−3.24	−0.68	−1.95	−1.26
42.50	4.43	1.19	−3.24	−0.68	−1.94	−1.26
43.00	4.43	1.17	−3.25	−0.67	−1.94	−1.26
43.50	4.42	1.16	−3.26	−0.67	−1.93	−1.26
44.00	4.42	1.15	−3.26	−0.67	−1.92	−1.26
44.50	4.41	1.14	−3.27	−0.66	−1.92	−1.26
45.00	4.41	1.13	−3.28	−0.66	−1.91	−1.26
45.50	4.41	1.12	−3.28	−0.65	−1.91	−1.26
46.00	4.40	1.11	−3.29	−0.65	−1.90	−1.26
46.50	4.40	1.10	−3.29	−0.65	−1.90	−1.26
47.00	4.40	1.10	−3.30	−0.64	−1.90	−1.25
47.50	4.39	1.09	−3.31	−0.64	−1.89	−1.25
48.00	4.39	1.08	−3.31	−0.63	−1.89	−1.25
48.50	4.39	1.07	−3.32	−0.63	−1.88	−1.25
49.00	4.39	1.06	−3.33	−0.63	−1.88	−1.25
49.50	4.38	1.05	−3.33	−0.62	−1.87	−1.25
50.00	4.38	1.04	−3.34	−0.62	−1.87	−1.25

TABLE 5.36

CARDINAL POINTS OF ASYMMETRICAL TWO-CYLINDER LENS OF RADII a AND b AND POTENTIALS V_2 AND V_1, RESPECTIVELY[a]

$a/b = 4$

V_2/V_1	f_i/b	z_{m_i}/b	z_{p_i}/b	f_o/b	z_{m_o}/b	z_{p_o}/b
1.50	143.27	129.92	−13.36	−116.89	−128.89	−12.00
2.00	51.37	43.57	−7.80	−36.30	−42.78	−6.48
2.50	30.53	24.59	−5.94	−19.30	−23.93	−4.64
3.00	21.97	16.95	−5.02	−12.68	−16.40	−3.72
3.50	17.43	12.95	−4.48	−9.31	−12.49	−3.18
4.00	14.65	10.53	−4.12	−7.32	−10.14	−2.82
4.50	12.78	8.91	−3.87	−6.02	−8.58	−2.56
5.00	11.44	7.76	−3.69	−5.12	−7.49	−2.37
5.50	10.45	6.90	−3.55	−4.45	−6.68	−2.22
6.00	9.67	6.23	−3.44	−3.95	−6.05	−2.11
6.50	9.05	5.70	−3.35	−3.55	−5.56	−2.01
7.00	8.55	5.26	−3.28	−3.23	−5.16	−1.93
7.50	8.13	4.90	−3.23	−2.97	−4.83	−1.87
8.00	7.77	4.59	−3.18	−2.75	−4.56	−1.81
8.50	7.47	4.33	−3.14	−2.56	−4.32	−1.76
9.00	7.21	4.10	−3.11	−2.40	−4.12	−1.72
9.50	6.98	3.90	−3.08	−2.26	−3.94	−1.68
10.00	6.78	3.72	−3.06	−2.14	−3.79	−1.65
10.50	6.60	3.56	−3.04	−2.04	−3.66	−1.62
11.00	6.44	3.41	−3.03	−1.94	−3.53	−1.59
11.50	6.30	3.28	−3.02	−1.86	−3.43	−1.57
12.00	6.17	3.17	−3.00	−1.78	−3.33	−1.55
12.50	6.06	3.06	−3.00	−1.71	−3.24	−1.53
13.00	5.95	2.96	−2.99	−1.65	−3.16	−1.51
13.50	5.85	2.87	−2.98	−1.59	−3.09	−1.49
14.00	5.76	2.79	−2.98	−1.54	3.02	−1.48
14.50	5.68	2.71	−2.97	−1.49	−2.96	−1.47
15.00	5.61	2.63	−2.97	−1.45	−2.90	−1.45
15.50	5.54	2.57	−2.97	−1.41	−2.85	−1.44
16.00	5.47	2.50	−2.97	−1.37	−2.80	−1.43
16.50	5.41	2.44	−2.97	−1.33	−2.75	−1.42
17.00	5.36	2.39	−2.97	−1.30	−2.71	−1.41
17.50	5.30	2.33	−2.97	−1.27	−2.67	−1.40
18.00	5.26	2.28	−2.97	−1.24	−2.63	−1.40
18.50	5.21	2.23	−2.97	−1.21	−2.60	−1.39
19.00	5.17	2.19	−2.98	−1.19	−2.57	−1.38
19.50	5.13	2.15	−2.98	−1.16	−2.54	−1.37
20.00	5.09	2.11	−2.98	−1.14	−2.51	−1.37
20.50	5.05	2.07	−2.99	−1.12	−2.48	−1.36

[a] Here, $a > b$.

TABLE 5.36 (continued)

$a/b = 4$

V_2/V_1	f_i/b	z_{m_i}/b	z_{p_i}/b	f_o/b	z_{m_o}/b	z_{p_o}/b
21.00	5.02	2.03	−2.99	−1.10	−2.45	−1.36
21.50	4.99	1.99	−2.99	−1.08	−2.43	−1.35
22.00	4.96	1.96	−3.00	−1.06	−2.40	−1.35
22.50	4.93	1.93	−3.00	−1.04	−2.38	−1.34
23.00	4.90	1.89	−3.01	−1.02	−2.36	−1.34
23.50	4.88	1.86	−3.01	−1.01	−2.34	−1.33
24.00	4.85	1.84	−3.02	−0.99	−2.32	−1.33
24.50	4.83	1.81	−3.02	−0.98	−2.30	−1.32
25.00	4.81	1.78	−3.03	−0.96	−2.28	−1.32
25.50	4.79	1.75	−3.03	−0.95	−2.27	−1.32
26.00	4.77	1.73	−3.04	−0.93	−2.25	−1.31
26.50	4.75	1.70	−3.04	−0.92	−2.23	−1.31
27.00	4.73	1.68	−3.05	−0.91	−2.22	−1.31
27.50	4.71	1.66	−3.05	−0.90	−2.20	−1.30
28.00	4.69	1.64	−3.06	−0.89	−2.19	−1.30
28.50	4.68	1.61	−3.06	−0.88	−2.18	−1.30
29.00	4.66	1.59	−3.07	−0.87	−2.16	−1.30
29.50	4.65	1.57	−3.08	−0.86	−2.15	−1.29
30.00	4.64	1.55	−3.08	−0.85	−2.14	−1.29
30.50	4.62	1.54	−3.09	−0.84	−2.13	−1.29
31.00	4.61	1.52	−3.09	−0.83	−2.12	−1.29
31.50	4.60	1.50	−3.10	−0.82	−2.11	−1.29
32.00	4.59	1.48	−3.11	−0.81	−2.09	−1.28
32.50	4.58	1.46	−3.11	−0.80	−2.08	−1.28
33.00	4.57	1.45	−3.12	−0.79	−2.07	−1.28
33.50	4.56	1.43	−3.12	−0.79	−2.07	−1.28
34.00	4.55	1.42	−3.13	−0.78	−2.06	−1.28
34.50	4.54	1.40	−3.14	−0.77	−2.05	−1.27
35.00	4.53	1.38	−3.14	−0.77	−2.04	−1.27
35.50	4.52	1.37	−3.15	−0.76	−2.03	−1.27
36.00	4.51	1.36	−3.16	−0.75	−2.02	−1.27
36.50	4.50	1.34	−3.16	−0.75	−2.01	−1.27
37.00	4.50	1.33	−3.17	−0.74	−2.01	−1.27
37.50	4.49	1.31	−3.17	−0.73	−2.00	−1.27
38.00	4.48	1.30	−3.18	−0.73	−1.99	−1.26
38.50	4.48	1.29	−3.19	−0.72	−1.98	−1.26
39.00	4.47	1.28	−3.19	−0.72	−1.98	−1.26
39.50	4.46	1.26	−3.20	−0.71	−1.97	−1.26
40.00	4.46	1.25	−3.21	−0.70	−1.97	−1.26
40.50	4.45	1.24	−3.21	−0.70	−1.96	−1.26
41.00	4.45	1.23	−3.22	−0.69	−1.95	−1.26
41.50	4.44	1.22	−3.22	−0.69	−1.95	−1.26
42.00	4.44	1.21	−3.23	−0.68	−1.94	−1.26

TABLE 5.36 (continued)

$a/b = 4$

V_2/V_1	f_i/b	z_{m_i}/b	z_{p_i}/b	f_o/b	z_{m_o}/b	z_{p_o}/b
42.50	4.43	1.19	−3.24	−0.68	−1.94	−1.26
43.00	4.43	1.18	−3.24	−0.68	−1.93	−1.25
43.50	4.42	1.17	−3.25	−0.67	−1.92	−1.25
44.00	4.42	1.16	−3.26	−0.67	−1.92	−1.25
44.50	4.41	1.15	−3.26	−0.66	−1.91	−1.25
45.00	4.41	1.14	−3.27	−0.66	−1.91	−1.25
45.50	4.41	1.13	−3.27	−0.65	−1.90	−1.25
46.00	4.40	1.12	−3.28	−0.65	−1.90	−1.25
46.50	4.40	1.11	−3.29	−0.65	−1.89	−1.25
47.00	4.40	1.10	−3.29	−0.64	−1.89	−1.25
47.50	4.39	1.10	−3.30	−0.64	−1.88	−1.25
48.00	4.39	1.09	−3.31	−0.63	−1.88	−1.25
48.50	4.39	1.08	−3.31	−0.63	−1.88	−1.25
49.00	4.39	1.07	−3.32	−0.63	−1.87	−1.25
49.50	4.38	1.06	−3.32	−0.62	−1.87	−1.24
50.00	4.38	1.05	−3.33	−0.62	−1.86	−1.24

TABLE 5.37

CARDINAL POINTS OF ASYMMETRICAL TWO-CYLINDER LENS OF RADII a AND b AND POTENTIALS V_2 AND V_1, RESPECTIVELY[a]

$a/b = 4.5$

V_2/V_1	f_i/b	z_{m_i}/b	z_{p_i}/b	f_o/b	z_{m_o}/b	z_{p_o}/b
1.50	158.36	143.57	−14.79	−129.17	−142.47	13.30
2.00	56.94	48.27	−8.67	−40.22	−47.43	−7.21
2.50	33.87	27.26	−6.62	−21.41	−26.58	−5.17
3.00	24.38	18.78	−5.59	−14.07	−18.22	−4.15
3.50	19.33	14.34	−4.99	−10.33	−13.88	−3.55
4.00	16.23	11.65	−4.58	−8.11	−11.26	−3.15
4.50	14.15	9.85	−4.30	−6.67	−9.53	−2.86
5.00	12.66	8.56	−4.09	−5.66	−8.30	−2.64
5.50	11.53	7.60	−3.93	−4.92	−7.39	−2.48
6.00	10.66	6.86	−3.81	−4.35	−6.70	−2.35
6.50	9.97	6.26	−3.71	−3.91	−6.15	−2.24
7.00	9.40	5.77	−3.63	−3.55	−5.70	−2.15
7.50	8.92	5.37	−3.56	−3.26	−5.33	−2.07
8.00	8.52	5.02	−3.50	−3.01	−5.02	−2.01

[a] Here, $a > b$.

TABLE 5.37 (continued)

$a/b = 4.5$

V_2/V_1	f_i/b	z_{m_i}/b	z_{p_i}/b	f_o/b	z_{m_o}/b	z_{p_o}/b
8.50	8.18	4.73	−3.45	−2.81	−4.76	−1.95
9.00	7.88	4.47	−3.41	−2.63	−4.53	−1.90
9.50	7.62	4.24	−3.38	−2.47	−4.33	−1.86
10.00	7.40	4.04	−3.35	−2.34	−4.16	−1.82
10.50	7.19	3.87	−3.33	−2.22	−4.01	−1.79
11.00	7.01	3.70	−3.31	−2.11	−3.87	−1.76
11.50	6.85	3.56	−3.29	−2.02	−3.75	−1.73
12.00	6.70	3.43	−3.27	−1.93	−3.64	−1.70
12.50	6.57	3.31	−3.26	−1.86	−3.54	−1.68
13.00	6.44	3.20	−3.25	−1.79	−3.45	−1.66
13.50	6.33	3.10	−3.24	−1.72	−3.36	−1.64
14.00	6.23	3.00	−3.23	−1.67	−3.29	−1.62
14.50	6.14	2.91	−3.22	−1.61	−3.22	−1.61
15.00	6.05	2.83	−3.21	−1.56	−3.15	−1.59
15.50	5.97	2.76	−3.21	−1.52	−3.10	−1.58
16.00	5.89	2.69	−3.21	−1.47	−3.04	−1.57
16.50	5.82	2.62	−3.20	−1.43	−2.99	−1.55
17.00	5.76	2.56	−3.20	−1.40	−2.94	−1.54
17.50	5.70	2.50	−3.20	−1.36	−2.89	−1.53
18.00	5.64	2.44	−3.20	−1.33	−2.85	−1.52
18.50	5.58	2.39	−3.20	−1.30	−2.81	−1.51
19.00	5.53	2.34	−3.20	−1.27	−2.78	−1.51
19.50	5.49	2.29	−3.20	−1.24	−2.74	−1.50
20.00	5.44	2.24	−3.20	−1.22	−2.71	−1.49
20.50	5.40	2.20	−3.20	−1.19	−2.68	−1.48
21.00	5.36	2.16	−3.20	−1.17	−2.65	−1.48
21.50	5.32	2.12	−3.20	−1.15	−2.62	−1.47
22.00	5.29	2.08	−3.20	−1.13	−2.59	−1.46
22.50	5.25	2.05	−3.21	−1.11	−2.56	−1.46
23.00	5.22	2.01	−3.21	−1.09	−2.54	−1.45
23.50	5.19	1.98	−3.21	−1.07	−2.52	−1.45
24.00	5.16	1.95	−3.21	−1.05	−2.49	−1.44
24.50	5.13	1.91	−3.22	−1.04	−2.47	−1.44
25.00	5.11	1.88	−3.22	−1.02	−2.45	−1.43
25.50	5.08	1.86	−3.22	−1.01	−2.43	−1.43
26.00	5.06	1.83	−3.23	−0.99	−2.42	−1.42
26.50	5.03	1.80	−3.23	−0.98	−2.40	−1.42
27.00	5.01	1.78	−3.24	−0.96	−2.38	−1.42
27.50	4.99	1.75	−3.24	−0.95	−2.36	−1.41
28.00	4.97	1.73	−3.24	−0.94	−2.35	−1.41
28.50	4.95	1.70	−3.25	−0.93	−2.33	−1.41
29.00	4.93	1.68	−3.25	−0.92	−2.32	−1.40
29.50	4.92	1.66	−3.26	−0.91	−2.30	−1.40

TABLE 5.37 (continued)

$a/b = 4.5$

V_2/V_1	f_i/b	z_{m_i}/b	z_{p_i}/b	f_o/b	z_{m_o}/b	z_{p_o}/b
30.00	4.90	1.64	−3.26	−0.89	−2.29	−1.40
30.50	4.88	1.62	−3.27	−0.88	−2.28	−1.39
31.00	4.87	1.59	−3.27	−0.87	−2.26	−1.39
31.50	4.85	1.58	−3.28	−0.86	−2.25	−1.39
32.00	4.84	1.56	−3.28	−0.86	−2.24	−1.39
32.50	4.82	1.54	−3.29	−0.85	−2.23	−1.38
33.00	4.81	1.52	−3.29	−0.84	−2.22	−1.38
33.50	4.80	1.50	−3.30	−0.83	−2.21	−1.38
34.00	4.78	1.48	−3.30	−0.82	−2.20	−1.38
34.50	4.77	1.47	−3.31	−0.81	−2.19	−1.37
35.00	4.76	1.45	−3.31	−0.81	−2.18	−1.37
35.50	4.75	1.43	−3.32	−0.80	−2.17	−1.37
36.00	4.74	1.42	−3.32	−0.79	−2.16	−1.37
36.50	4.73	1.40	−3.33	−0.78	−2.15	−1.37
37.00	4.72	1.39	−3.33	−0.78	−2.14	−1.36
37.50	4.71	1.37	−3.34	−0.77	−2.13	−1.36
38.00	4.70	1.36	−3.34	−0.76	−2.12	−1.36
38.50	4.69	1.35	−3.35	−0.76	−2.12	−1.36
39.00	4.69	1.33	−3.35	−0.75	−2.11	−1.36
39.50	4.68	1.32	−3.36	−0.74	−2.10	−1.36
40.00	4.67	1.31	−3.36	−0.74	−2.09	−1.36
40.50	4.66	1.29	−3.37	−0.73	−2.09	−1.35
41.00	4.65	1.28	−3.37	−0.73	−2.08	−1.35
41.50	4.65	1.27	−3.38	−0.72	−2.07	−1.35
42.00	4.64	1.26	−3.38	−0.72	−2.07	−1.35
42.50	4.63	1.24	−3.39	−0.71	−2.06	−1.35
43.00	4.63	1.23	−3.40	−0.71	−2.05	−1.35
43.50	4.62	1.22	3.40	−0.70	−2.05	−1.35
44.00	4.62	1.21	−3.41	−0.70	−2.04	−1.34
44.50	4.61	1.20	−3.41	−0.69	−2.04	−1.34
45.00	4.61	1.19	−3.42	0.69	−2.03	−1.34
45.50	4.60	1.18	−3.42	−0.68	−2.02	−1.34
46.00	4.60	1.17	−3.43	−0.68	−2.02	−1.34
46.50	4.59	1.16	−3.43	−0.67	−2.01	−1.34
47.00	4.59	1.15	−3.44	−0.67	−2.01	−1.34
47.50	4.58	1.14	−3.44	−0.66	−2.00	−1.34
48.00	4.58	1.13	−3.45	−0.66	−2.00	−1.34
48.50	4.57	1.12	−3.46	−0.66	−1.99	−1.34
49.00	4.57	1.11	−3.46	−0.65	−1.99	−1.34
49.50	4.57	1.10	−3.47	−0.65	−1.98	−1.33
50.00	4.56	1.09	−3.47	−0.65	−1.98	−1.33

TABLE 5.38

Cardinal Points of Asymmetrical Two-Cylinder Lens of Radii a and b and Potentials V_2 and V_1, Respectively[a]

$a/b = 5$

V_2/V_1	f_i/b	z_{m_i}/b	z_{p_i}/b	f_o/b	z_{m_o}/b	z_{p_o}/b
1.50	170.63	154.61	−16.02	−139.18	−153.58	−14.41
2.00	61.27	51.86	−9.41	−43.29	−51.12	−7.83
2.50	36.37	29.19	−7.18	−22.99	−28.61	−5.63
3.00	26.11	20.05	−6.06	−15.07	−19.59	−4.52
3.50	20.65	15.26	−5.40	−11.03	−14.90	−3.86
4.00	17.30	12.35	−4.95	−8.65	−12.07	−3.42
4.50	15.04	10.41	−4.64	−7.09	−10.20	−3.11
5.00	13.43	9.02	−4.41	−6.00	−8.88	−2.87
5.50	12.21	7.99	−4.23	−5.21	−7.90	−2.69
6.00	11.27	7.18	−4.08	−4.60	−7.14	−2.54
6.50	10.51	6.54	−3.97	−4.12	−6.55	−2.43
7.00	9.89	6.02	−3.87	−3.74	−6.06	−2.33
7.50	9.38	5.58	−3.80	−3.42	−5.66	−2.24
8.00	8.94	5.21	−3.73	−3.16	−5.33	−2.17
8.50	8.57	4.89	−3.67	−2.94	−5.05	−2.11
9.00	8.24	4.62	−3.63	−2.75	−4.80	−2.05
9.50	7.96	4.38	−3.59	−2.58	−4.59	−2.01
10.00	7.71	4.16	−3.55	−2.44	−4.40	−1.96
10.50	7.49	3.97	−3.52	−2.31	−4.24	−1.92
11.00	7.29	3.80	−3.49	−2.20	−4.09	−1.89
11.50	7.12	3.65	−3.47	−2.10	−3.96	−1.86
12.00	6.96	3.51	−3.45	−2.01	−3.84	−1.83
12.50	6.81	3.38	−3.43	−1.93	−3.73	−1.81
13.00	6.68	3.26	−3.41	−1.85	−3.64	−1.78
13.50	6.56	3.16	−3.40	−1.78	−3.55	−1.76
14.00	6.44	3.06	−3.39	−1.72	−3.46	−1.74
14.50	6.34	2.96	−3.38	−1.67	−3.39	−1.72
15.00	6.25	2.88	−3.37	−1.61	−3.32	−1.71
15.50	6.16	2.80	−3.36	−1.56	−3.26	−1.69
16.00	6.08	2.72	−3.35	−1.52	−3.20	−1.68
16.50	6.00	2.65	−3.35	−1.48	−3.14	−1.66
17.00	5.93	2.59	−3.34	−1.44	−3.09	−1.65
17.50	5.86	2.52	−3.34	−1.40	−3.04	−1.64
18.00	5.80	2.46	−3.33	−1.37	−2.99	−1.63
18.50	5.74	2.41	−3.33	−1.33	−2.95	−1.62
19.00	5.68	2.36	−3.33	−1.30	−2.91	−1.61
19.50	5.63	2.30	−3.33	−1.28	−2.87	−1.60
20.00	5.58	2.26	−3.33	−1.25	−2.84	−1.59
20.50	5.54	2.21	−3.32	−1.22	−2.80	−1.58

[a] Here, $a > b$.

TABLE 5.38 (continued)

$a/b = 5$

V_2/V_1	f_i/b	z_{m_i}/b	z_{p_i}/b	f_o/b	z_{m_o}/b	z_{p_o}/b
21.00	5.49	2.17	−3.32	−1.20	−2.77	−1.57
21.50	5.45	2.13	−3.32	−1.18	−2.74	−1.57
22.00	5.41	2.09	−3.32	−1.15	−2.71	−1.56
22.50	5.37	2.05	−3.32	−1.13	−2.69	−1.55
23.00	5.34	2.01	−3.33	−1.11	−2.66	−1.55
23.50	5.30	1.98	−3.33	−1.09	−2.63	−1.54
24.00	5.27	1.94	−3.33	−1.08	−2.61	−1.53
24.50	5.24	1.91	−3.33	−1.06	−2.59	−1.53
25.00	5.21	1.88	−3.33	−1.04	−2.57	−1.52
25.50	5.18	1.85	−3.33	−1.03	−2.55	−1.52
26.00	5.16	1.82	−3.34	−1.01	−2.53	−1.51
26.50	5.13	1.79	−3.34	−1.00	−2.51	−1.51
27.00	5.11	1.77	−3.34	−0.98	−2.49	−1.51
27.50	5.08	1.74	−3.34	−0.97	−2.47	−1.50
28.00	5.06	1.72	−3.35	−0.96	−2.45	−1.50
28.50	5.04	1.69	−3.35	−0.94	−2.44	−1.49
29.00	5.02	1.67	−3.35	−0.93	−2.42	−1.49
29.50	5.00	1.65	−3.36	−0.92	−2.41	−1.49
30.00	4.98	1.62	−3.36	−0.91	−2.39	−1.48
30.50	4.96	1.60	−3.36	−0.90	−2.38	−1.48
31.00	4.95	1.58	−3.37	−0.89	−2.36	−1.48
31.50	4.93	1.56	−3.37	−0.88	−2.35	−1.47
32.00	4.91	1.54	−3.37	−0.87	−2.34	−1.47
32.50	4.90	1.52	−3.38	−0.86	−2.33	−1.47
33.00	4.88	1.50	−3.38	−0.85	−2.31	−1.46
33.50	4.87	1.48	−3.38	−0.84	−2.30	−1.46
34.00	4.86	1.47	−3.39	−0.83	−2.29	−1.46
34.50	4.84	1.45	−3.39	−0.82	−2.28	−1.46
35.00	4.83	1.43	−3.40	−0.82	−2.27	−1.45
35.50	4.82	1.42	−3.40	−0.81	−2.26	−1.45
36.00	4.80	1.40	−3.41	−0.80	−2.25	−1.45
36.50	4.79	1.38	−3.41	−0.79	−2.24	−1.45
37.00	4.78	1.37	−3.41	−0.79	−2.23	−1.45
37.50	4.77	1.35	−3.42	−0.78	−2.22	−1.44
38.00	4.76	1.34	−3.42	−0.77	−2.21	−1.44
38.50	4.75	1.32	−3.43	−0.77	−2.21	−1.44
39.00	4.74	1.31	−3.43	−0.76	−2.20	−1.44
39.50	4.73	1.30	−3.44	−0.75	−2.19	−1.44
40.00	4.72	1.28	−3.44	−0.75	−2.18	−1.43
40.50	4.72	1.27	−3.45	−0.74	−2.17	−1.43
41.00	4.71	1.26	−3.45	−0.74	−2.17	−1.43
41.50	4.70	1.24	−3.45	−0.73	−2.16	−1.43
42.00	4.69	1.23	−3.46	−0.72	−2.15	−1.43

TABLE 5.38 (continued)

$a/b = 5$

V_2/V_1	f_i/b	z_{m_i}/b	z_{p_i}/b	f_o/b	z_{m_o}/b	z_{p_o}/b
42.50	4.68	1.22	−3.46	−0.72	−2.15	−1.43
43.00	4.68	1.21	−3.47	−0.71	−2.14	−1.43
43.50	4.67	1.20	−3.47	−0.71	−2.13	−1.42
44.00	4.66	1.18	−3.48	−0.70	−2.13	−1.42
44.50	4.66	1.17	−3.48	−0.70	−2.12	−1.42
45.00	4.65	1.16	−3.49	−0.69	−2.11	−1.42
45.50	4.64	1.15	−3.49	−0.69	−2.11	−1.42
46.00	4.64	1.14	−3.50	−0.68	−2.10	−1.42
46.50	4.63	1.13	−3.50	−0.68	−2.10	−1.42
47.00	4.63	1.12	−3.51	−0.67	−2.09	−1.41
47.50	4.62	1.11	−3.51	−0.67	−2.08	−1.41
48.00	4.62	1.10	−3.52	−0.67	−2.08	−1.41
48.50	4.61	1.09	−3.52	−0.66	−2.07	−1.41
49.00	4.61	1.08	−3.53	−0.66	−2.07	−1.41
49.50	4.60	1.07	−3.53	−0.65	−2.06	−1.41
50.00	4.60	1.06	−3.54	−0.65	−2.06	−1.41

VI

The Electrostatic Unipotential Lens

In Chapter IV we mentioned that the unipotential lens is characterized by the fact that the potentials on both sides of the lens are the same. This lens is of considerable interest in electron optics. Numerous geometries are available as shown in Fig. 4.7.

Two types of lens that are very frequently used are those consisting of three apertures, and those consisting of three coaxial tubes separated by certain distances. Normally, the separations between the electrodes are the same. In that case, we refer to the lens as a symmetrical unipotential lens. In some cases, however, the separations are not the same for reasons which we will explain in a subsequent chapter. The middle electrode is generally tied to the cathode. In that case the ratio of the potential of the outer electrodes to that of the inner one is infinite. Therefore, the refractive power of that lens is independent of the voltage of the outer electrodes. The cardinal points are then determined only by the geometry of the lens. If the position of the focus is critical, the dimensions of the lens have to be maintained very precisely. The advantage in this case is the fact that voltage regulation is unnecessary. If it should be desirable to control the focus with the potential of the lens, the central electrode should not be connected to the cathode. It may be at a lower or higher potential than that of the outer electrodes.

In what follows we shall analyze the lens shown in Fig. 6.1. Later on, we shall consider as a limiting case the condition in which the thickness of the central electrode is negligibly small. This latter is generally known as the three-aperture unipotential lens. We shall start by assuming that $V_2 > V_1$. We shall consider the case in which $V_1 > V_2$ later.

In the literature, the unipotential lens is sometimes referred to as the saddle-field lens or *einzel* lens, from the German word meaning single.

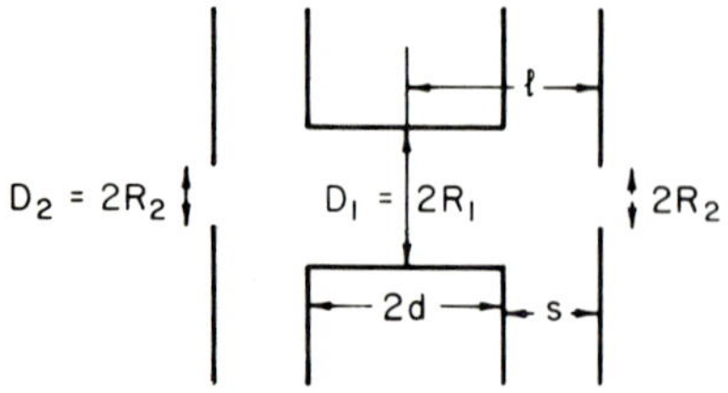

Fig. 6.1. The symmetrical unipotential lens with thick central electrode.

6.1 Axial Potential Distribution

From Chapter III, Eq. (3.158), we know that the axial potential distribution of an electrode with a circular aperture separating two uniform fields E_1 and E_2 is given by

$$\phi(z) = V - \frac{E_1 + E_2}{2} z + \frac{E_1 - E_2}{\pi} \left(z \arctan \frac{z}{R} + R \right) \tag{6.1}$$

Here

$$E_1 = -\left.\frac{\partial \phi}{\partial z}\right|_{z=-\infty} \qquad \text{and} \qquad E_2 = -\left.\frac{\partial \phi}{\partial z}\right|_{z=+\infty}$$

We shall assume that in Fig. 6.1 the lens consists of four apertures. Taking $z = 0$ at the geometrical center of the lens, the four apertures are located at $z = -l$, $z = -d$, $z = d$, and $z = l$. The following conditions are then valid at the various apertures:

$$\begin{array}{lll} z = -l & E_1 = 0 & E_2 = E \\ z = -d & E_1 = E & E_2 = 0 \\ z = d & E_1 = 0 & E_2 = -E \\ z = l & E_1 = -E & E_2 = 0 \end{array} \tag{6.2}$$

The radii of the apertures as shown in Fig. 6.1 are R_2 at $z = \pm l$ and R_1 at $z = \pm d$.

We shall now assume that the separation is sufficiently large, i.e., $l/R_2 \geq 3$ and $l/R_1 \geq 3$, to allow us to superimpose all the individual axial potential distributions in order to determine the distribution of the complete lens. We then have

$$\phi(z) = K - \frac{E}{2}(z + l) - \frac{E}{\pi}\left[(z + l) \arctan \frac{z + l}{R_2} + R_2\right]$$
$$- \frac{E}{2}(z + d) + \frac{E}{\pi}\left[(z + d) \arctan \frac{z + d}{R_1} + R_1\right]$$

$$+\frac{E}{2}(z-d)+\frac{E}{\pi}\left[(z-d)\arctan\frac{z-d}{R_1}+R_1\right]$$

$$+\frac{E}{2}(z-l)-\frac{E}{\pi}\left[(z-l)\arctan\frac{z-l}{R_2}+R_2\right] \tag{6.3}$$

The field E between the central and outer electrodes is,

$$E=\frac{V_2-V_1}{l-d} \tag{6.4}$$

At $z=\infty$, the potential should be V_2. Using this boundary condition and Eq. (6.4), we finally obtain

$$\phi(z)=V_2-\frac{V_2-V_1}{\pi(l-d)}\left[(z+l)\arctan\frac{z+l}{R_2}+(z-l)\arctan\frac{z-l}{R_2}\right.$$

$$\left.+2R_2-2R_1-(z+d)\arctan\frac{z+d}{R_1}-(z-d)\arctan\frac{z-d}{R_1}\right] \tag{6.5}$$

Equation (6.5) gives the axial potential distribution of a unipotential symmetrical lens for the case in which the ratio of the separation between the electrodes to the radii of the electrodes is larger than 3, for example.

In what follows, we shall see that the potential at the center of the lens and at the center of the outer electrodes play a very important part in the determination of the trajectories of the electron beam inside the lens.

6.2 Potentials at the Centers of the Inner and Outer Electrodes

The potential at the center of the lens is given by

$$\phi(0)=V_2-\frac{V_2-V_1}{\pi(l-d)}\left(2l\arctan\frac{l}{R_2}+2R_2-2R_1-2d\arctan\frac{d}{R_1}\right) \tag{6.6}$$

Since we assumed $l/R_2\geq 3$, we may expand, using the well-known series

$$\arctan\frac{l}{R_2}=\frac{\pi}{2}-\frac{R_2}{l}+\frac{{R_2}^3}{3l^3}-\cdots \tag{6.7}$$

and we obtain

$$\phi(0)=V_2-\frac{V_2-V_1}{\pi(l-d)}\left(\pi l-2R_1-2d\arctan\frac{d}{R_1}\right) \tag{6.8}$$

The potential at the center of the outer electrodes is given by

$$\phi(l) = V_2 - \frac{V_2 - V_1}{\pi(l-d)}\left[2l \arctan \frac{2l}{R_2} + 2R_2 - 2R_1 - (l+d)\arctan \frac{l+d}{R_1} - (l-d)\arctan \frac{l-d}{R_1}\right] \tag{6.9}$$

Again, expanding the term arctan $(2l/R_2)$, we obtain

$$\phi(l) = V_2 - \frac{V_2 - V_1}{\pi(l-d)}\left[\pi l + R_2 - 2R_1 - (l+d)\arctan \frac{l+d}{R_1} - (l-d)\arctan \frac{l-d}{R_1}\right] \tag{6.10}$$

The ratio of the potential at the center of the lens to that at the center of the outer electrodes is

$$u = \frac{\phi(0)}{\phi(l)} = \frac{V_2 - [(V_2 - V_1)/\pi(l-d)][\pi l - 2R_1 - 2d \arctan(d/R_1)]}{V_2 - [(V_2 - V_1)/\pi(l-d)][\pi l + R_2 - 2R_1 - (l+d)\arctan[(l+d)/R_1] - (l-d)\arctan[(l-d)/R_1]} \tag{6.11}$$

6.3 Potential Configuration along the Axis

At z close to the origin, we may expand the terms arctan $[(z+l)/R_2]$ and arctan $[(z+d)/R_1]$ as in Eq. (6.7), and we find that Eq. (6.5) reduces to an equation of a straight line. Next, we take the second derivative of Eq. (6.5) to determine the location of the points of inflection. By definition these occur where $\phi''(z) = 0$. We obtain

$$\frac{{R_2}^3}{[{R_2}^2 + (l+z)^2]^2} + \frac{{R_2}^3}{[{R_2}^2 + (l-z)^2]^2} = \frac{{R_1}^3}{[{R_1}^2 + (z+d)^2]^2} + \frac{{R_1}^3}{[{R_1}^2 + (z-d)^2]^2} \tag{6.12}$$

If we plot the potential distribution along the axis as given by Eq. (6.5) for various values of l, d, R_1, and R_2, we find that the general shape of the curve is as shown in Fig. 6.2. We shall divide this into several parts for the

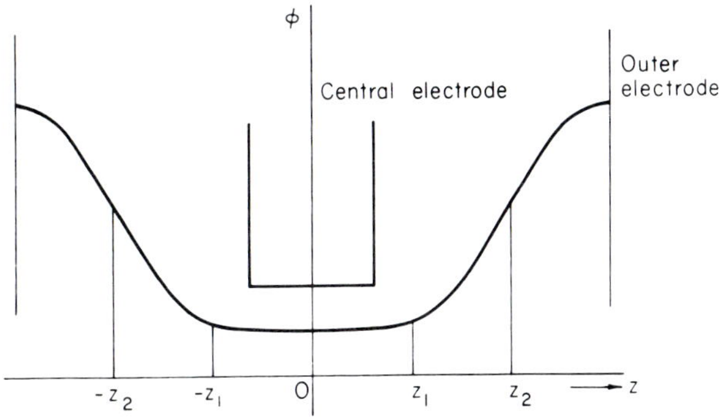

Fig. 6.2. General shape of the potential along the axis.

sake of analysis. First, as shown above, we have a linear region. Next, we have a series of parabolic regions. Let

$$\begin{aligned} \phi_1(z) &= \phi(0) & 0 \le |z| \le z_1 \\ \phi_2(z) &= \phi(0) + p(z \pm z_1)^2 & z_1 \le |z| \le z_2 \\ \phi_3(z) &= \phi(l) + q(z \pm l)^2 & z_2 \le |z| \le l \end{aligned} \tag{6.13}$$

In other words, we first have a linear region, next a parabola, until we reach the point of inflection z_2. After that, another parabola completes the model. In order to determine the coefficients p and q, we let

$$z_1 = g_1 l \qquad \text{and} \qquad z_2 - g_2 l \tag{6.14}$$

and since

$$\phi_2(g_2 l) = \phi_3(g_2 l), \qquad \phi_2{}'(g_2 l) = \phi_3{}'(g_2 l) \tag{6.15}$$

we obtain

$$p = \frac{1}{(1 - g_1)(g_2 - g_1)} \frac{\phi(l) - \phi(0)}{l^2} \tag{6.16}$$

$$q = \frac{-1}{(1 - g_1)(1 - g_2)} \frac{\phi(l) - \phi(0)}{l^2} \tag{6.17}$$

This oversimplification of the problem might indicate that the end results will be grossly inaccurate. It is true that the second derivative does not follow the shape that was assumed in our analysis. In reality, it looks more like curve 2 of Fig. 6.3. However, the potential distribution we are assuming depicts fairly well the situation at the vicinity of the center of the lens. As indicated in Chapter IV, the outer parts of the lens do not influence the course of the trajectory very much. In the central region the electrons are slowed down and are therefore strongly influenced by the field there. At this stage we therefore feel justified in pursuing the analysis with the above simplification. Later on, to verify our analysis, we shall compare our results with those obtained experimentally.

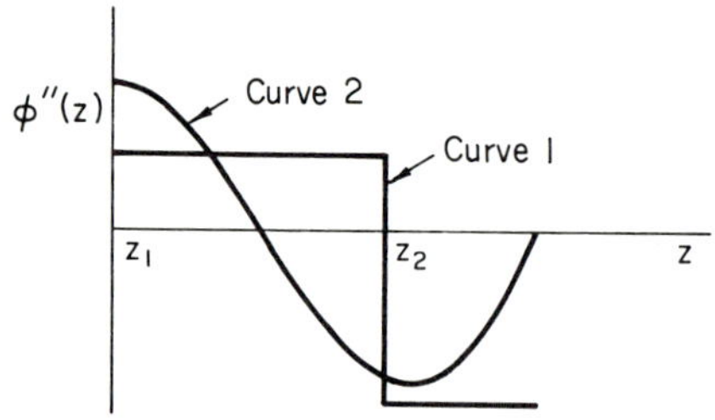

Fig. 6.3. The second derivative of the potential along the axis of a symmetrical unipotential lens. Curve 1 for the assumed parabolic distribution; curve 2 as obtained in an actual lens.

6.4 Ray Equations in the Regions of Interest

We shall consider three regions, namely: (a) the entrance region, (b) the central region, and (c) the exit region. In each region, the analysis will be divided as necessary to determine the ray equations.

(a) Entrance Region

From Chapter II, we know that the paraxial-ray equation is given by

$$4\phi(z)r'' + 2\phi'(z)r' + \phi''(z)r = 0 \tag{6.18}$$

If we now substitute for $\phi(z)$ and its derivatives, using Eq. (6.13), we obtain

$$4[\phi(l) + q(z+l)^2]r'' + 2[2q(z+l)]r' + 2qr = 0 \tag{6.19}$$

Let

$$\beta^2 = \frac{1}{\phi(l)}\frac{1}{(1-g_1)(1-g_2)}\frac{\phi(l)-\phi(0)}{l^2} \tag{6.20}$$

Since $\phi(l)$ is always greater than $\phi(0)$, we are allowed to write Eq. (6.20). Hence,

$$\beta = l^{-1}\{[(1-g_1)(1-g_2)]^{-1}(1-u)\}^{1/2} \tag{6.21}$$

and Eq. (6.19) becomes

$$[1 - \beta^2(z + l)^2]r'' - \beta^2(z + l)r' - (\beta^2/2)r = 0 \tag{6.22}$$

For simplicity, we shall make the following change in variable. We let $\zeta = z + l$, and

$$\beta\zeta = \sin \beta\mu \tag{6.23}$$

First, we must determine whether or not this substitution is valid. We have

$$\beta\zeta = \left[\frac{(z + z_2)^2(1 - u)}{l^2(1 - g_1)(1 - g_2)}\right]^{1/2} \tag{6.24}$$

The value of z in this region varies between $-l$ and $-z_2$. We shall see that the selection of g_1 and g_2 will always make Eq. (6.24) less than 1. We are therefore allowed to proceed with the substitution. We have

$$\frac{dr}{d\zeta} = \frac{dr}{d\mu}\frac{1}{\cos \beta\mu} \tag{6.25}$$

$$\frac{d^2r}{d\zeta^2} = \frac{d^2r}{d\mu^2}\frac{1}{\cos^2 \beta\mu} + \frac{dr}{d\mu}\frac{\beta \sin \beta\mu}{\cos^3 \beta\mu} \tag{6.26}$$

and Eq. (6.22) becomes

$$(-\sin^2 \beta\mu + 1)\left[\frac{d^2r}{d\mu^2}\frac{1}{\cos^2 \beta\mu} + \frac{dr}{d\mu}\frac{\beta \sin \beta\mu}{\cos^3 \beta\mu}\right] - \frac{\beta \sin \beta\mu}{\cos \beta\mu}\frac{dr}{d\mu} - \frac{\beta^2}{2}r = 0 \tag{6.27}$$

or

$$(d^2r/d\mu^2) - (\beta^2/2)r = 0 \tag{6.28}$$

The general solution of this differential equation is

$$r = C_1 \sinh\left(\frac{\beta}{\sqrt{2}}\mu\right) + C_2 \cosh\left(\frac{\beta}{\sqrt{2}}\mu\right) \tag{6.29}$$

Equation (6.29) can also be written as

$$r = D \sinh\left(\frac{\beta}{\sqrt{2}}\mu + \theta\right) \quad \text{or} \quad r = D \cosh\left(\frac{\beta}{\sqrt{2}}\mu + \theta\right) \tag{6.30}$$

where D and θ are arbitrary constants.

Is we consider incoming parallel beams at height r_o, we obtain

$$\mathrm{r}_1 = \mathrm{r}_o \cosh\left[\frac{1}{\sqrt{2}} \arcsin \beta(\mathrm{z}+l)\right] \tag{6.31}$$

where we have substituted for μ its value given by Eq. (6.23).

(b) Central Region

This region has three parts, two for $\phi_2(z)$, and one for $\phi_1(z)$ as given by Eq. (6.13). In the region $-z_2 < z < -z_1$, we have

$$\phi_2(z) = \phi(0) + p(z + z_1)^2 \tag{6.32}$$

If we let

$$\alpha = \frac{1}{l}\left\{[(1 - g_1)(g_2 - g_1)]^{-1}\frac{1 - u}{u}\right\}^{1/2} \tag{6.33}$$

we obtain

$$[1 + \alpha^2(z + z_1)^2]r'' + \alpha^2(z + z_1)r' + (\alpha^2/2)r = 0 \tag{6.34}$$

As previously, we make the substitution $\zeta = z + z_1$, and

$$\alpha\zeta = \alpha(z + z_1) = \sinh \alpha\mu \tag{6.35}$$

which is always permissible, and finally we obtain

$$(\sinh^2 \alpha\mu + 1)\left(\frac{d^2r}{d\mu^2}\frac{1}{\cosh^2 \alpha\mu} - \frac{dr}{d\mu}\frac{\alpha \sin \alpha\mu}{\cosh^3 \alpha\mu}\right) + \alpha\frac{\sinh \alpha\mu}{\cosh \alpha\mu} - \frac{dr}{d\mu} + \frac{\alpha^2}{2}r = 0 \tag{6.36}$$

or

$$(d^2r/d\mu^2) + (\alpha^2/2)r = 0 \tag{6.37}$$

or

$$r_2 = A \cos [(1/\sqrt{2}) \sinh^{-1} \alpha(z + z_1) - \varphi] \tag{6.38}$$

where A and φ are constants.

The second region is in the interval $-z_1 < z < z_1$. Here, $\phi_1(z) = \phi(0)$. The height of the ray and its slope at $z = -z_1$ are as in Eq. (6.38). We obtain

$$r_3 = (A\alpha/\sqrt{2})(z + z_1) + A \cos \phi \tag{6.39}$$

The third part is obtained exactly as before. We have

$$r_4 = A_1 \cos[(1/\sqrt{2}) \operatorname{arcsinh} \alpha(z - z_1) - \varphi_1] \tag{6.40}$$

(c) Exit Region

Here again, we have the same situation as in the entrance region. We can immediately write the ray equation

$$r_5 = B \cosh[(1/\sqrt{2}) \arcsin \beta(z - l) - \psi] \quad \text{or}$$

$$r_5 = B \sinh[(1/\sqrt{2}) \arcsin \beta(z - l) - \psi] \tag{6.41}$$

6.5 Evaluation of the Constants

To determine the value of the constants, we use the following boundary conditions:

$$\begin{aligned} r_1(-g_2 l) &= r_2(-g_2 l), & r_1'(-g_2 l) &= r_2'(-g_2 l) \\ r_2(-g_1 l) &= r_3(-g_1 l), & r_2'(-g_1 l) &= r_3'(-g_1 l) \\ r_3(g_1 l) &= r_4(g_1 l), & r_3'(g_1 l) &= r_4'(g_1 l) \\ r_4(g_2 l) &= r_5(g_2 l), & r_4'(g_2 l) &= r_5'(g_2 l) \end{aligned} \tag{6.42}$$

We define the following constants:

$$\lambda = [(g_2 - g_1)/(1 - g_2)]^{1/2} \tag{6.43}$$

$$\alpha_0 = (1/\sqrt{2}) \operatorname{arcsinh}[\alpha(g_2 - g_1)l] \tag{6.44}$$

$$\beta_0 = (1/\sqrt{2}) \arcsin[\beta(1 - g_2)l] \tag{6.45}$$

and after some simple computations, we obtain

$$A^2 = r_o^2(\cosh^2 \beta_0 + \lambda^2 \sinh^2 \beta_0) = r_o^2[1 + (1 + \lambda^2)\sinh^2 \beta_0]^{1/2} \tag{6.46}$$

$$\varphi = \arctan(\lambda \tanh \beta_0) - \alpha_0 \tag{6.47}$$

$$A_1^2 = A^2(1 + \sqrt{2}\alpha g_1 l \sin 2\varphi + 2\alpha^2 g_1^2 l^2 \sin^2 \varphi) \tag{6.48}$$

$$\varphi_1 = \operatorname{arccot}(\sqrt{2}\alpha g_1 l + \cot \varphi) \tag{6.49}$$

Here, A is always positive and φ is always negative for $u < 1$.

For the values of B and ψ, we can select either the cosh part or the sinh part of Eq. (6.41). However, we find that if $K = (1/\lambda)\tan(\alpha_0 - \varphi_1) < 1$, we must use the cosh term; and if $K > 1$, we must use the sinh term.

If $(1/\lambda)\tan(\alpha_0 - \varphi_1) < 1$,

$$\begin{aligned} B^2 &= A_1^2\left[\cos^2(\alpha_0 - \varphi_1) - \frac{1}{\lambda^2}\sin^2(\alpha_0 - \varphi_1)\right] \\ &= A_1^2\left[1 - \frac{\lambda^2 + 1}{\lambda^2}\sin^2(\alpha_0 - \varphi_1)\right] \end{aligned} \tag{6.50}$$

$$\psi = \operatorname{arctanh}\left[\frac{1}{\lambda}\tan(\alpha_0 - \varphi_1)\right] - \beta_0 \tag{6.51}$$

If $(1/\lambda)\tan(\alpha_0 - \varphi_1) > 1$, the sinh term should be used, and here

$$B^2 = A_1{}^2\left[\frac{1}{\lambda^2}\sin^2(\alpha_0 - \varphi_1) - \cos^2(\alpha_0 - \varphi_1)\right]$$

$$= A_1{}^2\left[\frac{\lambda^2 + 1}{\lambda^2}\sin^2(\alpha_0 - \varphi_1) - 1\right] \tag{6.52}$$

$$\psi = \operatorname{arctanh}[\lambda\cot(\alpha_0 - \varphi_1)] - \beta_0 \tag{6.53}$$

Here, B and ψ can be positive or negative. In the case of B, it depends on the value of $[(\lambda^2 + 1)/\lambda^2]\sin^2(\alpha_0 - \varphi_1) - 1$.

6.6 Focal Length

The focal length is determined by

$$f = -\frac{r_o}{(r')_{z=l}} \tag{6.54}$$

This is the projective focal length. It is equal to the objective focal length if the crossing with the axis takes place outside the lens. If $(1/\lambda)\tan(\alpha_0 - \varphi_1) < 1$, we obtain quite readily

$$f = \frac{r_o\sqrt{2}}{\beta B\sinh\psi} \tag{6.55}$$

Substituting for all the values in the expression for ψ, we obtain

$$\sinh\psi = \frac{\lambda^{-1}\sin(\alpha_0 - \varphi_1)\cosh\beta_0 - \cos(\alpha_0 - \varphi_1)\sinh\beta_0}{[\cos^2(\alpha_0 - \varphi_1) - \lambda^{-2}\sin^2(\alpha_0 - \varphi_1)]^{1/2}} \tag{6.56}$$

Then

$$B\sinh\psi = A_1[\lambda^{-1}\sin(\alpha_0 - \varphi_1)\cosh\beta_0 - \cos(\alpha_0 - \varphi_1)\sinh\beta_0] \tag{6.57}$$

$$B\sinh\psi = A_1[\cos\varphi_1(\lambda^{-1}\sin\alpha_0\cosh\beta_0 - \cos\alpha_0\sinh\beta_0) - \sin\varphi_1(\lambda^{-1}\cos\alpha_0\cosh\beta_0 + \sin\alpha_0\sinh\beta_0)] \tag{6.58}$$

But

$$A_1\sin\varphi_1 = A\sin\varphi \tag{6.59}$$

and

$$A_1\cos\varphi_1 = A(\cos\varphi + \sqrt{2}\,\alpha g_1 l\sin\varphi) \tag{6.60}$$

Then

$$B \sinh \psi = A[(\cos \varphi + \sqrt{2\alpha g_1 l} \sin \varphi)(\lambda^{-1} \sin \alpha_0 \cosh \beta_0 - \cos \alpha_0 \sinh \beta_0) - \sin \varphi(\lambda^{-1} \cos \alpha_0 \cosh \beta_0 + \sin \alpha_0 \sinh \beta_0)] \tag{6.61}$$

or

$$\begin{aligned} B \sinh \psi = A\{&\cos \varphi(\lambda^{-1} \sin \alpha_0 \cosh \beta_0 - \cos \alpha_0 \sinh \beta_0) \\ &+ \sin \varphi[\sqrt{2\alpha g_1 l}(\lambda^{-1} \sin \alpha_0 \cosh \beta_0 - \cos \alpha_0 \sinh \beta_0) \\ &- \lambda^{-1} \cos \alpha_0 \cosh \beta_0 - \sin \alpha_0 \sinh \beta_0]\} \end{aligned} \tag{6.62}$$

Now

$$\sin \varphi = \frac{\lambda \tanh \beta_0 \cos \alpha_0 - \sin \alpha_0}{(1 + \lambda^2 \tanh^2 \beta_0)^{1/2}} = \frac{\lambda \cos \alpha_0 \sinh \beta_0 - \sin \alpha_0 \cosh \beta_0}{(\cosh^2 \beta_0 + \lambda^2 \sinh^2 \beta_0)^{1/2}} \tag{6.63}$$

and

$$\cos \varphi = \frac{\cos \alpha_0 + \lambda \tanh \beta_0 \sin \alpha_0}{(1 + \lambda^2 \tanh^2 \beta_0)^{1/2}} = \frac{\cos \alpha_0 \cosh \beta_0 + \lambda \sin \alpha_0 \sinh \beta_0}{(\cosh^2 \beta_0 + \lambda^2 \sinh^2 \beta_0)^{1/2}} \tag{6.64}$$

From which we obtain

$$\begin{aligned} &B \sinh \psi \\ &= r_o\{(\cos \alpha_0 \cosh \beta_0 + \lambda \sin \alpha_0 \sinh \beta_0)(\lambda^{-1} \sin \alpha_0 \cosh \beta_0 - \cos \alpha_0 \sinh \beta_0) \\ &\quad + (\lambda \cos \alpha_0 \sinh \beta_0 - \sin \alpha_0 \cosh \beta_0)[\sqrt{2\alpha g_1 l}(\lambda^{-1} \sin \alpha_0 \cosh \beta_0 \\ &\quad - \cos \alpha_0 \sinh \beta_0) - \lambda^{-1} \cos \alpha_0 \cosh \beta_0 - \sin \alpha_0 \sinh \beta_0]\} \end{aligned} \tag{6.65}$$

or

$$\begin{aligned} B \sinh \psi = r_o\bigg\{&\sin 2\alpha_0 \left[\frac{\cosh^2 \beta_0}{\lambda} - \lambda \sinh^2 \beta_0\right] - \cos 2\alpha_0 \sinh 2\beta_0 \\ &\times \sqrt{2\alpha g_1 l}\left[\frac{1}{2} \sin 2\alpha_0 \sinh 2\beta_0 - \frac{\sin^2 \alpha_0 \cosh^2 \beta_0}{\lambda} \right. \\ &\left. - \lambda \cos^2 \alpha_0 \sinh^2 \beta_0\right]\bigg\} \end{aligned} \tag{6.66}$$

Now

$$\begin{aligned} &\frac{1}{2\lambda}(\lambda \sin 2\alpha_0 \sinh 2\beta_0 - 2 \sin^2 \alpha_0 \cosh^2 \beta_0 - 2\lambda^2 \cos^2 \alpha_0 \sinh^2 \beta_0) \\ &\quad = \frac{\cos^2 \alpha_0 \cosh^2 \beta_0}{2\lambda}(2\lambda \tan \alpha_0 2 \tanh \beta_0 - 2 \tan^2 \alpha_0 - 2\lambda^2 \tanh^2 \beta_0) \\ &\quad = -\frac{\cos^2 \alpha_0 \cosh^2 \beta_0}{\lambda}(\tan \alpha_0 - \lambda \tanh \beta_0)^2 \end{aligned} \tag{6.67}$$

Also

$$\sin 2\alpha_0(\cosh^2\beta_0 - \lambda^2\sinh^2\beta_0) - \lambda\cos 2\alpha_0\sinh 2\beta_0$$
$$= (\cosh^2\beta_0 + \lambda^2\sinh^2\beta_0)\sin\left(2\alpha_0 - \arctan\frac{2\lambda\tanh\beta_0}{1-\lambda^2\tanh^2\beta_0}\right) \quad (6.68)$$

We shall arbitrarily group the terms using Eqs. (6.67) and (6.68) in order to obtain

$$\frac{1}{f} = \frac{1}{f_1} + \frac{1}{f_2} \quad (6.69)$$

We shall see the significance of this grouping later. We obtain

$$\frac{1}{f_1} = \frac{\beta}{\lambda\sqrt{2}}[\cosh^2\beta_0 + \lambda^2\sinh^2\beta_0]\sin\left(2\alpha_0 - \arctan\frac{2\lambda\tanh\beta_0}{1-\lambda^2\tanh^2\beta_0}\right) \quad (6.70)$$

and

$$\frac{1}{f_2} = -\frac{\alpha\beta g_1 l}{\lambda}\cos^2\alpha_0\cosh^2\beta_0(\tan\alpha_0 - \lambda\tanh\beta_0)^2 \quad (6.71)$$

In other words, a lens with a thick central electrode can be considered to be a combination of two lenses with no separation. A similar result is obtained if $\lambda^{-1}\tan(\alpha_0 - \varphi_1) > 1$. We notice that the focal length will oscillate and can be either positive or negative. We shall discuss this later.

Now, Eq. (6.71) can be written as follows:

$$\frac{1}{f_2} = -\frac{g_1(1-u)}{l(1-g_1)(g_2-g_1)u^{1/2}}\cos^2\alpha_0\cosh^2\beta_0(\tan\alpha_0 - \lambda\tanh\beta_0)^2 \quad (6.72)$$

We are dealing with the case in which the potential of the central electrode is lower than that of the outer electrodes. Also, we are considering the system as a lens and not as a mirror. Therefore, u is always positive and, at most, equal to 1; g_2 is always greater than g_1. We can therefore conclude that $1/f_2$ is always negative.

6.7 Midfocal Length

The midfocal length is given by

$$z_m = l - (r/r')_{z=l} \quad (6.73)$$

For $|\lambda^{-1}\tan(\alpha_0 - \varphi_1)| < 1$, we obtain

$$z_m = l + \frac{\sqrt{2}}{\beta}\coth\psi \quad (6.74)$$

Now

$$\coth \psi = \coth[\operatorname{arctanh} \lambda^{-1} \tan(\alpha_0 - \varphi_1) - \beta_0] \tag{6.75}$$

$$\coth \psi = \frac{\lambda \cos(\alpha_0 - \varphi_1)\cosh \beta_0 - \sin(\alpha_0 - \varphi_1)\sinh \beta_0}{\sin(\alpha_0 - \varphi_1)\cosh \beta_0 - \lambda \cos(\alpha_0 - \varphi_1)\sinh \beta_0} \tag{6.76}$$

Also, we know that

$$\sin \varphi_1 = (A/A_1) \sin \varphi \tag{6.77}$$

$$\cos \varphi_1 = (A/A_1)(\cos \varphi + \sqrt{2\alpha g_1 l} \sin \varphi) \tag{6.78}$$

$$\varphi = \arctan \lambda \tanh \beta_0 - \alpha_0 \tag{6.79}$$

Hence,

$$\frac{z_m}{l} = 1 + \frac{1}{\sqrt{2\beta l}} \frac{\sqrt{2\alpha g_1 l} I + J}{\sqrt{2\alpha g_1 l} M + N} \tag{6.80}$$

where

$$I = (\sin^2 \alpha_0 - \lambda^2 \cos^2 \alpha_0) \sinh\left(2\beta_0 - \operatorname{arctanh} \frac{2\lambda \tan \alpha_0}{\lambda^2 + \tan^2 \alpha_0}\right) \tag{6.81}$$

$$J = [4\lambda^2 + (\lambda^2 + 1)^2 \sinh^2 2\beta_0]^{1/2} \sin\left[2\alpha_0 + \arctan\left(\frac{2\lambda}{\lambda^2 - 1} \coth 2\beta_0\right)\right] \tag{6.82}$$

$$M = -\cos^2 \alpha_0 \cosh^2 \beta_0 (\tan \alpha_0 - \lambda \tanh \beta_0)^2 \tag{6.83}$$

and

$$N = \lambda\sqrt{2/\beta f_1} \tag{6.84}$$

Similar results are obtained if $|\lambda^{-1} \tan(\alpha_0 - \varphi_1)| > 1$.

6.8 Principal Plane

The principal plane is given by

$$z_p = z_m - f \tag{6.85}$$

which after some simple manipulations gives

$$\frac{z_p}{l} = 1 + \frac{1}{\sqrt{2\beta l}} \frac{\sqrt{2 l \alpha g_1} I + J - 2\lambda}{\sqrt{2\alpha g_1 l} M + N} \tag{6.86}$$

where I, J, M, and N have the same meanings as before.

So far, we have said nothing about z_1 and z_2. By examining field plots for a range of lenses where $R_2 \leq R_1$ and where $(l/R_2) \geq 3$, it appears that a good approximation is attained if we choose the following values:

$$z_1 = \frac{d}{2} \quad \text{and} \quad z_2 = d + \frac{2}{3}(l - d) \quad \text{or} \quad g_1 = \frac{d}{2l}$$

$$\text{and} \quad g_2 = \frac{2}{3} + \frac{d}{3l} \tag{6.87}$$

6.9 Trajectories in the Three-Aperture Lens

The three-aperture lens as shown in Fig. 6.4 is a special case of Fig. 6.1 where $d = 0$. Here, $z_1 = 0$ and all the relations derived earlier are valid if the linear part of the axial potential distribution is removed. In other words, $g_1 = 0$.

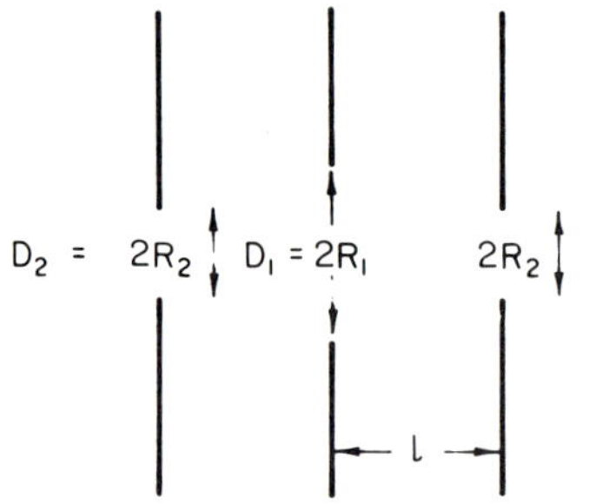

Fig. 6.4. The three-aperture symmetrical unipotential lens. This is a special case of Fig. 6.1 when $d \to 0$.

The trajectory equations can readily be written:

(a) *Entrance region*

$$r_1 = r_o \cosh[(1/\sqrt{2}) \arcsin \beta(z + l)] \tag{6.88}$$

(b) *Central region*

$$r_2 = r_4 = A \cos[(1/\sqrt{2}) \operatorname{arcsinh} \alpha z - \varphi] \tag{6.89}$$

(c) *Exit region*

$$r_5 = B \cosh[(1/\sqrt{2}) \arcsin \beta(z - l) - \psi] \quad \text{or}$$

$$r_5 = B \sinh[(1/\sqrt{2}) \arcsin \beta(z - l) - \psi] \tag{6.90}$$

Also, the values of α, β, φ, and ψ are readily obtained by setting $d = 0$ and $g_1 = 0$, where they were defined. Here, $\lambda = \sqrt{2}$.

$$\beta = \frac{1}{l}\left(\frac{1-u}{1-g_2}\right)^{1/2} = \frac{1}{l}\,[3(1-u)]^{1/2} \tag{6.91}$$

$$\alpha = \frac{1}{l}\left(\frac{1-u}{u}\frac{1}{g_2}\right)^{1/2} = \frac{1}{l}\left(\frac{3}{2}\frac{1-u}{u}\right)^{1/2} \tag{6.92}$$

$$\alpha_0 = \frac{1}{\sqrt{2}}\operatorname{arcsinh}\left(\frac{2}{3}\frac{1-u}{u}\right)^{1/2} \tag{6.93}$$

$$\beta_0 = \frac{1}{\sqrt{2}}\arcsin\left(\frac{1-u}{3}\right)^{1/2} \tag{6.94}$$

$$A^2 = r_o{}^2(\cosh^2\beta_0 + 2\sinh^2\beta_0) = r_o{}^2(1 + 3\sinh^2\beta_0) \tag{6.95}$$

$$\varphi = \arctan(\sqrt{2}\tanh\beta_0) - \alpha_0 \tag{6.96}$$

If $K = (1/\sqrt{2})\tan(\alpha_0 - \varphi) < 1$, the cosh term is used, and

$$B^2 = A^2[1 - (3/2)\sin^2(\alpha_0 - \varphi)] \tag{6.97}$$

$$\psi = \operatorname{arctanh}[(1/\sqrt{2})\tan(\alpha_0 - \varphi)] - \beta_0 \tag{6.98}$$

and if $K > 1$, the sinh term should be used, and we obtain

$$B^2 = A^2[(3/2)\sin^2(\alpha_0 - \varphi) - 1] \tag{6.99}$$

$$\psi = \operatorname{arctanh}[\sqrt{2}\cot(\alpha_0 - \varphi)] - \beta_0 \tag{6.100}$$

The cardinal points are also obtained the same way. The focal length is given by

$$\frac{1}{f} = \beta\left(\frac{3}{2}\cosh^2\beta_0 - 1\right)\sin\left(2\alpha_0 - \arctan\frac{2\sqrt{2}\tanh\beta_0}{1 - 2\tanh^2\beta_0}\right) \tag{6.101}$$

which was obtained from Eq. (6.70). We can readily see from Eq. (6.71) that $1/f_2 = 0$. Similarly, we obtain

$$\frac{z_m}{l} = 1 + \frac{1}{[6(1-u)]^{1/2}}\,\frac{(9\cosh^2 2\beta_0 - 1)^{1/2}\sin[2\alpha_0 + \arctan(2\sqrt{2}\coth 2\beta_0)]}{(3\cosh^2\beta_0 - 2)\sin\left(2\alpha_0 - \arctan\dfrac{2\sqrt{2}\tanh\beta_0}{1 - 2\tanh^2\beta_0}\right)} \tag{6.102}$$

and

$$\frac{z_p}{l} = 1 + \frac{1}{[6(1-u)]^{1/2}}$$
$$\times \frac{(9\cosh^2\beta_0 - 1)^{1/2}\sin(2\alpha_0 + \arctan 2\sqrt{2}\coth 2\beta_0) - 2\sqrt{2}}{(3\cosh^2\beta_0 - 2)\sin\left(2\alpha_0 - \arctan\dfrac{2\sqrt{2}\tanh\beta_0}{1 - 2\tanh^2\beta_0}\right)} \tag{6.103}$$

If a comparison is made of the results obtained here with those for the lens with thick central electrode, we can draw the following conclusions:

(a) The $1/f_2$ term as given by Eq. (6.71) is the additional term due to the thickness of the central electrode. Therefore, if $1/f_1$ is positive, the lens is weaker in the thick electrode case than in the thin electrode case. If $1/f_1$ is negative, the thick electrode will obviously make it stronger. Generally speaking, we consider the first intersection as the most interesting mode of operation. In that case the increase in the width of the central electrode will decrease the converging power of the lens.

(b) It is obvious that the influence of the thickness of the central electrode becomes more noticeable as the thickness increases.

(c) As is expected, the influence of the thickness becomes negligible as $u \to 1$. This is also the case if the central aperture is large. On the other hand, if $u \to 0$, the $1/f_2$ term becomes very important, and it is possible in this case to obtain converging powers much stronger than those with thin central electrodes.

6.10 Course of the Trajectories in the Regions of Interest of the Three-Aperture Lens

(a) Entrance Region

Here, we have

$$r_1 = r_o \cosh\left[\frac{1}{\sqrt{2}} \arcsin \beta(z + l)\right] \tag{6.104}$$

for the region

$$-l < z < -\tfrac{2}{3}l \tag{6.105}$$

We notice that whatever the value of $u = \phi(0)/\phi(l)$ the value of r_1 increases as the beam moves in that region. Its maximum value occurs at $z = -\frac{2}{3}l$, and here

$$r_1 = r_o \cosh\left[\frac{1}{\sqrt{2}} \arcsin\left(\frac{1-u}{3}\right)^{1/2}\right] \tag{6.106}$$

When $u = 1$, there is no lens action, and as expected $r_1 = r_o$. The minimum value of u is zero, and then $r_1 = 1.096\, r_o$. It is clear that the region of incidence affects the trajectory very little.

(b) Central Region

Here, we have

$$r_2 = A \cos\left(\frac{1}{\sqrt{2}} \operatorname{arcsinh} \alpha z - \varphi\right) \tag{6.107}$$

in the region

$$-\tfrac{2}{3}l < z < \tfrac{2}{3}l \tag{6.108}$$

From the boundary condition

$$r_1'(-\tfrac{2}{3}l) = r_2'(-\tfrac{2}{3}l), \tag{6.109}$$

we know that the height of the trajectory will initially increase here, and shortly thereafter it will decrease. Depending on the value of u, it might cross the axis once or several times or even emerge from this region without ever crossing the axis. It crosses the axis either when $r_2 = 0$ or when

$$\frac{1}{\sqrt{2}} \operatorname{arcsinh} \alpha z - \varphi = (2n + 1)\frac{\pi}{2} \tag{6.110}$$

where n is an integer.

Solving for z in Eq. (6.110) and substituting for α and φ, we obtain the value of z where the ray intersects the axis.

$$z = l\left(\frac{2}{3}\frac{u}{1-u}\right)^{1/2} \sinh\left\{\sqrt{2} \arctan\left\{\sqrt{2} \tanh\left[\frac{1}{\sqrt{2}} \arcsin\left(\frac{1-u}{3}\right)^{1/2}\right]\right\}\right.$$

$$\left. -\operatorname{arcsinh}\left(\frac{2}{3}\frac{1-u}{u}\right)^{1/2} + (2n+1)\frac{\pi}{\sqrt{2}}\right\} \tag{6.111}$$

Let us now determine the value of the ratio u for which the trajectory will intersect the axis just at the end of the central region, i.e., at $z = \frac{2}{3}l$. Equation (6.111) can then be written as

$$\operatorname{arcsinh}\left(\frac{2}{3}\frac{1-u}{u}\right)^{1/2} - (2n+1)\frac{\pi}{2\sqrt{2}}$$

$$= \frac{1}{\sqrt{2}} \arctan\left\{\sqrt{2} \tanh\left[\frac{1}{\sqrt{2}} \arcsin\left(\frac{1-u}{3}\right)^{1/2}\right]\right\} \tag{6.112}$$

The roots of this transcendental equation, as found with the help of a digital computer, are

$$n = 0 \qquad u = 1.38 \times 10^{-1}$$
$$n = 1 \qquad u = 1.61 \times 10^{-3}$$
$$n = 2 \qquad u = 1.90 \times 10^{-5}$$

It is clear therefore that for $u > 0.138$, the ray does not intersect the axis in the central region. For values of $u < 0.138$, there can be one or more intersections depending on the value of u.

Next we shall obtain the limit of the value of z at the point of intersection when $u \to 0$. We have from Eq. (6.111)

$$\lim_{u \to o} \left(\frac{z}{l}\right) = \left(\frac{2}{3}\frac{u}{1-u}\right)^{1/2} \sinh\left(0.743 - \operatorname{arcsinh}\left(\frac{2}{3}\frac{1-u}{u}\right)^{1/2} + n\pi\sqrt{2} + 2.222\right) \tag{6.113}$$

or

$$\lim_{u \to o} \left(\frac{z}{l}\right) = \left(\frac{2}{3}\frac{u}{1-u}\right)^{1/2} \left[\left(\frac{2+u}{3u}\right)^{1/2} \sinh(2.965 + n\pi\sqrt{2}) - \left(\frac{2}{3}\frac{1-u}{u}\right)^{1/2} \cosh(2.965 + n\pi\sqrt{2})\right] \tag{6.114}$$

$$\lim_{u \to o} (z/l) = \tfrac{2}{3}(\sinh 2.965 - \cosh 2.965)(\cosh n\pi\sqrt{2} - \sinh n\pi\sqrt{2}) \tag{6.115}$$

For $n = 0$,

$$\lim_{u \to o} (z/l) = -0.035 \tag{6.116}$$

and for $n = 1, 2, 3$, etc.

$$\lim_{u \to o} (z/l) \tag{6.117}$$

is negligible.

For small values of u, Eq. (6.111) can be written as follows:

$$\frac{z}{l} = \left(\frac{2}{9}\frac{2+x}{1+x}\right)^{1/2} \sinh(2.965 + n\pi\sqrt{2}) - \frac{2}{3}\cosh(2.965 + n\pi\sqrt{2})$$
$$\approx \left(\frac{2}{3} + \frac{u}{2}\right) \sinh(2.965 + n\pi\sqrt{2}) - \frac{2}{3}\cosh(2.965 + n\pi\sqrt{2}) \tag{6.118}$$

which simplifies to

$$z/l \cong (u/4)\exp(2.965 + n\pi\sqrt{2}) \tag{6.119}$$

It is therefore clear that for small values of u, the z coordinate of the point of intersection of the ray with the axis in the central region varies linearly with u.

The ray reaches a maximum or a minimum in the central region when

$$(1/\sqrt{2}) \operatorname{arcsinh} \alpha z - \varphi = n\pi \tag{6.120}$$

The absolute value of those maxima or minima is

$$A = r_0\left\{1 + 3 \sinh^2\left[\frac{1}{\sqrt{2}} \arcsin\left(\frac{1-u}{3}\right)^{1/2}\right]\right\}^{1/2} \tag{6.121}$$

(c) Exit Region

$$r_5 = B \sinh\left[\frac{1}{\sqrt{2}} \arcsin \beta(z - l) - \psi\right] \tag{6.122}$$

or the cosh term, in the interval $\frac{2}{3}l < z < l$. For r_5 to be zero we must use the sinh term because the cosh term can never be equal to zero. We have

$$\sinh[(1/\sqrt{2}) \arcsin \beta(z - l) - \psi] = 0 \tag{6.123}$$

or

$$z = l + \beta^{-1} \sin \psi\sqrt{2} \tag{6.124}$$

If the intersection is to occur in the interval of the exit region, we must have

$$\tfrac{2}{3}l \leq l + \beta^{-1} \sin \psi\sqrt{2} \leq l \tag{6.125}$$

Let us first assume that the intersection will take place in the beginning of the exit region. We then have

$$\tfrac{2}{3}l = l + \beta^{-1} \sin \psi\sqrt{2} \tag{6.126}$$

Or, using Eq. (6.53), corresponding to the sinh term,

$$\operatorname{arctanh}[\sqrt{2} \cot(\alpha_0 - \varphi)] - \beta_0 = \frac{1}{\sqrt{2}} \arcsin\left(-\frac{\beta l}{3}\right) \tag{6.127}$$

This finally reduces to

$$\frac{1}{\sqrt{2}} \operatorname{arcsinh}\left(\frac{2}{3}\frac{1-u}{u}\right)^{1/2} = \varphi + (2n + 1)\frac{\pi}{2} \tag{6.128}$$

This equation is similar to Eq. (6.110), which we have already analyzed.

If the intersection is to occur at the end of the exit region, we have

$$l = l + \beta^{-1} \sin \psi\sqrt{2} \tag{6.129}$$

which obviously leads to $\psi = 0$, or, using Eq. (6.100)

$$\operatorname{arctanh}[\sqrt{2}\cot(\alpha_0 - \varphi)] - \beta_0 = 0 \tag{6.130}$$

Substituting for α_0, β_0 and φ, we obtain

$$\operatorname{arctanh}\left(\sqrt{2}\cot\left\{\sqrt{2}\operatorname{arcsinh}\left[\frac{2}{3}\left(\frac{1-u}{u}\right)\right]^{1/2} - \arctan\left[\sqrt{2}\tanh\left(\frac{1}{\sqrt{2}}\arcsin\left(\frac{1-u}{3}\right)^{1/2}\right)\right]\right\}\right) - \frac{1}{\sqrt{2}}\arcsin\left(\frac{1-u}{3}\right)^{1/2} = 0 \tag{6.131}$$

The zeros of this transcendental equation as obtained with the help of a digital computer are

$$u_1 = 1.99 \times 10^{-1}$$
$$u_2 = 2.31 \times 10^{-3}$$
$$u_3 = 2.92 \times 10^{-5} \qquad \text{etc.}$$

We can easily show that when

$$(1/\sqrt{2})\tan(\alpha_0 - \varphi) = \pm 1 \tag{6.132}$$

there can be no intersection in the exit region.

(d) Intersection outside the Lens Region

Here, we have $z > l$. In this region the second derivative of the paraxial ray is zero. The slope is constant beyond the lens. We have

$$r = r_5(l) + [d(r_5)/dz]\,(z - l) \tag{6.133}$$

Letting $r = 0$, we obtain

$$z = l - [r_5(l)/r_5'(l)] \tag{6.134}$$

Using Eq. (6.41), we obtain

$$\frac{z}{l} = 1 + \left[\frac{2}{3(1-u)}\right]^{1/2}\coth\psi \quad \text{or} \quad 1 + \left[\frac{2}{3(1-u)}\right]^{1/2}\tanh\psi \tag{6.135}$$

For the special case in which

$$(1/\sqrt{2})\tan(\alpha_0 - \varphi) = -1 \tag{6.136}$$

it is readily seen that no intersection occurs outside the lens. When

$$(1/\sqrt{2})\tan(\alpha_0 - \varphi) = 1 \tag{6.137}$$

Eq. (6.135) becomes

$$\frac{z}{l} = 1 + \left[\frac{2}{3(1-u)}\right]^{1/2} \tag{6.138}$$

If the relation for α_0 is substituted in Eq. (6.137), we obtain

$$\frac{1}{\sqrt{2}} \tan\left(\frac{1}{\sqrt{2}} \operatorname{arcsinh} \frac{2\alpha l}{3} - \varphi\right) = 1 \tag{6.139}$$

and when we substitute for φ, we get:

$$\sqrt{2} \operatorname{arcsinh}\left[\frac{2}{3}\left(\frac{1}{u} - 1\right)\right]^{1/2} - \arctan\sqrt{2} - n\pi$$
$$= \arctan\left\{\sqrt{2}\tanh\left[\frac{1}{\sqrt{2}} \arcsin\left(\frac{1-u}{3}\right)^{1/2}\right]\right\} \tag{6.140}$$

The roots are obtained with a computer. We have

$$n = 0 \qquad u_0 = 0.3143$$
$$n = 1 \qquad u_1 = 3.86 \times 10^{-3}$$
$$n = 2 \qquad u_2 = 4.54 \times 10^{-5}$$

Therefore, for $u_o = 0.3143$, $z/l = 2$, and for all other u_1, u_2, etc., we have approximately $z/l = 1.867$. Table 6.1[1] summarizes the complete analysis.

A step by step analysis of the trajectory for a lens with $u = 10^{-2}$ is given here as an example. For simplicity, assuming $r_o = 1$, we have

$$\alpha = 12.19, \qquad \beta = 1.72, \qquad A = 1.26, \qquad \varphi = -1.45, \qquad B = -1.19$$

and

$$\psi = -0.22, \qquad \alpha_0 = 1.97, \qquad \beta_0 = 0.43, \qquad |K| = 0.41$$

Table 6.2 lists the values of r normalized with respect to l for equal z intervals. Trajectories for different values of u are plotted in Figs. 6.5–6.7.

The focal length, the midfocal length, and the position of the principal plane for the three-aperture lens (i.e., with thin central electrode) as shown in Fig. 6.4 are plotted as a function of u in Figs. 6.8–6.11. The oscillation of the focal length for small values of u is clearly shown in Fig. 6.11. Table 6.3 lists the values obtained with the computer.

Similar curves are given in Figs. 6.12–6.14 for the thick central electrode lens shown in Fig. 6.1. The curves are drawn for various values of d/l.

The values are also listed in Tables 6.4–6.12 for better accuracy.

In Fig. 6.15 the maximum value of l/f is plotted as a function of d/l. The corresponding values of u are also indicated.

[1] All tables for this chapter will be found at the end of the chapter.

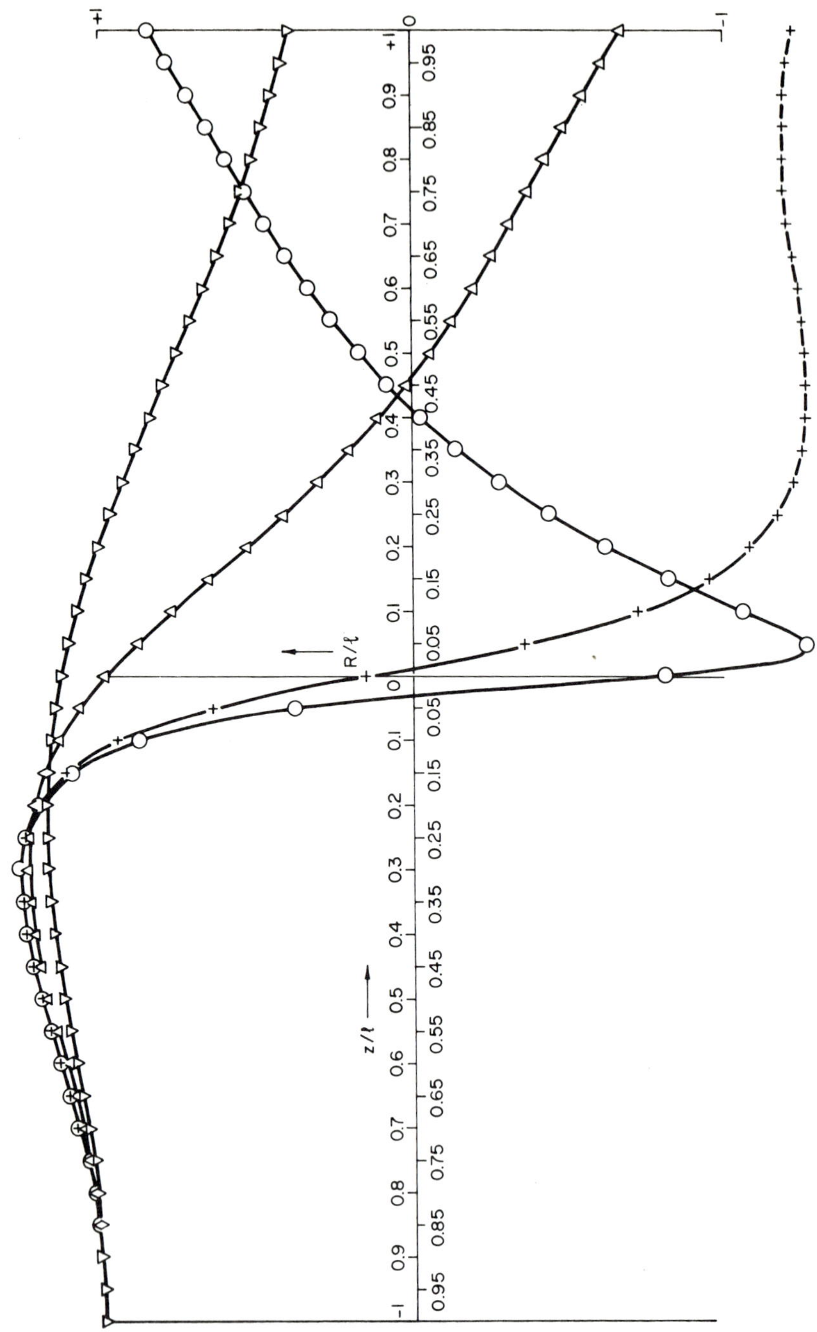

Fig. 6.5. Trajectories for different values of u (○, $u = 0.001$; +, $u = 0.01$; △, $u = 0.1$; ▽, $u = 0.3$).

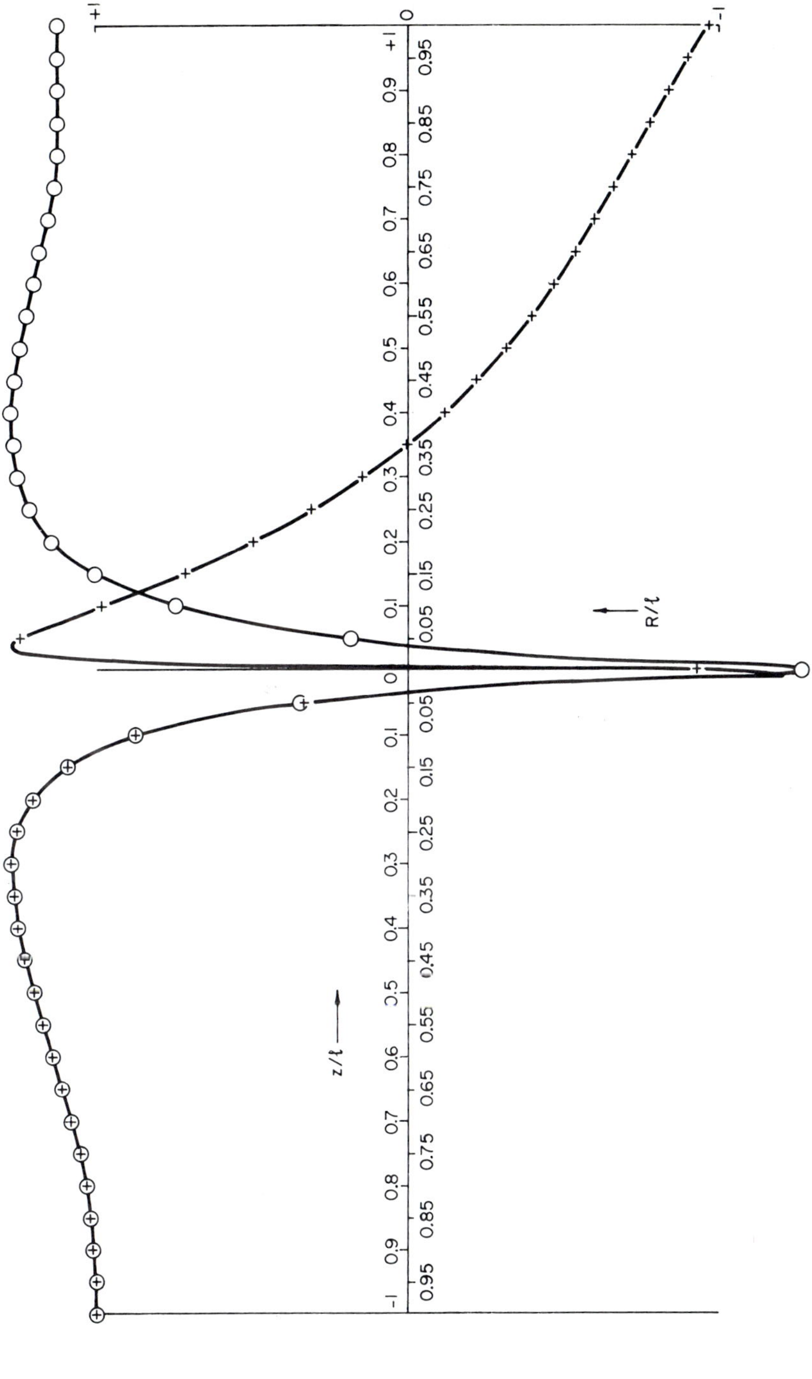

Fig. 6.5 Trajectories for different values of u (○, $u = 0.0001$; +, $u = 0.00001$).

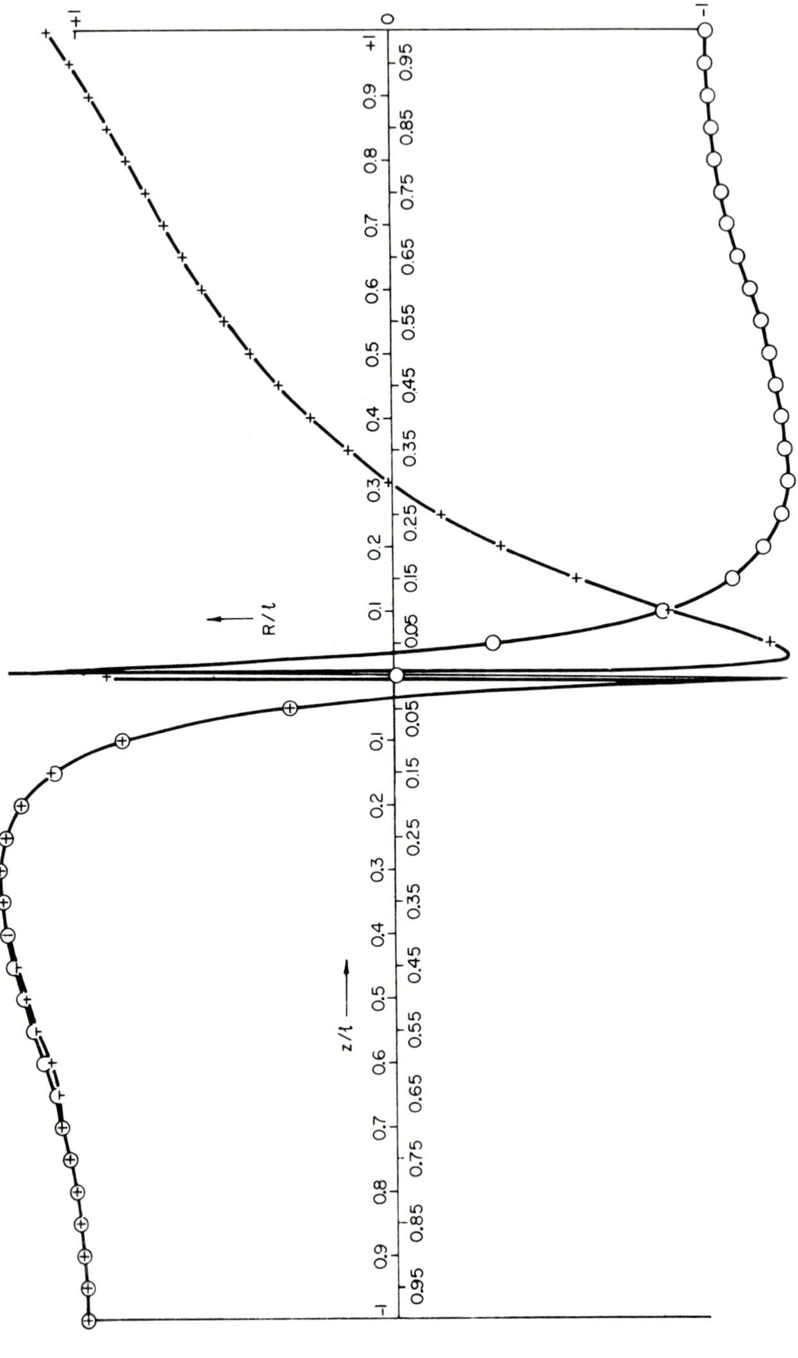

Fig. 6.7. Trajectories for different values of u (○, $u = 0.000001$; +, $u = 0.0000001$).

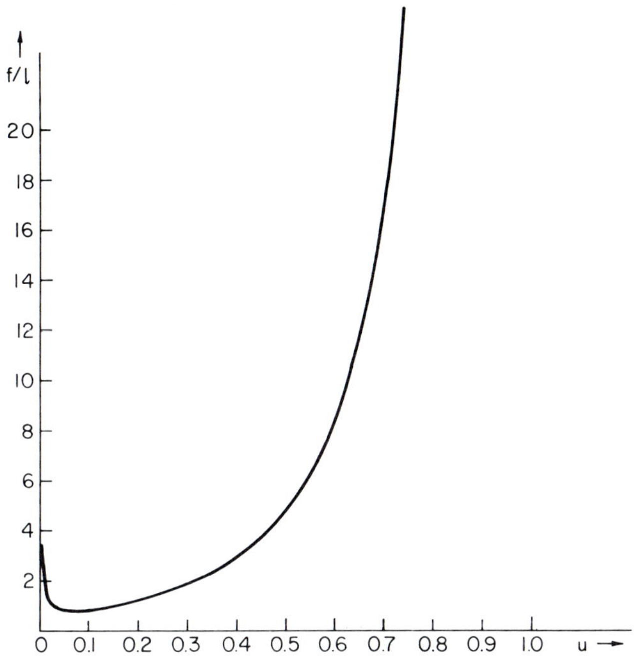

Fig. 6.8. The focal length as a function of *u* for the three-aperture lens.

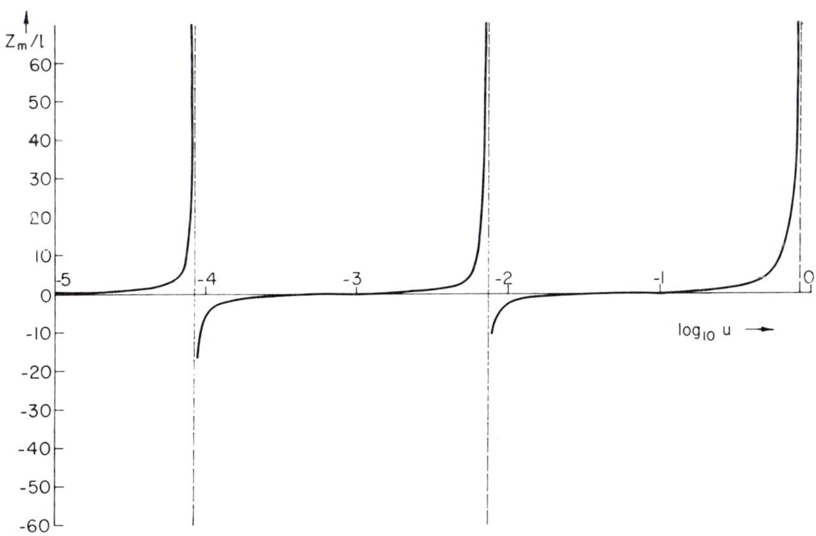

Fig. 6.9. The midfocal length as a function of *u* for the three-aperture lens.

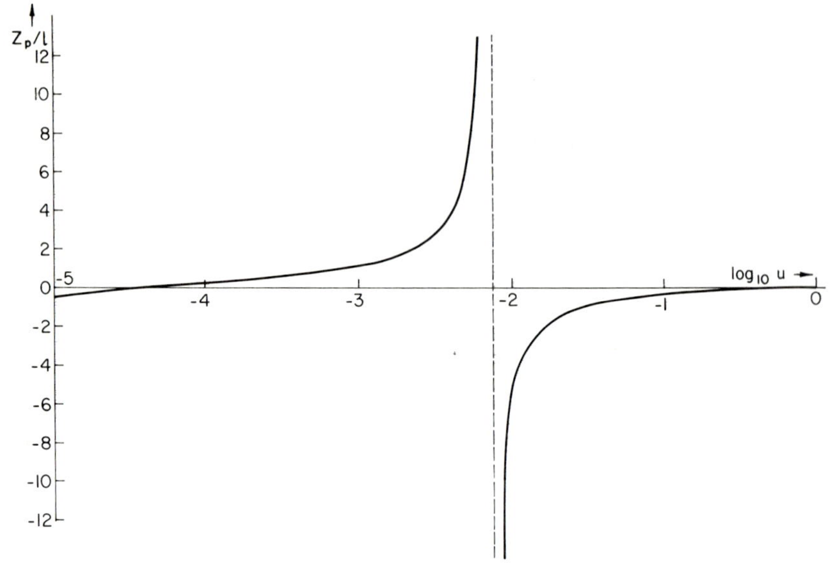

Fig. 6.10. The position of the principal plane as a function of u for the three-aperture lens.

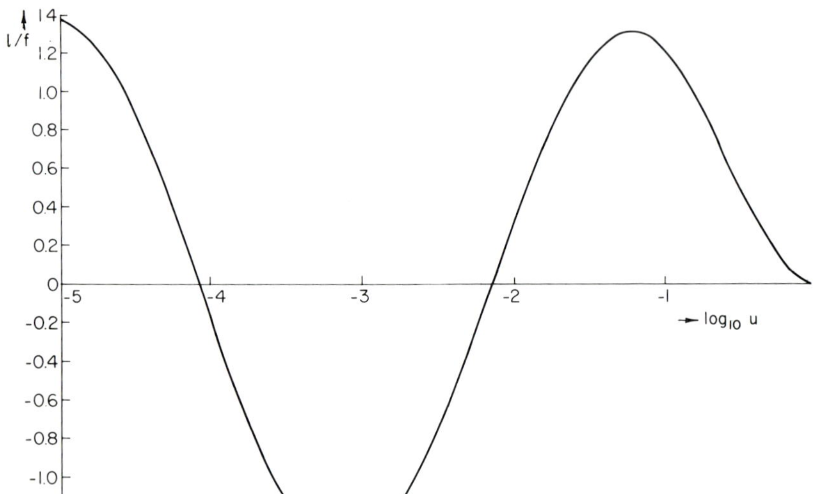

Fig. 6.11. The strength of the lens l/f as a function of u as u tends to zero. The oscillation is expected from Eq. (6.101).

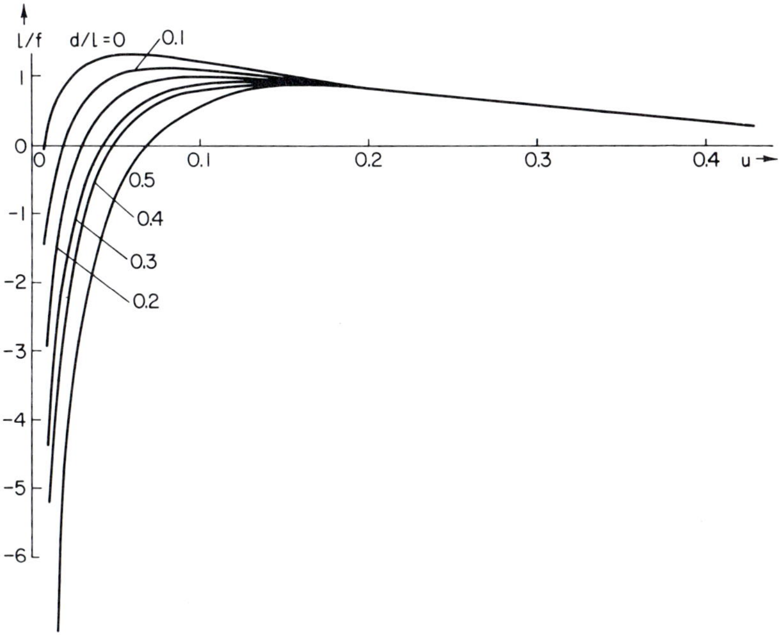

Fig. 6.12. Curve showing l/f as a function of u for the *einzel* lens with thick central electrode.

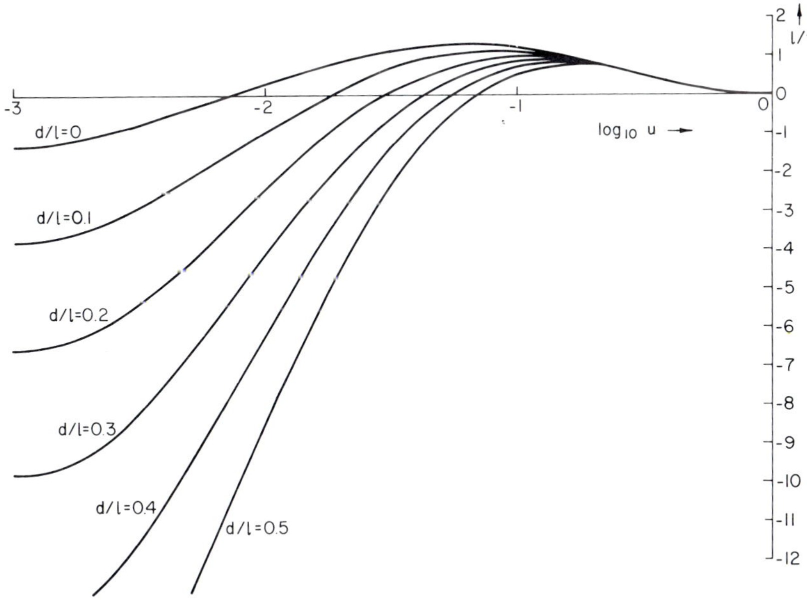

Fig. 6.13. The value of l/f as a function of u as u tends to zero for different values of d/l.

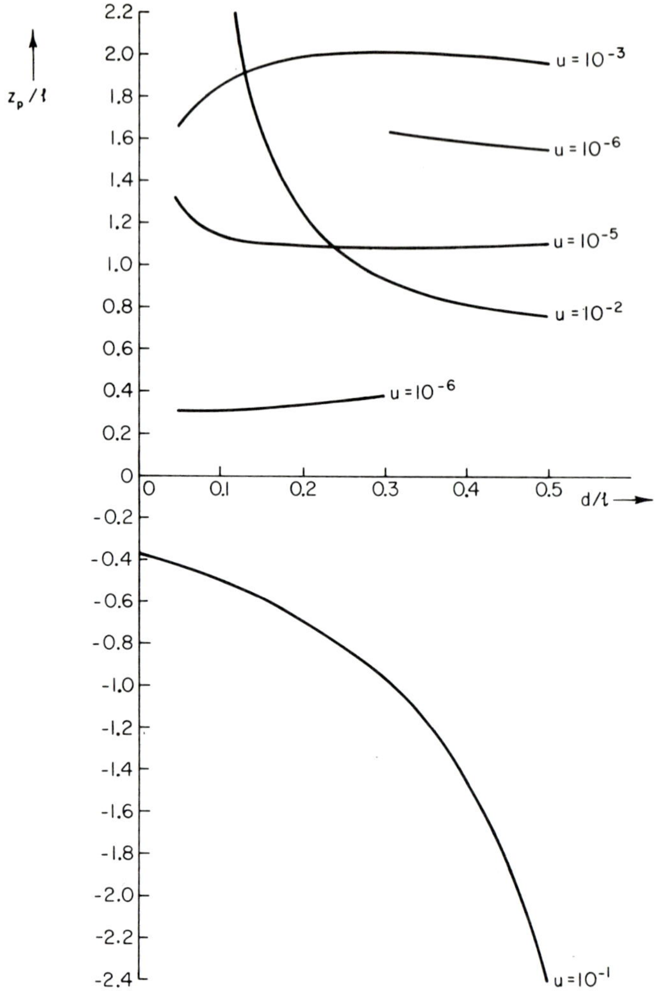

Fig. 6.14. The variation of the principal plane position with d/l for fixed values of u.

6.11 Design Examples

The preceding analysis will now be applied to the design of unipotential lenses.

Example 1

Let us design a unipotential thin central electrode lens with a reasonably short focal length. Let $z_m/l = 0.45$. Reference to Fig. 6.9 or Table 6.3 gives the corresponding value of $u = 0.1$.

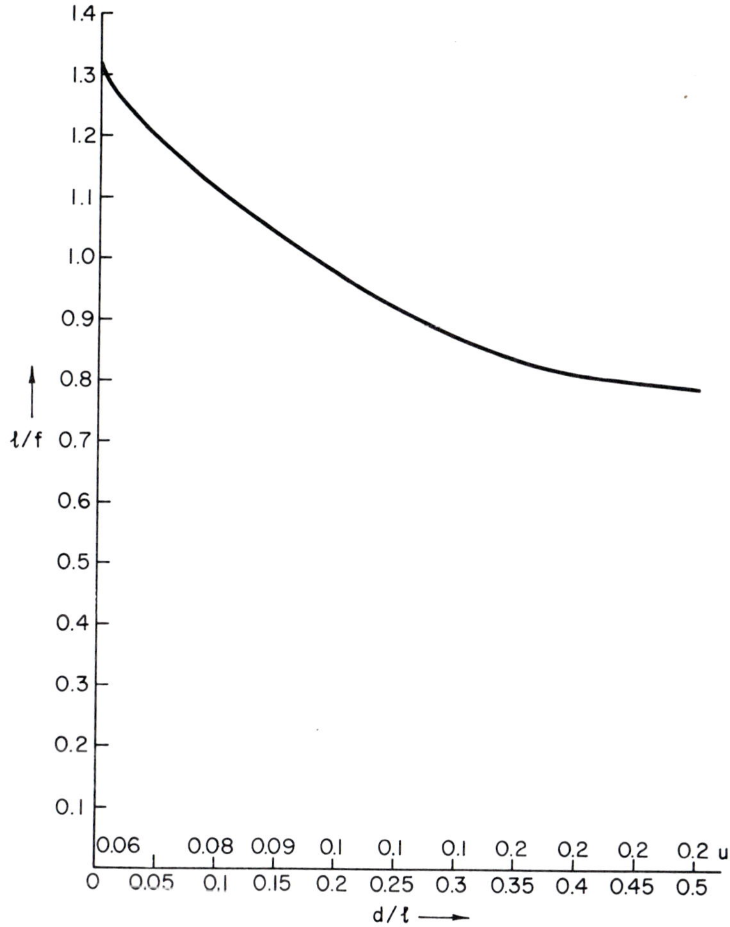

Fig. 6.15. Maximum value of l/f as a function of d/l. The corresponding values of u are indicated on the same axis. In this case, we are considering the positive values of f before it turns negative as u decreases.

From Fig. 6.8 or Table 6.3 we obtain $f/l = 0.82$. To begin, we shall assume that the central electrode is connected to the cathode. From Eq. (6.11), we obtain

$$u = \left(1 + \frac{l}{R_1} \arctan \frac{l}{R_1} - \frac{R_2}{2R_1}\right)^{-1} = 0.1 \tag{6.141}$$

We shall select $R_2/R_1 = 1/2$. Then $l/R_1 = 6.52$. Let $l = 8.15$ mm. Hence $R_1 = 1.25$ mm, and $R_2 = 0.625$ mm. The complete lens is now designed.

Example 2

Let us assume that, for reasons which we will explain in a subsequent chapter, we need to double R_1 and R_2 while retaining the same spacing and the same cardinal points. In that case the value of u should be retained. The only way out is to disconnect the central electrode from the cathode and adjust the ratio V_1/V_2. We now have, from Eq. (6.11)

$$u = \frac{\phi(0)}{\phi(l)} = \frac{V_2 - [(V_2 - V_1)/\pi l][\pi l - 2R_1]}{V_2 - [(V_2 - V_1)/\pi l][\pi l + R_2 - 2R_1 - 2l \arctan(l/R_1)]} = 0.1 \tag{6.142}$$

For $R_1 = 2.5$ mm, $R_2 = 1.25$ mm, and $l = 8.15$ mm, we find $V_1/V_2 = -0.12$. In other words, if the outer electrodes are connected to ground, and the cathode is made, e.g., 50 kV negative with respect to ground, the central electrode should then be 56 kV negative with respect to ground or 6 kV negative with respect to the cathode. This can easily be arranged with a potentiometer.

6.12 Case in Which $V_1 > V_2$ or $u > 1$

The complete analysis which has been presented above is valid for the case in which $u < 1$. If the central electrode is made positive with respect to the outer electrodes, $u > 1$.

As long as the formulas which were derived earlier were not restricted to the condition $u \leq 1$, we can safely assume that they are valid for any u. The potential representation along the axis, and its derivatives are valid for $u > 1$.

If reference is made to Eqs. (6.21) and (6.33), we find that in this case

$$\alpha = j\alpha, \qquad \text{and} \qquad \beta = j\beta \tag{6.143}$$

where $j = \sqrt{-1}$. Here

$$\alpha = \frac{1}{l}\left[\frac{1}{(1 - g_1)(g_2 - g_1)}\left(\frac{u - 1}{u}\right)\right]^{1/2} \tag{6.144}$$

and

$$\beta = \frac{1}{l}\left[\frac{1}{(1-g_1)(1-g_2)}(u-1)\right]^{1/2} \tag{6.145}$$

It is therefore easy to obtain the results for $u > 1$. We simply replace $r = r(\alpha, \beta, z)$ in $u < 1$ by $r = r(j\alpha, j\beta, jz)$ for $u > 1$. Remembering that $j \sin x = \sinh jx$ and $\cos x = \cosh jx$, etc., we obtain from Eqs. (6.44) and (6.45) new values for α_0 and β_0

$$\alpha_1 = \frac{1}{\sqrt{2}} \arcsin\left(\frac{g_2 - g_1}{1 - g_1}\,\frac{u-1}{u}\right)^{1/2} \tag{6.146}$$

$$\beta_1 = \frac{1}{\sqrt{2}} \operatorname{arcsinh}\left[\frac{1-g_2}{1-g_1}(u-1)\right]^{1/2} \tag{6.147}$$

The focal length is then

$$\frac{1}{f_1} = -\frac{\beta}{\lambda\sqrt{2}}(\cos^2\beta_1 - \lambda^2 \sin^2\beta_1)\sinh\left(2\alpha_1 - \operatorname{arctanh}\frac{2\lambda \tan\beta_1}{1 + \lambda^2 \tan^2\beta_1}\right) \tag{6.148}$$

and

$$\frac{1}{f_2} = -\frac{\alpha\beta g_1 l}{\lambda}\cosh^2\alpha_1 \cos^2\beta_1(\tanh\alpha_1 - \lambda\tan\beta_1)^2 \tag{6.149}$$

For the special case of a thin central electrode, we obtain

$$\frac{1}{f} = -\beta\left(\frac{3}{2}\cos^2\beta_1 - 1\right)\sinh\left(2\alpha_1 - \operatorname{arctanh}\frac{2\sqrt{2}\tan\beta_1}{1 + 2\tan^2\beta_1}\right) \tag{6.150}$$

The midfocal length is given by

$$\frac{z_m}{l} = 1 + \frac{1}{\sqrt{2}j\beta l}\,\frac{\sqrt{2}j\alpha g_1 lI + J}{\sqrt{2}j\alpha g_1 lM + N} \tag{6.151}$$

where

$$I = j(-\sinh^2\alpha_1 - \lambda^2\cosh^2\alpha_1)\sin\left[2\beta_1 - \arctan\left(\frac{2\lambda\tanh\alpha_1}{\lambda^2 - \tanh^2\alpha_1}\right)\right] \tag{6.152}$$

$$J = -[(\lambda^2+1)^2\sin^2 2\beta_1 - 4\lambda^2]^{1/2}\sinh\left[2\alpha_1 - \operatorname{arctanh}\left(\frac{2\lambda}{\lambda^2 - 1}\cot 2\beta_1\right)\right] \tag{6.153}$$

$$M = \cosh^2\alpha_1\cosh^2\beta_1[\tanh\alpha_1 - \lambda\tan\beta_1]^2 \tag{6.154}$$

and

$$N = \frac{\lambda\sqrt{2}}{j\beta f_1} \tag{6.155}$$

and the principal plane is

$$\frac{z_{\mathrm{P}}}{l} = 1 + \frac{1}{\sqrt{2}j\beta l}\,\frac{\sqrt{2}j\alpha g_1 lI + J - 2\lambda}{\sqrt{2}j\alpha g_1 lM + N} \tag{6.156}$$

Upon substitution for *I*, *J*, *M*, and *N*, we obtain real numbers for the cardinal points.

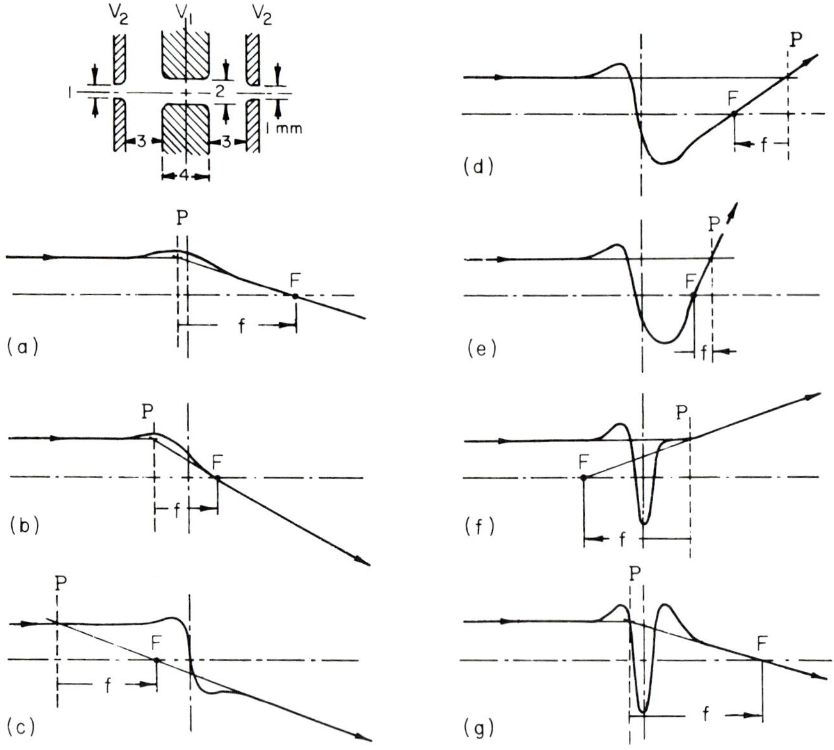

Fig. 6.16. The beam intersects the axis once, twice, cr three times depending on the voltage ratio. (a) $V_1/V_2 = 0.3$; (b) $V_1/V_2 = 0.11$; (c) $V_1/V_2 = 0.05$; (d) $V_1/V_2 = 0.01$; (e) $V_1/V_2 = 0$; (f) $V_1/V_2 = -0.0028$. In (*g*) V_1/V_2 is small enough to cause three intersections. The ratio V_1/V_2 should not be confused with $u = \phi(0)/\phi(l)$ (from Heise and Rang, 1949).

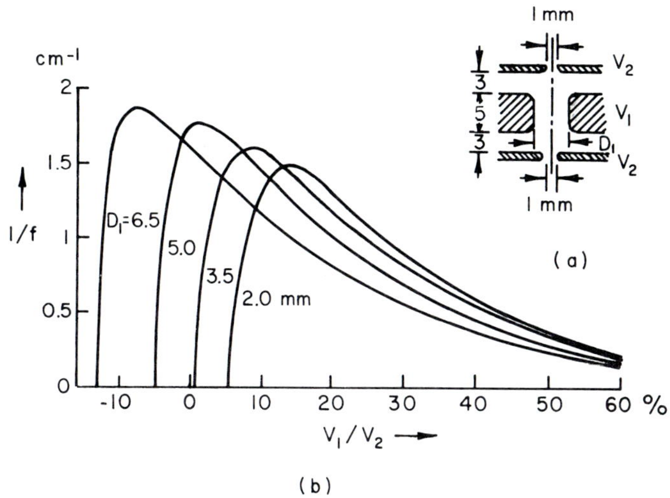

Fig. 6.17. The focal length as a function of V_1/V_2 for various diameters D_1 of the central electrode (from Heise and Rang, 1949).

6.13 Experimental Results

The detailed theory which was presented in all the preceding sections was first developed by Regenstreif. It is based on numerous simplifying assumptions. The final results are, therefore, valid only in very special cases. Nevertheless with the help of this theory, it was possible to understand the detailed operation of the lens. We will show that while the quantitative results are not quite accurate, the qualitative results are reliable. In what follows we shall present experimental results which were obtained by numerous researchers using different methods.

We have seen in the previous analysis that the trajectory can intersect the axis several times depending on the strength of the lens. As an example, Fig. 6.16 shows the course of the trajectory as the ratio V_1/V_2 is decreased. In general a lens is operated in such a way that only one intersection occurs outside the lens. We are, therefore, primarily interested in the projective and not in the objective focal point. It is easy to see that for a given geometry, it is possible to obtain two values of V_1/V_2 which will give the same focal length when only one intersection occurs. In Figs. 6.17 and 6.18 the focal length is given as a function of V_1/V_2 for various values of D_1, the diameter of the central electrode. The results were obtained experimentally for the lens whose geometry is shown in Fig. 6.17a. As the diameter D_1 gets smaller, the ratio $u = \phi(0)/\phi(l)$ gets smaller, and therefore the lens gets stronger. This means that to retain the same focal length, the voltage of the central

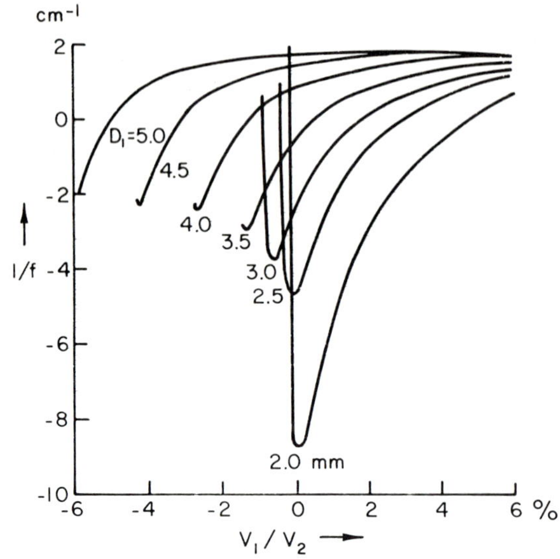

Fig. 6.18. The focal length as a function of V_1/V_2 for various values of D_1 for the geometry shown in Fig. 6.17a. Here, two intersections occur (from Heise and Rang,1949).

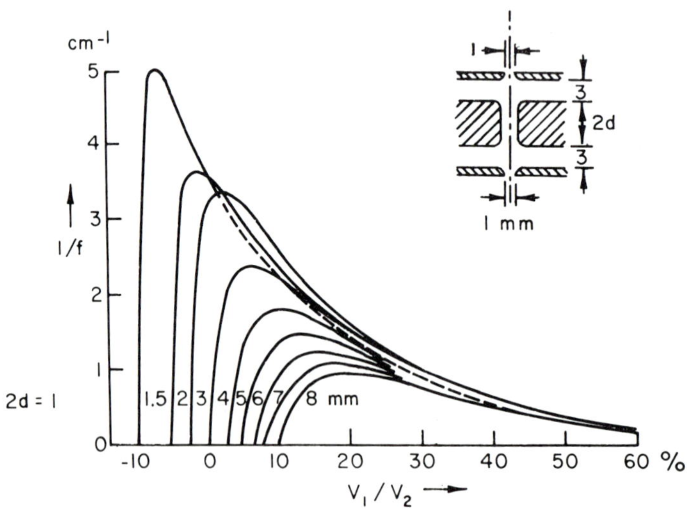

Fig. 6.19. Focal length as a function of V_1/V_2 for various values of $2d$, the thickness of the central electrode (from Heise and Rang, 1949).

electrode should be made more negative as D_1 increases. However, in the range shown, the minimum focal length decreases as D_1 increases.

In Figs. 6.19 and 6.20 similar curves were obtained with $2d$, the thickness of the central electrode, as the variable. We find here that for a given voltage ratio, the maximum strength of the lens (minimum focal length) increases as the thickness decreases which verifies the theory where it was shown that for the first intersection, the three-aperture lens can actually be stronger than the lens with thick central electrode. As V_1/V_2 decreases, u decreases and here the $1/f_2$ term in Eq. (6.69) predominates. We should, therefore, expect an increase in lens strength with thickness of central electrode. This is clearly verified in Fig. 6.20. The curves for $2d = 1$ and 1.5 do not follow the same pattern. This

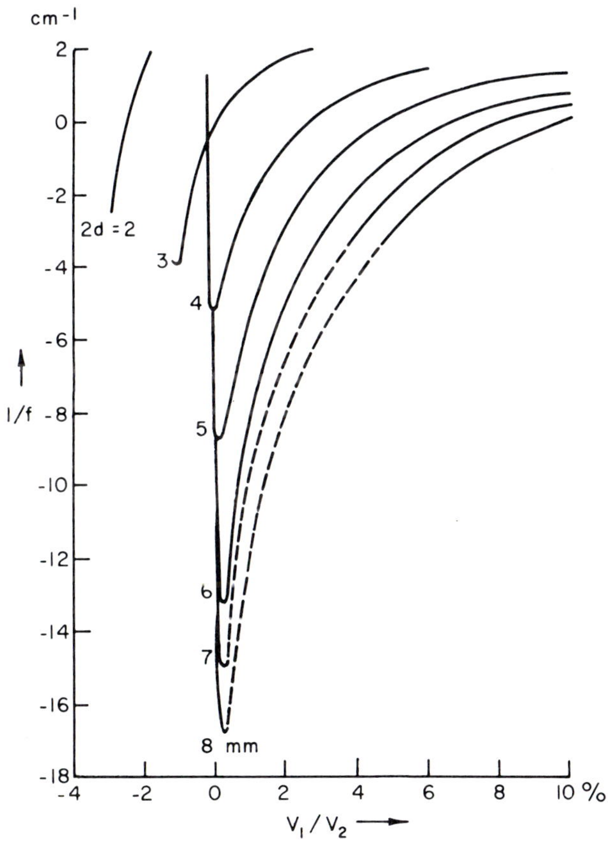

Fig. 6.20. Focal length as a function of V_1/V_2 for various values of $2d$, the thickness of the central electrode, for the geometry shown in Fig. 6.19. Here, two intersections occur (from Heise and Rang, 1949).

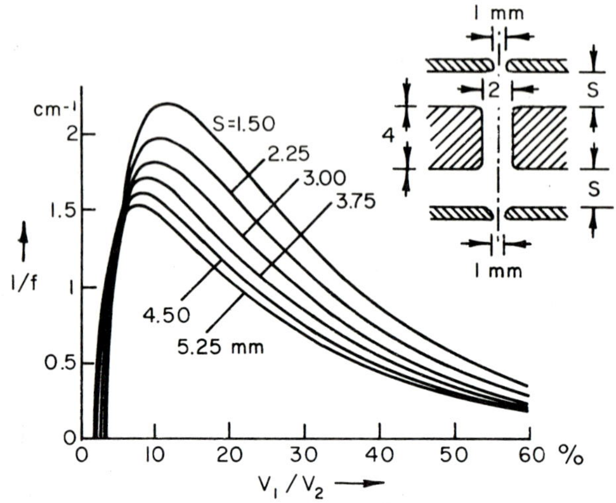

Fig. 6.21. Influence of the spacing S. Focal length as a function of V_1/V_2 for various values of S (from Heise and Rang, 1949).

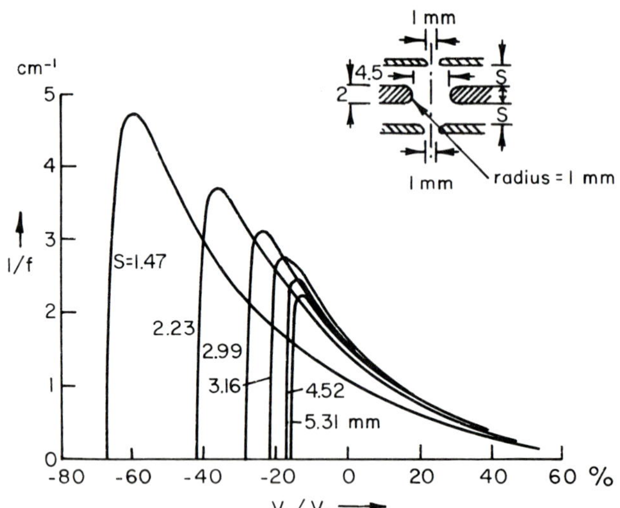

Fig. 6.22. Influence of the spacing S. Focal length as a function of V_1/V_2 for a wider range of V_1/V_2 (from Heise and Rang, 1949).

is due to some rounding of the edges which results in an apparent increase in D_1.

The effect of the spacing S between the central and outer electrodes is shown in Figs. 6.21 and 6.22 for two different values of D_1, the diameter of the central electrode. Obviously, the lens becomes weaker as the spacing is increased.

Experimental values for a unipotential lens where the central electrode is tied to the cathode, i.e., $V_1/V_2 = 0$, were obtained for various values of the

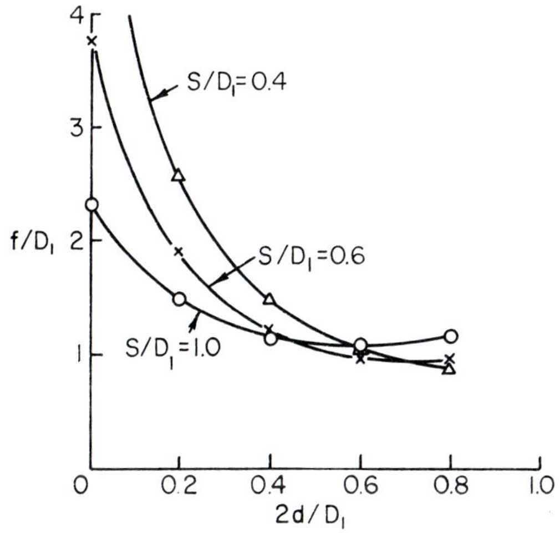

Fig. 6.23. Focal length as a function of $2d/D_1$ for various values of S/D_1. Here, $V_1/V_2=0$ and $D_1 = D_2$ (from Vine, 1960).

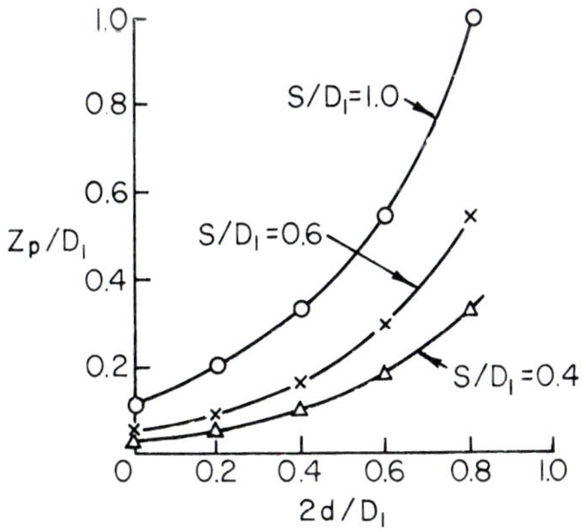

Fig. 6.24. Position of principal plane z_p/D_1 as a function of $2d/D_1$ for various values of S/D_1. Here, $V_1/V_2 = 0$, and $D_1 = D_2$ (from Vine, 1960).

ratio $2d/D$. Here $R_1 = R_2 = R = D/2$. Figures 6.23 and 6.24 give the focal length and location of principal points. At first glance, the reader might think that the fact that the focal length decreases with increasing thickness of the central electrode violates the statement made earlier in this chapter. For given voltage ratios V_1/V_2, the values of $u = \phi(0)/\phi(l)$ change, as $2d$ changes. When $V_1/V_2 = 0$, u decreases as $2d$ increases, and obviously the focal length decreases. If a plot of $2d/l$ as a function of D_1/l is made keeping l/f constant, contours as shown in Fig. 6.25 will be obtained. Similar contours are shown for the midfocal length in Fig. 6.26. The dotted lines represent estimated curves and not measured values. Positive contours correspond to the range

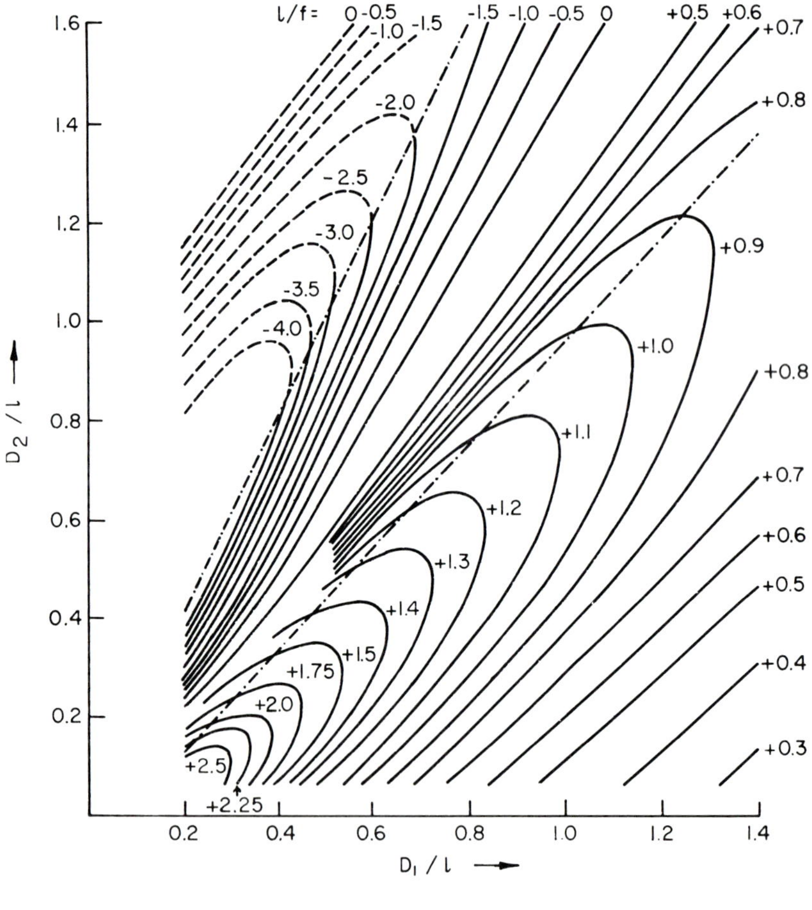

Fig. 6.25. Contours of fixed focal length; D_2 as a function of D_1 keeping f constant (from Von Lippert and Pohlit, 1952).

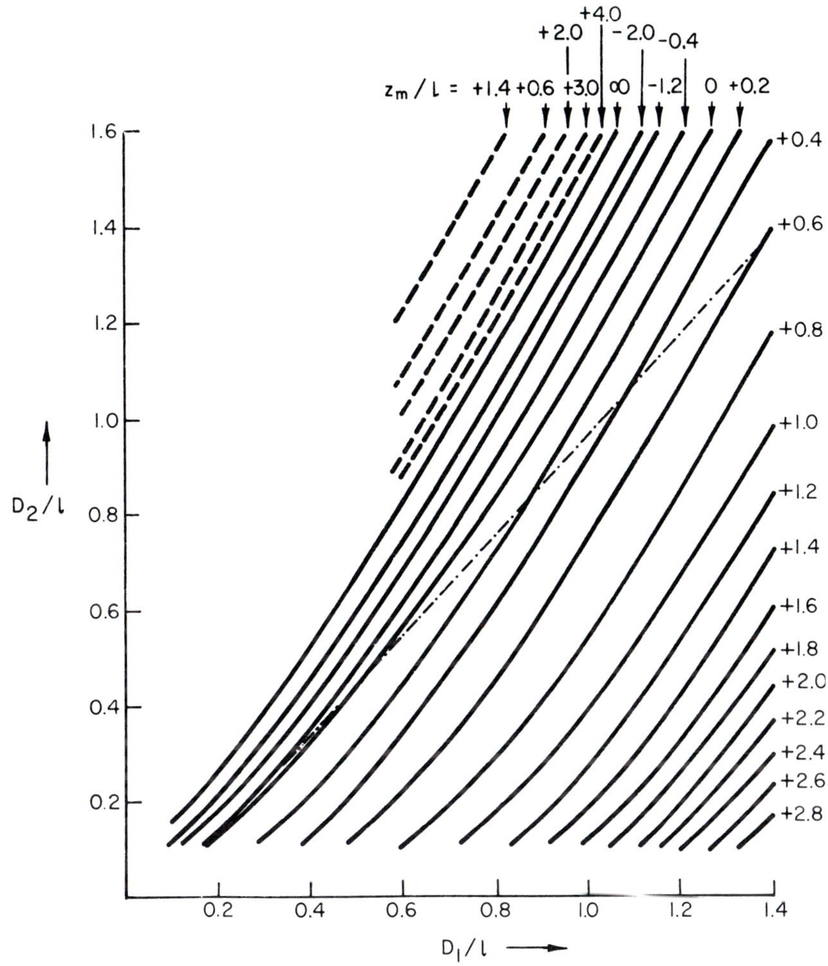

Fig. 6.26. Contours of fixed value of midfocal length (from Von Lippert and Pohlit, 1952).

where one intersection occurs while the negative contours are for the second intersection.

A clear comparison between the symmetrical *einzel* lens and the asymmetrical one is made in Figs. 6.27 and Figs. 6.28. Figure 6.29 also gives the focal length of an asymmetrical lens in which not only the spacing is different but also the opening on each side of the central electrode. In the cases shown, the asymmetrical lens seems to be somewhat stronger than its symmetrical

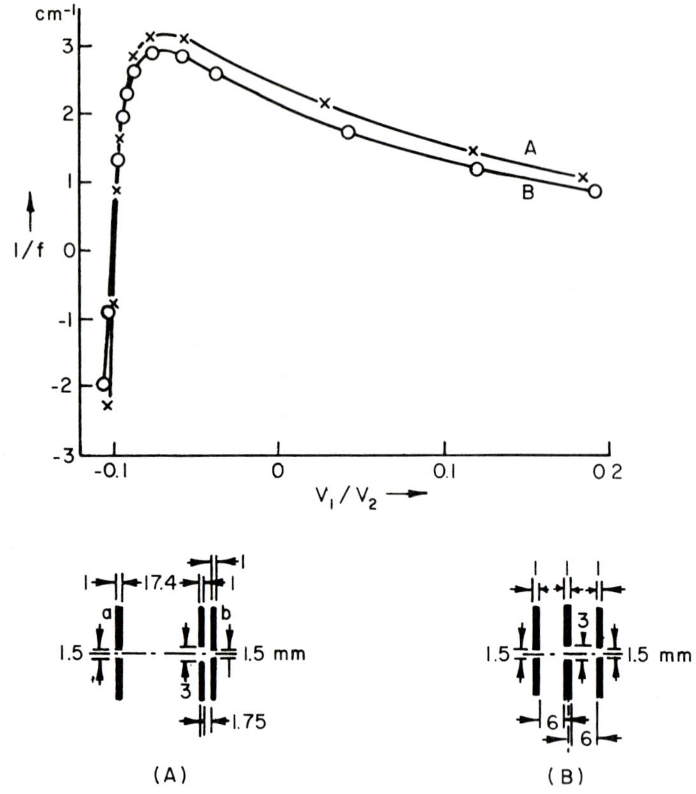

Fig. 6.27. Comparison between a symmetrical and an asymmetrical *einzel* lens (from Everitt and Hanszen, 1956).

counterpart. However, this is not always true. In asymmetrical lenses the object-side and image-side cardinal points are no longer equal.

So far we have considered unipotential lenses consisting of either three apertures or two apertures and a central thick electrode. In Volume 2, Fig. 10.13 of Chapter X, a series of lenses consisting of cylinders is shown. Table 6.13 lists values for the focal length and distance of principal point for each lens.

The choice of a unipotential lens depends on several factors. In Volume 2, Chapter X, we shall discuss spherical aberration of several types of *einzel* lenses, and we shall see how the geometry of the lens plays a very important role in this case.

Finally, Figs. 6.30 and 6.31 show the lens-power characteristics obtained experimentally by Liebmann (1949).

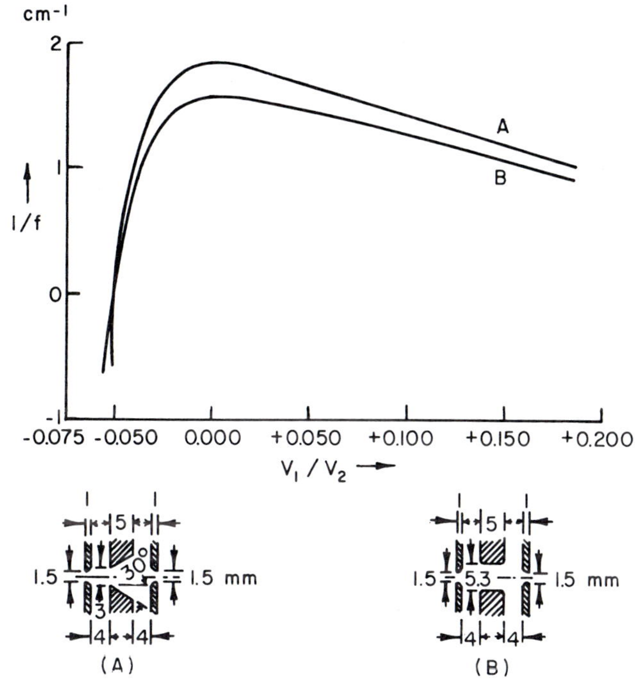

Fig. 6.28. Comparison of a symmetrical and an asymmetrical *einzel* lens (from Hanszen, 1958).

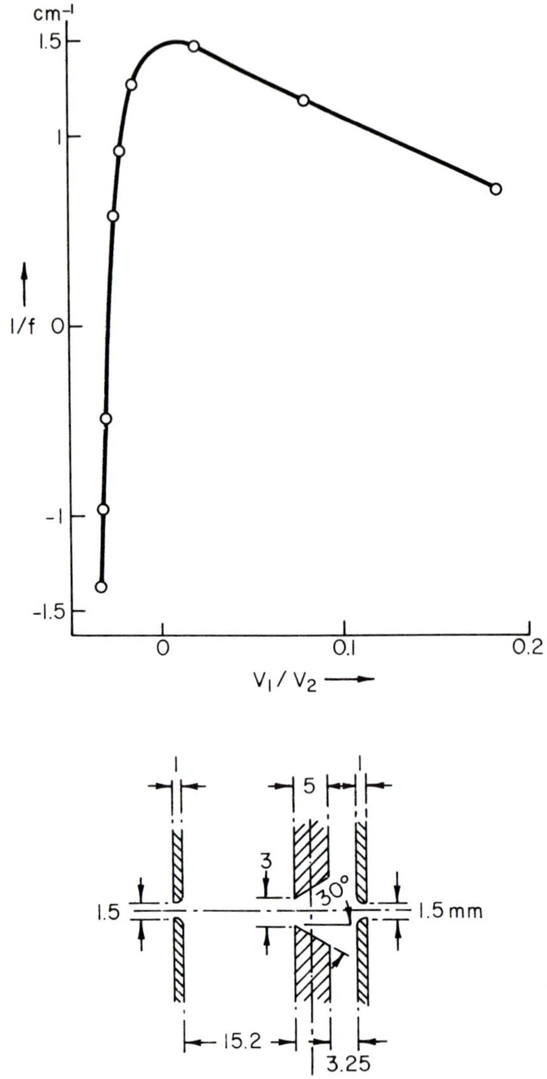

Fig. 6.29. Focal length of an asymmetrical *einzel* lens as a function of V_1/V_2 (from Everitt and Hanszen, 1956).

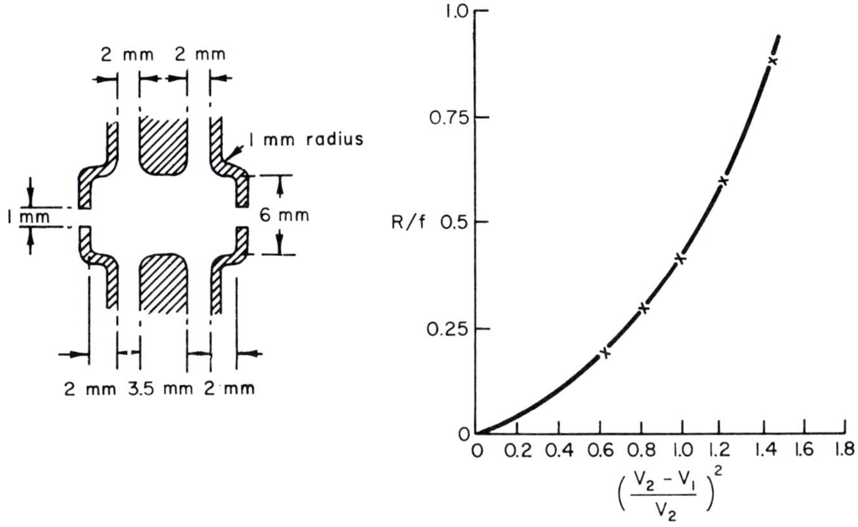

Fig. 6.30. Lens power of an *einzel* lens (from Liebmann, 1949).

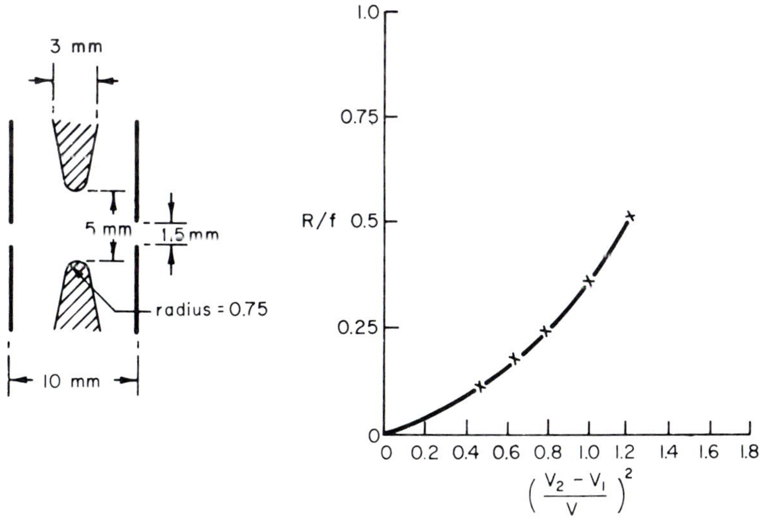

Fig. 6.31. Lens power of an *einzel* lens with rounded inner electrode (from Liebmann, 1949).

FURTHER READING:

Everitt, C. W. F., and Hanszen, K. J. (1956). *Optik* **13**, 385.
Hanszen, K. J. (1958). *Optik* **15**, 304.
Heise, F., and Rang, O. (1949). *Optik* **5**, 201.
Liebmann, G. (1949). *Proc. Phys. Soc.* (*London*) **B62**, 213.
Regenstrief, E. (1951). *Ann. Radioélectricité* **6**, 51, 164.
Vine, J. (1960). *Brit. J. Appl. Phys.* **11**, 408.
Von Lippert, W., and Pohlit, W. (1952). *Optik* **9**, 456.

TABLE 6.1

Interval of u	Location of the intersections
$1 \geq u \geq 0.199$	There will be one intersection outside the lens. In the limit, at $u = 0.199$, the intersection will take place at $z = l$ or just at the boundary of the lens.
$0.199 \geq u \geq 0.138$	One intersection will occur inside the lens region between $z = \frac{2}{3}l$ and $z = l$. In the limit, at $u = 0.138$, the intersection will occur at $z = \frac{2}{3}l$.
$0.138 \geq u \geq 0.007$	Only one intersection will occur in the region $0 \leq z \leq \frac{2}{3}l$.
$u = 7.014 \times 10^{-3}$	The ray will intersect the axis at $z = 0.008l$ (practically in the geometrical center) and will emerge parallel to the axis. We can therefore say that two intersections occur, one inside the lens and the other at infinity.
$7.014 \times 10^{-3} \geq u \geq 8.51 \times 10^{-5}$	In this interval, two intersections occur. At $u = 2.31 \times 10^{-3}$, there will be an intersection at $z = 0.06l$ and at $z = l$. At $u = 1.61 \times 10^{-3}$, there will be an intersection at $z = 0.0725$ and at $z = \frac{2}{3}l$.
$u = 8.5 \times 10^{-5}$	Here, the ray intersects the axis first at $z = -0.022l$, next at $z = 0.035l$, and then emerges parallel to the axis.
$u < 8.5 \times 10^{-5}$	Three or more intersections occur. For example, at $u = 2.92 \times 10^{-5}$, first intersection at $z = -0.03l$, second at $z = 0.012l$, and third at $z = l$.

TABLE 6.2

VALUES OF r NORMALIZED WITH RESPECT TO l AS A FUNCTION OF z/l FOR THE THREE-APERTURE ELECTRODE WHERE $u = 10^{-2}$

z	r	z	r	z	r	z	r
−1.00	1.00	−0.45	1.23	0.10	−0.72	0.65	−1.22
−0.95	1.00	−0.40	1.25	0.15	−0.95	0.70	−1.20
−0.90	1.01	−0.35	1.26	0.20	−1.08	0.75	−1.19
−0.85	1.02	−0.30	1.26	0.25	−1.17	0.80	−1.19
−0.80	1.03	−0.25	1.25	0.30	−1.22	0.85	−1.19
−0.75	1.05	−0.20	1.21	0.35	−1.25	0.90	−1.19
−0.70	1.07	−0.15	1.12	0.40	−1.26	0.95	−1.20
−0.65	1.11	−0.10	0.95	0.45	−1.26	1.00	−1.22
−0.60	1.14	−0.05	0.64	0.50	−1.26		
−0.55	1.17	0	0.15	0.55	−1.25		
−0.50	1.20	0.05	−0.36	0.60	−1.24		

TABLE 6.3
CARDINAL POINTS OF THE THREE-APERTURE UNIPOTENTIAL LENS IN UNITS OF l, THE SPACING BETWEEN THE ELECTRODES

u	l/f	f/l	z_m/l	z_p/l
$d/l = 0$				
0.000001	0.02	58.09	−57.84	−115.92
0.000002	0.67	1.49	−1.06	−2.55
0.000003	0.99	1.01	−0.47	−1.48
0.000004	1.16	0.86	−0.22	−1.08
0.000005	1.27	0.79	−0.07	−0.86
0.000006	1.33	0.75	0.03	−0.72
0.000007	1.37	0.73	0.12	−0.62
0.000008	1.38	0.72	0.19	−0.54
0.000009	1.39	0.72	0.25	−0.47
0.00001	1.39	0.72	0.30	−0.42
0.00002	1.18	0.85	0.70	−0.15
0.00003	0.92	1.09	1.06	−0.02
0.00004	0.69	1.45	1.50	0.06
0.00005	0.49	2.03	2.14	0.12
0.00006	0.32	3.10	3.27	0.16
0.00007	0.17	5.77	5.97	0.20
0.00008	0.04	23.46	23.70	0.24
0.00009	−0.07	−13.68	−13.41	0.27
0.0001	−0.18	−5.68	−5.38	0.29
0.0002	−0.80	−1.24	−0.77	0.48
0.0003	−1.09	−0.92	−0.32	0.60
0.0004	−1.24	−0.80	−0.11	0.69
0.0005	−1.32	−0.75	0.02	0.78
0.0006	−1.37	−0.73	0.12	0.85
0.0007	−1.39	−0.72	0.20	0.92
0.0008	−1.39	−0.72	0.27	0.99
0.0009	−1.38	−0.72	0.33	1.05
0.001	−1.37	−0.73	0.38	1.12
0.002	−1.08	−0.92	0.82	1.75
0.003	−0.79	−1.26	1.28	2.54
0.004	−0.55	−1.82	1.92	3.75
0.005	−0.34	−2.93	3.08	6.01
0.006	−0.17	−6.00	6.20	12.20
0.007	−0.02	−59.56	59.80	119.36
0.008	0.11	8.85	−8.57	−17.43
0.009	0.23	4.42	−4.11	−8.53
0.01	0.33	3.07	−2.73	−5.80
0.02	0.91	1.10	−0.58	−1.68
0.03	1.14	0.87	−0.22	−1.10
0.04	1.25	0.80	−0.04	−0.84
0.05	1.30	0.77	0.08	−0.69

TABLE 6.3 (continued)

u	l/f	f/l	z_m/l	z_p/l
$d/l = 0$				
0.06	1.31	0.77	0.18	−0.59
0.07	1.30	0.77	0.26	−0.51
0.08	1.28	0.78	0.33	−0.46
0.09	1.25	0.80	0.39	−0.41
0.10	1.21	0.82	0.45	−0.37
0.20	0.84	1.19	1.01	−0.18
0.30	0.55	1.81	1.70	−0.11
0.40	0.35	2.84	2.78	−0.06
0.50	0.21	4.66	4.63	−0.04
0.60	0.12	8.24	8.21	−0.02
0.70	0.06	16.41	16.40	−0.01
0.80	0.02	41.08	41.07	0.00
0.90	0.01	181.74	181.74	0.00

TABLE 6.4

CARDINAL POINTS OF THE UNIPOTENTIAL LENS EXPRESSED IN UNITS OF l, THE SPACING BETWEEN THE ELECTRODES

u	l/f_1	l/f_2	l/f	f/l	z_m/l	z_p/l
$d/l = 0.1$						
0.000001	0.01	−129.40	−129.39	−0.01	0.30	0.31
0.000002	0.68	−86.01	−85.32	−0.01	0.48	0.49
0.000003	1.01	−63.86	−62.85	−0.02	0.59	0.61
0.000004	1.20	−50.23	−49.03	−0.02	0.68	0.71
0.000005	1.31	−40.94	−39.63	−0.03	0.76	0.79
0.000006	1.37	−34.19	−32.82	−0.03	0.83	0.87
0.000007	1.41	−29.07	−27.66	−0.04	0.90	0.94
0.000008	1.43	−25.05	23.62	−0.04	0.97	1.01
0.000009	1.44	−21.83	−20.39	−0.05	1.03	1.08
0.00001	1.44	−19.18	−17.75	−0.06	1.09	1.14
0.00002	1.22	−6.87	−5.65	−0.18	1.71	1.89
0.00003	0.96	−3.02	−2.06	−0.48	−0.65	−0.17
0.00004	0.72	−1.39	−0.67	−1.49	−3.40	−1.91
0.00005	0.52	−0.62	−0.10	−10.00	−23.27	−13.27
0.00006	0.34	−0.24	0.10	9.85	20.15	10.30
0.00007	0.19	−0.07	0.12	8.18	13.43	5.25
0.00008	0.05	−0.01	0.05	20.20	24.20	3.99
0.00009	−0.06	−0.01	−0.07	−13.90	−10.48	3.42
0.0001	−0.17	−0.05	−0.22	−4.59	−1.50	3.09
0.0002	−0.82	−0.82	−1.65	−0.61	1.61	2.22

TABLE 6.4 (continued)

u	l/f_1	l/f_2	l/f	f/l	z_m/l	z_p/l
$d/l=0.1$						
0.0003	−1.12	−1.40	−2.53	−0.40	1.63	2.03
0.0004	−1.28	−1.77	−3.05	−0.33	1.62	1.95
0.0005	−1.37	−2.00	−3.37	−0.30	1.61	1.91
0.0006	−1.41	−2.16	−3.57	−0.28	1.60	1.88
0.0007	−1.43	−2.26	−3.69	−0.27	1.60	1.87
0.0008	−1.44	−2.33	−3.77	−0.27	1.60	1.86
0.0009	−1.43	−2.37	−3.80	−0.26	1.60	1.86
0.001	−1.41	−2.40	−3.82	−0.26	1.60	1.86
0.002	−1.13	−2.33	−3.46	−0.29	1.65	1.94
0.003	−0.83	−2.13	−2.97	−0.34	1.75	2.08
0.004	−0.58	−1.95	−2.52	−0.40	1.86	2.26
0.005	−0.36	−1.79	−2.15	−0.46	1.99	2.46
0.006	−0.18	−1.65	−1.83	−0.55	0.79	1.34
0.007	−0.03	−1.53	−1.56	−0.64	0.92	1.57
0.008	0.11	−1.43	−1.32	−0.76	1.07	1.83
0.009	0.22	−1.34	−1.11	−0.90	1.24	2.14
0.01	0.33	−1.26	−0.93	−1.07	1.45	2.52
0.02	0.93	−0.78	0.15	6.61	−6.06	−12.67
0.03	1.18	−0.55	0.63	1.59	−0.91	−2.50
0.04	1.29	−0.41	0.88	1.14	−0.36	−1.50
0.05	1.34	−0.33	1.01	0.99	−0.11	−1.10
0.06	1.35	−0.26	1.09	0.92	0.04	−0.88
0.07	1.34	−0.22	1.13	0.89	0.15	−0.73
0.08	1.32	−0.18	1.14	0.88	0.24	−0.63
0.09	1.29	−0.15	1.14	0.88	0.32	−0.56
0.10	1.26	−0.13	1.13	0.89	0.39	−0.50
0.20	0.88	−0.04	0.84	1.19	1.06	−0.13
0.30	0.58	−0.01	0.57	1.77	1.74	−0.02
0.40	0.37	0	0.36	2.75	2.77	0.03
0.50	0.22	0	0.22	4.48	4.54	0.06
0.60	0.13	0	0.13	7.88	7.96	0.08
0.70	0.06	0	0.06	15.65	15.74	0.09
0.80	0.03	0	0.03	39.11	39.21	0.10
0.90	0.01	0	0.01	172.81	172.91	0.10

TABLE 6.5

CARDINAL POINTS OF THE UNIPOTENTIAL LENS EXPRESSED IN UNITS OF l, THE SPACING BETWEEN THE ELECTRODES

u	l/f_1	l/f_2	l/f	f/l	z_m/l	z_p/l
$d/l = 0.15$						
0.000001	0.00	−201.47	−201.47	−0.00	0.32	0.32
0.000002	0.69	−134.04	−133.35	−0.01	0.49	0.50
0.000003	1.03	−99.59	−98.56	−0.01	0.60	0.61
0.000004	1.22	−78.37	−77.15	−0.01	0.69	0.70
0.000005	1.33	−63.90	−62.57	−0.02	0.77	0.78
0.000006	1.40	−53.39	−51.99	−0.02	0.83	0.85
0.000007	1.44	−45.41	−43.97	−0.02	0.90	0.92
0.000008	1.46	−39.15	−37.69	−0.03	0.96	0.98
0.000009	1.46	−34.12	−32.65	−0.03	1.02	1.05
0.00001	1.46	−29.99	−28.53	−0.04	1.07	1.11
0.00002	1.25	−10.78	−9.53	−0.10	1.64	1.74
0.00003	0.98	−4.75	−3.77	−0.27	−0.39	−0.12
0.00004	0.74	−2.20	−1.46	−0.69	−1.99	−1.30
0.00005	0.54	−0.99	−0.45	−2.22	−6.63	−4.42
0.00006	0.36	−0.39	−0.03	−29.36	−75.81	−46.45
0.00007	0.20	−0.11	0.09	11.47	22.75	11.28
0.00008	0.06	−0.01	0.05	19.43	25.73	6.30
0.00009	−0.06	−0.01	−0.07	−14.39	−9.57	4.82
0.0001	−0.17	−0.07	−0.24	−4.23	−0.13	4.10
0.0002	−0.83	−1.27	−2.10	−0.48	2.07	2.55
0.0003	−1.14	−2.17	−3.31	−0.30	1.96	2.26
0.0004	−1.30	−2.73	−4.03	−0.25	1.89	2.14
0.0005	−1.39	−3.10	−4.49	−0.22	1.84	2.07
0.0006	−1.44	−3.34	−4.78	−0.21	1.81	2.02
0.0007	−1.46	−3.50	−4.96	−0.20	1.79	1.99
0.0008	−1.46	−3.61	5.07	−0.20	1.77	1.97
0.0009	−1.46	−3.68	−5.14	−0.19	1.76	1.96
0.001	−1.44	−3.73	−5.17	−0.19	1.75	1.95
0.002	−1.15	−3.63	−4.78	−0.21	1.74	1.95
0.003	−0.85	−3.32	−4.17	−0.24	1.78	2.02
0.004	−0.59	−3.03	−3.62	−0.28	1.84	2.11
0.005	−0.38	−2.78	−3.16	−0.32	1.91	2.23
0.006	−0.19	−2.57	−2.76	−0.36	1.99	2.36
0.007	−0.03	−2.38	−2.42	−0.41	0.72	1.13
0.008	0.10	−2.22	−2.12	−0.47	0.81	1.28
0.009	0.22	−2.08	−1.86	−0.54	0.90	1.44
0.01	0.33	−1.96	−1.63	−0.61	1.00	1.61
0.02	0.94	−1.21	−0.27	−3.68	4.25	7.93
0.03	1.20	−0.86	0.34	2.95	−2.26	−5.20
0.04	1.31	−0.65	0.67	1.50	−0.71	−2.21
0.05	1.36	−0.51	0.86	1.17	−0.29	−1.46

TABLE 6.5 (continued)

u	l/f_1	l/f_2	l/f	f/l	z_m/l	z_p/l
$d/l = 0.15$						
0.06	1.38	−0.41	0.97	1.03	−0.07	−1.11
0.07	1.37	−0.34	1.03	0.97	0.07	−0.90
0.08	1.35	−0.28	1.06	0.94	0.18	−0.76
0.09	1.32	−0.24	1.08	0.93	0.27	−0.65
0.10	1.28	−0.20	1.08	0.93	0.35	−0.58
0.20	0.89	−0.06	0.84	1.19	1.09	−0.10
0.30	0.59	−0.02	0.57	1.75	1.77	0.02
0.40	0.38	−0.01	0.37	2.70	2.77	0.07
0.50	0.23	−0.00	0.23	4.39	4.50	0.11
0.60	0.13	−0.00	0.13	7.70	7.83	0.13
0.70	0.07	−0.00	0.07	15.28	15.41	0.14
0.80	0.03	−0.00	0.03	38.13	38.27	0.15
0.90	0.01	−0.00	0.01	168.34	168.49	0.15

TABLE 6.6

CARDINAL POINTS OF THE UNIPOTENTIAL LENS EXPRESSED IN UNITS OF l, THE SPACING BETWEEN THE ELECTRODES

u	l/f_1	l/f_2	l/f	f/l	z_m/l	z_p/l
$d/l = 0.2$						
0.000001	0.00	−279.06	−279.06	−0.00	0.34	0.34
0.000002	0.70	−185.84	−185.15	−0.01	0.51	0.51
0.000003	1.04	−138.17	−137.13	−0.01	0.61	0.62
0.000004	1.24	−108.78	−107.55	−0.01	0.70	0.71
0.000005	1.35	−88.73	−87.38	−0.01	0.77	0.78
0.000006	1.42	−74.16	−72.74	−0.01	0.84	0.85
0.000007	1.46	−63.10	−61.64	−0.02	0.90	0.92
0.000008	1.48	−54.42	−52.94	−0.02	0.96	0.98
0.000009	1.49	−47.44	−45.95	−0.02	1.01	1.03
0.00001	1.49	−41.72	−40.23	−0.02	1.07	1.09
0.00002	1.27	−15.04	−13.76	−0.07	1.60	1.67
0.00003	1.00	−6.65	−5.65	−0.18	−0.27	−0.09
0.00004	0.76	−3.10	−2.33	−0.43	−1.51	−1.08
0.00005	0.55	−1.40	−0.85	−1.18	−4.35	−3.17
0.00006	0.37	−0.56	−0.19	−5.27	−16.60	−11.32
0.00007	0.21	−0.16	+0.04	22.54	53.03	30.49
0.00008	0.07	−0.02	+0.05	19.23	28.38	9.15
0.00009	−0.06	−0.01	−0.07	−15.21	−9.06	6.15
0.0001	−0.17	−0.09	−0.25	−3.95	0.99	4.94

TABLE 6.6 (continued)

u	l/f_1	l/f_2	l/f	f/l	z_m/l	z_p/l
$d/l=0.2$						
0.0002	−0.84	−1.73	−2.58	−0.39	2.37	2.75
0.0003	−1.16	−2.98	−4.13	−0.24	2.16	2.40
0.0004	−1.32	−3.76	−5.08	−0.20	2.05	2.24
0.0005	−1.41	−4.27	−5.68	−0.18	1.98	2.15
0.0006	−1.46	−4.60	−6.07	−0.16	1.93	2.10
0.0007	−1.48	−4.83	−6.31	−0.16	1.90	2.06
0.0008	−1.49	−4.98	−6.47	−0.15	1.87	2.03
0.0009	−1.48	−5.08	−6.56	−0.15	1.85	2.01
0.001	−1.47	−5.15	−6.61	−0.15	1.84	1.99
0.002	−1.18	−5.02	−6.19	−0.16	1.78	1.94
0.003	−0.87	−4.59	−5.46	−0.18	1.79	1.97
0.004	−0.61	−4.19	−4.80	−0.21	1.81	2.02
0.005	−0.39	−3.85	−4.24	−0.24	1.85	2.09
0.006	−0.20	−3.56	−3.76	−0.27	1.90	2.17
0.007	−0.04	−3.30	−3.34	−0.30	0.62	0.92
0.008	0.10	−3.08	−2.98	−0.34	0.69	1.03
0.009	0.22	−2.89	−2.67	−0.38	0.76	1.13
0.01	0.33	−2.72	−2.39	−0.42	0.83	1.25
0.02	0.95	−1.68	−0.73	−1.37	1.96	3.33
0.03	1.22	−1.19	0.03	39.77	−39.07	−78.84
0.04	1.34	−0.90	0.44	2.29	−1.49	−3.78
0.05	1.39	−0.71	0.68	1.47	−0.58	−2.05
0.06	1.40	−0.57	0.83	1.20	−0.23	−1.43
0.07	1.39	−0.47	0.92	1.08	−0.03	−1.11
0.08	1.37	−0.39	0.98	1.02	0.11	−0.91
0.09	1.34	−0.33	1.01	0.99	0.21	−0.78
0.10	1.31	−0.28	1.02	0.98	0.30	−0.68
0.20	0.91	−0.08	0.84	1.20	1.13	−0.07
0.30	0.60	−0.03	0.58	1.73	1.79	0.06
0.40	0.39	−0.01	0.38	2.65	2.77	0.12
0.50	0.24	0.00	0.23	4.30	4.45	0.15
0.60	0.13	−0.00	0.13	7.52	7.70	0.17
0.70	0.07	−0.00	0.07	14.90	15.09	0.19
0.80	0.03	−0.00	0.03	37.15	37.34	0.20
0.90	0.01	−0.00	0.01	163.88	164.07	0.20

TABLE 6.7

CARDINAL POINTS OF THE UNIPOTENTIAL LENS EXPRESSED IN UNITS OF l, THE SPACING BETWEEN THE ELECTRODES

u	l/f_1	l/f_2	l/f	f/l	z_m/l	z_p/l
$d/l = 0.25$						
0.000001	−0.01	−362.67	−362.68	−0.00	0.35	0.36
0.000002	0.70	−241.78	−241.07	−0.00	0.52	0.53
0.000003	1.06	−179.87	−178.82	−0.01	0.63	0.63
0.000004	1.25	−141.69	−140.43	−0.01	0.71	0.72
0.000005	1.37	−115.63	−114.25	−0.01	0.78	0.79
0.000006	1.44	−96.68	−95.23	−0.01	0.85	0.86
0.000007	1.49	−82.28	−80.80	−0.01	0.90	0.92
0.000008	1.51	−70.99	−69.48	−0.01	0.96	0.98
0.000009	1.52	−61.91	−60.39	−0.02	1.01	1.03
0.00001	1.52	−54.46	−52.95	−0.02	1.07	1.09
0.00002	1.30	−19.69	−18.39	−0.05	1.57	1.63
0.00003	1.03	−8.74	−7.71	−0.13	−0.19	−0.06
0.00004	0.78	−4.08	−3.30	−0.30	−1.25	−0.95
0.00005	0.57	−1.85	−1.29	−0.78	−3.41	−2.63
0.00006	0.38	−0.75	−0.37	−2.72	−10.17	−7.45
0.00007	0.22	−0.23	−0.01	−132.67	−365.28	−232.60
0.00008	0.08	−0.03	0.05	19.78	32.70	12.93
0.00009	−0.05	−0.01	−0.06	−16.47	−9.01	7.46
0.0001	−0.16	−0.10	−0.27	−3.73	1.94	5.67
0.0002	−0.85	−2.22	−3.08	−0.33	2.56	2.89
0.0003	−1.17	−3.83	−5.01	−0.20	2.28	2.48
0.0004	−1.34	−4.85	−6.20	−0.16	2.14	2.31
0.0005	−1.44	−5.51	−6.95	−0.14	2.06	2.21
0.0006	−1.49	−5.95	−7.44	−0.13	2.00	2.14
0.0007	−1.51	−6.25	−7.76	−0.13	1.96	2.09
0.0008	−1.52	−6.45	−7.96	−0.13	1.93	2.06
0.0009	−1.51	−6.58	−8.09	−0.12	1.91	2.03
0.001	−1.50	−6.67	−8.16	−0.12	1.89	2.01
0.002	−1.20	−6.51	−7.71	−0.13	1.80	1.93
0.003	−0.89	−5.96	−6.85	−0.15	1.78	1.93
0.004	−0.63	−5.45	−6.07	−0.16	1.79	1.96
0.005	−0.40	−5.00	−5.40	−0.19	1.81	2.00
0.006	−0.21	−4.62	−4.83	−0.21	1.84	2.05
0.007	−0.05	−4.29	−4.34	−0.23	0.57	0.80
0.008	0.09	−4.00	−3.91	−0.26	0.63	0.89
0.009	0.22	−3.75	−3.54	−0.28	0.68	0.97
0.01	0.33	−3.53	−3.20	−0.31	0.74	1.05
0.02	0.97	−2.19	−1.22	−0.82	1.42	2.23
0.03	1.23	−1.55	−0.32	−3.17	3.88	7.05
0.04	1.36	−1.17	0.19	5.34	−4.53	−9.87
0.05	1.41	−0.92	0.49	2.04	−1.14	−3.17

TABLE 6.7 (continued)

u	l/f_1	l/f_2	l/f	f/l	z_m/l	z_p/l
$d/l = 0.25$						
0.06	1.43	−0.74	0.68	1.46	−0.48	−1.95
0.07	1.42	−0.61	0.81	1.24	−0.18	−1.42
0.08	1.40	−0.51	0.89	1.13	0.00	−1.13
0.09	1.37	−0.43	0.94	1.07	0.14	−0.93
0.10	1.33	−0.37	0.96	1.04	0.24	−0.80
0.20	0.93	−0.10	0.83	1.20	1.16	−0.04
0.30	0.62	−0.03	0.58	1.71	1.82	0.10
0.40	0.40	−0.01	0.38	2.61	2.78	0.17
0.50	0.24	−0.00	0.24	4.21	4.41	0.20
0.60	0.14	−0.00	0.14	7.35	7.57	0.22
0.70	0.07	−0.00	0.07	14.52	14.76	0.24
0.80	0.03	−0.00	0.03	36.16	36.41	0.24
0.90	0.00	−0.00	0.01	159.41	159.66	0.25

TABLE 6.8

CARDINAL POINTS OF THE UNIPOTENTIAL LENS EXPRESSED IN UNITS OF l, THE SPACING BETWEEN THE ELECTRODES

u	l/f_1	l/f_2	l/f	f/l	z_m/l	z_p/l
$d/l = 0.3$						
0.000001	−0.02	−452.87	−452.88	−0.00	0.37	0.38
0.000002	0.71	−302.24	−301.52	−0.00	0.54	0.54
0.000003	1.07	−225.01	−223.94	−0.00	0.64	0.64
0.000004	1.27	−177.33	−176.06	−0.01	0.72	0.73
0.000005	1.40	−144.78	−143.39	−0.01	0.79	0.80
0.000006	1.47	−121.11	−119.64	−0.01	0.85	0.86
0.000007	1.51	−103.11	−101.60	−0.01	0.91	0.92
0.000008	1.54	−88.99	−87.46	−0.01	0.97	0.98
0.000009	1.55	−77.63	−76.09	−0.01	1.02	1.03
0.00001	1.54	−68.32	−66.77	−0.01	1.07	1.08
0.00002	1.33	−24.77	−23.44	−0.04	1.56	1.60
0.00003	1.05	−11.04	−9.98	−0.10	−0.14	−0.04
0.00004	0.80	−5.18	−4.38	−0.23	−1.08	−0.85
0.00005	0.58	−2.37	−1.78	−0.56	−2.87	−2.31
0.00006	0.39	−0.97	−0.57	−1.75	−7.63	−5.88
0.00007	0.23	−0.30	−0.70	−14.26	−45.42	−31.16
0.00008	0.08	−0.04	0.05	21.46	39.93	18.47
0.00009	−0.04	−0.01	−0.05	−18.38	−9.59	8.78
0.0001	−0.16	−0.12	−0.28	−3.57	2.76	6.33

TABLE 6.8 (continued)

u	l/f_1	l/f_2	l/f	f/l	z_m/l	z_p/l
$d/l=0.3$						
0.0002	−0.86	−2.74	−3.60	−0.28	2.70	2.98
0.0003	−1.19	−4.74	−5.93	−0.17	2.36	2.53
0.0004	−1.37	−6.02	−7.38	−0.14	2.21	2.34
0.0005	−1.46	−6.84	−8.31	−0.12	2.11	2.23
0.0006	−1.51	−7.40	−8.91	−0.11	2.05	2.16
0.0007	−1.54	−7.77	−9.31	−0.11	2.00	2.11
0.0008	−1.54	−8.02	−9.56	−0.10	1.96	2.07
0.0009	−1.54	−8.19	−9.72	−0.10	1.94	2.04
0.001	−1.52	−8.29	−9.82	−0.10	1.91	2.01
0.002	−1.23	−8.11	−9.33	−0.11	1.80	1.91
0 003	−0.91	−7.43	−8.35	−0.12	1.77	1.89
0.004	−0.64	−6.79	−7.44	−0.13	1.77	1.90
0.005	−0.42	−6.24	−6.66	−0.15	1.78	1.93
0.006	−0.22	−5.77	−5.99	−0.17	1.79	1.96
0.007	−0.05	−5.36	−5.41	−0.18	1.81	1.99
0.008	0.09	−5.00	−4.91	−0.20	0.60	0.80
0.009	0.22	−4.69	−4.47	−0.22	0.64	0.87
0.01	0.33	−4.41	−4.08	−0.24	0.69	0.93
0.02	0.98	−2.74	−1.76	−0.57	1.18	1.75
0.03	1.25	−1.94	−0.69	−1.46	2.19	3.64
0.04	1.38	−1.47	−0.08	−11.80	12.62	24.42
0.05	1.44	−1.15	0.28	3.52	−2.61	−6.13
0.06	1.45	−0.93	0.52	1.92	−0.93	−2.85
0.07	1.45	−0.77	0.68	1.47	−0.41	−1.88
0.08	1.43	−0.64	0.78	1.28	−0.14	−1.42
0.09	1.40	−0.54	0.85	1.17	0.03	−1.14
0.10	1.36	−0.46	0.90	1.12	0.16	−0.95
0.20	0.95	−0.13	0.83	1.21	1.20	−0.01
0.30	0.63	−0.04	0.59	1.70	1.84	0.15
0.40	0.41	−0.02	0.39	2.56	2.78	0.21
0.50	0.25	−0.01	0.24	4.12	4.37	0.25
0.60	0.14	−0.00	0.14	7.17	7.44	0.27
0.70	0.07	−0.00	0.07	14.15	14.43	0.29
0.80	0.03	−0.00	0.03	35.18	35.48	0.29
0.90	0.01	−0.00	0.01	154.94	155.24	0.30

TABLE 6.9

CARDINAL POINTS OF THE UNIPOTENTIAL LENS EXPRESSED IN UNITS OF l, THE SPACING BETWEEN THE ELECTRODES

u	l/f_1	l/f_2	l/f	f/l	z_m/l	z_p/l
$d/l = 0.35$						
0.000001	−0.03	−550.30	−550.33	−0.00	1.61	1.61
0.000002	0.72	−367.67	−366.95	−0.00	0.55	0.56
0.000003	1.09	−273.92	−272.83	−0.00	0.65	0.66
0.000004	1.29	−216.00	−214.70	−0.00	0.73	0.74
0.000005	1.42	−176.43	−175.02	−0.01	0.80	0.81
0.000006	1.50	−147.64	−146.15	−0.01	0.86	0.87
0.000007	1.54	−125.76	−124.22	−0.01	0.92	0.93
0.000008	1.56	−108.58	−107.01	−0.01	0.97	0.98
0.000009	1.57	−94.75	−93.18	−0.01	1.02	1.03
0.00001	1.57	−83.41	−81.83	−0.01	1.07	1.08
0.00002	1.36	−30.34	−28.98	−0.03	1.54	1.58
0.00003	1.08	−13.56	−12.49	−0.08	−0.10	−0.02
0.00004	0.82	−6.39	−5.57	−0.18	−0.96	−0.78
0.00005	0.60	−2.94	−2.34	−0.43	−2.51	−2.08
0.00006	0.41	−1.22	−0.81	−1.24	−6.23	−4.99
0.00007	0.24	−0.39	−0.15	−6.88	−25.14	−18.26
0.00008	0.09	−0.05	+0.04	25.40	53.33	27.94
0.00009	−0.04	−0.01	−0.05	−21.31	−11.14	10.17
0.0001	−0.16	−0.13	−0.29	−3.45	3.47	6.92
0.0002	−0.87	−3.28	−4.15	−0.24	2.80	3.04
0.0003	−1.21	−5.71	−6.92	−0.14	2.42	2.56
0.0004	−1.39	−7.26	−8.65	−0.12	2.25	2.36
0.0005	−1.49	−8.27	−9.75	−0.10	2.14	2.24
0.0006	−1.54	−8.94	−10.48	−0.10	2.07	2.17
0.0007	−1.57	−9.39	−10.96	−0.09	2.02	2.11
0.0008	−1.57	−9.70	−11.27	−0.09	1.98	2.07
0.0009	−1.57	−9.91	−11.48	−0.09	1.95	2.04
0.001	−1.55	−10.04	−11.59	−0.09	1.93	2.01
0.002	−1.26	9.83	−11.08	−0.09	1.80	1.89
0.003	−0.94	−9.02	−9.96	−0.10	1.76	1.86
0.004	−0.66	−8.25	−8.91	−0.11	1.75	1.86
0.005	−0.43	−7.58	−8.01	−0.12	1.74	1.87
0.006	−0.23	−7.01	−7.24	−0.14	1.75	1.89
0.007	−0.06	−6.51	−6.57	−0.15	1.76	1.91
0.008	0.08	−6.08	−5.99	−0.17	0.58	0.75
0.009	0.21	−5.70	−5.49	−0.18	0.62	0.80
0.01	0.33	−5.36	−5.04	−0.20	0.66	0.86
0.02	1.00	−3.34	−2.34	−0.43	1.06	1.48
0.03	1.28	−2.36	−1.09	−0.92	1.66	2.58
0.04	1.41	−1.79	−0.38	−2.62	3.46	6.08
0.05	1.46	−1.41	0.06	17.31	−16.40	−33.71
0.06	1.48	−1.14	0.34	2.90	−1.91	−4.80

TABLE 6.9 (continued)

u	l/f_1	l/f_2	l/f	f/l	z_m/l	z_p/l
$d/l = 0.35$						
0.07	1.47	−0.94	0.54	1.86	−0.79	−2.65
0.08	1.45	−0.78	0.67	1.49	−0.35	−1.84
0.09	1.42	−0.66	0.76	1.31	−0.11	−1.42
0.10	1.39	−0.57	0.82	1.22	0.06	−1.16
0.20	0.98	−0.16	0.82	1.22	1.24	0.02
0.30	0.65	−0.05	0.60	1.68	1.87	0.19
0.40	0.42	−0.02	0.40	2.52	2.78	0.26
0.50	0.25	−0.01	0.25	4.03	4.33	0.30
0.60	0.15	−0.00	0.14	6.99	7.31	0.32
0.70	0.07	−0.00	0.07	13.77	14.11	0.34
0.80	0.03	−0.00	0.03	34.20	34.54	0.34
0.90	0.01	−0.00	0.01	150.47	150.82	0.35

TABLE 6.10

CARDINAL POINTS OF THE UNIPOTENTIAL LENS EXPRESSED IN UNITS OF l, THE SPACING BETWEEN THE ELECTRODES

u	l/f_1	l/f_2	l/f	f/l	z_m/l	z_p/l
$d/l = 0.4$						
0.000001	−0.03	−655.69	−655.72	−0.00	1.59	1.59
0.000002	0.73	−438.59	−437.87	−0.00	0.57	0.57
0.000003	1.10	−326.99	−325.89	−0.00	0.67	0.67
0.000004	1.32	−258.00	−256.68	−0.00	0.75	0.75
0.000005	1.44	−210.84	−209.40	−0.00	0.81	0.82
0.000006	1.52	−176.52	−174.99	−0.01	0.87	0.88
0.000007	1.57	−150.41	−148.84	−0.01	0.93	0.93
0.000008	1.59	−129.91	−128.31	−0.01	0.98	0.98
0.000009	1.61	−113.41	−111.80	−0.01	1.03	1.04
0.00001	1.61	−99.86	−98.26	−0.01	1.07	1.08
0.00002	1.39	−36.44	−35.06	−0.03	1.53	1.56
0.00003	1.10	−16.35	−15.25	−0.07	−0.06	0.01
0.00004	0.84	−7.75	−6.90	−0.14	−0.86	−0.71
0.00005	0.62	−3.59	−2.97	−0.34	−2.24	−1.90
0.00006	0.42	−1.50	−1.08	−0.93	−5.32	−4.39
0.00007	0.25	−0.49	−0.24	−4.25	−17.67	−13.42
0.00008	0.10	−0.07	0.03	35.73	84.81	49.08
0.00009	−0.03	−0.01	−0.04	−26.08	−14.40	11.68
0.0001	−0.15	−0.15	−0.30	−3.37	4.11	7.48
0.0002	−0.89	−3.85	−4.73	−0.21	2.87	3.08

TABLE 6.10 (continued)

u	l/f_1	l/f_2	l/f	f/l	z_m/l	z_p/l
$d/l = 0.4$						
0.0003	−1.23	−6.73	−7.96	−0.13	2.45	2.58
0.0004	−1.41	−8.58	−9.99	−0.10	2.27	2.37
0.0005	−1.51	−9.79	−11.30	−0.09	2.16	2.25
0.0006	−1.57	−10.59	−12.16	−0.08	2.08	2.17
0.0007	−1.60	−11.14	−12.73	−0.08	2.03	2.11
0.0008	−1.61	−11.51	−13.11	−0.08	1.99	2.07
0.0009	−1.60	−11.75	−13.35	−0.07	1.96	2.03
0.001	−1.59	−11.92	−13.50	−0.07	1.93	2.00
0.002	−1.28	−11.68	−12.97	−0.08	1.80	1.87
0.003	−0.96	−10.73	−11.69	−0.09	1.75	1.83
0.004	−0.68	−9.82	−10.50	−0.10	1.72	1.82
0.005	−0.45	−9.03	−9.47	−0.11	1.71	1.82
0.006	−0.24	−8.34	−8.59	−0.12	1.71	1.83
0.007	−0.07	−7.76	−7.83	−0.13	1.71	1.84
0.008	0.08	−7.24	−7.17	−0.14	1.78	1.92
0.009	0.21	−6.79	−6.58	−0.15	0.61	0.76
0.01	0.33	−6.40	−6.07	−0.16	0.65	0.81
0.02	1.01	−3.98	−2.97	−0.34	0.98	1.32
0.03	1.30	−2.82	−1.53	−0.65	1.41	2.06
0.04	1.43	−2.14	−0.70	−1.42	2.26	3.68
0.05	1.49	−1.68	−0.19	−5.26	6.19	11.45
0.06	1.51	−1.36	0.15	6.63	−5.64	−12.27
0.07	1.50	−1.12	0.38	2.61	−1.54	−4.14
0.08	1.48	−0.94	0.55	1.83	−0.69	−2.52
0.09	1.45	−0.79	0.66	1.52	−0.31	−1.82
0.10	1.42	−0.68	0.74	1.35	−0.07	−1.43
0.20	1.00	−0.19	0.81	1.23	1.29	0.05
0.30	0.66	−0.06	0.60	1.66	1.90	0.24
0.40	0.43	−0.02	0.40	2.47	2.78	0.31
0.50	0.26	−0.01	0.25	3.94	4.29	0.35
0.60	0.15	0.00	0.15	6.81	7.19	0.37
0.70	0.08	−0.00	0.07	13.39	13.78	0.39
0.80	0.03	−0.00	0.03	33.22	33.61	0.39
0.90	0.01	−0.00	0.01	146.01	146.41	0.40

TABLE 6.11

CARDINAL POINTS OF THE UNIPOTENTIAL LENS EXPRESSED IN UNITS OF l, THE SPACING BETWEEN THE ELECTRODES

u	l/f_1	l/f_2	l/f	f/l	z_m/l	z_p/l
$d/l = 0.45$						
0.000001	−0.04	−769.84	−769.88	−0.00	1.57	1.57
0.000002	0.73	−515.57	−514.83	−0.00	0.59	0.59
0.000003	1.12	−384.68	−383.56	−0.00	0.68	0.68
0.000004	1.34	−303.69	−302.35	−0.00	0.76	0.76
0.000005	1.47	−248.31	−246.84	−0.00	0.82	0.83
0.000006	1.55	−207.97	−206.42	−0.00	0.88	0.88
0.000007	1.60	−177.29	−175.69	−0.01	0.93	0.94
0.000008	1.63	−153.18	−151.56	−0.01	0.98	0.99
0.000009	1.64	−133.77	−132.14	−0.01	1.03	1.04
0.00001	1.64	−117.84	−116.20	−0.01	1.08	1.09
0.00002	1.42	−43.15	−41.73	−0.02	1.52	1.54
0.00003	1.13	−19.43	−18.30	−0.05	−0.03	0.03
0.00004	0.87	−9.25	−8.38	−0.12	−0.77	−0.65
0.00005	0.64	−4.31	−3.67	−0.27	−2.02	−1.75
0.00006	0.44	−1.82	−1.38	−0.72	−4.66	−3.94
0.00007	0.27	−0.61	−0.34	−2.92	−13.72	−10.81
0.00008	0.11	−0.10	0.01	88.11	236.57	148.46
0.00009	−0.02	−0.00	−0.03	−34.62	−21.24	13.38
0.0001	−0.15	−0.15	−0.30	−3.33	4.69	8.02
0.0002	−0.90	−4.44	−5.34	−0.19	2.92	3.11
0.0003	−1.25	−7.83	−9.07	−0.11	2.47	2.58
0.0004	−1.44	−9.99	−11.43	−0.09	2.28	2.37
0.0005	−1.54	−11.41	−12.95	−0.08	2.17	2.24
0.0006	−1.60	−12.36	−13.96	−0.07	2.09	2.16
0.0007	−1.63	−13.00	−14.63	−0.07	2.03	2.10
0.0008	−1.64	−13.44	−15.08	−0.07	1.99	2.06
0.0009	−1.63	−13.74	−15.37	−0.07	1.95	2.02
0.001	−1.62	−13.93	−15.55	−0.06	1.93	1.99
0.002	−1.32	−13.69	−15.00	−0.07	1.78	1.85
0.003	−0.99	−12.58	−13.57	−0.07	1.73	1.80
0.004	−0.70	−11.52	−12.22	−0.08	1.70	1.78
0.005	−0.46	−10.59	−11.05	−0.09	1.69	1.78
0.006	−0.26	−9.79	−10.05	−0.10	1.68	1.78
0.007	−0.08	−9.11	−9.19	−0.11	1.68	1.79
0.008	0.07	−8.51	−8.43	−0.12	1.68	1.80
0.009	0.21	−7.98	−7.77	−0.13	0.61	0.73
0.01	0.33	−7.51	−7.19	−0.14	0.64	0.78
0.02	1.02	−4.69	−3.66	−0.27	0.93	1.21
0.03	1.32	−3.32	−2.01	−0.50	1.27	1.76
0.04	1.46	−2.52	−1.06	−0.94	1.80	2.74
0.05	1.52	−1.98	−0.46	−2.17	3.10	5.27

TABLE 6.11 (continued)

u	l/f_1	l/f_2	l/f	f/l	z_m/l	z_p/l
$d/l=0.45$						
0.06	1.54	−1.60	−0.06	−15.96	16.96	32.92
0.07	1.53	−1.32	0.21	4.69	−3.61	−8.30
0.08	1.51	−1.11	0.41	2.45	−1.31	−3.75
0.09	1.48	−0.94	0.55	1.83	−0.62	−2.44
0.10	1.45	−0.80	0.65	1.54	−0.27	−1.81
0.20	1.02	−0.22	0.80	1.25	1.33	0.09
0.30	0.68	−0.08	0.61	1.65	1.93	0.28
0.40	0.44	−0.03	0.41	2.43	2.79	0.36
0.50	0.27	−0.01	0.26	3.85	4.25	0.40
0.60	0.15	−0.00	0.15	6.64	7.06	0.42
0.70	0.08	−0.00	0.08	13.01	13.45	0.44
0.80	0.03	−0.00	0.03	32.23	32.68	0.44
0.90	0.01	−0.00	0.01	141.54	141.99	0.45

TABLE 6.12

CARDINAL POINTS OF THE UNIPOTENTIAL LENS EXPRESSED IN UNITS OF l, THE SPACING BETWEEN THE ELECTRODES

u	l/f_1	l/f_2	l/f	f/l	z_m/l	z_p/l
$d/l=0.5$						
0.000001	−0.05	−893.68	−893.68	−0.00	1.55	1.55
0.000002	0.74	−599.25	598.51	−0.00	0.60	0.60
0.000003	1.14	−447.46	−446.33	−0.00	0.70	0.70
0.000004	1.36	−353.47	−352.11	−0.00	0.77	0.77
0.000005	1.50	−289.17	−287.67	−0.00	0.83	0.84
0.000006	1.58	−242.31	−240.73	−0.00	0.89	0.89
0.000007	1.63	−206.64	−205.01	−0.00	0.94	0.95
0.000008	1.66	−178.62	−176.96	−0.01	0.99	1.00
0.000009	1.67	−156.05	−154.38	−0.01	−1.04	1.04
0.00001	1.67	−137.51	−135.84	−0.01	1.08	1.09
0.00002	1.45	−50.53	−49.07	−0.02	1.51	1.53
0.00003	1.16	−22.84	−21.68	−0.05	0.00	0.05
0.00004	0.89	−10.93	−10.03	−0.10	−0.70	−0.60
0.00005	0.66	−5.13	−4.47	−0.22	−1.84	−1.62
0.00006	0.46	−2.20	−1.74	−0.57	−4.15	−3.57
0.00007	0.28	−0.75	−0.47	−2.12	−11.24	−9.12
0.00008	0.12	−0.13	−0.01	−84.09	−255.44	−171.34
0.00009	−0.02	−0.00	−0.02	−53.08	−37.70	15.38
0.0001	−0.14	−0.16	−0.30	−3.33	5.22	8.55

TABLE 6.12 (continued)

u	l/f_1	l/f_2	l/f	f/l	z_m/l	z_p/l
$d/l=0.5$						
0.0002	−0.91	−5.07	−5.98	−0.17	2.95	3.12
0.0003	−1.27	−8.99	−10.26	−0.10	2.48	2.58
0.0004	−1.46	−11.51	−12.97	−0.08	2.28	2.36
0.0005	−1.57	−13.15	−14.73	−0.07	2.16	2.23
0.0006	−1.63	−14.26	−15.89	−0.06	2.09	2.15
0.0007	−1.66	−15.01	−16.68	−0.06	2.03	2.09
0.0008	−1.67	−15.53	−17.20	−0.06	1.98	2.04
0.0009	−1.67	−15.88	−17.54	−0.06	1.95	2.00
0.001	−1.65	−16.11	−17.76	−0.06	1.92	1.97
0.002	−1.35	−15.85	−17.20	−0.06	1.77	1.83
0.003	−1.01	−14.58	−15.59	−0.06	1.71	1.78
0.004	−0.72	−13.36	−14.08	−0.07	1.68	1.75
0.005	−0.48	−12.29	−12.77	−0.08	1.66	1.74
0.006	−0.27	−11.37	−11.64	−0.09	1.65	1.73
0.007	−0.09	−10.57	−10.66	−0.09	1.64	1.74
0.008	0.07	−9.88	−9.81	−0.10	1.64	1.74
0.009	0.20	−9.27	−9.06	−0.11	1.64	1.75
0.01	0.32	−8.73	−8.40	−0.12	0.64	0.76
0.02	1.04	−5.45	−4.41	−0.23	0.90	1.13
0.03	1.34	−3.87	−2.53	−0.40	1.17	1.57
0.04	1.48	−2.93	−1.45	−0.69	1.56	2.25
0.05	1.55	−2.31	−0.76	−1.32	2.26	3.57
0.06	1.57	−1.87	−0.30	−3.36	4.38	7.74
0.07	1.57	−1.54	0.03	39.53	−38.45	−77.99
0.08	1.55	−1.29	0.26	3.91	−2.76	−6.67
0.09	1.52	−1.09	0.42	2.36	−1.15	−3.51
0.10	1.48	−0.94	0.55	1.83	−0.56	−2.39
0.20	1.05	−0.26	0.79	1.27	1.39	0.12
0.30	0.70	−0.09	0.61	1.64	1.97	0.33
0.40	0.45	−0.03	0.42	2.39	2.80	0.41
0.50	0.28	−0.01	0.27	3.76	4.21	0.45
0.60	0.16	−0.00	0.15	6.46	6.93	0.47
0.70	0.08	−0.00	0.08	12.64	13.12	0.49
0.80	0.03	−0.00	0.03	31.25	31.74	0.49
0.90	0.01	−0.00	0.01	137.07	137.57	0.50

TABLE 6.13

Lens number[a]	V_1/V_2	f_o/R accuracy ±1%	z_p/R accuracy ±20%
Fig. 10.13a	0	6.20	−0.22
Fig. 10.13b	0	3.84	−0.20
Fig. 10.13c	0	2.78	−0.18
Fig. 10.13d	0	1.84	−0.18
Fig. 10.13e	0	1.94	−0.18
Fig. 10.13f	0	1.72	−0.22
Fig. 10.13g	0	3.30	−0.08
Fig. 10.13h	0	1.74	−0.08
Fig. 10.13i	0	1.56	−0.18
Fig. 10.13j	0	2.30	−0.12
Fig. 10.13k	0	1.86	−0.18
Fig. 10.13l	0	1.86	−0.18
Fig. 6.30	+0.20	4.86	−0.22
Fig. 6.30	+0.10	3.40	−0.22
Fig. 6.30	0	2.44	−0.22
Fig. 6.30	−0.10	1.66	−0.22
Fig. 6.30	−0.20	7.16	−0.22
Fig. 6.31	+0.30	8.76	−0.24
Fig. 6.31	+0.20	5.64	−0.24
Fig. 6.31	+0.10	4.04	−0.24
Fig. 6.31	0	2.76	−0.24
Fig. 6.31	−0.10	1.88	−0.24

[a] Fig. 10.13 appears in Volume 2.

VII

Optical Image Formation by Rotationally Symmetrical Magnetic Fields

The analogy between the propagation of light and the motion of charged particles inside a magnetic field is not as obvious as that for electrostatic fields. Nevertheless, the study of electron optics for the case of magnetic fields will follow the procedure adopted in the case of electrostatic fields very closely.

From electron ballistics we already know that in a magnetic field the charged particle is accelerated in a direction perpendicular to both the field direction and the velocity direction. Therefore, a constant magnetic field does not change the total energy of the particle, but alters the curvature of its path.

7.1 The Homogeneous Magnetic Field

We shall start our analysis with the simplest case, namely, the homogeneous magnetic field. This can be very closely approximated by a long solenoid. A coil is referred to as "long" when its length is much greater than its diameter. In other words, we are neglecting fringe fields, and we assume an abrupt start and termination of the field at the entrance and exit of the coil, respectively. Of course, we can show that such a field can never exist in practice, but for a long solenoid the approximation is valid.

The force developed by the magnetic field is centripetal and is balanced by the centrifugal force developed by the circular motion. We have

$$Bev = mv^2/\rho \tag{7.1}$$

where B is the flux density in teslas, v the velocity in meters per second, e and m the charge and mass of the electron in coulombs and kilograms, respectively, and ρ the radius of curvature in meters. Here, B is assumed constant and v is generally constant.

$$\rho = (m/e)(v/B) \tag{7.2}$$

The period τ which is required to cover one full circumference of the circle $2\pi\rho$ is

$$\tau = 2\pi\rho/v = (2\pi m/e)(1/B) \tag{7.3}$$

The period is therefore independent of the velocity of the electrons and of the radius of the trajectory.

Let us now consider a point H_o from which electrons originate with velocity v and at angle θ with the axis of the magnetic field, which coincides with the axis of the solenoid. Since we have assumed that the field starts abruptly, the point H_o can be either inside or outside the solenoid. The longitudinal component of the velocity is

$$v_z = v \cos\theta \tag{7.4}$$

along the axis of the coil, and the transversal motion is perpendicular to the axis

$$v_r = v \sin\theta \tag{7.5}$$

The longitudinal component is parallel to the magnetic field, and hence is not affected by it.

The transversal motion will follow a circle of radius

$$\rho = (mv/eB)\sin\theta \tag{7.6}$$

The path of the electron is, of course, a helix. If we consider the trajectories of electrons starting from H_o with various initial angles $\theta_1, \theta_2, \ldots,$ they will intersect the line through H_o parallel to the axis at various distances from H_o given by

$$l_1 = (v\cos\theta_1)\tau, \qquad l_2 = (v\cos\theta_2)\tau, \quad \text{etc.} \tag{7.7}$$

The condition for focusing of the electron trajectories originating from one point H_o is that

$$l = v_z\tau = (2\pi mv/eB)\cos\theta \tag{7.8}$$

be the same for all electrons. Now, if we expand $\cos\theta$ into its series and consider only paraxial rays, $\cos\theta \approx 1$ and

$$l_0 = 2\pi mv/eB = \text{constant} \tag{7.9}$$

for all small angles θ. The long solenoid has therefore the capability of forming images only for paraxial rays. Since initially parallel rays emerge parallel due to the lack of interaction with the homogeneous magnetic field, the focal length is here infinite. The image is always erect since a full 2π-rotation took place and the magnification is unity because the image and object sizes are equal. The projection of the helical path on a transverse plane is shown in Fig. 7.1.

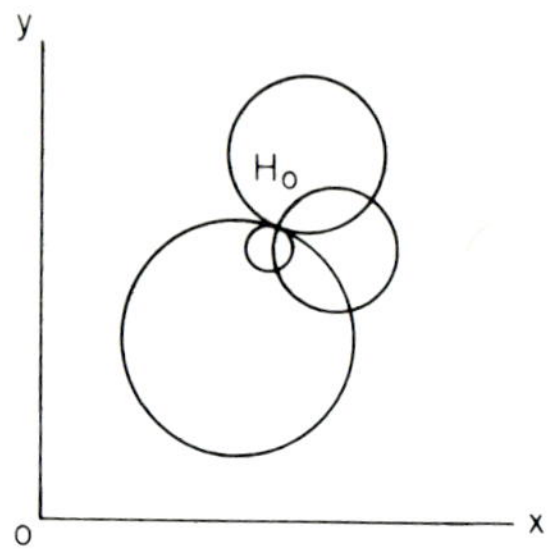

Fig. 7.1. The projection of the electron paths on an *xy* plane, transverse to the axis of symmetry.

The use of the long solenoid is not restricted to focusing of charged particles. If B and v are known, and if the angle is fixed by means of a circular opening, the ratio e/m can be measured. This was done by various researchers. It is also possible to determine the variation in velocity of the electrons emanating from H_0 if all the remaining parameters of Eq. (7.8) are kept constant. This is the principle of velocity spectrographs.

7.2 The Inhomogeneous Magnetic Field

If the length of the coil is not large compared with the radius, the field will have not only a longitudinal component but also a radial component, and both components of the field will depend on the axial as well as the radial coordinates.

In general, for negative charges, i.e., electrons, we have

$$m\mathbf{a} = e\mathbf{B} \times \mathbf{v} \tag{7.10}$$

In cylindrical coordinates the components of the velocity of the electron are

$$v_r = \dot{r}, \qquad v_\theta = r\dot{\theta}, \qquad v_z = \dot{z} \tag{7.11}$$

in which the dots above the variables represent differentiation with respect to time. The components of B are B_r, B_θ, and B_z. Expanding Eq. (7.10), we have

$$F_r = eB_\theta \dot{z} - eB_z r\dot{\theta} \tag{7.12}$$

$$F_\theta = eB_z \dot{r} - eB_r \dot{z} \tag{7.13}$$

$$F_z = eB_r r\dot{\theta} - eB_\theta \dot{r} \tag{7.14}$$

where F_r, F_θ, and F_z are the forces in the r, θ, and z directions, respectively.

Now the radial force F_r is the net result of the difference between the inward force $m\ddot{r}$ and the centrifugal force $mr\dot{\theta}^2$

$$F_r = m\ddot{r} - mr\dot{\theta}^2 \tag{7.15}$$

Next, the time rate of change of the angular momentum $mrv_\theta = mr^2\dot{\theta}$ is equal to the moment of the force with respect to the axis of rotation

$$\frac{d}{dt}(mr^2\dot{\theta}) = rF_\theta \tag{7.16}$$

or

$$F_\theta = \frac{1}{r}\frac{d}{dt}(mr^2\dot{\theta}) \tag{7.17}$$

and finally the equation for the longitudinal motion is

$$m\ddot{z} = F_z \tag{7.18}$$

Since we have rotational symmetry, the circular component of the magnetic field B_θ, is zero. Keeping that in mind, and substituting Eqs. (7.16)–(7.18) in Eqs. (7.12)–(7.14) we finally have

$$m\ddot{r} = -eB_z r\dot{\theta} + mr\dot{\theta}^2 \tag{7.19}$$

$$\frac{d}{dt}(mr^2\dot{\theta}) = er\dot{r}B_z - erB_r\dot{z} \tag{7.20}$$

$$m\ddot{z} = eB_r r\dot{\theta} \tag{7.21}$$

7.3 Series Expansion of the Magnetic Field

The magnetic scalar potential Ψ obeys Laplace's equation as does the electric potential ϕ. Therefore, in rotationally symmetrical systems, we can express the potential $\Psi(r, z)$ as a function of its variation along the axis. We can thus immediately write the same series for Ψ as for ϕ in Eq. (2.30),

$$\Psi(r, z) = \Psi(z) - \frac{r^2}{4}\Psi''(z) + \frac{r^4}{64}\Psi^{\mathrm{IV}}(z) \cdots \tag{7.22}$$

or

$$\Psi(r, z) = \sum_{n=0}^{\infty} \frac{(-1)^n \Psi^{(2n)}}{(n!)^2} \left(\frac{r}{2}\right)^{2n} \tag{7.23}$$

where the exponent between parentheses represents an order of derivation and not a power.

$$B_z = -\mu(\partial\Psi/\partial z) \qquad \text{and} \qquad B_r = -\mu(\partial\Psi/\partial r) \tag{7.24}$$

We therefore have

$$B_z(r, z) = B(z) - \frac{r^2}{4} B''(z) + \frac{r^4}{64} B^{\text{IV}}(z) \cdots \tag{7.25}$$

or

$$B_z(r, z) = \sum_{n=0}^{\infty} \frac{(-1)^n B^{(2n)}}{(n!)^2} \left(\frac{r}{2}\right)^{2n} \tag{7.26}$$

and similarly

$$B_r(r, z) = -\frac{r}{2} B'(z) + \frac{r^3}{16} B'''(z) \cdots \tag{7.27}$$

or

$$B_r(r, z) = \sum_{n=1}^{\infty} \frac{(-1)^n B^{(2n-1)}}{n!(n-1)!} \left(\frac{r}{2}\right)^{2n-1} \tag{7.28}$$

We shall now define a magnetic vector potential **A** so that

$$\mathbf{B} = \text{curl}\, \mathbf{A} \tag{7.29}$$

Let us expand Eq. (7.29) in cylindrical coordinates

$$B_r = \frac{1}{r}\frac{\partial A_z}{\partial \theta} - \frac{\partial A_\theta}{\partial z} \tag{7.30}$$

$$B_\theta = \frac{\partial A_r}{\partial z} - \frac{\partial A_z}{\partial r} \tag{7.31}$$

$$B_z = \frac{1}{r}\frac{\partial}{\partial r}(rA_\theta) - \frac{1}{r}\frac{\partial A_r}{\partial \theta} \tag{7.32}$$

Since we are dealing with rotationally symmetrical systems, all variations with respect to θ vanish. Hence, $B_\theta = 0$, $(\partial A_z/\partial\theta)$, and $(\partial A_r/\partial\theta)$ are also equal to zero. We therefore have

$$B_z = \frac{1}{r}\frac{\partial}{\partial r}(rA_\theta) \tag{7.33}$$

If we now substitute for B_z the series given by Eq. (7.26) and perform the integration, we obtain

$$A_\theta = \frac{r}{2} B(z) - \frac{r^3}{16} B''(z) + \frac{r^5}{384} B^{IV}(z) \cdots \tag{7.34}$$

or

$$A_\theta = \sum_{n=0}^{\infty} \frac{(-1)^n B^{(2n)}}{n!(n+1)!} \left(\frac{r}{2}\right)^{2n+1} \tag{7.35}$$

It should be borne in mind that all the above relations are only valid for the case where the region under consideration does not include any current flow. Strictly speaking, the electron beam itself is a current flow but the magnetic field it generates is so weak that it can be ignored.

7.4 Paraxial-Ray Equation

We shall confine our analysis to paraxial rays. In that case we can limit our development to first order terms, and Eq. (7.27) can now be written as

$$B_r = -\frac{r}{2} B'(z) \tag{7.36}$$

If we substitute this in Eq. (7.20), we have

$$\frac{d}{dt}(mr^2\dot{\theta}) = eB_z r\dot{r} + e\frac{r^2}{2}\frac{\partial B_z}{\partial z}\dot{z} = \frac{d}{dt}\left(\frac{e}{2} r^2 B_z\right) \tag{7.37}$$

Equation (7.37) can now be integrated to give

$$mr^2\dot{\theta} = \frac{e}{2} r^2 B_z + C \tag{7.38}$$

where C is a constant.

If we assume that $\dot{\theta} = 0$ outside the influence of the field, i.e., where $B_z = 0$, then C, the constant of integration, is zero. This means we are only considering meridional rays, that is, rays whose original direction in the field free space is situated in a plane through the original point and the axis. We will deal with skew rays later.

We therefore have

$$\dot{\theta} = (e/2m)B_z \tag{7.39}$$

This indicates that the angular velocity is dependent only on the axial component of the magnetic field. If we substitute Eqs. (7.36) and (7.39) in Eq. (7.21), we have

$$m\ddot{z} = -\frac{e}{2}\frac{\partial B_z}{\partial z} r^2 \frac{e}{2m} B(z) \tag{7.40}$$

or

$$\ddot{z} = -(e/2m)^2 r^2 B_z \, \partial B_z/\partial z \tag{7.41}$$

Since we are limiting our analysis to first order terms, Eq. (7.41) is equal to zero. This means that the axial velocity of the electrons is constant if we confine ourselves to paraxial rays.

If we now substitute Eq. (7.39) in Eq. (7.19) ,we have

$$\ddot{r} = -(e^2/4m^2) B_z{}^2 r \tag{7.42}$$

This is an important conclusion similar to the one we reached in the case of electrostatic fields. Equation (7.42) tells us that in the case of magnetic fields, as with electrostatic fields, the force that drives the electrons toward the axis (the direction is inward because of the negative sign) is directly proportional to the distance r from the axis. This, as we already know, is the principle of a focusing field.

It is advisable to obtain Eq. (7.42) without the time variable.

$$v = v_z = dz/dt = (2eV/m)^{1/2} \tag{7.43}$$

where V is the accelerating potential.

$$\frac{d^2r}{dt^2} = \frac{d}{dz}\left(\frac{dr}{dt}\right)\frac{dz}{dt} = \left(\frac{2eV}{m}\right)^{1/2} \frac{d}{dz}\left(\frac{dr}{dz}\frac{dz}{dt}\right) = \frac{2eV}{m}\frac{d^2r}{dz^2} \tag{7.44}$$

Hence,

$$\frac{2eV}{m}\frac{d^2r}{dz^2} = -\frac{e^2}{4m^2} B_z{}^2 r \tag{7.45}$$

or

$$r'' = -(e/8mV) B_z{}^2 r \tag{7.46}$$

where the primes indicate differentiation with respect to z.

We have obtained a differential equation for the paraxial electron trajectories in a rotationally symmetrical magnetic field which is of second order, linear, and homogeneous. All the conclusions arrived at in the case of electrostatic lenses are therefore applicable here.

An important observation can be made from Eq. (7.42). The force is proportional to the square of the magnetic field. This means that if we reverse the direction of the magnetic field by reversing the current, the direction of the force toward the axis should not change. In other words, there will be no change in the focus.

7.5 Image Rotation

Equation (7.39) indicates that a rotation of the image occurs in a magnetic field. To eliminate the time variable, we write

$$\frac{d\theta}{dt} = \left(\frac{2eV}{m}\right)^{1/2} \frac{d\theta}{dz} \tag{7.47}$$

or using Eq. (7.39),

$$\frac{d\theta}{dz} = \left(\frac{e}{8mV}\right)^{1/2} B_z(z) \tag{7.48}$$

Starting from an initial point $z = z_o$, the total angle of rotation at a point $z = z_1$, is given by

$$\theta = \left(\frac{e}{8mV}\right)^{1/2} \int_{z_o}^{z_1} B_z(z)\, dz \tag{7.49}$$

Also, since the angle of rotation is a function of B_z, a reversal of the magnetic field will result in a reversal of the direction of rotation. One way to obtain zero rotation is to place two coaxial and identical magnetic lenses side by side and reverse the current of one with respect to the other.

Let us now integrate Eq. (7.46):

$$r_i' - r_o' = -(e/8mV) \int_{-\infty}^{+\infty} rB_z^2\, dz \tag{7.50}$$

7.6 Busch's Equation for a Weak Lens

If the rays in the object space are parallel to the axis, $r_o' = 0$ and, since we are dealing with a thin lens, we can assume that r inside the lens does not vary appreciably (as we have done in the case of the weak electrostatic lens), and we have

$$\frac{r_i'}{r_o} = -\frac{e}{8mV} \int_{-\infty}^{+\infty} B_z^2\, dz = \frac{-1}{f_i} \tag{7.51}$$

Since $f_i = f_o = f$ because of the symmetry of the lens, we have for the thin lens:

$$\frac{1}{f} = \frac{e}{8mV} \int_{-\infty}^{+\infty} B_z^2 \, dz \tag{7.52}$$

This is Busch's formula for a weak lens.

Two major differences between magnetic and electrostatic lenses are immediately apparent. First, the charge to mass ratio e/m appears in Eq. (7.52), whereas it did not appear in the relation for the focal length in the electrostatic lens. Hence, in magnetic lenses particles of different e/m ratio will follow different trajectories. Next, Eq. (7.52) is not homogeneous with respect to the voltage V as was the case in electrostatic lenses. A high degree of voltage stability is necessary in this case. However, the trajectory will remain constant if B^2/V is constant, or if

$$B = C\sqrt{V} \tag{7.53}$$

where C is a constant.

All the above relations are valid only for slow electrons. If the voltage is raised to the point where the speed of the electrons is an appreciable fraction of the speed of light, we must substitute for V the corrected value

$$V_r = V\left[1 + \frac{eV}{2m_0 c^2}\right] = V(1 + 0.978 \times 10^{-6}\ V) \tag{7.54}$$

where c is the speed of light. Figure 7.2 shows some shapes of the axial distribution of the magnetic field in a lens where the fringe cannot be ignored, in other words, a thin lens.

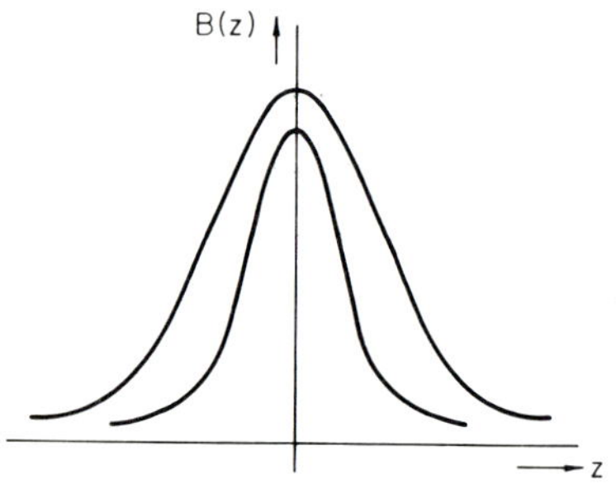

Fig. 7.2. Distribution of the axial magnetic field of a rotationally symmetrical system.

Depending upon whether or not the lens is shielded, and how the pole pieces are shaped, the curves can be either peaked at the center or spread out along the axis. If the area is evaluated for each of the curves, very little variation is observed within practical ranges. Since this expression appears in the relation for the angular rotation, Eq. (7.49), we can expect very little variation in the net rotation of the image as the shape of the field is varied. If the integral of the square of the magnetic field intensity $\int_{-\infty}^{+\infty} B^2(z)\,dz$ is evaluated, a wide variation results as the shapes of the curves change in Fig. 7.2.

The variation of the magnetic field intensity along the axis is a function of the magnetic lens geometry and design. The focal length of the lens depends very much on the axial distribution of the field. The most logical way to handle the problem is to map the field distribution experimentally and calculate the focal length. If we attempt to design a lens for a given focal length and other parameters, it would necessitate the tedious labor of plotting the fields for a large number of cases. In order to avoid this, attempts have been made to obtain empirical relations between the number of ampere-turns and the focal lengths for various lens geometries. A better insight can be obtained if we can express the distribution of the field analytically to fit the curves which are obtained experimentally. However, it is also necessary that the equation relating B as a function of z be of such a nature as to lead to some meaningful results when substituted in the paraxial equation for the focal length given by Eq. (7.52). The two requirements are not entirely incompatible. Several attempts have been made, and we propose to discuss these with the hope of learning more about the magnetic lens. The limitations of each case will be clearly indicated.

7.7 The Glaser Model

The field distribution along the axis for a single circular turn or for a close-wound coil of large diameter which can be approximated as a simple linear conductor, can be calculated with the use of Ampere's law to be

$$B_z = \frac{B_0}{[1 + (z/R)^2]^{3/2}} \tag{7.55}$$

where B_0 is the maximum flux density in the center, and R is the radius of the loop or the coil. Glaser proposed the following field distribution

$$B_z = \frac{B_0}{[1 + (z/a)^2]^{\mu}} \tag{7.56}$$

in which B_0 is the maximum value as shown in Fig. 7.3 and a is a parameter related to the half-width d (the value of z for which $B_z = \frac{1}{2}B_0$) by

$$d = a(2^{1/\mu} - 1)^{1/2} \tag{7.57}$$

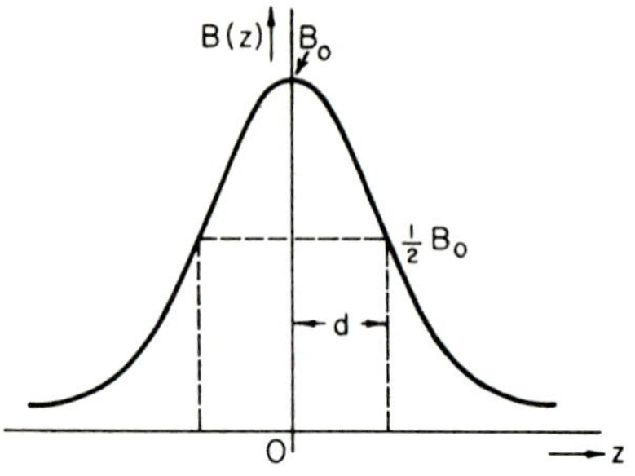

Fig. 7.3. Glaser's bell-shaped curve.

where μ is a constant depending on the coil shape and shielding used. In particular, Glaser investigated the case where $\mu = 1$, namely,

$$B_z = \frac{B_0}{1 + (z/a)^2} \tag{7.58}$$

where $a = d$.

This is generally referred to as the bell-shaped curve. It is possible to select the various parameters of Eq. (7.56) to fit any magnetic field distribution very closely. However, the particular case where $\mu = 1$ given by Eq. (7.58) fits only the case where the pole pieces are close to saturation. This is clearly shown in Fig. 7.4.

The curves are drawn for five different cases. We can readily see that Eq. (7.58) very closely approximates the distribution for saturated pole

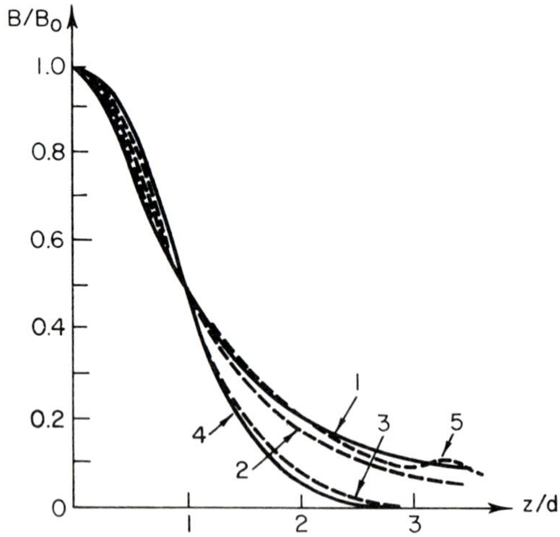

Fig. 7.4. Distribution of the axial magnetic field for various lenses. (1) $B = B_0/[1 + (z^2/a^2)]$; (2) $B = B_0/[1 + (z^2/a^2)]^{3/2}$; (3) $B = B_0 \exp(-z^2/a^2)$; (4) nonsaturated pole pieces; (5) saturated pole pieces.

pieces. This situation is normally encountered in the objective lens of an electron miscroscope.

Although the correspondence of Eq. (7.58) to practical magnetic lenses is very restricted, we shall nevertheless carry out the analysis using this particular field distribution, since it happens to be one of the rare cases where one can obtain the necessary relations, mathematically, to the end. If we substitute Eq. (7.58) in Eq. (7.46), we have

$$\frac{d^2r}{dz^2} = -\frac{eB_0{}^2}{8mV}\frac{r}{[1+(z/a)^2]^2} \tag{7.59}$$

If we let

$$y = r/a, \qquad x = z/a, \qquad \text{and} \qquad K^2 = (eB_0{}^2a^2/8mV) \tag{7.60}$$

then Eq. (7.59) changes to

$$\frac{d^2y}{dx^2} = -\frac{K^2y}{(1+x^2)^2} \tag{7.61}$$

Furthermore, if we let

$$x = \cot\varphi \tag{7.62}$$

then

$$\frac{dx}{d\varphi} = -\frac{1}{\sin^2\varphi} \qquad \text{and} \qquad 1 + x^2 = \frac{1}{\sin^2\varphi} \tag{7.63}$$

$$\frac{dy}{dx} = -\sin^2\varphi\,\frac{dy}{d\varphi} \tag{7.64}$$

and

$$\frac{d^2y}{dx^2} = \sin^4\varphi\,\frac{d^2y}{d\varphi^2} + 2\sin^3\varphi\cos\varphi\,\frac{dy}{d\varphi} \tag{7.65}$$

and Eq. (7.61) reduces to

$$\frac{d^2y}{d\varphi^2} + 2\cot\varphi\,\frac{dy}{d\varphi} + K^2y = 0 \tag{7.66}$$

In order to solve Eq. (7.66), we will attempt to eliminate the middle term, i.e., the first derivative of y. With this in mind, we let

$$y = u(\varphi)v(\varphi) \tag{7.67}$$

Then

$$\frac{dy}{d\varphi} = u\,\frac{dv}{d\varphi} + v\,\frac{du}{d\varphi} \tag{7.68}$$

and

$$\frac{d^2y}{d\varphi^2} = u\frac{d^2v}{d\varphi^2} + 2\frac{du}{d\varphi}\frac{dv}{d\varphi} + v\frac{d^2u}{d\varphi^2} \tag{7.69}$$

If we now substitute this in Eq. (7.66), we get

$$u\frac{d^2v}{d\varphi^2} + \left(2\frac{du}{d\varphi} + 2u\cot\varphi\right)\frac{dv}{d\varphi} + \left[\frac{d^2u}{d\varphi^2} + 2\frac{du}{d\varphi}\cot\varphi + uK^2\right]v = 0 \tag{7.70}$$

Now let us set

$$2\frac{du}{d\varphi} + 2u\cot\varphi = 0 \tag{7.71}$$

hence,

$$u = \exp\left(-\int\cot\varphi\, d\varphi\right) = \exp(-\log\sin\varphi) = [\exp(\log\sin\varphi)]^{-1} = (\sin\varphi)^{-1} \tag{7.72}$$

and

$$y = uv(\varphi) = [v(\varphi)/\sin\varphi] = (1 + x^2)^{1/2}\, v(\varphi) \tag{7.73}$$

$$\frac{du}{d\varphi} = -\frac{\cos\varphi}{\sin^2\varphi} \tag{7.74}$$

and

$$\frac{d^2u}{d\varphi^2} = \frac{1 + \cos^2\varphi}{\sin^3\varphi} \tag{7.75}$$

Equation (7.70) then reduces to

$$\frac{d^2v}{d\varphi^2} + (1 + K^2)v = 0 \tag{7.76}$$

If we had started with Eq. (7.56) rather than Eq. (7.58), we would have obtained

$$\frac{d^2y}{d\varphi^2} + 2\cot\varphi\frac{dy}{d\varphi} + K^2\sin^{4(\mu-1)}\varphi\cdot y = 0 \tag{7.77}$$

and

$$\frac{d^2v(\varphi)}{d\varphi^2} + [1 + K^2\sin^{4(\mu-1)}\phi]v(\varphi) = 0 \tag{7.78}$$

which, for the particular case $\mu = 1$, reduce to Eqs. (7.66) and (7.76), respectively. Two particular solutions of Eq. (7.76) are

$$v = \sin(1 + K^2)^{1/2}\varphi \qquad \text{and} \qquad v = \cos(1 + K^2)^{1/2}\varphi \tag{7.79}$$

and the general solution can now be written, if reference is made to Eqs. (7.62) and (7.73),

$$y = (1 + x^2)^{1/2}\{C_1 \sin[(1 + K^2)^{1/2} \operatorname{arccot} x] + C_2 \cos[(1 + K^2)^{1/2} \operatorname{arccot} x]\} \tag{7.80}$$

where C_1 and C_2 are constants.

For simplification, let

$$\omega^2 = K^2 + 1 = (eB_0{}^2a^2/8mV) + 1 \tag{7.81}$$

Equation (7.80) can now be written as

$$y = (\sin\varphi)^{-1}[C_1 \sin\omega\varphi + C_2 \cos\omega\varphi] \tag{7.82}$$

which can be simplified to

$$y = A\,\frac{\sin\omega(\varphi + \psi)}{\sin\varphi} \tag{7.83}$$

where A and ψ are constants.

Since $z = a\cot\varphi$, a variation of z from $-\infty$ to $+\infty$ corresponds to a variation of φ from π to 0.

7.8 The Focal Points

Since we are considering a symmetrical lens, the direction of the incoming beam is immaterial. Let us consider the case of a parallel incoming beam situated at $z = +\infty$ or $\varphi = 0$. The same results will be obtained with opposite signs if the incoming beam arrives from $z = -\infty$. Now if $\varphi = 0$, y in Eq. (7.83) must have a finite value, it cannot go to infinity. Thus, the constant Ψ must necessarily be equal to zero. We thus have

$$y = A\,(\sin\omega\varphi/\sin\varphi) \tag{7.84}$$

For the limit $\varphi = 0$, we use L'Hôpital's rule and obtain

$$y_{z=\infty} = \omega A \tag{7.85}$$

Next, the slope of y at $z = \infty$ is given by

$$y' = \frac{dy}{d\varphi}\frac{d\varphi}{dz} = \frac{A}{a}(\sin\omega\varphi\cos\varphi - \omega\sin\varphi\cos\omega\varphi) \tag{7.86}$$

For $z = \infty$, $\varphi = 0$, $y' = 0$. Therefore, this indeed verifies that the beam is parallel to the axis at $z = \infty$ as originally stipulated.

By definition, the point where this incoming parallel beam crosses the axis is the focal point. This occurs when

$$\sin \omega\varphi = 0 \tag{7.87}$$

or

$$\omega\varphi = n\pi \qquad (n = 1, 2, 3, \text{etc.}) \tag{7.88}$$

Since $z/a = \cot \varphi$ and $\varphi = n\pi/\omega$, the midfocal length, i.e., the distance z of the focal point from the center of the lens, is then

$$z_{\mathrm{m_o}} = a \cot(n\pi/\omega) \qquad (n = 1, 2, 3, \ldots) \tag{7.89}$$

As expected, we can readily see that there can be numerous intersections inside the lens, depending on the various parameters involved. We shall investigate this in more detail.

Following exactly the same kind of reasoning, we consider an incoming ray from $-\infty$ and obtain

$$z_{\mathrm{m_i}} = -a \cot(n\pi/\omega) \tag{7.90}$$

Since we are presently dealing with a symmetrical magnetic lens, we notice that the focal lengths are equal for rays coming from the left as well as from the right. This is also verified in practice; n cannot take an unlimited number of values. We have seen that for z varying from $-\infty$ to $+\infty$ this corresponds to a variation in φ from π to 0.

Hence

$$n \leq (K^2 + 1)^{1/2} \tag{7.91}$$

and n increases as K increases. Remembering that $K^2 = (eB_0{}^2a^2/8mV)$, we see that the number of crossings on the axis, that is, the number of focal points, increases as B increases, V decreases, and a increases. This is exactly as expected in a magnetic lens. For a fixed voltage and B_0, an increase in a corresponds to an increase in the thickness of the lens.

We can now consider the first crossing when $n = 1$

$$(z_{\mathrm{m_o}})_1 = a \cot \frac{\pi}{(K^2 + 1)^{1/2}} \tag{7.92}$$

and, in general,

$$(z_{\mathrm{m_o}})_n = a \cot \frac{n\pi}{(K^2 + 1)^{1/2}} \tag{7.93}$$

The number of focal points, or crossings of the axis, is clearly shown for various values of K^2 in Fig. 7.5.

These paths are not on the same plane. However, in Fig. 7.5 their projections on the same meridional plane are shown. The field distribution has been drawn, too, to indicate the location of the trajectories relative to the field. From Eq. (7.93), we can see that for $K^2 < 3$ the lens has one focal point, for $3 < K^2 < 8$, two focal points, and for $8 < K^2 < 15$, three focal points, etc. In all cases the focal length can never be less than a.

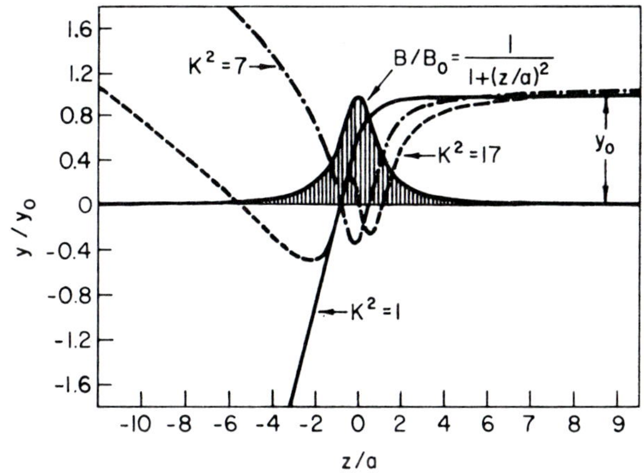

Fig. 7.5. Beam trajectories in a rotationally symmetrical magnetic lens as a function of the parameter K^2 (from Zworykin *et al.*, 1945[a]).

[a] See reference list at end of this volume.

7.9 The Focal Lengths

Let us now refer to Fig. 7.6. By definition $f_o = -F_o P_o$.

$$f_o = -y/\tan\alpha \tag{7.94}$$

The value of y at $z = \infty$ is given by Eq. (7.85).

$$\tan\alpha = dy/dz \tag{7.95}$$

For simplicity, we shall first consider the case in which $n = 1$. We can generalize later on. If $n = 1$, $\varphi = \pi/\omega$. From Eq. (7.86), we now have

$$dy/dz = (A\omega/a)\sin(\pi/\omega) \tag{7.96}$$

hence

$$\left(\frac{1}{f_o}\right)_{n=1} = \frac{1}{f_o} = -\frac{1}{a}\sin\frac{\pi}{(K^2+1)^{1/2}} \tag{7.97}$$

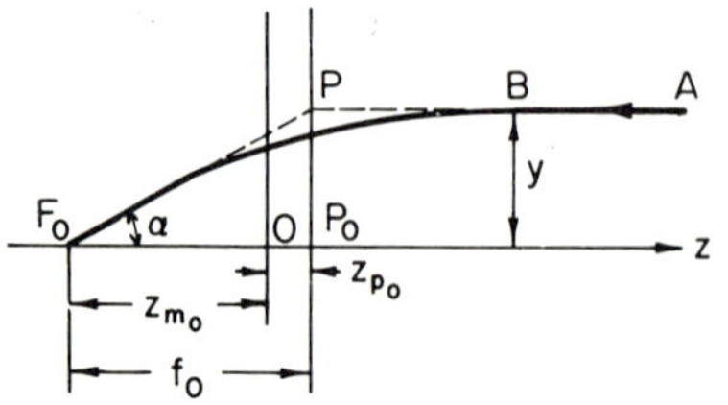

Fig. 7.6. Cardinal points of a magnetic lens. The incoming beam from A is parallel the axis; F_o is the focal point; PP_o is the principal plane; z_{m_o} is the midfocal length.

For the focal distance of order n, we have, following exactly the same procedure,

$$\left(\frac{1}{f_o}\right)_n = \frac{1}{a}(-1)^n \sin \frac{n\pi}{(K^2+1)^{1/2}} \tag{7.98}$$

The smallest value of the focal length f_o occurs when

$$\sin \frac{\pi}{(K^2+1)^{1/2}} = 1 \tag{7.99}$$

or when $K^2 = 3$

The focal length f is then equal to a, which is equal to the value of z for which $B = \frac{1}{2}B_0$. If we refer to Eq. (7.92), we find that the minimum focal point occurs at the center of the lens since $z_{m_o} = 0$. Figure 7.7 shows the variation of the focal length as a function of K^2 as well as the distance of the focal point from the center of the lens. We can see that

$$f = a \qquad \text{for} \qquad K^2 = 3 \tag{7.100}$$

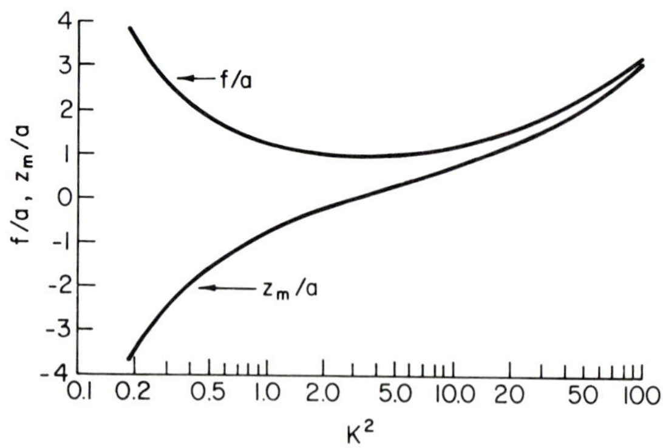

Fig. 7.7. Variation of the focal length and midfocal length with K^2 (from Glaser, 1941).

7.10 The Principal Planes

The position of the principal planes can easily be calculated if reference is made to Fig. 7.6 where, following our sign convention, we have, for $n = 1$

$$z_{p_o} = z_{m_o} - f_o = a \cot \frac{\pi}{\omega} + \frac{a}{\sin(\pi/\omega)} \tag{7.101}$$

or

$$z_{p_o} = a \cot \frac{\pi}{2\omega} = a \cot \frac{\pi}{2(K^2 + 1)^{1/2}} \tag{7.102}$$

Similarly, the second principal plane is

$$z_{p_i} = -a \cot \frac{\pi}{2(K^2 + 1)^{1/2}} \tag{7.103}$$

In general, for the focal point of order n, the principal plane is

$$(z_{p_o})_n = a \cot \frac{n\pi}{\omega} - (-1)^n \frac{a}{\sin(n\pi/\omega)} \tag{7.104}$$

When more than one crossing occurs, our sign convention is no longer always valid.

Figure 7.8 shows the positions of the first focal points and principal points as functions of K^2.

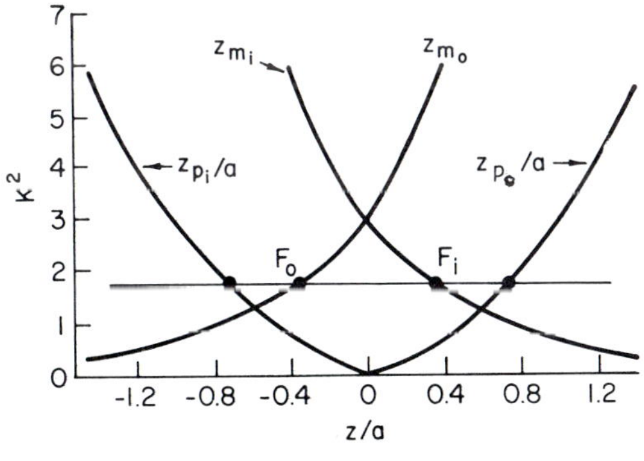

Fig. 7.8. Coordinates of the principal points and focal points as functions of K^2 (from Glaser, 1941).

The separation between the two principal planes is therefore

$$\overline{P_i P_o} = 2z_p = 2a \cot \frac{1}{2} \frac{\pi}{(K^2 + 1)^{1/2}} \tag{7.105}$$

In strong lenses, say for $K^2 > 1$, the separation increases, and for weak lenses it tends to zero.

Figure 7.9 shows the constructional details for which normal optical procedures have been used. Strictly speaking, these are valid only for the case in which object and image are outside the influence of the field. However, Glaser has shown that these procedures are also valid for the case in which object or image is deeply immersed in the field. This will be shown in the next section. The dotted lines indicate the projection of the actual paths of the electrons on the meridional plane.

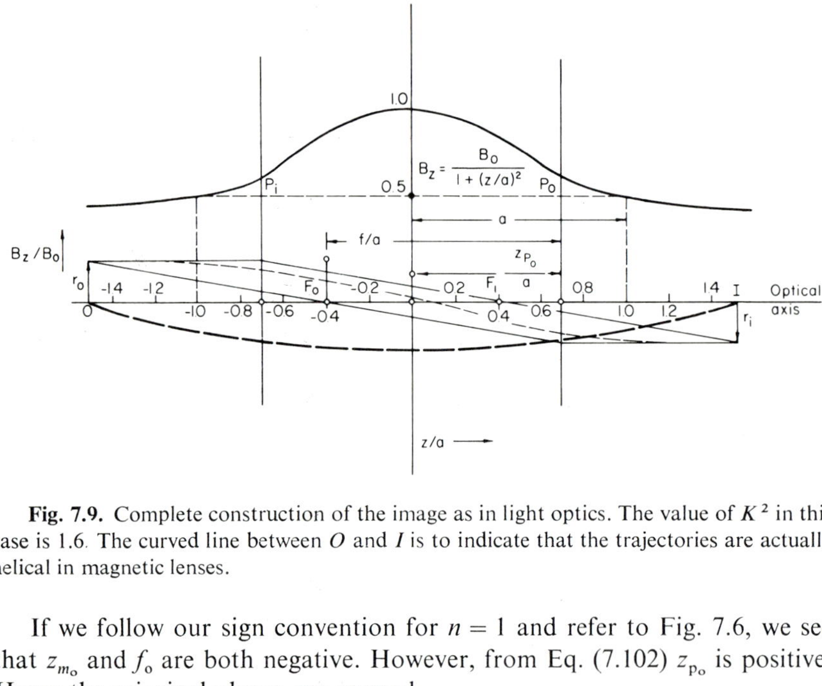

Fig. 7.9. Complete construction of the image as in light optics. The value of K^2 in this case is 1.6. The curved line between O and I is to indicate that the trajectories are actually helical in magnetic lenses.

If we follow our sign convention for $n = 1$ and refer to Fig. 7.6, we see that z_{m_o} and f_o are both negative. However, from Eq. (7.102) z_{p_o} is positive. Hence the principal planes are crossed.

Actually, we can easily show that the principal planes are crossed for any value of K^2; however, when more than one focus occurs, this fact is of little value.

7.11 Object and Image Relationship

Since Eq. (7.46) is a linear second order differential equation, we know that it has two independent solutions, and that the general solution is a linear combination of these two particular solutions. As we did in the case of the electrostatic field, we consider $m(z)$ and $n(z)$ as the two particular solutions. Hence,

$$r(z) = C_1 m(z) + C_2 n(z) \tag{7.106}$$

Now let us consider the trajectories which pass through an object point H_o of coordinate z_o. Then

$$r(z_o) = C_1 m(z_o) + C_2 n(z_o) \tag{7.107}$$

Next, we consider the case of one trajectory which crosses the axis at H_i where $z = z_i$. Then

$$r(z_i) = C_1 m(z_i) + C_2 n(z_i) \tag{7.108}$$

If we cancel out the constant C_2 in the two equations, we have

$$r(z_i) = C_1 \left[m(z_i) - \frac{n(z_i)}{n(z_o)} m(z_o) \right] + \frac{n(z_i)}{n(z_o)} r(z_o) \tag{7.109}$$

In order to obtain an image of H_o at H_i, all the rays emanating from H_o must converge to H_i. For this to happen, the first term in the right-hand side of Eq. (7.109) must vanish. Thus,

$$m(z_i)n(z_o) - n(z_i)m(z_o) = 0 \tag{7.110}$$

In our particular case, referring to Eq. (7.82), we have

$$m(z) = \sin \omega\varphi / \sin \varphi \qquad \text{and} \qquad n(z) = \cos \omega\varphi / \sin \varphi \tag{7.111}$$

Substituting these relations in Eq. (7.110) and remembering the trigonometric identity

$$\sin a \cos b - \sin b \cos a = \sin(a - b) \tag{7.112}$$

we obtain

$$\sin \omega(\varphi_i - \varphi_o) = 0 \tag{7.113}$$

or

$$\varphi_o - \varphi_i = n\pi/\omega \qquad (n = 1, 2, 3 \quad \text{etc.}) \tag{7.114}$$

If we refer to Eqs. (7.60) and (7.62), we have

$$z_o = a \cot \varphi_o = a \cot[\varphi_i + n\pi/\omega] \tag{7.115}$$

Remembering the trigonometric identity

$$\cot(a + b) = \frac{\cot a \cot b - 1}{\cot a + \cot b} \tag{7.116}$$

we can write

$$z_o = \frac{a \cot \varphi_i \cot(n\pi/\omega) - a}{\cot \varphi_i + \cot(n\pi/\omega)} = \frac{a z_i \cot(n\pi/\omega) - a^2}{z_i + a \cot(n\pi/\omega)} \tag{7.117}$$

where we have multiplied the numerator and denominator by a. Furthermore, by subtracting left and right the term $a \cot(n\pi/\omega)$, we have

$$z_o - a \cot n \frac{\pi}{\omega} = -a^2 \frac{1 + \cot^2(n\pi/\omega)}{z_i + a \cot(n\pi/\omega)} \tag{7.118}$$

which finally gives

$$\left(z_o - a \cot n \frac{\pi}{\omega}\right)\left(z_i + a \cot n \frac{\pi}{\omega}\right) = \frac{-a^2}{\sin^2(n\pi/\omega)} \tag{7.119}$$

From Eqs. (7.93) and (7.98), we have

$$\begin{aligned} z_{m_o} &= a \cot(n\pi/\omega), & f_o &= \frac{(-1)^n a}{\sin(n\pi/\omega)}, \\ z_{m_i} &= -a \cot n \frac{\pi}{\omega}, & f_i &= \frac{(-1)^n(-a)}{\sin(n\pi/\omega)} \end{aligned} \tag{7.120}$$

and if we let

$$x_i = z_i - z_{m_i} \quad \text{and} \quad x_o = z_o - z_{m_o} \tag{7.121}$$

we will find that

$$x_i x_o = f_o f_i \tag{7.122}$$

which is Newton's well-known formula derived in light optics. We have thus shown that there is, indeed, an image formation even though the object or image is immersed inside the field.

The focal length, the position of the focal point, and the principal plane as a ratio of a are given for various values of K^2 in Table 7.1.

TABLE 7.1[a]

K^2	a/f_o	z_m/a	z_p/a	K^2	a/f_o	z_m/a	z_p/a
0.1	−0.146	−6.791	0.073	1.0	−0.796	−0.761	0.496
0.2	−0.270	−3.562	0.138	1.1	−0.827	−0.680	0.529
0.3	−0.377	−2.459	0.196	1.2	−0.854	−0.609	0.562
0.4	−0.468	−1.891	0.248	1.3	−0.877	−0.547	0.593
0.5	−0.545	−1.538	0.297	1.4	−0.897	−0.492	0.623
0.6	−0.611	−1.290	0.341	1.5	−0.915	−0.442	0.651
0.7	−0.668	−1.113	0.383	1.6	−0.930	−0.397	0.679
0.8	−0.717	−0.971	0.423	1.7	−0.942	−0.355	0.706
0.9	−0.759	−0.857	0.460	1.8	−0.953	−0.317	0.732

TABLE 7.1 (continued)

K^2	a/f_o	z_m/a	z_p/a	K^2	a/f_o	z_m/a	z_p/a
1.9	−0.963	−0.281	0.758	6.0	−0.927	0.403	1.482
2.0	−0.971	−0.248	0.782	6.1	−0.924	0.413	1.495
2.1	−0.977	−0.217	0.806	6.2	−0.921	0.423	1.508
2.2	−0.983	−0.188	0.830	6.3	−0.918	0.432	1.522
2.3	−0.987	−0.160	0.853	6.4	−0.915	0.442	1.535
2.4	−0.991	−0.134	0.875	6.5	−0.912	0.451	1.548
2.5	−0.994	−0.109	0.897	6.6	−0.908	0.460	1.561
2.6	−0.996	−0.085	0.918	6.7	−0.905	0.469	1.574
2.7	−0.998	−0.063	0.939	6.8	−0.902	0.478	1.586
2.8	−0.999	−0.041	0.960	6.9	−0.899	0.487	1.599
2.9	−1.000	−0.020	0.980	7.0	−0.896	0.496	1.612
3.0	−1.000	0.000	1.000	7.1	−0.893	0.504	1.624
3.1	−1.000	0.019	1.019	7.2	−0.890	0.513	1.636
3.2	−0.999	0.038	1.039	7.3	−0.887	0.521	1.649
3.3	−0.998	0.056	1.057	7.4	−0.884	0.529	1.661
3.4	−0.997	0.073	1.076	7.5	−0.881	0.538	1.673
3.5	−0.996	0.090	1.094	7.6	−0.878	0.546	1.685
3.6	−0.994	0.106	1.112	7.7	−0.875	0.554	1.697
3.7	−0.993	0.122	1.130	7.8	−0.872	0.562	1.709
3.8	−0.991	0.138	1.147	7.9	−0.869	0.570	1.720
3.9	−0.989	0.153	1.164	8.0	−0.866	0.577	1.732
4.0	−0.986	0.167	1.181	8.1	−0.863	0.585	1.744
4.1	−0.984	0.182	1.198	8.2	−0.860	0.593	1.755
4.2	−0.981	0.196	1.215	8.3	−0.857	0.600	1.767
4.3	−0.979	0.209	1.231	8.4	−0.856	0.608	1.778
4.4	−0.976	0.222	1.247	8.5	−0.852	0.615	1.789
4.5	−0.973	0.235	1.263	8.6	−0.849	0.623	1.801
4.6	−0.971	0.248	1.278	8.7	−0.846	0.630	1.812
4.7	−0.968	0.261	1.294	8.8	−0.843	0.637	1.823
4.8	−0.965	0.273	1.309	8.9	−0.841	0.644	1.834
4.9	−0.962	0.285	1.325	9.0	−0.838	0.651	1.845
5.0	−0.959	0.297	1.340	9.1	−0.835	0.658	1.856
5.1	−0.956	0.308	1.354	9.2	−0.833	0.665	1.867
5.2	−0.953	0.319	1.369	9.3	−0.830	0.672	1.877
5.3	−0.950	0.330	1.384	9.4	−0.827	0.679	1.888
5.4	−0.946	0.341	1.398	9.5	−0.825	0.686	1.899
5.5	−0.943	0.352	1.412	9.6	−0.822	0.693	1.909
5.6	−9.940	0.363	1.426	9.7	−0.819	0.699	1.920
5.7	−0.937	0.373	1.440	9.8	−0.817	0.706	1.930
5.8	−0.934	0.383	1.454	9.9	−0.814	0.713	1.941
5.9	−0.931	0.393	1.468	10.0	−0.812	0.719	1.951

[a] The values for the focal length, the midfocal length, and the distance of the principal point from the center of the lens are given for $n = 1$ in units of a for an electron beam coming from the right, at $z = +\infty$.

7.12 Magnification

Reference to Eq. (7.109) in which we have eliminated the first term on the right-hand side shows that the magnification is given by

$$M = r(z_i)/r(z_o) = n(z_i)/n(z_o) \tag{7.123}$$

Substituting for $n(z)$ the value given by Eq. (7.111), we have

$$M = (\cos \omega\varphi_i/\sin \varphi_i)\,(\sin \varphi_o/\cos \omega\varphi_o) \tag{7.124}$$

Using Eq. (7.114), we can write

$$M = \frac{\cos(\omega\varphi_o - n\pi)}{\sin[\varphi_o - (n\pi/\omega)]}\,\frac{\sin \varphi_o}{\cos \omega\varphi_o}$$

$$= (-1)^n \frac{\sin \varphi_o}{\sin \varphi_o \cos(n\pi/\omega) - \cos \varphi_o \sin(n\pi/\omega)} \tag{7.125}$$

which, using Eqs. (7.60), and (7.62), can be written as

$$M = (-1)^{n-1} \frac{1}{z_o - a \cot(n\pi/\omega)}\,\frac{a}{\sin(n\pi/\omega)} \tag{7.126}$$

which, when reference is made to Eqs. (7.120) and (7.121) reduces to

$$M = -(f_o/x_o) \tag{7.127}$$

or using Eq. (7.122)

$$M = -f_o/x_o = -x_i/f_i \tag{7.128}$$

7.13 Image Rotation

We have already established that a trajectory of a particle inside a magnetic lens undergoes an inward force which focuses it and a transversal force which rotates it. If we substitute for B the bell-shaped curve given by Eq. (7.58) in Eq. (7.49), we have

$$\theta = \left(\frac{e}{8mV} B_0\right)^{1/2} \int_{z_o}^{z_i} \frac{dz}{1 + (z^2/a^2)} \tag{7.129}$$

If we remember that $(z/a) = \cot \varphi$ and $K^2 = (eB_0{}^2 a^2/8mV)$ and that $dz = (-a/\sin^2 \varphi)\,d\varphi$, Eq. (7.129) reduces to

$$\theta = K(\varphi_o - \varphi_i) \tag{7.130}$$

If z varies from $z_i = \infty$ to $z_o = z_{m_o}$, φ varies from 0 to $[n\pi/(K^2 + 1)^{1/2}]$. Hence,

$$\theta = nK\pi/(K^2 + 1)^{1/2} \tag{7.131}$$

For large K, $\theta \approx n\pi$. For example, if $K^2 = 2$, then $n = 1$ and $\theta = \pi(2/3)^{1/2}$ rad $= 146°$. Equation (7.130) gives a physical meaning to the variable φ. It is now possible, for example, to determine K by measuring the rotation of the image for an incoming parallel beam.

7.14 The Parameter K^2

We have seen from the above analysis that all the cardinal points of a magnetic lens in which the field distribution was assumed to be given by Eq. (7.58) depend on the parameter K^2. It is therefore useful to investigate the role which this parameter plays in the overall results.

We already know that the minimum focal length (distance between the focal point and corresponding principal point) is $f = a$ for $K^2 = 3$. The focal point is then at the center of the lens, and the principal point is a distance a from the center of the lens. Let us calculate the magnetic field necessary to obtain the minimum focal length.

$$K^2 = \frac{eB_0{}^2a^2}{8mV} = \frac{1.76 \times 10^{11}}{8}\frac{B_0{}^2a^2}{V} = 3 \tag{7.132}$$

TABLE 7.2[a]

Voltage V (in kV)	$a = 2$mm $= 2 \times 10^3$ m B_0 (in teslas)	$a = 5$ mm $= 5 \times 10^{-3}$ m B_0 (in teslas)	$a = 10$ mm $= 10^{-2}$ m B_0 (in teslas)
5	0.414	0.166	0.083
10	0.587	0.235	0.117
20	0.834	0.334	0.167
30	1.026	0.411	0.205
50	1.338	0.535	0.267
75	1.657	0.663	0.331
100	1.935	0.774	0.387
150	2.422	0.969	0.484
200	2.856	1.142	0.571
300	3.638	1.455	0.728
400	4.357	1.743	0.871
500	5.040	2.016	1.008
600	5.699	2.280	1.140
700	6.343	2.537	1.269
1000	8.214	3.286	1.643

[a] Maximum flux density in teslas necessary to obtain minimum focal length, i.e., for $K^2 = 3$. Correction for relativity is made in all cases.

Table 7.2 gives the values of B for three values of a for various voltages in which a correction for relativity was made with the use of Eq. (7.54).

Even in the objectives of the electron microscope which are generally strong lenses, K^2 is less than 3. Therefore, we can immediately see that the first focus of a magnetic lens is the one normally used.

Some of the values of B_0 given in Table 7.2 are extremely difficult, if not impossible, to obtain without pole-piece saturation.

7.15 Comparison of Glaser's Result with Busch's Equation

If K^2 is small compared to 1, Eq. (7.97) for the focal length can be written as

$$\frac{a}{f} = \sin \frac{\pi}{(K^2 + 1)^{1/2}} \approx \sin \pi \frac{K^2}{2} \approx \pi \frac{K^2}{2} \tag{7.133}$$

or

$$\frac{1}{f} = \frac{\pi e a}{16m} \frac{{B_0}^2}{V} \tag{7.134}$$

If we substitute Eq. (7.58) in Eq. (7.52) ,we get

$$\frac{1}{f} = \frac{e}{8mV} \int_{-\infty}^{+\infty} \frac{{B_0}^2}{[1 + (z/a)^2]^2}\, dz \tag{7.135}$$

From tables of integrals, we know that

$$\int \frac{dx}{(x^2 + 1)^2} = \frac{x}{2(x^2 + 1)} + \frac{1}{2} \arctan x \tag{7.136}$$

and Eq. (7.135) reduces to

$$\frac{1}{f} = \frac{e}{8mV} \frac{{B_0}^2 a\pi}{2} = \frac{\pi e a}{16m} \frac{{B_0}^2}{V} \tag{7.137}$$

which is identical to Eq. (7.134).

One can therefore observe that Busch's formula for the focal length as given by Eq. (7.52) is only valid for very weak fields. Indeed, it is only there that the distance of the electron from the axis does not change appreciably inside the lens.

Figure 7.10 shows the variation of the focal length a/f as a function of K^2. One curve was obtained according to Eq. (7.97) derived by Glaser and the other form Eq. (7.52) derived by Busch. From these curves, we can see that for $K^2 = 0.2$ (which is almost 10 times smaller than the value normally used in practice), the discrepancy is already about 16 %. If we consider, say $K^2 = 2$, as, for instance, in the objective of an electron microscope, the error is almost 300 %! Since we are dealing with paraxial rays, we know that the force on the electron is proportional to its height from the axis. The assumption that the radius is constant inside the lens leads to a stronger force on the electron than is actually the case because normally the trajectory will bend toward the axis, and the force on the particle decreases accordingly. We would therefore expect the focal length as given by Busch's formula, Eq. (7.52), to be smaller than the true focal length. This is confirmed by Fig. 7.10.

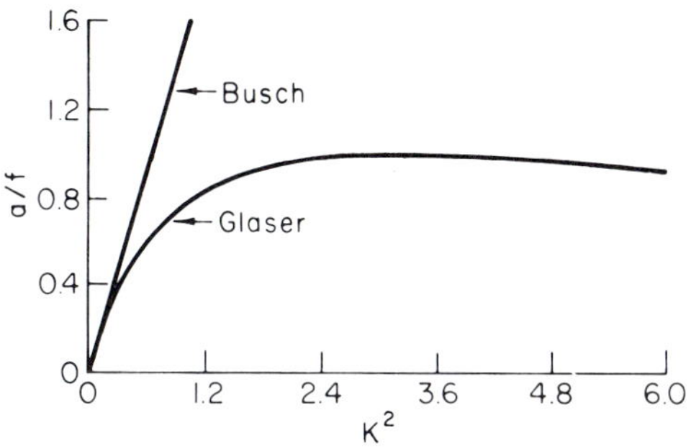

Fig. 7.10. The focal length as a function of K^2 as computed according to Glaser's and Busch's formulas. We can readily see that the error is considerable for large values of K^2.

7.16 The Magnetic Lens with Unsymmetrical Pole Pieces

So far we have assumed that the pole pieces were perfectly symmetrical and that the curve representing the axial distribution of the magnetic field is symmetrically bell-shaped. Very often the pole pieces are not symmetrical. This is frequently the case in magnetic electron microscopes. In that case the field distribution is no longer symmetrical. We shall show in what follows that even though we have an unsymmetrical field, we can still apply all the previous analysis by considering the field as split in two parts at $z = 0$, each part having the same maximum field at $z = 0$, but a different value of d, i.e., the value for z for which $B = \frac{1}{2}B_0$.

We can thus consider the case where

$$B(z) = \frac{B_0}{1 + (z^2/a^2)} \tag{7.138}$$

for $z \leq 0$ where $d = a$, and for $z \geq 0$

$$B(z) = \frac{B_0}{1 + (z^2/b^2)} \tag{7.139}$$

where $d = b$.

We can thus assume that

$$z = a \cot \varphi \quad \text{for} \quad z \leq 0, \quad \text{and} \quad z = b \cot \varphi \quad \text{for} \quad z \geq 0 \tag{7.140}$$

and as before, we let

$$\omega_1 = (K_1{}^2 + 1)^{1/2} = \left(1 + \frac{e}{8m}\frac{B_0{}^2a^2}{V}\right)^{1/2}$$

and (7.141)

$$\omega_2 = (K_2{}^2 + 1)^{1/2} = \left(1 + \frac{eB_0{}^2b^2}{8mV}\right)^{1/2}$$

We consider a point at $z = z_1$ where $y_1(z_1) = 0$. Let the corresponding value for φ be φ_1, i.e., $z_1 = a \cot \varphi_1$. From Eq. (7.83) $\psi = \varphi_1$. We obtain

$$y_1(z) = C_1 \frac{\sin \omega_1(\varphi - \varphi_1)}{\sin \varphi} \tag{7.142}$$

with $z = a \cot \varphi$, and the trajectory on the right-hand side can similarly be written as

$$y_2(z) = C_2 \frac{\sin \omega_2(\varphi - \varphi_2)}{\sin \varphi} \tag{7.143}$$

with $z = b \cot \varphi$.

The general trajectory equation can now be obtained by combining the two solutions. It must however be borne in mind that where the two trajectories meet at $z = 0$

$$y_1(0) = y_2(0) \tag{7.144}$$

and

$$y_1'(0) = y_2'(0) \tag{7.145}$$

If we take the derivative of Eq. (7.142), remembering that $dz/d\varphi = -(a/\sin^2 \varphi)$, we obtain

$$y_1'(z) = -C_1 \frac{\omega \sin \varphi \cos \omega(\varphi - \varphi_1) - \sin \omega(\varphi - \varphi_1) \cos \varphi}{\sin^2 \varphi} \frac{\sin^2 \varphi}{a} \tag{7.146}$$

At $z = z_1$, $\varphi = \varphi_1$, and $y_1'(z) = 1$. The constant C_1 is then

$$C_1 = -a/\omega \sin \varphi_1 \tag{7.147}$$

Hence

$$y_1(z) = \frac{-a}{\omega_1} \frac{1}{\sin \varphi_1} \frac{\sin \omega_1(\varphi - \varphi_1)}{\sin \varphi} \tag{7.148}$$

Now at $z = 0$, $\varphi = \pi/2$ hence

$$y_1(\tfrac{1}{2}\pi) = y_2(\tfrac{1}{2}\pi) \qquad \text{and} \qquad y_1'(\tfrac{1}{2}\pi) = y_2'(\tfrac{1}{2}\pi) \tag{7.149}$$

We can now equate Eqs. (7.142) and (7.143) and their derivatives at $\varphi = \pi/2$, and we have

$$C_2 \sin \omega_2\left(\frac{\pi}{2} - \varphi_2\right) = \frac{a}{\omega_1} \frac{1}{\sin \varphi_1} \sin \omega_1(\varphi_1 - \pi/2) \tag{7.150}$$

$$C_2 \cos \omega_2\left(\frac{\pi}{2} - \varphi_2\right) = \frac{-b}{\omega_2} \frac{1}{\sin \varphi_1} \cos \omega_1(\varphi_1 - \pi/2) \tag{7.151}$$

Dividing Eq. (7.150) by Eq. (7.151), we have

$$\tan \omega_2\left(\frac{\pi}{2} - \varphi_2\right) = \frac{a}{b} \frac{\omega_2}{\omega_1} \tan \omega_1\left(\frac{\pi}{2} - \varphi_1\right) \tag{7.152}$$

We now have a valuable relation between the two trajectories. To check our results, let us assume that $a = b$ and $\omega_1 = \omega_2$; in other words, we are returning to the symmetrical case. Then

$$\tan \omega\ (\tfrac{1}{2}\pi - \varphi_2) = \tan \omega(\tfrac{1}{2}\pi - \varphi_1) \tag{7.153}$$

or

$$\omega(\varphi_1 - \varphi_2) = n\pi \qquad n = 1, 2, 3, \qquad \text{etc.} \tag{7.154}$$

which is identical to Eq. (7.114), if we let $\varphi_1 = \varphi_o$ and $\varphi_2 = \varphi_i$.

Equation (7.153) can be solved graphically as in Fig. 7.11. On a horizontal line we position points A and B so that $OA = a\omega_2$ and $OB = b\omega_1$. At A we draw a line making an angle θ_1 so that $\theta_1 = \omega_1(\tfrac{1}{2}\pi - \varphi_1)$ which intersects the perpendicular at point C. Then the angle θ_2 which CB makes with OB is equal to $\omega_2(\tfrac{1}{2}\pi - \varphi_2)$.

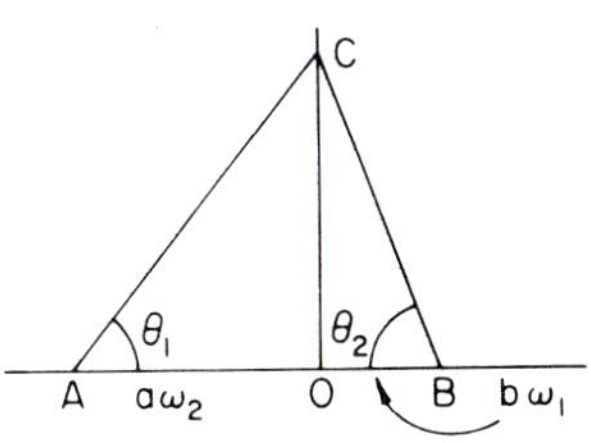

Fig. 7.11. Graphical calculation of the asymmetrical magnetic lens.

7.17 Accuracy of the Glaser Model

From Fig. 7.4 we can see that the Glaser model which has been so useful in understanding some of the essential properties of magnetic lenses does not fit all magnetic lenses. For example, in case the pole pieces are far from saturation, a field representation given by

$$B = B_0 \exp[-(z/a)^2] \tag{7.155}$$

fits the curve more closely. Unfortunately, the computation of the cardinal points in this case is difficult. Figure 7.12 shows a comparison of the values for focal length for various field distributions. We can see that a marked difference exists between Glaser's distribution and that given by Eq. (7.155).

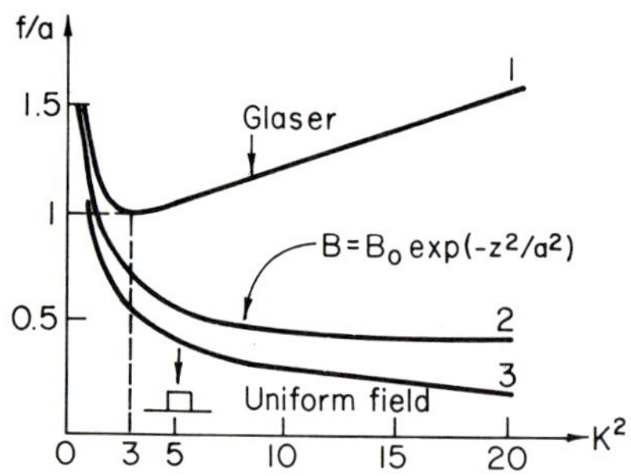

Fig. 7.12. The focal length as a function of K^2 for various field configurations.

7.18 Numerical Example Using Glaser's Formulas

Given an electron gun with a crossover of 25 μ (1 $\mu = 10^{-6}$ m), positioned in an electron optical system 50 cm from the geometrical center of a symmetrical magnetic lens. The electrons are accelerated by 100 kV.

With the use of an accurate Gauss meter and an infinitely small probe, the flux density at the center of the lens was found to be 0.1 T. The distance a from the center of the lens at which the flux density drops to 0.05 T is 1 cm. Assuming no aberrations exist in the system, compute (a) the location of the screen where a sharp spot can be obtained, (b) the size of the spot.

Solution

Let us first compute K^2

$$K^2 = \frac{eB_0{}^2a^2}{8mV} = \frac{1.76 \times 10^{11}(0.1)^2(1 \times 10^{-2})^2}{8 \times 10^5(1 + 0.978 \times 10^{-6} \times 10^5)} = 0.20$$

[The additional term in the denominator is a relativity correction; see Eq. (7.54).]

We can either refer to the curves or compute the values of f, z_m, and z_p. We shall compute the values in order to show how the equations are used:

$$|z_m| = 0.01 \left| \cot \frac{\pi}{(0.20 + 1)^{1/2}} \right| \quad \text{m}$$
$$= 3.56 \quad \text{cm}$$

$$f = \frac{0.01}{\sin[\pi/(0.20 + 1)^{1/2}]} \quad \text{m}$$
$$= 3.70 \quad \text{cm}$$

$$z_p = 0.01 \cot \frac{\pi}{2(0.20 + 1)^{1/2}} \quad \text{m}$$
$$= 0.14 \quad \text{cm}$$

In other words, the principal plane is not very far from the center of the lens. The lens is symmetrical so these values are the same for both sides of the lens. We shall use the signs in accordance with our convention.

In order to determine the distance of the image we use the well-known formula

$$\frac{1}{u_i} + \frac{1}{u_o} = \frac{1}{f}$$

(u_o object distance, u_i image distance). Here, $f = 3.70$ and $u_o = 50 + 0.14 = 50.14$

$$u_i = \frac{3.70 \times 50.14}{50.14 - 3.70} \approx 4.00$$

and the distance from the screen to the center of the lens is

$$4.00 - 0.14 = 3.86 \quad \text{(cm)}$$

Since there are no aberrations, the magnification

$$M = \frac{u_i}{u_o} = \frac{4.00}{50.14}$$

and the spot size is $25(4.00/50.14) \approx 2\ \mu$.

7.19 Combined Electric and Magnetic Fields in Rotationally Symmetrical Systems

So far we have dealt with the case of a pure electric field and a pure magnetic field through which an electron was moving in a region of constant electric potential. In the case where a combination of the two fields exists, the analysis is somewhat more complicated, but the general method of attack is the same.

The force on a charged particle in a combined electric and magnetic field is given by

$$\mathbf{F} = m\mathbf{a} = e[\nabla\phi + \mathbf{B} \times \mathbf{v}] \tag{7.156}$$

where $\nabla\phi$ is the gradient of electric potential, $\mathbf{B}$ is the magnetic flux density, and $\mathbf{v}$ the velocity. As before, e and m are the charge and mass of the electron, respectively.

The components of $\mathbf{B}$ are given by Eq. (7.30), (7.31), and (7.32); and following the same reasoning as with pure magnetic fields in which we took advantage of the rotational symmetry, we obtain

$$\frac{m}{e}(\ddot{r} - r\dot{\theta}^2) = -\dot{\theta}\frac{\partial}{\partial r}(rA_\theta) + \frac{\partial\phi}{\partial r} \tag{7.157}$$

$$\frac{m}{e}\frac{1}{r}\frac{d}{dt}(r^2\dot{\theta}) = \dot{z}\frac{\partial A_\theta}{\partial z} + \frac{\dot{r}}{r}\frac{\partial(rA_\theta)}{\partial r} = \frac{1}{r}\frac{d}{dt}(rA_\theta) \tag{7.158}$$

$$\frac{m}{e}\ddot{z} = -r\dot{\theta}\frac{\partial A_\theta}{\partial z} + \frac{\partial\phi}{\partial z} \tag{7.159}$$

As expected, Eq. (7.158) has no electric potential term since in electric fields of rotational symmetry there is no force in the θ direction. Only magnetic fields rotate electrons about the axis.

Equation (7.158) can be integrated to give

$$mr^2\dot{\theta} = erA_\theta + C \tag{7.160}$$

where C is a constant of integration. If we limit ourselves to paraxial rays and use Eq. (7.34), Eq. (7.160) reduces to

$$mr^2\dot{\theta} = \frac{er^2B(z)}{2} + C \tag{7.161}$$

Let us first assume that there is no initial rotation of the electron beam before it comes under the influence of the magnetic field. In other words, the constant of integration will be assumed equal to zero, i.e., $C = 0$.

Substituting the value of $\dot{\theta}$ from Eq. (7.160) in Eqs. (7.157) and (7.159), we obtain

$$m\ddot{r} = e\frac{\partial}{\partial r}\left(\phi - \frac{1}{2}\frac{e}{m}A_\theta{}^2\right) \tag{7.162}$$

and

$$m\ddot{z} = e\frac{\partial}{\partial z}\left(\phi - \frac{1}{2}\frac{e}{m}A_\theta{}^2\right) \tag{7.163}$$

If we multiply Eq. (7.162) by $\dot{r}$ and Eq. (7.163) by $\dot{z}$, add the two equations, integrate, and use Eq. (7.161), we obtain:

$$\tfrac{1}{2}m[\dot{r}^2 + (r\dot{\theta})^2 + \dot{z}^2] = e\phi. \tag{7.164}$$

This is the energy equation. Since neither A nor B appears in this equation, it verifies the fact that the energy of a beam cannot be changed by a magnetic field.

As with the case of pure magnetic fields, we can obtain the rotation of the electron beam from Eq. (7.161)

$$\theta = \left(\frac{e}{8m}\right)^{1/2}\int\frac{B}{\sqrt{\phi}}\,dz \tag{7.165}$$

This was to be expected since we started with no rotation and the electric field does not rotate the beam.

Since we are limiting ourselves to paraxial rays, we use only the first-order terms in the series expansion for ϕ and A_θ as given by Eqs. (2.30) and (7.34), and using these in Eqs. (7.162) and (7.163), we obtain

$$\ddot{r} = -\frac{re}{2m}\left\{\phi''(z) + \frac{1}{2}\frac{e}{m}B^2(z)\right\} \tag{7.166}$$

and

$$\ddot{z} - (e/m)\phi'(z) \tag{7.167}$$

Using the following relations

$$\dot{z}^2 = \frac{2e\phi(z)}{m}, \qquad \dot{r} = \frac{dr}{dz}\dot{z}, \qquad \text{and} \qquad \ddot{r} = \frac{d}{dz}(\dot{r})\dot{z} \tag{7.168}$$

we eliminate the time variable and obtain

$$\frac{d^2r}{dz^2} + \frac{\phi'(z)}{2\phi(z)}\frac{dr}{dz} + \frac{r}{4\phi(z)}\left(\phi''(z) + \frac{eB^2(z)}{2m}\right) = 0 \tag{7.169}$$

We can readily see that if no magnetic field is present, Eq. (7.169) reduces to the paraxial-ray equation for purely electrostatic fields, whereas if no electric fields are present, we obtain Eq. (7.46).

7.20 Theorem of Larmor

So far we have always assumed that the initial conditions were such that the constant of integration $C = 0$. This really means that $\dot{\theta} = 0$ initially, or, in other words, there is no initial angular velocity of the beam. If we integrate Eq. (7.158) between the limits z_o and z_i such that z_o is in a region far away from the magnetic field, then $\dot{\theta} = 0$ and $C = 0$. However, if z_o is selected somewhere where the magnetic flux density B has a finite value, then $C \neq 0$, and it is logical to ask whether the above paraxial-ray equation (7.169) is still valid. In other words, do we still obtain stigmatic image formation in that case? The answer to this question can easily be found if we consider a new set of coordinates with an angular velocity

$$\dot{\theta}_o = eB/2m \tag{7.170}$$

Here, B is the value of the magnetic flux density at z_o. Reference to Eq. (7.161) clearly shows that $C = 0$ when $\dot{\theta}$ has the value given by Eq. (7.170) We have thus reduced our problem to the original conditions and all the above considerations will be valid. The final rotation of the beam as given by Eq. (7.165) will have to be modified to include the initial rotation prior to integration.

With the help of Larmor's theorem, we are thus capable of reducing a problem involving a rotation of the trajectory to one in which the trajectory moves in a meridional plane.

7.21 Special Case: *B* is Constant

Let us consider the special case in which we have a homogeneous magnetic field ($B = B_z = \text{constant}$, $B_r = 0$) to which is superimposed an arbitrary rotationally symmetrical electric field with a coincident axis. If a new set of axes is again selected which rotate with an angular velocity as given by Eq. (7.170), the motion of the electrons now takes place in only one plane. In fact, if reference is made to Eq. (7.169), it is evident that in this case we can analyze the problem exactly as if it were a purely electric field, provided the following equivalent potential is used:

$$\phi_e = \phi - (e/m)B^2r^2 \tag{7.171}$$

In many applications in atomic physics, the additional term $(e/m)B^2r^2$ can be neglected with respect to ϕ. Larmor's theorem is mostly referred to in its special form which states that the action of a homogeneous magnetic field can be nullified by introducing a coordinate system which rotates with the Larmor frequency given by Eq. (7.170). However, as we have already seen, Larmor's theorem can also be used very effectively even when the magnetic field is not uniform.

Further Reading

Dosse, J. (1941). *Z. Phys.* **117**, 722.
Glaser, W. (1941). *Z. Phys.*, **117**, 285.

VIII

The Symmetrical Magnetic Lens

8.1 The Short Air Coil

Starting from Ampere's law, the magnetic flux density at a point located on the axis due to a circular conductor can be calculated to be

$$B = \frac{\mu_0 I R^2}{2(R^2 + z^2)^{3/2}} \tag{8.1}$$

Here, I is the current, R the radius of the conductor, z the distance along the axis (which is perpendicular to the plane of the circular conductor), and μ is the permeability. Equation (8.1) is valid for a single turn. When hundreds of turns are wound as in a torus, some corrective factor must be used. Lyle developed empirical relations, some of which are included here. Figure 8.1 shows two special cases. For the case shown in Fig. 8.1a, a torus of square cross section, the radius R in Eq. (8.1) should be

$$R = R_{\mathrm{m}}[1 + (a^2/24R_{\mathrm{m}}^{\ 2})] \tag{8.2}$$

where R_{m} is the mean radius. For the case shown in Fig. 8.1b, a rectangular torus, B is obtained by adding the individual fields given by two circular coplanar turns, each carrying half the total current and of radii

$$R_1 = R + d \qquad \text{and} \qquad R_2 = R - d \tag{8.3}$$

where

$$R = R_{\mathrm{m}}[1 + (d^2/24R_{\mathrm{m}}^{\ 2})] \qquad \text{and} \qquad d = (a^2 - b^2)/12 \tag{8.4}$$

This is true for $a > b$.

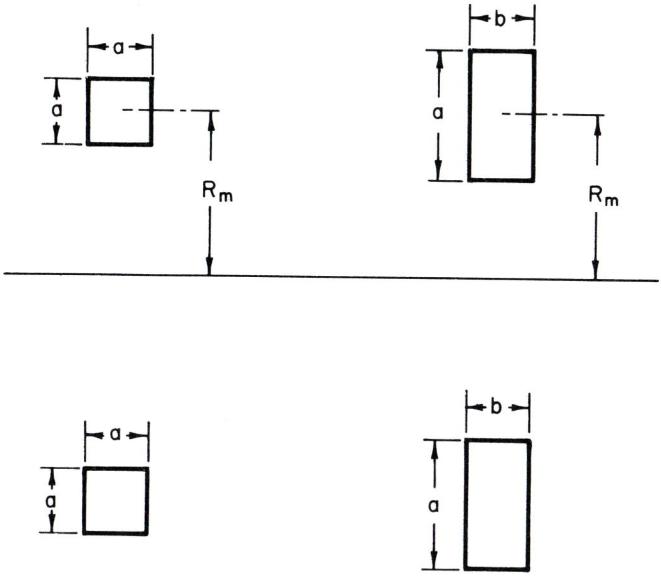

Fig. 8.1. Two types of magnetic lens with no iron core.

8.2 Focal Length of an Air Coil

From Chapter VII, we know that the focal length of a weak magnetic lens is given approximately by

$$\frac{1}{f} = \frac{e}{8mV_r} \int_{-\infty}^{+\infty} B_z^2 \, dz \tag{8.5}$$

where e and m are the charge and mass of the particle, respectively, and V_r is the voltage corrected for relativity.

$$\int_{-\infty}^{+\infty} B_z^2 \, dz \rightarrow 8mV_r/ef \tag{8.6}$$

For a circular loop of one conductor,

$$\int_{-\infty}^{+\infty} B_z^2 \, dz = \int_{-\infty}^{+\infty} \frac{\mu_0 I^2 R^4}{4(R^2 + z^2)^3} \, dz = \frac{\mu_0{}^2 R^4 I^2}{4} \int_{-\infty}^{+\infty} \frac{1}{(R^2 + z^2)^3} \, dz \tag{8.7}$$

where we have used Eq. (8.1). By performing the integration, we obtain

$$\int_{-\infty}^{+\infty} B_z^2 \, dz = 3\pi\mu_0{}^2 I^2/16D \tag{8.8}$$

where $D = 2R$ is the diameter of the coil. Then

$$(3\pi\mu_0{}^2/16D)I^2 = 8mV_r/ef \tag{8.9}$$

and if we express V_r in kilovolts,

$$I = \left[\frac{16 \times 8 \times 10^3}{(4\pi \times 10^{-7})^2\, 1.76 \times 10^{11} \times 3\pi}\right]^{1/2} \left(\frac{D}{f} V_b\right)^{1/2} = 220\left(\frac{D}{f} V_b\right)^{1/2} \tag{8.10}$$

where $V_b = V_r/1000$; or,

$$\frac{D}{f} = \frac{1}{(220)^2} \frac{I^2}{V_b} \tag{8.11}$$

This formula is based on the assumption that the lens is weak (long focal length) and is valid for paraxial rays only. In addition, we have considered the case of only one turn. Normally, thousands of turns are used and a corrective factor is necessary in this case. We must then replace I by

$$NI = 200G(V_b\, D/f)^{1/2} \tag{8.12}$$

where G is a coil form factor of the order of 1.2, approximately.

8.3 The Shielded Magnetic Lens

Most electron optical devices use lenses with relatively short focal lengths. When a coil is shielded except for a narrow air gap in the center, the field on the axis is sharply concentrated and is at least an order of magnitude higher than that obtained with an air core lens. This is the reason why most lenses are shielded.

In Chapter VII we dealt with the general properties of magnetic lenses. Starting with the bell-shaped field assumed by Glaser we were able to obtain analytically the complete data for the magnetic lens. We have also seen how the analysis can be adapted for asymmetrical lenses. With the Glaser bell-shaped field a complete qualitative analysis of the lens was possible. Unfortunately, the quantitative results are not accurate, since they are based on a field distribution which approximates the real case only when the lens is highly saturated. Glaser's distribution is given by

$$B = \frac{B_0}{1 + (z/a)^2} \tag{8.13}$$

Here, B_0 is the maximum flux density at the center of the lens, and a is the axial position z where $B = \frac{1}{2}B_0$. Another distribution given by Ramberg is

$$B = B_0 \operatorname{sech}^2 bz \tag{8.14}$$

Here, $b = 2.63/D$ and D is the bore diameter. This field distribution was determined the same way that the axial potential distribution of an equidiameter two-cylinder electrostatic lens was obtained in Chapter III. Thus, it is only valid in cases where the gap width is extremely small.

Most of the results available in the literature were obtained either experimentally or with a resistance network. In what follows, we shall develop an analytical expression from which the axial field distribution can be obtained for reasonable ratios of S/D where S is the gap width and D the diameter of the bore.

8.4 Magnetic Scalar Potential of a Symmetrical Lens

The magnetic field intensity H and the magnetic flux density B satisfy Maxwell's equations

$$\nabla \cdot \mathbf{B} = 0 \tag{8.15}$$

and

$$\nabla \times \mathbf{H} = \mathbf{J} \tag{8.16}$$

Here, $\mathbf{J}$ is the current density producing the magnetic field intensity $\mathbf{H}$. Since $\mathbf{H}$ is conservative, in current-free regions it is possible to define a magnetic scalar potential ψ such that

$$\mathbf{H} = \nabla\psi \tag{8.17}$$

We can therefore write

$$\nabla \cdot \mathbf{B} = \nabla \cdot (\mu \nabla\psi) = \nabla\mu \nabla\psi + \mu \nabla^2\psi = 0 \tag{8.18}$$

In regions of constant permeability, this reduces to Laplace's equation

$$\nabla^2\psi = 0 \tag{8.19}$$

We shall now apply the scalar magnetic potential to the symmetrical magnetic lens shown in Fig. 8.2. This is a cross section of a rotationally symmetrical lens about the z axis. Laplace's equation can then conveniently be expressed in cylindrical coordinates

$$\nabla^2\psi = \frac{1}{r}\frac{\partial}{\partial r}\left(r\frac{\partial\psi}{\partial r}\right) + \frac{\partial^2\psi}{\partial z^2} = 0 \tag{8.20}$$

The term involving θ is, of course, zero because of the rotational symmetry. We have seen in Chapter III that a solution to this equation is given by

$$\psi(r, z) = \int_{-\infty}^{+\infty} f(\alpha)e^{j\alpha z}I_0(\alpha r)\,d\alpha \tag{8.21}$$

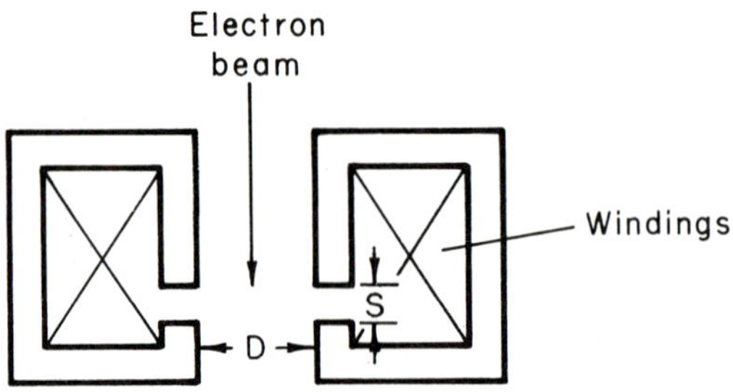

Fig. 8.2. Typical geometry of a symmetrical magnetic lens of air gap diameter D and width S. Lenses of this kind are characterized by the ratio S/D.

where $I_0(\alpha r)$ is the modified Bessel function of first kind, of order zero. See Eq. (3.50).

To evaluate the function $f(\alpha)$, we shall follow exactly the same procedure as in Chapter III. Along the bore radius R, we have

$$\psi(R, z) = \int_{-\infty}^{\infty} f(\alpha)e^{j\alpha z}I_0(\alpha R)\, d\alpha \tag{8.22}$$

which is a function of z only. The Fourier inversion integral

$$g(z) = (1/2\pi)\int_{-\infty}^{+\infty} G(\alpha)e^{j\alpha z}\, d\alpha \tag{8.23}$$

is used to obtain $g(z)$ from its Fourier transform $G(\alpha)$. Thus,

$$f(\alpha) = G(\alpha)/2\pi I_0(\alpha R) \tag{8.24}$$

The function $f(\alpha)$ can, therefore, be defined in terms of the Fourier integral of the potential variation along the bore radius of the lens. Since the magnetic material of the core has a very large permeability, and since we are operating in the region below saturation, we can safely assume that

$$NI = \sum H\, \Delta S \tag{8.25}$$

where NI are the ampere-turns, H is the magnetic field intensity in the gap, and S is the gap width. In regions far away from the center where we are sure no fringing occurs, we can write

$$H = NI/S \tag{8.26}$$

Along the bore radius, the magnetic field intensity does not vary linearly Some curvature exists due to fringe effects. We have here exactly the same condition as in the case of the electrostatic two-cylinder lens discussed in detail in Chapter III. We can assume that the magnetic potential varies linearly along the bore radius as shown in Fig. 8.3. While this approximation is not

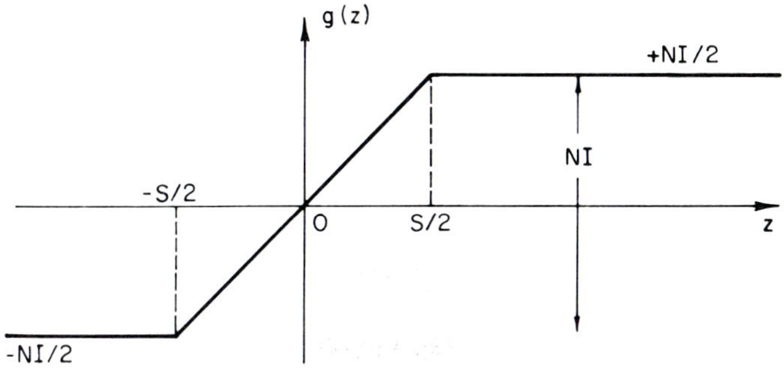

Fig. 8.3. Assumed magnetic scalar potential distribution along the radius R in the air gap.

correct, we shall see that the field distribution along the axis is, nevertheless, fairly accurate for reasonable ratios of S/D, as was the case in Chapter III.

The following boundary conditions can now be used

$$\psi(R, z) = g(z) = \begin{cases} -\frac{1}{2}NI & z \leq -\frac{1}{2}S \\ (NI/S)z & -\frac{1}{2}S \leq z \leq \frac{1}{2}S \\ +\frac{1}{2}NI & z \geq \frac{1}{2}S \end{cases} \tag{8.27}$$

We are now in a position to evaluate the function $G(\alpha)$ given by

$$G(\alpha) = \int_{-\infty}^{+\infty} g(z)e^{-j\alpha z}\, dz \tag{8.28}$$

where $g(z)$ is given by Eq. (8.27). The function $g(z)$ does not obey the condition for existence of the Fourier transform pair defined by Eqs. (8.28) and (8.23) given by

$$\int_{-\infty}^{+\infty} |g(z)|\, dz < \infty \tag{8.29}$$

We can modify this function by letting

$$g_1(z) = (\exp -|\tau z|)g(z) \tag{8.30}$$

where τ is a small positive constant.

We can now write

$$G_1(\alpha) = \int g_1(z)e^{-j\alpha z}\, dz \tag{8.31}$$

and we obtain

$$G(\alpha) = \lim_{\tau \to 0} G_1(\alpha) \tag{8.32}$$

We then have

$$G_1(\alpha) = \int_{-\infty}^{-S/2} -\tfrac{1}{2}NI \exp(-j\alpha z - |\tau z|)\, dz$$
$$+ \int_{-S/2}^{S/2} (NI/S)z \exp(-j\alpha z - |\tau z|)\, dz$$
$$+ \int_{S/2}^{\infty} \tfrac{1}{2}NI \exp(-j\alpha z - |\tau z|)\, dz \tag{8.33}$$

We consider the absolute value of z to be

$$|z| = \begin{cases} -z & z \le 0 \\ +z & z \ge 0 \end{cases} \tag{8.34}$$

We then have

$$G_1(\alpha) = NI\Bigg[-\frac{1}{2}\int_{-\infty}^{-S/2} \exp[(\tau - j\alpha)z]\, dz + S^{-1}\int_{-S/2}^{0} z\exp[(\tau - j\alpha)z]\, dz$$
$$+ S^{-1}\int_{0}^{+S/2} z\exp[-(\tau + j\alpha)z]\, dz + \frac{1}{2}\int_{+S/2}^{\infty} \exp[-(\tau + j\alpha)z]\, dz\Bigg] \tag{8.35}$$

Performing the integrations, we obtain

$$G_1(\alpha) = \frac{NI}{2}\left[\frac{\exp[-(\tau + j\alpha)S/2]}{\tau + j\alpha} - \frac{\exp[-(\tau - j\alpha)S/2]}{\tau - j\alpha}\right]$$
$$+ \frac{NI}{S}\left\{\frac{\exp[-(\tau - j\alpha)S/2]}{(\tau - j\alpha)^2}\{[(\tau - j\alpha)S/2] + 1\}\right.$$
$$- \frac{\exp[-(\tau + j\alpha)S/2]}{(\tau + j\alpha)^2}\{[(\tau + j\alpha)S/2] + 1\}$$
$$\left. + \frac{1}{(\tau + j\alpha)^2} - \frac{1}{(\tau - j\alpha)^2}\right\} \tag{8.36}$$

Taking the limit as τ approaches zero and rearranging, we obtain

$$G(\alpha) = \frac{NI}{S}\,\frac{\exp[j\alpha S/2] - \exp[-j\alpha S/2]}{(j\alpha)^2} = \frac{2NI}{jS\alpha^2}\,\frac{\exp[j\alpha S/2] - \exp[-j\alpha S/2]}{2j} \tag{8.37}$$

or

$$G(\alpha) = (2NI/jS\alpha^2)\sin(S\alpha/2) \tag{8.38}$$

We can now evaluate $f(\alpha)$ in Eq. (8.24)

$$f(\alpha) = (NI/jS\alpha^2)\sin(S\alpha/2)[\pi I_0(\alpha R)]^{-1} \tag{8.39}$$

Substituting for $f(\alpha)$ in Eq. (8.21), we obtain

$$\psi(r, z) = \frac{NI}{\pi S}\int_{-\infty}^{+\infty} \frac{\sin(S\alpha/2)}{j\alpha^2} e^{j\alpha z} \frac{I_0(\alpha r)}{I_0(\alpha R)}\, d\alpha \tag{8.40}$$

We can simplify this equation by writing

$$\psi(r, z) = \frac{NI}{\pi S}\left[\int_{-\infty}^{0} \frac{\sin(S\alpha/2)}{j\alpha^2} e^{j\alpha z} \frac{I_0(\alpha r)}{I_0(\alpha R)}\, d\alpha + \int_{0}^{\infty} \frac{\sin(S\alpha/2)}{j\alpha^2} e^{j\alpha z} \frac{I_0(\alpha r)}{I_0(\alpha R)}\right] d\alpha \tag{8.41}$$

Next, let $\alpha = -\alpha'$, then $d\alpha = -d\alpha'$. Since $\sin(-x) = -\sin(x)$ and $I_0(-x) = I_0(x)$, we obtain

$$\psi(r, z) = \frac{NI}{\pi S}\left[\int_{0}^{\infty} \frac{-\sin(S\alpha'/2)\, e^{-j\alpha' z}}{j\alpha'^2} \frac{I_0(\alpha' r)}{I_0(\alpha' R)}\, d\alpha' + \int_{0}^{\infty} \frac{\sin(S\alpha/2)}{j\alpha^2} e^{j\alpha z} \frac{I_0(\alpha r)}{I_0(\alpha R)}\, d\alpha\right] \tag{8.42}$$

Here, α and α' are dummy variables. Therefore,

$$\psi(r, z) = \frac{NI}{\pi}\int_{0}^{\infty} \frac{\sin(S\alpha/2)}{S\alpha/2} \frac{\sin(\alpha z)}{\alpha} \frac{I_0(\alpha r)}{I_0(\alpha R)}\, d\alpha \tag{8.43}$$

This is the solution for the scalar magnetic potential of the symmetrical magnetic lens using the approximate boundary condition for unsaturated pole pieces.

8.5 The Flux Density along the z Axis

The scalar potential along the z axis of the lens can be obtained by letting $r = 0$ in Eq. (8.43). We obtain

$$\psi(0, z) = \frac{NI}{\pi}\int_{0}^{\infty} \frac{\sin(S\alpha/2)}{S\alpha/2} \frac{\sin(\alpha z)}{\alpha} \frac{1}{I_0(\alpha R)}\, d\alpha \tag{8.44}$$

We shall now change the variable of integration in order to express the magnetic field in terms of known lens dimensions. We let

$$\alpha = x/R, \qquad \text{then} \qquad d\alpha = dx/R \tag{8.45}$$

and we obtain

$$\psi(0, z) = \frac{NI}{\pi} \int_0^\infty \frac{\sin(Sx/2R)\sin(xz/R)}{(S/2R)x^2 I_0(x)}\, dx \tag{8.46}$$

Using Eq. (8.17) and remembering the relationship between B and H, we can write

$$B(z) = \mu_0 \frac{NI}{\pi} \frac{d}{dz} \int_0^\infty \frac{\sin[(S/D)x]\sin(2xz/D)}{(S/D)x^2 I_0(x)}\, dx \tag{8.47}$$

where we have replaced the radius by the diameter D of the bore. Performing the differentiation and simplifying, we get

$$B(z) = \frac{2\mu_0 NI}{\pi S} \int_0^\infty \frac{\sin[(S/D)x]\cos(2xz/D)}{xI_0(x)}\, dx \tag{8.48}$$

This is the magnetic flux density along the axis of the symmetrical magnetic lens of gap width S and diameter D. It was first obtained by Lenz. It should be remembered that this relation was obtained with the following assumptions:

(a) No saturation occurs. The pole pieces are equipotentials; i.e., the permeability of the material is infinite.

(b) The magnetic scalar potential varies linearly along the radius R between the pole pieces.

The first assumption is realistic, in practice. In general, the material chosen for the core of magnetic lenses is of high permeability, and numerous experimental and theoretical results were obtained which justify assumption (a). If the lens is operated very close to saturation or beyond saturation, the mathematical model is no longer valid, and the only way to obtain the magnetic field distribution is either by relaxation technique or by experimental measurements.

Assumption (b) is closely approximated in cases where S/D does not exceed 2. The linear distribution is certainly valid for $S/D \to 0$ or for $S/D \to \infty$. However, for values used in practice the linear relationship does not always hold. However, for $S/D \leq 2$, the potential distribution along the axis is accurate enough for our purpose.

The first and second derivatives of $B(z)$ are

$$B'(z) = \frac{-4\mu_0 NI}{\pi SD} \int_0^\infty \frac{\sin[(S/D)x]\sin(2xz/D)}{I_0(x)}\, dx \tag{8.49}$$

and

$$B''(z) = -\frac{8\mu_0 NI}{\pi SD^2} \int_0^\infty \frac{x\sin[(S/D)x]\cos(2xz/D)}{I_0(x)}\, dx \tag{8.50}$$

Equation (8.48) was used in obtaining the cardinal points for a series of values of S/D listed in the tables. Equations (8.49) and (8.50) will be used in Volume 2, Chapter X when aberrations of magnetic lenses are discussed.

8.6 Considerations in the Design of the Pole Pieces

A symmetrical magnetic lens is generally as shown in Fig. 8.4a. Figure 8.4b is an asymmetrical lens which is sometimes used in practice. In the case of the symmetrical lens, there exists mirror symmetry about the midplane of the lens and only half of it need be studied. Since the midplane is an equipotential surface, it can be replaced by a plane of infinite permeability which will thus reduce the symmetrical lens to an asymmetrical one. We shall therefore study the geometry shown in Fig. 8.4b, and from it obtain the necessary information for the symmetrical case.

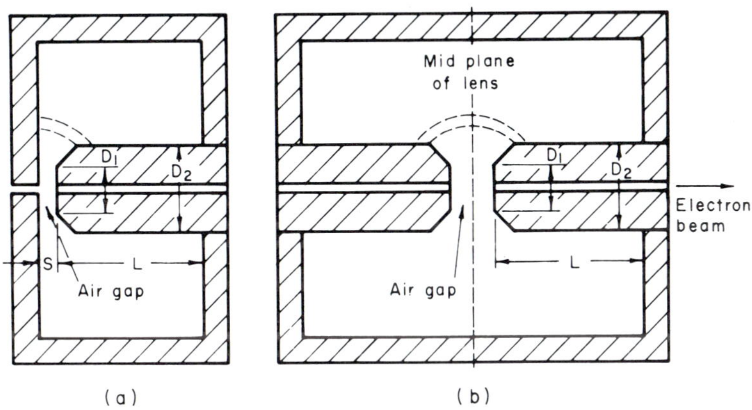

Fig. 8.4. Magnetic circuits of (a) asymmetrical and (b) symmetrical lenses.

The lens dimensions are expressed here in units of D_1, the pole-face diameter; D is the bore diameter, which is not considered in what follows. Its effect will be considered later. The relative flux density B/B_p is defined as the ratio between the flux density B at any point in the iron to the flux density B_p in the parallel part of the air gap.

For a core material of very high permeability and for so long as we are below saturation, we have

$$B_p \approx 4\pi \cdot 10^{-7} NI/S \tag{8.51}$$

where S is the air gap, and NI the ampere-turns.

An enlarged diagram of the pole piece of the asymmetrical lens is shown in Fig. 8.5a. The bore is not shown. The flux entering the pole face near the axis is approximately parallel to the axis. Further away from the axis, the flux lines are curved as shown in the figure. In Fig. 8.5b, the relative flux

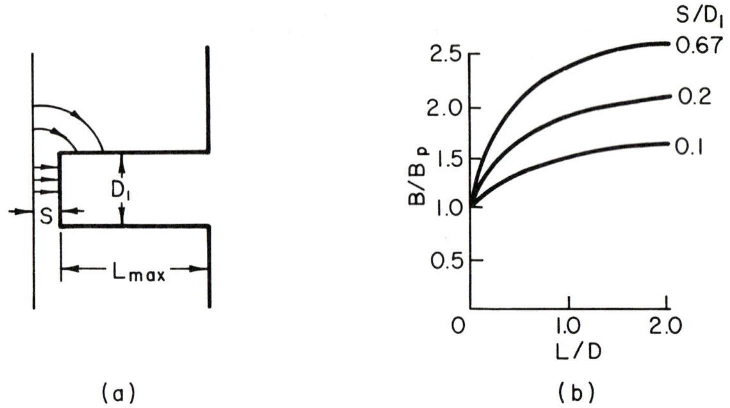

Fig. 8.5. (a) An enlarged diagram of the pole piece of the asymmetrical lens; (b) relative flux density B/B_p as a function of L/D_1 for three values of S/D_1 (from Mulvey, 1953.)

density B/B_p is plotted as a function of L/D_1 for three values of the ratio of gap width to pole-face diameter S/D_1; L is measured from the pole face. These results were obtained experimentally by Mulvey. It is evident that the flux density in the cylinder increases with increasing values of L/D_1. In order to avoid saturation, the flux density in the core must not exceed a certain limit. This, in turn, limits the flux density at the air gap. To avoid this difficulty, the pole pieces are tapered as shown in Fig. 8.6a in order to increase the cross section beyond the pole face and accommodate the leakage flux without danger of saturation. Figure 8.6b shows the relative flux density B/B_p at a distance L/D_1 along the pole piece for various taper angles. Here, $S/D_1 = 0.2$.

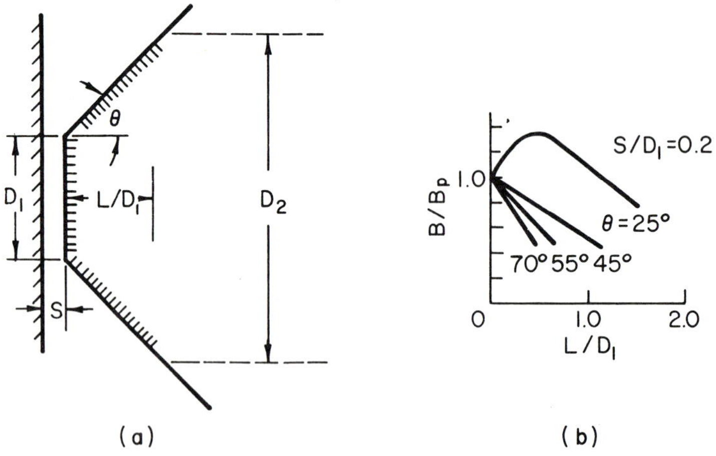

Fig. 8.6. (a) Pole piece with taper angle (the bore is not shown); (b) variation of flux density along the pole piece for various values of the taper angle θ (from Mulvey, 1953).

For small taper angles (less than 25°), the increase in iron cross section is not enough to prevent the increase in flux density at the core. On the other hand, we are not interested in overdesigning the magnetic circuit. Therefore, angles of 70° or more are also undesirable. Beyond a certain distance L, the taper can be stopped at a bore diameter D_2 where the cross section is enough to avoid saturation.

Figure 8.7 shows the variation of the flux density for various values of S/D_1 and D_2/D_1. The solid lines are for the case $L_{max}/D_1 = 3$, and the dotted lines for $L_{max}/D_1 = 6$.

In all the above analysis, we have ignored the influence of the bore diameter D. As a general rule, it is necessary to design the lens so that

$$D_1 \geq 3D$$

Two numerical examples which apply Mulvey's measurements in designing pole pieces follow.

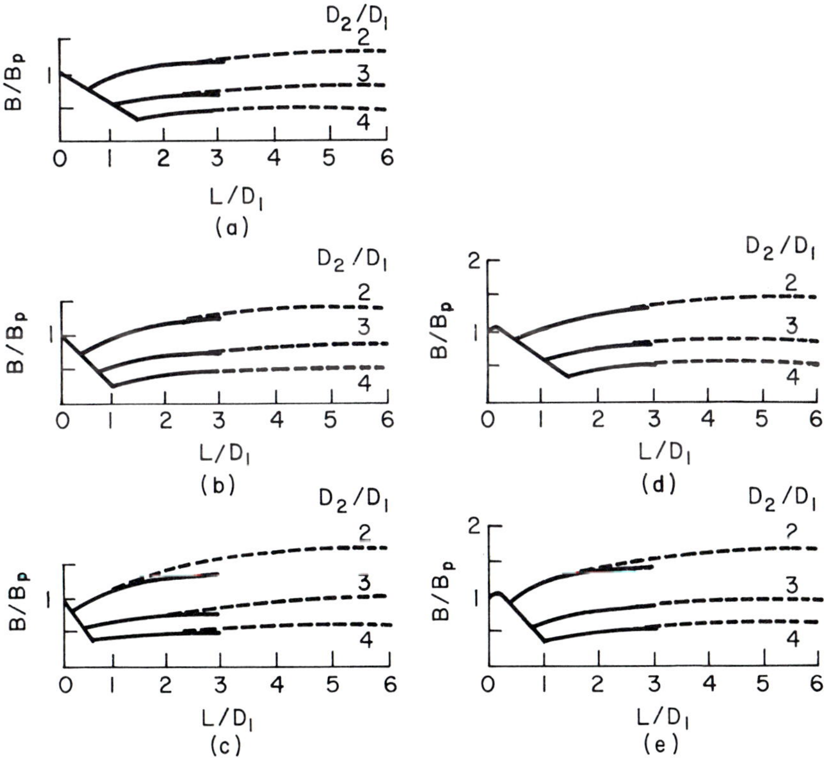

Fig. 8.7. The variation of the flux density as a function of the distance from the pole face, for various values of S/D_1 and D_2/D_1: (a) $S/D_1 = 0.2$, $\theta = 45°$; (b) $S/D_1 = 0.2$, $\theta = 55°$; (c) $S/D_1 = 0.2$, $\theta = 70°$; (d) $S/D_1 = 0.3$, $\theta = 45°$; (e) $S/D_1 = 0.3$, $\theta = 55°$ (from Mulvey, 1953).

Example 1

Let us consider the case of an asymmetrical lens which is to be used to focus a beam of 50 kV. Suppose that after other design considerations have been taken into account, we find that $B_p = 1.9$ T (tesla = Wb/m^2) and $S/D = 1$, $NI = 6000$ ampere turns and $D = 0.4$ cm. In order to minimize saturation at the pole face, a value of $S/D_1 = 0.3$ and a taper angle of 55° are chosen. Then $D_1 = 1.33$ cm. Allowing a total core length of at least 10 cm to accommodate the winding, $L/D \geq 7.5$. Referring to Fig. 8.7e, for $S/D_1 = 0.3$, we see that for $D_2/D_1 = 4$ or $D_2 = 5.3$ cm, the relative flux density in the cylindrical core does not rise above $0.6B_p$ or 1.14 T.

Example 2

In this example we shall consider the case of a symmetrical lens. We shall assume that $D = 0.2$ cm, $S/D = 1.5$, and $NI = 4700$ ampere turns. Then $S = 0.3$ cm and $B_p = 1.97$ *T*. The design proceeds exactly as in Example 1. Here, however, $S = \frac{1}{2}(0.3) = 0.15$ cm and B_p remains unchanged. By making $D_1 = 0.75$ cm and $\theta = 55°$, and choosing $D_2/D_1 = 4$ or $D_2 = 3$ cm, we find that the design is adequate.

8.7 Analysis of the Symmetrical Magnetic Lens

Using the axial field distribution given by Eq. (8.48), the cardinal points and the other design parameters of a series of magnetic lenses were obtained using a digital computer. It was convenient to express the field distribution and the z and r coordinates in relative units rather than mks units. The lens bore radius R and the maximum axial field B_0 are used as units of measure. We define

$$k^2 = (e/8mV_r)B_0^{\,2}R^2 \tag{8.52}$$

Here, V_r is the accelerating potential corrected for relativity as given by

$$V_r = V(1 + 0.978 \times 10^{-6}V) \tag{8.53}$$

and e and m are the charge and mass of the electron, respectively.

By writing $\bar{z} = z/R$, we obtain

$$B(\bar{z}) = \frac{2\mu_0 NI}{\pi S}\int_0^\infty \frac{\sin(St/D)\cos(t\bar{z})}{tI_0(t)}\,dt \tag{8.54}$$

and similarly for $B'(\bar{z})$ and $B''(\bar{z})$.

At $\bar{z} = 0$, the field is given by

$$B_0 = \frac{2\mu_0 NI}{\pi S} \int_0^\infty \frac{\sin(St/D)}{tI_0(t)}\, dt \tag{8.55}$$

Using Eq. (8.51), we obtain the ratio of the maximum axial field B_0 to the field in the pole-piece gap B_p

$$\frac{B_0}{B_p} = \frac{2}{\pi} \int_0^\infty \frac{\sin(St/D)}{tI_0(t)}\, dt \tag{8.56}$$

The data on the magnetic lenses were obtained for a series of values of S/D and for a whole range of k^2. The latter represents the strength of the lens. Magnetic lenses of various values of S and D, but of equal values of S/D, have the same characteristics as long as no saturation occurs.

The numerical integration of the ray equation was started for $|\bar{z}| \geq 6.0$ with an increment of $\Delta\bar{z} = 0.01$. This large interval was needed in order to guarantee that the starting and ending points of $\bar{z}$ would lie in a field-free region for the range of values of S/D that were studied.

As we mentioned in Chapter IV, there are two sets of cardinal points in a strong lens; the objective and the projective. For small values of k^2, the two sets are equal. In the numerical analysis program, when the ray crossed the axis at an intermediate point where the exact value of $\bar{z}$ is not known, the value was calculated by interpolation.

The values of $B(z)/B_0$ are tabulated as a function of $\bar{z}$ in Tables 8.1a–h[1] for all the values of S/D analyzed. From Fig. 8.8, where a few representative values of S/D are plotted, we can see how the axial field distribution broadens and flattens out near the center of the gap as S/D increases. Because of the symmetry of the lens, only half of the complete curve is shown.

In Chapter VII we used a different value of K^2 as defined by Glaser. We have the relation

$$(K^2)_{\text{Glaser}} = (a/R)^2 k^2 \tag{8.57}$$

Here, a is the value of z where the field attains half the maximum value, as defined in Chapter VII.

The field half-value width, in relative units, is plotted as a function of S/D in Fig. 8.9. From the figure, we can see that a/R asymptotically approaches S/D as S/D becomes large.

The ratio of the maximum axial field to the field in the gap at the pole pieces is plotted as a function of S/D in Fig. 8.10.

In order to relate the number of ampere-turns to the constant k^2, we define

$$\beta = k^2 V_r/(NI)^2 \tag{8.58}$$

[1] All tables for this chapter will be found at the end of the chapter.

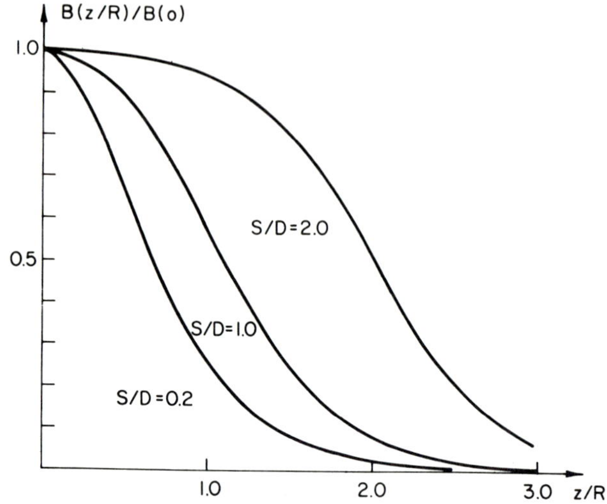

Fig. 8.8. Variation of the normalized axial field as a function of z for various values of S/D.

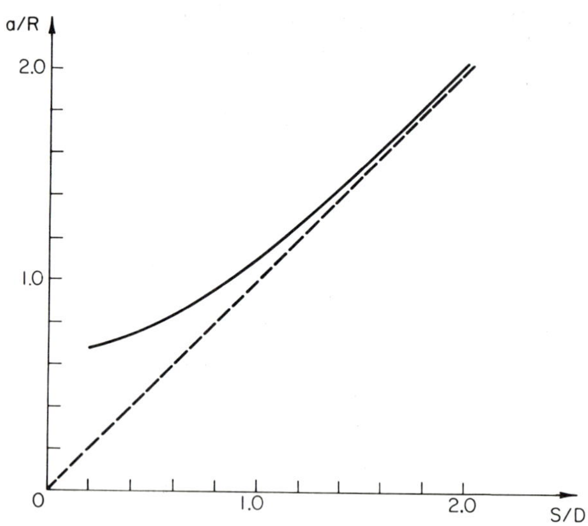

Fig. 8.9. Half-value width of $B(z)$ as a function of S/D.

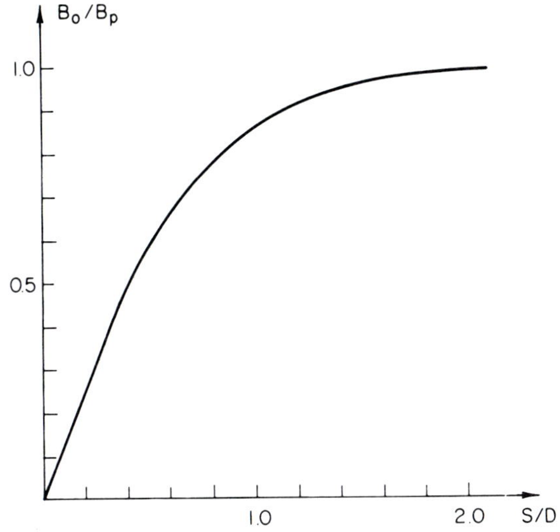

Fig. 8.10. Graph showing B_0/B_p as a function of S/D.

We can write

$$k^2 = \frac{e}{8mV_r}\left(\frac{B_0}{B_p}\right)^2 B_p{}^2R^2 \tag{8.59}$$

or

$$k^2 = \frac{e\mu_0{}^2}{32m}\left(\frac{B_0}{B_p}\right)^2 (D/S)^2\,\frac{(NI)^2}{V_r} \tag{8.60}$$

and

$$\beta = 0.0087(B_0/B_p)^2\,(D/S)^2 \tag{8.61}$$

With the above relations, it is possible to calculate the necessary number of ampere-turns for a given lens when all the other parameters have been fixed.

Figure 8.11 shows the variation of β as a function of S/D. The values of β, B_0/B_p, and a/R are listed in Table 8.2a. The number of ampere turns as a function of S/D for four values of the accelerating potential for a fixed objective focal length $f/R = 3.00$ is given in Table 8.2b.

The objective optical properties f_o/R and z_o/R are plotted for the lenses with $S/D = 0.1$, 0.5, 1.0, and 2.0 as a function of k^2 in Figs. 8.12 and 8.13. At small values of k^2 the focal length is larger for the smaller values of S/D. As k^2 increases, f_o/R decreases monotonically. The focal lengths for lenses with smaller values of S/D decrease more rapidly than for lenses with larger

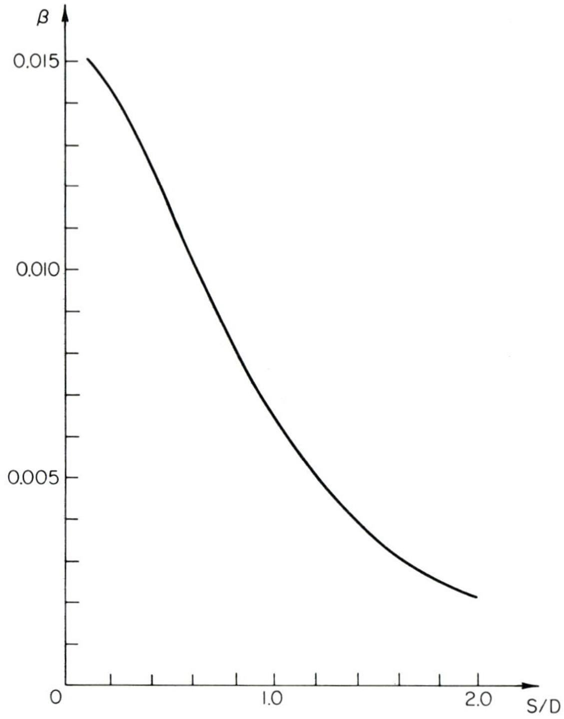

Fig. 8.11. β as a function of S/D. This relates the ampere turns, accelerating potential, and k^2 for any value of S/D.

values of S/D. This results in a crossover of the focal lengths in the region $0.8 \le k^2 \le 3.3$ for the range of lenses considered. After the crossover region, the focal lengths are longer for larger values of S/D. As k^2 increases, the focal lengths continue to decrease at a decreasing rate. For large values of k^2, the focal lengths do not vary appreciably for different values of S/D. For example, for $k^2 = 100$, the focal lengths are 0.36, 0.37, 0.38, and 0.38 for S/D equal to 0.1, 0.5, 1.0, and 2.0, respectively.

The midfocal length z_o/R is drawn as a function of k^2 in Fig. 8.13. Again we see a monotonic decrease as k^2 increases with z_o/R being smaller for larger values of S/D. The midfocal lengths, unlike f_o/R, remain separated as k^2 increases, becoming negative in the range $0.8 \le k^2 \le 8.0$. This separation is still evident at extremely high values of k^2. For example, at $k^2 = 100$, z_o/R is -0.65, -0.79, -1.14, and -2.08 for $S/D = 0.1$, 0.5, 1.0, and 2.0, respectively.

The objective optical properties discussed above are of particular interest when the image or object is immersed in the lens field, for example, in the electron microscope objective lens. In most applications, the projective optical properties are of interest. The magnetically focused cathode-ray tube and the projective lens of the electron microscope are examples of some devices in which the projective data are meaningful.

The projective focal lengths are of an oscillatory nature as k^2 increases. This is a result of the multiple crossings of the optical axis by the beam for strong lenses and the definition of f_p as discussed in Chapter IV. For practical purposes, only beams that cross the axis once are of interest. In this region, the projective focal lengths initially decrease, reaching minimum values and then increase again as k^2 increases. This is shown in Fig. 8.14 for the series of lenses $S/D = 0.1$, 0.5, 1.0, and 2.0. As with the objective focal length, the projective focal length is largest for small values of S/D for small values of k^2. Here, of course, there is very little difference in the values of the projective and objective quantities. This correspondence increases with k^2 for smaller values of S/D up to approximately $k^2 = 1.0$ for $S/D = 0.1$.

The minimum value of f_p is of particular interest, since it represents the minimum value of focal length obtainable when the lens is used as a projective lens. Furthermore, operation of the lens at or near this point makes f_p invariant with small fluctuations in k^2 due to the broad shallow nature of the minima.

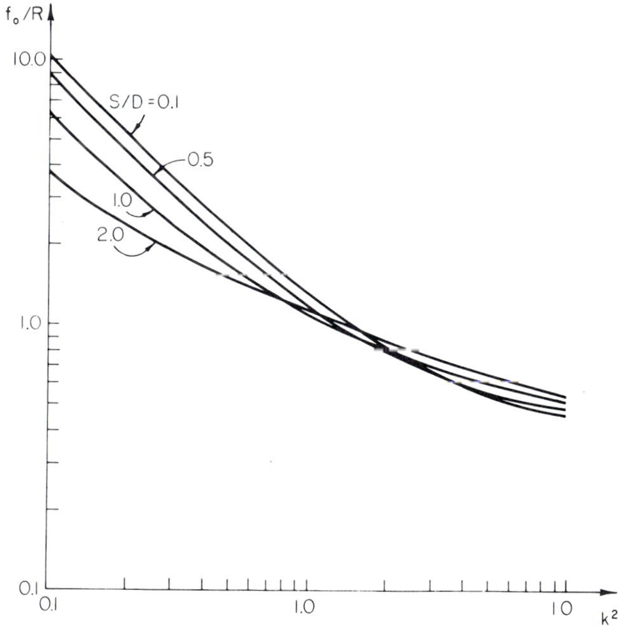

Fig. 8.12. Objective focal length as a function of lens excitation.

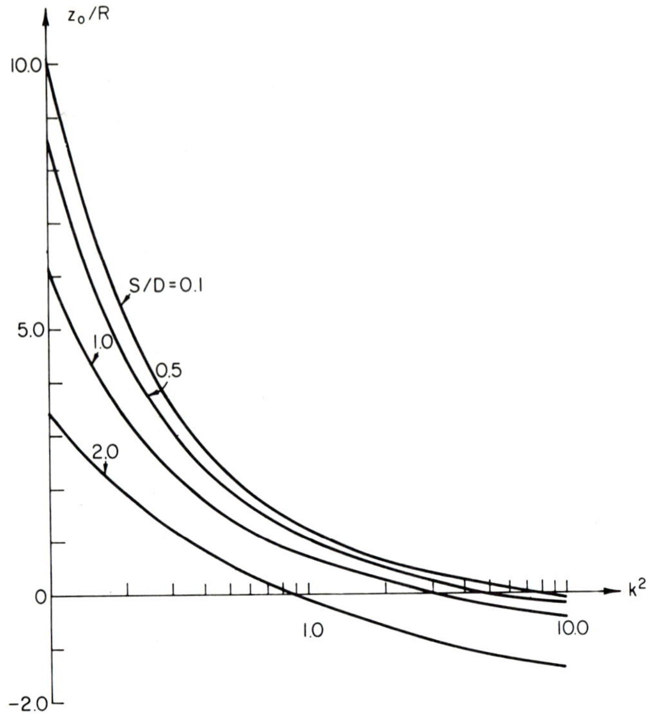

Fig. 8.13. Objective midfocal length as a function of lens excitation.

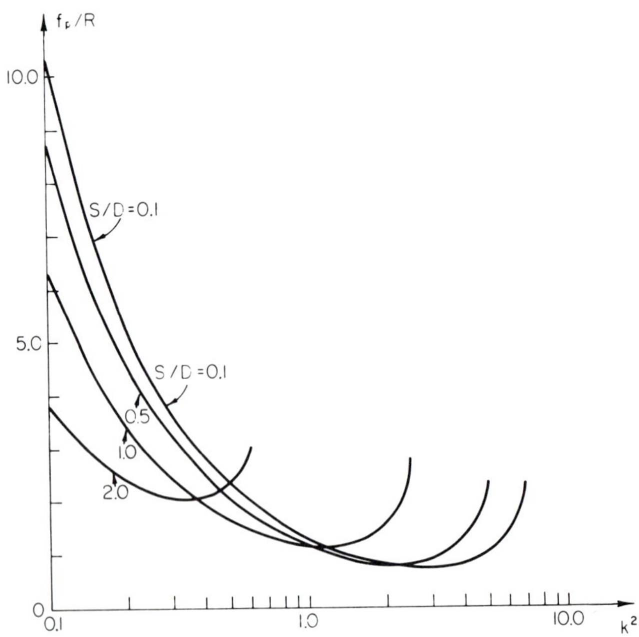

Fig. 8.14. Projective focal length as a function of lens excitation.

For the range of lenses considered, the minima fall between $0.35 \leq k^2 \leq 3.0$. After the minimum, further increase in k^2 results in a rapid increase of f_p which asymptotically goes to infinity. This corresponds to the value of k^2 for which the beam leaves the lens field parallel to the optical axis.

The projective midfocal length z_p/R as a function of k^2 is shown in Fig. 8.15. As k^2 increases, z_p/R decreases monotonically, becomes negative, and then asymptotically approaches minus infinity as the beam leaves the lens parallel to the optical axis. The complete lens data are listed in Tables 8.3–8.31.

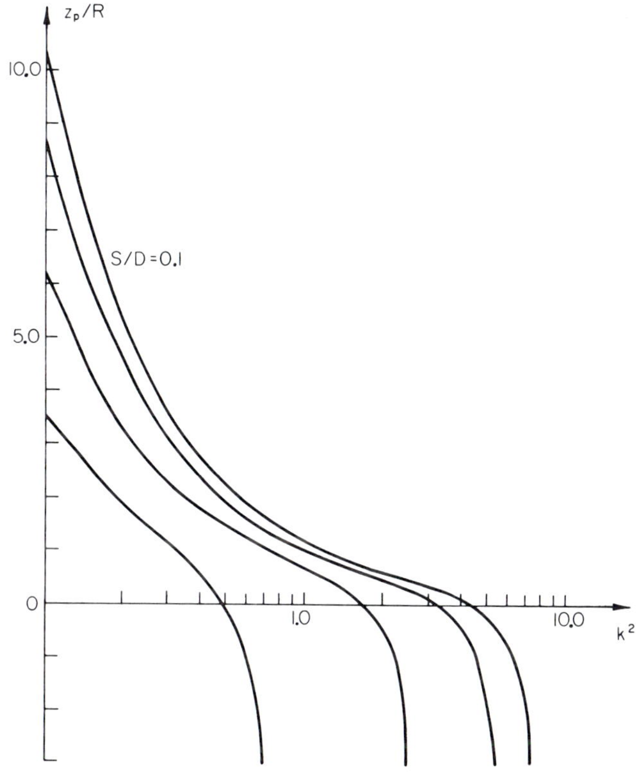

Fig. 8.15. Projective midfocal length as a function of lens excitation.

FURTHER READING

Lenz, F. (1950). *Optik*, **17**, 243.
Lenz, F. (1950). *Ann. Physik* **9**, 245.
Mulvey, T. (1953). *Proc. Phys. Soc.* (*London*) **B66**, 441.
Parks, H. G. (1969). M.Sc. Dissertation. Syracuse Univ., Syracuse, New York.

TABLES 8.1a–h

NORMALIZED AXIAL MAGNETIC FLUX DENSITY AS A FUNCTION OF z/R

TABLE 8.1a

	$B(z)/B_0$			
z/R	$S/D = 0.01$	$S/D = 0.02$	$S/D = 0.03$	$S/D = 0.04$
0	1.00	1.00	1.00	1.00
0.20	0.93	0.93	0.93	0.93
0.40	0.75	0.75	0.75	0.75
0.60	0.55	0.55	0.55	0.55
0.80	0.38	0.38	0.38	0.38
1.00	0.25	0.25	0.25	0.25
1.20	0.16	0.16	0.16	0.16
1.40	0.10	0.10	0.10	0.10
1.60	0.06	0.06	0.06	0.06
1.80	0.04	0.04	0.04	0.04
2.00	0.02	0.02	0.02	0.02
2.20	0.01	0.01	0.01	0.01
2.40	0.01	0.01	0.01	0.01
2.60	0.01	0.01	0.01	0.01
2.80	0.00	0.00	0.00	0.00
3.00	0.00	0.00	0.00	0.00

TABLE 8.1b

	$B(z)/B_0$			
z/R	$S/D = 0.05$	$S/D = 0.06$	$S/D = 0.07$	$S/D = 0.08$
0	1.00	1.00	1.00	1.00
0.20	0.93	0.93	0.93	0.93
0.40	0.75	0.75	0.76	0.76
0.60	0.55	0.55	0.55	0.55
0.80	0.38	0.38	0.38	0.38
1.00	0.25	0.25	0.25	0.25
1.20	0.16	0.16	0.16	0.16
1.40	0.10	0.10	0.10	0.10
1.60	0.06	0.06	0.06	0.06
1.80	0.04	0.04	0.04	0.04
2.00	0.02	0.02	0.02	0.02
2.20	0.01	0.01	0.01	0.01
2.40	0.01	0.01	0.01	0.01
2.60	0.01	0.01	0.01	0.01
2.80	0.00	0.00	0.00	0.00
3.00	0.00	0.00	0.00	0.00

TABLE 8.1c

z/R	$B(z)/B_0$			
	$S/D = 0.09$	$S/D = 0.10$	$S/D = 0.20$	$S/D = 0.30$
0	1.00	1.00	1.00	1.00
0.20	0.93	0.93	0.93	0.94
0.40	0.76	0.76	0.77	0.78
0.60	0.55	0.56	0.57	0.59
0.80	0.38	0.38	0.39	0.41
1.00	0.25	0.25	0.26	0.28
1.20	0.16	0.16	0.17	0.18
1.40	0.10	0.10	0.10	0.11
1.60	0.06	0.06	0.07	0.07
1.80	0.04	0.04	0.04	0.04
2.00	0.02	0.02	0.03	0.03
2.20	0.01	0.01	0.02	0.02
2.40	0.01	0.01	0.01	0.01
2.60	0.01	0.01	0.01	0.01
2.80	0.00	0.00	0.00	0.00
3.00	0.00	0.00	0.00	0.00

TABLE 8.1d

z/R	$B(z)/B_0$	
	$S/D = 0.40$	$S/D = 0.50$
0	1.00	1.00
0.20	0.95	0.95
0.40	0.80	0.83
0.60	0.62	0.66
0.80	0.44	0.48
1.00	0.30	0.33
1.20	0.19	0.22
1.40	0.12	0.14
1.60	0.08	0.09
1.80	0.05	0.05
2.00	0.03	0.03
2.20	0.02	0.02
2.40	0.01	0.01
2.60	0.01	0.01
2.80	0.00	0.00
3.00	0.00	0.00

TABLE 8.1e

z/R	$B(z)/B_0$			
	$S/D = 0.60$	$S/D = 0.70$	$S/D = 0.80$	$S/D = 0.90$
0	1.00	1.00	1.00	1.00
0.20	0.96	0.97	0.97	0.98
0.40	0.85	0.88	0.90	0.92
0.60	0.69	0.73	0.77	0.81
0.80	0.52	0.57	0.62	0.67
1.00	0.37	0.41	0.46	0.52
1.20	0.25	0.28	0.32	0.37
1.40	0.16	0.18	0.22	0.25
1.60	0.10	0.12	0.14	0.17
1.80	0.06	0.07	0.09	0.11
2.00	0.04	0.05	0.06	0.07
2.20	0.02	0.03	0.03	0.04
2.40	0.02	0.02	0.02	0.03
2.60	0.01	0.01	0.01	0.02
2.80	0.01	0.01	0.01	0.01
3.00	0.00	0.00	0.01	0.01

TABLE 8.1f

z/R	$B(z)/B_0$			
	$S/D = 1.0$	$S/D = 1.1$	$S/D = 1.2$	$S/D = 1.3$
0	1.00	1.00	1.00	1.00
0.20	0.98	0.99	0.99	0.99
0.40	0.93	0.95	0.96	0.97
0.60	0.84	0.87	0.90	0.92
0.80	0.72	0.77	0.81	0.84
1.00	0.57	0.63	0.69	0.74
1.20	0.43	0.48	0.55	0.61
1.40	0.30	0.35	0.40	0.47
1.60	0.20	0.24	0.28	0.33
1.80	0.13	0.16	0.19	0.23
2.00	0.08	0.10	0.12	0.15
2.20	0.05	0.06	0.08	0.09
2.40	0.03	0.04	0.05	0.06
2.60	0.02	0.02	0.03	0.04
2.80	0.01	0.01	0.02	0.02
3.00	0.01	0.01	0.01	0.01
3.20	0.00	0.01	0.01	0.01
3.40	0.00	0.00	0.00	0.01
3.60	0.00	0.00	0.00	0.00
3.80	0.00	0.00	0.00	0.00
4.00	0.00	0.00	0.00	0.00

TABLE 8.1g

z/R	$B(z)/B_0$			
	$S/D = 1.4$	$S/D = 1.5$	$S/D = 1.6$	$S/D = 1.7$
0	1.00	1.00	1.00	1.00
0.20	0.99	0.99	1.00	1.00
0.40	0.97	0.98	0.98	0.99
0.60	0.93	0.95	0.96	0.97
0.80	0.87	0.90	0.92	0.93
1.00	0.78	0.82	0.86	0.88
1.20	0.66	0.72	0.77	0.81
1.40	0.53	0.59	0.65	0.71
1.60	0.39	0.45	0.52	0.58
1.80	0.27	0.33	0.38	0.45
2.00	0.18	0.22	0.27	0.32
2.20	0.12	0.14	0.18	0.22
2.40	0.07	0.09	0.11	0.14
2.60	0.05	0.06	0.07	0.09
2.80	0.03	0.04	0.05	0.06
3.00	0.02	0.02	0.03	0.04
3.20	0.01	0.01	0.02	0.02
3.40	0.01	0.01	0.01	0.01
3.60	0.00	0.01	0.01	0.01
3.80	0.00	0.00	0.00	0.01
4.00	0.00	0.00	0.00	0.00

TABLE 8.1h

z/R	$B(z)/B_0$		
	$S/D = 1.8$	$S/D = 1.9$	$S/D = 2.0$
0	1.00	1.00	1.00
0.20	1.00	1.00	1.00
0.40	0.99	0.99	0.99
0.60	0.97	0.98	0.98
0.80	0.95	0.96	0.97
1.00	0.91	0.93	0.94
1.20	0.85	0.88	0.90
1.40	0.76	0.80	0.84
1.60	0.64	0.70	0.75
1.80	0.51	0.58	0.64
2.00	0.38	0.44	0.51
2.20	0.26	0.32	0.38
2.40	0.18	0.22	0.26
2.60	0.11	0.14	0.17
2.80	0.07	0.09	0.11
3.00	0.04	0.06	0.07
3.20	0.03	0.04	0.04
3.40	0.02	0.02	0.03
3.60	0.01	0.01	0.02
3.80	0.01	0.01	0.01
4.00	0.00	0.01	0.01

TABLE 8.2a
NORMALIZED VALUES OF B_0 AND a, AND β AS A FUNCTION OF S/D

S/D	B_0/B_p	a/R	β
0.01	0.013		0.0147
0.02	0.027		0.0159
0.03	0.040		0.0155
0.04	0.053		0.0153
0.05	0.066		0.0152
0.06	0.079		0.0152
0.07	0.093		0.0152
0.08	0.106		0.0152
0.09	0.119		0.0152
0.10	0.132	0.667	0.0152
0.20	0.259	0.674	0.0146
0.30	0.377	0.699	0.0137
0.40	0.483	0.733	0.0127
0.50	0.577	0.776	0.0116
0.60	0.656	0.828	0.0104
0.70	0.723	0.886	0.0093
0.80	0.778	0.952	0.0082
0.90	0.823	1.023	0.0073
1.0	0.860	1.099	0.0064
1.1	0.889	1.179	0.0057
1.2	0.912	1.263	0.0050
1.3	0.931	1.350	0.0046
1.4	0.945	1.440	0.0040
1.5	0.957	1.532	0.0035
1.6	0.966	1.625	0.0032
1.7	0.973	1.720	0.0029
1.8	0.979	1.816	0.0026
1.9	0.983	1.912	0.0023
2.0	0.987	2.010	0.0021
3.0	0.999	3.001	0.0010

TABLE 8.2b

AMPERE TURNS AS A FUNCTION OF S/D FOR VARIOUS ACCELERATING POTENTIALS, FOR FIXED OBJECTIVE FOCAL LENGTH $f/R = 3.0$

S/D	10 kV	50 kV	100 kV	1000 kV
0.01	508	1157	1673	7103
0.02	487	1109	1605	6812
0.03	492	1120	1621	6881
0.04	494	1126	1629	6916
0.05	496	1130	1635	6939
0.06	496	1130	1635	6939
0.07	496	1130	1635	6939
0.08	496	1130	1635	6939
0.09	496	1130	1635	6939
0.10	496	1130	1635	6939
0.20	499	1137	1645	6984
0.30	508	1158	1675	7109
0.40	512	1167	1689	7169
0.50	528	1203	1740	7387
0.60	540	1230	1780	7554
0.70	551	1257	1818	7717
0.80	566	1290	1866	7919
0.90	576	1313	1900	8064
1.00	600	1365	1970	8430
1.10	610	1390	2011	8537
1.20	619	1412	2042	8670
1.30	629	1433	2073	8798
1.40	655	1493	2160	9169
1.50	679	1548	2240	9509
1.60	711	1619	2343	9945
1.70	746	1701	2461	10447
1.80	788	1796	2599	11033
1.90	838	1910	2763	11730
2.00	849	1935	2800	11886

TABLE 8.3
RATIO OF GAP WIDTH TO BORE DIAMETER $S/D = 0.01$

	Objective			Projective		
k^2	f/R	z_m/R	z_p/R	f/R	z_m/R	z_p/R
0.02	48.87	48.86	−0.01	48.87	48.86	−0.01
0.04	24.62	24.61	−0.02	24.62	24.61	−0.02
0.06	16.54	16.52	−0.03	16.54	16.52	−0.03
0.08	12.51	12.47	−0.04	12.51	12.47	−0.04
0.10	10.08	10.04	−0.04	10.08	10.04	−0.04
0.12	8.47	8.42	−0.05	8.47	8.42	−0.05
0.14	7.32	7.25	−0.06	7.32	7.25	−0.06
0.16	6.45	6.38	−0.07	6.45	6.38	−0.07
0.18	5.78	5.70	−0.08	5.78	5.70	−0.08
0.20	5.24	5.15	−0.09	5.24	5.15	−0.09
0.22	4.80	4.71	−0.10	4.81	4.71	−0.10
0.24	4.43	4.33	−0.10	4.44	4.33	−0.11
0.26	4.12	4.02	−0.10	4.13	4.01	−0.12
0.28	3.85	3.75	−0.11	3.87	3.74	−0.13
0.30	3.62	3.51	−0.11	3.64	3.50	−0.14
0.32	3.42	3.30	−0.11	3.44	3.29	−0.15
0.34	3.24	3.12	−0.12	3.26	3.11	−0.15
0.36	3.08	2.96	−0.12	3.10	2.94	−0.16
0.38	2.94	2.81	−0.13	2.97	2.79	−0.17
0.40	2.81	2.68	−0.13	2.84	2.66	−0.18
0.42	2.70	2.56	−0.13	2.73	2.53	−0.19
0.44	2.59	2.45	−0.14	2.62	2.42	−0.20
0.46	2.49	2.35	−0.14	2.53	2.32	−0.21
0.48	2.41	2.26	−0.14	2.44	2.22	−0.22
0.50	2.32	2.18	−0.15	2.37	2.13	−0.23
0.52	2.25	2.10	−0.15	2.29	2.05	−0.24
0.54	2.18	2.03	−0.15	2.23	1.97	−0.25
0.56	2.12	1.96	−0.15	2.16	1.90	−0.26
0.58	2.06	1.90	−0.16	2.11	1.83	−0.27
0.60	2.00	1.84	−0.16	2.05	1.77	−0.28
0.62	1.95	1.79	−0.16	2.00	1.71	−0.29
0.64	1.90	1.74	−0.16	1.96	1.65	−0.30
0.66	1.85	1.69	−0.17	1.91	1.60	−0.31
0.68	1.81	1.64	−0.17	1.87	1.55	−0.32
0.70	1.77	1.60	−0.17	1.83	1.50	−0.33
0.72	1.73	1.56	−0.18	1.80	1.45	−0.34
0.74	1.70	1.52	−0.18	1.76	1.41	−0.35
0.76	1.66	1.48	−0.18	1.73	1.37	−0.36
0.78	1.63	1.45	−0.18	1.70	1.33	−0.37
0.80	1.60	1.42	−0.19	1.67	1.29	−0.39
0.82	1.57	1.38	−0.19	1.64	1.25	−0.40
0.84	1.55	1.35	−0.19	1.62	1.21	−0.41

TABLE 8.3 (continued)

	Objective			Projective		
k^2	f/R	z_m/R	z_p/R	f/R	z_m/R	z_p/R
0.86	1.52	1.32	−0.20	1.59	1.18	−0.42
0.88	1.50	1.30	−0.20	1.57	1.14	−0.43
0.90	1.47	1.27	−0.20	1.55	1.11	−0.44
0.92	1.45	1.25	−0.20	1.53	1.08	−0.45
0.94	1.43	1.22	−0.21	1.51	1.05	−0.46
0.96	1.41	1.20	−0.21	1.49	1.02	−0.47
0.98	1.39	1.18	−0.21	1.47	0.99	−0.48
1.00	1.37	1.15	−0.22	1.45	0.96	−0.49
1.02	1.35	1.13	−0.22	1.44	0.93	−0.51
1.04	1.33	1.11	−0.22	1.42	0.91	−0.52
1.06	1.32	1.09	−0.22	1.41	0.88	−0.53
1.08	1.30	1.07	−0.23	1.39	0.85	−0.54
1.10	1.29	1.06	−0.23	1.38	0.83	−0.55
1.12	1.27	1.04	−0.23	1.37	0.80	−0.56
1.14	1.26	1.02	−0.24	1.35	0.78	−0.57
1.16	1.24	1.01	−0.24	1.34	0.76	−0.59
1.18	1.23	0.99	−0.24	1.33	0.73	−0.60
1.20	1.22	0.97	−0.25	1.32	0.71	−0.61
1.22	1.21	0.96	−0.25	1.31	0.69	−0.62
1.24	1.20	0.94	−0.25	1.30	0.67	−0.63
1.26	1.18	0.93	−0.25	1.29	0.65	−0.65
1.28	1.17	0.92	−0.26	1.28	0.62	−0.66
1.30	1.16	0.90	−0.26	1.27	0.60	−0.67
1.32	1.15	0.89	−0.26	1.26	0.58	−0.68
1.34	1.14	0.88	−0.26	1.26	0.56	−0.69
1.36	1.13	0.87	−0.27	1.25	0.54	−0.71
1.38	1.12	0.85	−0.27	1.24	0.52	−0.72
1.40	1.12	0.84	−0.27	1.23	0.50	−0.73
1.42	1.11	0.83	−0.28	1.23	0.48	−0.74
1.44	1.10	0.82	−0.28	1.22	0.46	−0.76
1.46	1.09	0.81	−0.28	1.22	0.45	−0.77
1.48	1.08	0.80	−0.28	1.21	0.43	−0.78
1.50	1.08	0.79	−0.29	1.20	0.41	−0.80
1.52	1.07	0.78	−0.29	1.20	0.39	−0.81
1.54	1.06	0.77	−0.29	1.19	0.37	−0.82
1.56	1.05	0.76	−0.30	1.19	0.35	−0.83
1.58	1.05	0.75	−0.30	1.19	0.34	−0.85
1.60	1.04	0.74	−0.30	1.18	0.32	−0.86
1.62	1.03	0.73	−0.30	1.18	0.30	−0.88
1.64	1.03	0.72	−0.31	1.17	0.28	−0.89
1.66	1.02	0.71	−0.31	1.17	0.27	−0.90
1.68	1.02	0.70	−0.31	1.17	0.25	−0.92
1.70	1.01	0.70	−0.31	1.16	0.23	−0.93

TABLE 8.3 (continued)

k^2	Objective f/R	Objective z_m/R	Objective z_p/R	Projective f/R	Projective z_m/R	Projective z_p/R
1.72	1.01	0.69	−0.32	1.16	0.22	−0.94
1.74	1.00	0.68	−0.32	1.16	0.20	−0.96
1.76	1.00	0.67	−0.32	1.15	0.18	−0.97
1.78	0.99	0.66	−0.33	1.15	0.17	−0.99
1.80	0.99	0.66	−0.33	1.15	0.15	−1.00
1.82	0.98	0.65	−0.33	1.15	0.13	−1.01
1.84	0.98	0.64	−0.33	1.15	0.12	−1.03
1.86	0.97	0.64	−0.34	1.14	0.10	−1.04
1.88	0.97	0.63	−0.34	1.14	0.08	−1.06
1.90	0.96	0.62	−0.34	1.14	0.07	−1.07
1.92	0.96	0.61	−0.34	1.14	0.05	−1.09
1.94	0.95	0.61	−0.35	1.14	0.04	−1.10
1.96	0.95	0.60	−0.35	1.14	0.02	−1.12
1.98	0.95	0.60	−0.35	1.14	0.00	−1.13
2.00	0.94	0.59	−0.35	1.14	−0.01	−1.15
2.02	0.94	0.58	−0.36	1.13	−0.03	−1.16
2.04	0.94	0.58	−0.36	1.13	−0.04	−1.18
2.06	0.93	0.57	−0.36	1.13	−0.06	−1.19
2.08	0.93	0.56	−0.36	1.13	−0.08	−1.21
2.10	0.93	0.56	−0.37	1.13	−0.09	−1.22
2.12	0.92	0.55	−0.37	1.13	−0.11	−1.24
2.14	0.92	0.55	−0.37	1.13	−0.12	−1.26
2.16	0.92	0.54	−0.37	1.13	−0.14	−1.27
2.18	0.91	0.54	−0.38	1.13	−0.15	−1.29
2.20	0.91	0.53	−0.38	1.13	−0.17	−1.31
2.22	0.91	0.53	−0.38	1.14	−0.19	−1.32
2.24	0.91	0.52	−0.38	1.14	−0.20	−1.34
2.26	0.90	0.52	−0.39	1.14	−0.22	−1.35
2.28	0.90	0.51	−0.39	1.14	−0.23	−1.37
2.30	0.90	0.51	−0.39	1.14	−0.25	−1.39
2.32	0.89	0.50	−0.39	1.14	−0.26	−1.41
2.34	0.89	0.50	−0.40	1.14	−0.28	−1.42
2.36	0.89	0.49	−0.40	1.14	−0.30	−1.44
2.38	0.89	0.49	−0.40	1.14	−0.31	−1.46
2.40	0.89	0.48	−0.40	1.15	−0.33	−1.47
2.42	0.88	0.48	−0.41	1.15	−0.34	−1.49
2.44	0.88	0.47	−0.41	1.15	−0.36	−1.51
2.46	0.88	0.47	−0.41	1.15	−0.38	−1.53
2.48	0.88	0.46	−0.41	1.15	−0.39	−1.55
2.50	0.88	0.46	−0.42	1.16	−0.41	−1.56

TABLE 8.4

RATIO OF GAP WIDTH TO BORE DIAMETER $S/D = 0.02$

	Objective			Projective		
k^2	f/R	z_m/R	z_p/R	f/R	z_m/R	z_p/R
0.02	49.29	49.28	−0.01	49.29	49.28	−0.01
0.04	24.82	24.81	−0.01	24.82	24.81	−0.01
0.06	16.66	16.64	−0.02	16.66	16.64	−0.02
0.08	12.59	12.56	−0.03	12.59	12.56	−0.03
0.10	10.14	10.11	−0.04	10.14	10.11	−0.04
0.12	8.51	8.47	−0.04	8.51	8.47	−0.04
0.14	7.35	7.30	−0.05	7.35	7.30	−0.05
0.16	6.47	6.42	−0.06	6.47	6.42	−0.06
0.18	5.80	5.73	−0.06	5.80	5.73	−0.06
0.20	5.25	5.18	−0.07	5.25	5.18	−0.07
0.22	4.81	4.73	−0.08	4.81	4.73	−0.08
0.24	4.44	4.36	−0.08	4.44	4.35	−0.09
0.26	4.12	4.04	−0.08	4.13	4.04	−0.09
0.28	3.85	3.76	−0.09	3.86	3.76	−0.10
0.30	3.62	3.53	−0.09	3.63	3.52	−0.11
0.32	3.41	3.32	−0.09	3.43	3.31	−0.12
0.34	3.23	3.14	−0.10	3.25	3.13	−0.12
0.36	3.07	2.97	−0.10	3.09	2.96	−0.13
0.38	2.93	2.83	−0.10	2.95	2.81	−0.14
0.40	2.80	2.69	−0.11	2.82	2.68	−0.15
0.42	2.68	2.57	−0.11	2.71	2.55	−0.15
0.44	2.58	2.47	−0.11	2.60	2.44	−0.16
0.46	2.48	2.37	−0.12	2.51	2.34	−0.17
0.48	2.39	2.27	−0.12	2.42	2.24	−0.18
0.50	2.31	2.19	−0.12	2.34	2.16	−0.19
0.52	2.24	2.11	−0.12	2.27	2.08	−0.19
0.54	2.17	2.04	−0.13	2.20	2.00	−0.20
0.56	2.10	1.97	−0.13	2.14	1.93	−0.21
0.58	2.04	1.91	−0.13	2.08	1.86	−0.22
0.60	1.99	1.85	−0.13	2.02	1.80	−0.22
0.62	1.93	1.80	−0.14	1.97	1.74	−0.23
0.64	1.88	1.75	−0.14	1.93	1.69	−0.24
0.66	1.84	1.70	−0.14	1.88	1.63	−0.25
0.68	1.80	1.65	−0.14	1.84	1.58	−0.26
0.70	1.76	1.61	−0.15	1.80	1.54	−0.26
0.72	1.72	1.57	−0.15	1.76	1.49	−0.27
0.74	1.68	1.53	−0.15	1.73	1.45	−0.28
0.76	1.65	1.49	−0.15	1.70	1.41	−0.29
0.78	1.61	1.46	−0.16	1.66	1.37	−0.30
0.80	1.58	1.42	−0.16	1.64	1.33	−0.31
0.82	1.56	1.39	−0.16	1.61	1.29	−0.31
0.84	1.53	1.36	−0.17	1.58	1.26	−0.32

TABLE 8.4 (continued)

	Objective			Projective		
k^2	f/R	z_m/R	z_p/R	f/R	z_m/R	z_p/R
0.86	1.50	1.33	−0.17	1.56	1.22	−0.33
0.88	1.48	1.31	−0.17	1.53	1.19	−0.34
0.90	1.45	1.28	−0.17	1.51	1.16	−0.35
0.92	1.43	1.25	−0.18	1.49	1.13	−0.36
0.94	1.41	1.23	−0.18	1.47	1.10	−0.37
0.96	1.39	1.21	−0.18	1.45	1.07	−0.37
0.98	1.37	1.18	−0.18	1.43	1.05	−0.38
1.00	1.35	1.16	−0.19	1.41	1.02	−0.39
1.02	1.33	1.14	−0.19	1.39	0.99	−0.40
1.04	1.31	1.12	−0.19	1.38	0.97	−0.41
1.06	1.30	1.10	−0.19	1.36	0.94	−0.42
1.08	1.28	1.08	−0.20	1.35	0.92	−0.43
1.10	1.26	1.06	−0.20	1.33	0.90	−0.44
1.12	1.25	1.05	−0.20	1.32	0.87	−0.45
1.14	1.24	1.03	−0.21	1.31	0.85	−0.45
1.16	1.22	1.01	−0.21	1.29	0.83	−0.46
1.18	1.21	1.00	−0.21	1.28	0.81	−0.47
1.20	1.20	0.98	−0.21	1.27	0.79	−0.48
1.22	1.18	0.97	−0.22	1.26	0.77	−0.49
1.24	1.17	0.95	−0.22	1.25	0.75	−0.50
1.26	1.16	0.94	−0.22	1.24	0.73	−0.51
1.28	1.15	0.92	−0.22	1.23	0.71	−0.52
1.30	1.14	0.91	−0.23	1.22	0.69	−0.53
1.32	1.13	0.90	−0.23	1.21	0.67	−0.54
1.34	1.12	0.89	−0.23	1.20	0.65	−0.55
1.36	1.11	0.87	−0.23	1.19	0.64	−0.56
1.38	1.10	0.86	−0.24	1.18	0.62	−0.57
1.40	1.09	0.85	−0.24	1.18	0.60	−0.58
1.42	1.08	0.84	−0.24	1.17	0.58	−0.59
1.44	1.07	0.83	−0.24	1.16	0.57	−0.60
1.46	1.06	0.82	−0.25	1.15	0.55	−0.61
1.48	1.06	0.81	−0.25	1.15	0.53	−0.62
1.50	1.05	0.80	−0.25	1.14	0.52	−0.63
1.52	1.04	0.79	−0.25	1.14	0.50	−0.64
1.54	1.03	0.78	−0.26	1.13	0.49	−0.65
1.56	1.03	0.77	−0.26	1.12	0.47	−0.66
1.58	1.02	0.76	−0.26	1.12	0.45	−0.67
1.60	1.01	0.75	−0.26	1.11	0.44	−0.68
1.62	1.00	0.74	−0.27	1.11	0.42	−0.69
1.64	1.00	0.73	−0.27	1.10	0.41	−0.70
1.66	0.99	0.72	−0.27	1.10	0.39	−0.71
1.68	0.99	0.71	−0.27	1.10	0.38	−0.72
1.70	0.98	0.70	−0.28	1.09	0.36	−0.73

TABLE 8.4 (continued)

	Objective			Projective		
k^2	f/R	z_m/R	z_p/R	f/R	z_m/R	z_p/R
1.72	0.97	0.70	−0.28	1.09	0.35	−0.74
1.74	0.97	0.69	−0.28	1.08	0.34	−0.75
1.76	0.96	0.68	−0.28	1.08	0.32	−0.76
1.78	0.96	0.67	−0.29	1.08	0.31	−0.77
1.80	0.95	0.67	−0.29	1.07	0.29	−0.78
1.82	0.95	0.66	−0.29	1.07	0.28	−0.79
1.84	0.94	0.65	−0.29	1.07	0.27	−0.80
1.86	0.94	0.64	−0.29	1.07	0.25	−0.81
1.88	0.93	0.64	−0.30	1.06	0.24	−0.82
1.90	0.93	0.63	−0.30	1.06	0.22	−0.84
1.92	0.93	0.62	−0.30	1.06	0.21	−0.85
1.94	0.92	0.62	−0.30	1.06	0.20	−0.86
1.96	0.92	0.61	−0.31	1.05	0.18	−0.87
1.98	0.91	0.60	−0.31	1.05	0.17	−0.88
2.00	0.91	0.60	−0.31	1.05	0.16	−0.89
2.02	0.90	0.59	−0.31	1.05	0.14	−0.90
2.04	0.90	0.58	−0.32	1.05	0.13	−0.92
2.06	0.90	0.58	−0.32	1.05	0.12	−0.93
2.08	0.89	0.57	−0.32	1.04	0.11	−0.94
2.10	0.89	0.57	−0.32	1.04	0.09	−0.95
2.12	0.89	0.56	−0.32	1.04	0.08	−0.96
2.14	0.88	0.56	−0.33	1.04	0.07	−0.97
2.16	0.88	0.55	−0.33	1.04	0.05	−0.99
2.18	0.88	0.54	−0.33	1.04	0.04	−1.00
2.20	0.87	0.54	−0.33	1.04	0.03	−1.01
2.22	0.87	0.53	−0.34	1.04	0.02	−1.02
2.24	0.87	0.53	−0.34	1.04	0.00	−1.04
2.26	0.86	0.52	−0.34	1.04	−0.01	−1.05
2.28	0.86	0.52	−0.34	1.04	−0.02	−1.06
2.30	0.86	0.51	−0.34	1.04	−0.04	−1.07
2.32	0.86	0.51	−0.35	1.04	−0.05	−1.09
2.34	0.85	0.50	−0.35	1.04	−0.06	−1.10
2.36	0.85	0.50	−0.35	1.04	−0.07	−1.11
2.38	0.85	0.49	−0.35	1.04	−0.09	−1.12
2.40	0.84	0.49	−0.36	1.04	−0.10	−1.14
2.42	0.84	0.49	−0.36	1.04	−0.11	−1.15
2.44	0.84	0.48	−0.36	1.04	−0.12	−1.16
2.46	0.84	0.48	−0.36	1.04	−0.14	−1.18
2.48	0.84	0.47	−0.36	1.04	−0.15	−1.19
2.50	0.83	0.47	−0.37	1.04	−0.16	−1.20

TABLE 8.5
RATIO OF GAP WIDTH TO BORE DIAMETER $S/D = 0.03$

	Objective			Projective		
k^2	f/R	z_m/R	z_p/R	f/R	z_m/R	z_p/R
0.02	49.18	49.17	−0.01	49.18	49.17	−0.01
0.04	24.76	24.75	−0.01	24.76	24.75	−0.01
0.06	16.63	16.60	−0.02	16.63	16.60	−0.02
0.08	12.56	12.53	−0.03	12.56	12.53	−0.03
0.10	10.12	10.08	−0.04	10.12	10.08	−0.04
0.12	8.49	8.45	−0.04	8.49	8.45	−0.04
0.14	7.33	7.28	−0.05	7.33	7.28	−0.05
0.16	6.46	6.40	−0.06	6.46	6.40	−0.06
0.18	5.78	5.72	−0.06	5.78	5.72	−0.06
0.20	5.24	5.17	−0.07	5.24	5.17	−0.07
0.22	4.80	4.72	−0.08	4.80	4.72	−0.08
0.24	4.43	4.35	−0.08	4.43	4.34	−0.09
0.26	4.11	4.03	−0.08	4.12	4.03	−0.09
0.28	3.84	3.76	−0.09	3.85	3.75	−0.10
0.30	3.61	3.52	−0.09	3.62	3.51	−0.11
0.32	3.41	3.31	−0.09	3.42	3.30	−0.12
0.34	3.23	3.13	−0.10	3.24	3.12	−0.12
0.36	3.07	2.97	−0.10	3.08	2.95	−0.13
0.38	2.92	2.82	−0.10	2.94	2.80	−0.14
0.40	2.80	2.69	−0.11	2.82	2.67	−0.15
0.42	2.68	2.57	−0.11	2.70	2.55	−0.15
0.44	2.57	2.46	−0.11	2.60	2.44	−0.16
0.46	2.48	2.36	−0.12	2.50	2.33	−0.17
0.48	2.39	2.27	−0.12	2.42	2.24	−0.18
0.50	2.31	2.19	−0.12	2.34	2.15	−0.19
0.52	2.23	2.11	−0.12	2.26	2.07	−0.19
0.54	2.16	2.04	−0.13	2.20	1.99	−0.20
0.56	2.10	1.97	−0.13	2.13	1.92	−0.21
0.58	2.04	1.91	−0.13	2.07	1.86	−0.22
0.60	1.98	1.85	−0.13	2.02	1.80	−0.22
0.62	1.93	1.79	−0.14	1.97	1.74	−0.23
0.64	1.88	1.74	−0.14	1.92	1.68	−0.24
0.66	1.84	1.69	−0.14	1.88	1.63	−0.25
0.68	1.79	1.65	−0.14	1.84	1.58	−0.26
0.70	1.75	1.61	−0.15	1.80	1.53	−0.26
0.72	1.71	1.56	−0.15	1.76	1.49	−0.27
0.74	1.68	1.53	−0.15	1.73	1.44	−0.28
0.76	1.64	1.49	−0.15	1.69	1.40	−0.29
0.78	1.61	1.45	−0.16	1.66	1.36	−0.30
0.80	1.58	1.42	−0.16	1.63	1.33	−0.31
0.82	1.55	1.39	−0.16	1.60	1.29	−0.31
0.84	1.52	1.36	−0.17	1.58	1.25	−0.32

TABLE 8.5 (continued)

	Objective			Projective		
k^2	f/R	z_m/R	z_p/R	f/R	z_m/R	z_p/R
0.86	1.50	1.33	−0.17	1.55	1.22	−0.33
0.88	1.47	1.30	−0.17	1.53	1.19	−0.34
0.90	1.45	1.28	−0.17	1.51	1.16	−0.35
0.92	1.43	1.25	−0.18	1.48	1.13	−0.36
0.94	1.41	1.23	−0.18	1.46	1.10	−0.37
0.96	1.38	1.20	−0.18	1.44	1.07	−0.37
0.98	1.36	1.18	−0.18	1.43	1.04	−0.38
1.00	1.35	1.16	−0.19	1.41	1.02	−0.39
1.02	1.33	1.14	−0.19	1.39	0.99	−0.40
1.04	1.31	1.12	−0.19	1.37	0.97	−0.41
1.06	1.29	1.10	−0.19	1.36	0.94	−0.42
1.08	1.28	1.08	−0.20	1.34	0.92	−0.43
1.10	1.26	1.06	−0.20	1.33	0.89	−0.44
1.12	1.25	1.04	−0.20	1.32	0.87	−0.45
1.14	1.23	1.03	−0.20	1.30	0.85	−0.45
1.16	1.22	1.01	−0.21	1.29	0.83	−0.46
1.18	1.21	1.00	−0.21	1.28	0.81	−0.47
1.20	1.19	0.98	−0.21	1.27	0.79	−0.48
1.22	1.18	0.97	−0.22	1.26	0.76	−0.49
1.24	1.17	0.95	−0.22	1.25	0.74	−0.50
1.26	1.16	0.94	−0.22	1.23	0.73	−0.51
1.28	1.15	0.92	−0.22	1.23	0.71	−0.52
1.30	1.14	0.91	−0.23	1.22	0.69	−0.53
1.32	1.12	0.90	−0.23	1.21	0.67	−0.54
1.34	1.11	0.88	−0.23	1.20	0.65	−0.55
1.36	1.11	0.87	−0.23	1.19	0.63	−0.56
1.38	1.10	0.86	−0.24	1.18	0.62	−0.57
1.40	1.09	0.85	−0.24	1.17	0.60	−0.58
1.42	1.08	0.84	−0.24	1.17	0.58	−0.59
1.44	1.07	0.83	−0.24	1.16	0.56	−0.60
1.46	1.06	0.81	−0.25	1.15	0.55	−0.61
1.48	1.05	0.80	−0.25	1.15	0.53	−0.62
1.50	1.05	0.79	0.25	1.14	0.51	−0.63
1.52	1.04	0.78	−0.25	1.13	0.50	−0.64
1.54	1.03	0.77	−0.26	1.13	0.48	−0.65
1.56	1.02	0.76	−0.26	1.12	0.47	−0.66
1.58	1.02	0.76	−0.26	1.12	0.45	−0.67
1.60	1.01	0.75	−0.26	1.11	0.44	−0.68
1.62	1.00	0.74	−0.27	1.11	0.42	−0.69
1.64	1.00	0.73	−0.27	1.10	0.41	−0.70
1.66	0.99	0.72	−0.27	1.10	0.39	−0.71
1.68	0.98	0.71	−0.27	1.09	0.38	−0.72
1.70	0.98	0.70	−0.28	1.09	0.36	−0.73

TABLE 8.5 (continued)

	Objective			Projective		
k^2	f/R	z_m/R	z_p/R	f/R	z_m/R	z_p/R
1.72	0.97	0.69	−0.28	1.09	0.35	−0.74
1.74	0.97	0.69	−0.28	1.08	0.33	−0.75
1.76	0.96	0.68	−0.28	1.08	0.32	−0.76
1.78	0.96	0.67	−0.29	1.08	0.31	−0.77
1.80	0.95	0.66	−0.29	1.07	0.29	−0.78
1.82	0.95	0.66	−0.29	1.07	0.28	−0.79
1.84	0.94	0.65	−0.29	1.07	0.26	−0.80
1.86	0.94	0.64	−0.29	1.06	0.25	−0.81
1.88	0.93	0.64	−0.30	1.06	0.24	−0.82
1.90	0.93	0.63	−0.30	1.06	0.22	−0.84
1.92	0.92	0.62	−0.30	1.06	0.21	−0.85
1.94	0.92	0.61	−0.30	1.05	0.20	−0.86
1.96	0.91	0.61	−0.31	1.05	0.18	−0.87
1.98	0.91	0.60	−0.31	1.05	0.17	−0.88
2.00	0.91	0.60	−0.31	1.05	0.16	−0.89
2.02	0.90	0.59	−0.31	1.05	0.14	−0.90
2.04	0.90	0.58	−0.32	1.05	0.13	−0.92
2.06	0.90	0.58	−0.32	1.04	0.12	−0.93
2.08	0.89	0.57	−0.32	1.04	0.10	−0.94
2.10	0.89	0.57	−0.32	1.04	0.09	−0.95
2.12	0.88	0.56	−0.32	1.04	0.08	−0.96
2.14	0.88	0.55	−0.33	1.04	0.06	−0.97
2.16	0.88	0.55	−0.33	1.04	0.05	−0.99
2.18	0.87	0.54	−0.33	1.04	0.04	−1.00
2.20	0.87	0.54	−0.33	1.04	0.03	−1.01
2.22	0.87	0.53	−0.34	1.04	0.01	−1.02
2.24	0.87	0.53	−0.34	1.04	0.00	−1.04
2.26	0.86	0.52	−0.34	1.04	−0.01	−1.05
2.28	0.86	0.52	−0.34	1.04	−0.02	−1.06
2.30	0.86	0.51	−0.34	1.04	−0.04	−1.07
2.32	0.85	0.51	−0.35	1.04	−0.05	−1.09
2.34	0.85	0.50	−0.35	1.04	−0.06	−1.10
2.36	0.85	0.50	−0.35	1.04	−0.08	−1.11
2.38	0.85	0.49	−0.35	1.04	−0.09	−1.12
2.40	0.84	0.49	−0.36	1.04	−0.10	−1.14
2.42	0.84	0.48	−0.36	1.04	−0.11	−1.15
2.44	0.84	0.48	−0.36	1.04	−0.13	−1.16
2.46	0.84	0.47	−0.36	1.04	−0.14	−1.18
2.48	0.83	0.47	−0.36	1.04	−0.15	−1.19
2.50	0.83	0.47	−0.37	1.04	−0.16	−1.20

TABLE 8.6
RATIO OF GAP WIDTH TO BORE DIAMETER $S/D = 0.04$

	Objective			Projective		
k^2	f/R	z_m/R	z_p/R	f/R	z_m/R	z_p/R
0.02	50.22	50.21	−0.00	50.22	50.21	−0.00
0.04	25.26	25.25	−0.01	25.26	25.25	−0.01
0.06	16.93	16.92	−0.01	16.93	16.92	−0.01
0.08	12.77	12.76	−0.02	12.77	12.76	−0.02
0.10	10.28	10.26	−0.02	10.28	10.26	−0.02
0.12	8.62	8.59	−0.03	8.62	8.59	−0.03
0.14	7.43	7.40	−0.03	7.43	7.40	−0.03
0.16	6.54	6.50	−0.04	6.54	6.50	−0.04
0.18	5.85	5.80	−0.04	5.85	5.80	−0.04
0.20	5.29	5.25	−0.05	5.29	5.25	−0.05
0.22	4.84	4.79	−0.05	4.84	4.79	−0.05
0.24	4.46	4.41	−0.05	4.46	4.41	−0.06
0.26	4.14	4.08	−0.05	4.14	4.08	−0.06
0.28	3.86	3.81	−0.06	3.87	3.80	−0.06
0.30	3.62	3.57	−0.06	3.63	3.56	−0.07
0.32	3.42	3.36	−0.06	3.43	3.35	−0.07
0.34	3.23	3.17	−0.06	3.24	3.16	−0.08
0.36	3.07	3.00	−0.06	3.08	3.00	−0.08
0.38	2.92	2.86	−0.07	2.94	2.85	−0.09
0.40	2.79	2.72	−0.07	2.81	2.71	−0.09
0.42	2.67	2.60	−0.07	2.69	2.59	−0.10
0.44	2.57	2.49	−0.07	2.58	2.48	−0.10
0.46	2.47	2.39	−0.08	2.48	2.38	−0.11
0.48	2.38	2.30	−0.08	2.39	2.28	−0.11
0.50	2.30	2.22	−0.08	2.31	2.19	−0.12
0.52	2.22	2.14	−0.08	2.24	2.11	−0 12
0.54	2.15	2.06	−0.08	2.17	2.04	−0.13
0.56	2.08	2.00	−0.09	2.10	1.97	−0.13
0.58	2.02	1.93	−0.09	2.04	1.90	−0.14
0.60	1.97	1.87	−0.09	1.99	1.84	−0.14
0.62	1.91	1.82	0.09	1.93	1.79	−0.15
0.64	1.86	1.77	−0.10	1.88	1.73	−0.15
0.66	1.82	1.72	−0.10	1.84	1.68	−0.16
0.68	1.77	1.67	−0.10	1.80	1.63	−0.16
0.70	1.73	1.63	−0.10	1.75	1.59	−0.17
0.72	1.69	1.59	−0.11	1.72	1.54	−0.17
0.74	1.66	1.55	−0.11	1.68	1.50	−0.18
0.76	1.62	1.51	−0.11	1.65	1.46	−0.18
0.78	1.59	1.48	−0.11	1.61	1.42	−0.19
0.80	1.56	1.44	−0.11	1.58	1.39	−0.19
0.82	1.53	1.41	−0.12	1.55	1.35	−0.20
0.84	1.50	1.38	−0.12	1.53	1.32	−0.20

TABLE 8.6 (continued)

	Objective			Projective		
k^2	f/R	z_m/R	z_p/R	f/R	z_m/R	z_p/R
0.86	1.47	1.35	−0.12	1.50	1.29	−0.21
0.88	1.45	1.32	−0.12	1.47	1.26	−0.22
0.90	1.42	1.30	−0.13	1.45	1.23	−0.22
0.92	1.40	1.27	−0.13	1.43	1.20	−0.23
0.94	1.38	1.25	−0.13	1.41	1.17	−0.23
0.96	1.36	1.22	−0.13	1.39	1.15	−0.24
0.98	1.34	1.20	−0.14	1.37	1.12	−0.24
1.00	1.32	1.18	−0.14	1.35	1.10	−0.25
1.02	1.30	1.16	−0.14	1.33	1.08	−0.25
1.04	1.28	1.14	−0.14	1.31	1.05	−0.26
1.06	1.26	1.12	−0.14	1.29	1.03	−0.26
1.08	1.25	1.10	−0.15	1.28	1.01	−0.27
1.10	1.23	1.08	−0.15	1.26	0.99	−0.28
1.12	1.22	1.06	−0.15	1.25	0.97	−0.28
1.14	1.20	1.05	−0.15	1.23	0.95	−0.29
1.16	1.19	1.03	−0.16	1.22	0.93	−0.29
1.18	1.17	1.01	−0.16	1.21	0.91	−0.30
1.20	1.16	1.00	−0.16	1.19	0.89	−0.30
1.22	1.15	0.98	−0.16	1.18	0.87	−0.31
1.24	1.13	0.97	−0.17	1.17	0.86	−0.31
1.26	1.12	0.95	−0.17	1.16	0.84	−0.32
1.28	1.11	0.94	−0.17	1.15	0.82	−0.33
1.30	1.10	0.93	−0.17	1.14	0.81	−0.33
1.32	1.09	0.91	−0.17	1.13	0.79	−0.34
1.34	1.08	0.90	−0.18	1.12	0.77	−0.34
1.36	1.07	0.89	−0.18	1.11	0.76	−0.35
1.38	1.06	0.88	−0.18	1.10	0.74	−0.36
1.40	1.05	0.87	−0.18	1.09	0.73	−0.36
1.42	1.04	0.85	−0.19	1.08	0.72	−0.37
1.44	1.03	0.84	−0.19	1.07	0.70	−0.37
1.46	1.02	0.83	−0.19	1.07	0.69	−0.38
1.48	1.01	0.82	−0.19	1.06	0.67	−0.38
1.50	1.01	0.81	−0.19	1.05	0.66	−0.39
1.52	1.00	0.80	−0.20	1.04	0.65	−0.40
1.54	0.99	0.79	−0.20	1.04	0.63	−0.40
1.56	0.98	0.78	−0.20	1.03	0.62	−0.41
1.58	0.97	0.77	−0.20	1.02	0.61	−0.42
1.60	0.97	0.76	−0.20	1.02	0.60	−0.42
1.62	0.96	0.75	−0.21	1.01	0.58	−0.43
1.64	0.95	0.75	−0.21	1.01	0.57	−0.43
1.66	0.95	0.74	−0.21	1.00	0.56	−0.44
1.68	0.94	0.73	−0.21	1.00	0.55	−0.45
1.70	0.93	0.72	−0.21	0.99	0.54	−0.45

TABLE 8.6 (continued)

	Objective			Projective		
k^2	f/R	z_m/R	z_p/R	f/R	z_m/R	z_p/R
1.72	0.93	0.71	−0.22	0.99	0.53	−0.46
1.74	0.92	0.70	−0.22	0.98	0.51	−0.47
1.76	0.92	0.70	−0.22	0.98	0.50	−0.47
1.78	0.91	0.69	−0.22	0.97	0.49	−0.48
1.80	0.91	0.68	−0.22	0.97	0.48	−0.48
1.82	0.90	0.67	−0.23	0.96	0.47	−0.49
1.84	0.89	0.67	−0.23	0.96	0.46	−0.50
1.86	0.89	0.66	−0.23	0.95	0.45	−0.50
1.88	0.88	0.65	−0.23	0.95	0.44	−0.51
1.90	0.88	0.64	−0.23	0.95	0.43	−0.52
1.92	0.87	0.64	−0.24	0.94	0.42	−0.52
1.94	0.87	0.63	−0.24	0.94	0.41	−0.53
1.96	0.87	0.63	−0.24	0.94	0.40	−0.54
1.98	0.86	0.62	−0.24	0.93	0.39	−0.54
2.00	0.86	0.61	−0.24	0.93	0.38	−0.55
2.02	0.85	0.61	−0.25	0.93	0.37	−0.56
2.04	0.85	0.60	−0.25	0.92	0.36	−0.56
2.06	0.84	0.59	−0.25	0.92	0.35	−0.57
2.08	0.84	0.59	−0.25	0.92	0.34	−0.58
2.10	0.84	0.58	−0.25	0.92	0.33	−0.59
2.12	0.83	0.58	−0.26	0.91	0.32	−0.59
2.14	0.83	0.57	−0.26	0.91	0.31	−0.60
2.16	0.82	0.57	−0.26	0.91	0.30	−0.61
2.18	0.82	0.56	−0.26	0.91	0.29	−0.61
2.20	0.82	0.55	−0.26	0.91	0.28	−0.62
2.22	0.81	0.55	−0.26	0.90	0.28	−0.63
2.24	0.81	0.54	−0.27	0.90	0.27	−0.64
2.26	0.81	0.54	−0.27	0.90	0.26	−0.64
2.28	0.80	0.53	−0.27	0.90	0.25	−0.65
2.30	0.80	0.53	−0.27	0.90	0.24	−0.66
2.32	0.80	0.52	−0.27	0.89	0.23	−0.66
2.34	0.79	0.52	−0.28	0.89	0.22	−0.67
2.36	0.79	0.51	−0.28	0.89	0.21	−0.68
2.38	0.79	0.51	−0.28	0.89	0.20	−0.69
2.40	0.79	0.51	−0.28	0.89	0.20	−0.69
2.42	0.78	0.50	−0.28	0.89	0.19	−0.70
2.44	0.78	0.50	−0.28	0.89	0.18	−0.71
2.46	0.78	0.49	−0.29	0.89	0.17	−0.72
2.48	0.77	0.49	−0.29	0.89	0.16	−0.72
2.50	0.77	0.48	−0.29	0.88	0.15	−0.73

TABLE 8.7
RATIO OF GAP WIDTH TO BORE DIAMETER $S/D = 0.05$

	Objective			Projective		
k^2	f/R	z_m/R	z_p/R	f/R	z_m/R	z_p/R
0.02	49.61	49.60	−0.01	49.61	49.60	−0.01
0.04	24.96	24.95	−0.01	24.96	24.95	−0.01
0.06	16.75	16.73	−0.02	17.75	16.73	−0.02
0.08	12.64	12.62	−0.02	12.64	12.62	−0.02
0.10	10.18	10.15	−0.03	10.18	10.15	−0.03
0.12	8.53	8.50	−0.03	8.53	8.50	−0.03
0.14	7.36	7.32	−0.04	7.36	7.32	−0.04
0.16	6.48	6.44	−0.04	6.48	6.44	−0.04
0.18	5.80	5.75	−0.05	5.80	5.75	−0.05
0.20	5.25	5.20	−0.05	5.25	5.20	−0.05
0.22	4.80	4.75	−0.06	4.81	4.75	−0.06
0.24	4.43	4.37	−0.06	4.43	4.37	−0.07
0.26	4.11	4.05	−0.06	4.12	4.05	−0.07
0.28	3.84	3.77	−0.07	3.85	3.77	−0.08
0.30	3.61	3.54	−0.07	3.62	3.53	−0.08
0.32	3.40	3.33	−0.07	3.41	3.32	−0.09
0.34	3.22	3.14	−0.08	3.23	3.14	−0.09
0.36	3.06	2.98	−0.08	3.07	2.97	−0.10
0.38	2.91	2.83	−0.08	2.93	2.82	−0.11
0.40	2.79	2.70	−0.08	2.80	2.69	−0.11
0.42	2.67	2.58	−0.09	2.68	2.57	−0.12
0.44	2.56	2.47	−0.09	2.58	2.46	−0.12
0.46	2.47	2.37	−0.09	2.48	2.35	−0.13
0.48	2.38	2.28	−0.09	2.40	2.26	−0.13
0.50	2.29	2.20	−0.10	2.31	2.17	−0.14
0.52	2.22	2.12	−0.10	2.24	2.09	−0.15
0.54	2.15	2.05	−0.10	2.17	2.02	−0.15
0.56	2.08	1.98	−0.10	2.11	1.95	−0.16
0.58	2.02	1.92	−0.11	2.05	1.88	−0.16
0.60	1.97	1.86	−0.11	1.99	1.82	−0.17
0.62	1.92	1.80	−0.11	1.94	1.76	−0.18
0.64	1.87	1.75	−0.11	1.89	1.71	−0.18
0.66	1.82	1.70	−0.12	1.85	1.66	−0.19
0.68	1.78	1.66	−0.12	1.81	1.61	−0.19
0.70	1.74	1.61	−0.12	1.77	1.56	−0.20
0.72	1.70	1.57	−0.12	1.73	1.52	−0.21
0.74	1.66	1.54	−0.13	1.69	1.48	−0.21
0.76	1.63	1.50	−0.13	1.66	1.44	−0.22
0.78	1.59	1.46	−0.13	1.63	1.40	−0.23
0.80	1.56	1.43	−0.13	1.60	1.37	−0.23
0.82	1.53	1.40	−0.14	1.57	1.33	−0.24
0.84	1.51	1.37	−0.14	1.54	1.30	−0.24

TABLE 8.7 (continued)

	Objective			Projective		
k^2	f/R	z_m/R	z_p/R	f/R	z_m/R	z_p/R
0.86	1.48	1.34	−0.14	1.52	1.26	−0.25
0.88	1.45	1.31	−0.14	1.49	1.23	−0.26
0.90	1.43	1.29	−0.15	1.47	1.20	−0.26
0.92	1.41	1.26	−0.15	1.45	1.18	−0.27
0.94	1.39	1.24	−0.15	1.42	1.15	−0.28
0.96	1.36	1.21	−0.15	1.40	1.12	−0.28
0.98	1.34	1.19	−0.16	1.38	1.10	−0.29
1.00	1.33	1.17	−0.16	1.37	1.07	−0.30
1.02	1.31	1.15	−0.16	1.35	1.05	−0.30
1.04	1.29	1.13	−0.16	1.33	1.02	−0.31
1.06	1.27	1.11	−0.17	1.32	1.00	−0.32
1.08	1.26	1.09	−0.17	1.30	0.98	−0.32
1.10	1.24	1.07	−0.17	1.29	0.96	−0.33
1.12	1.23	1.05	−0.17	1.27	0.94	−0.34
1.14	1.21	1.04	−0.17	1.26	0.91	−0.34
1.16	1.20	1.02	−0.18	1.24	0.89	−0.35
1.18	1.18	1.00	−0.18	1.23	0.88	−0.36
1.20	1.17	0.99	−0.18	1.22	0.86	−0.36
1.22	1.16	0.97	−0.18	1.21	0.84	−0.37
1.24	1.15	0.96	−0.19	1.20	0.82	−0.38
1.26	1.13	0.94	−0.19	1.19	0.80	−0.38
1.28	1.12	0.93	−0.19	1.17	0.78	−0.39
1.30	1.11	0.92	−0.19	1.16	0.77	−0.40
1.32	1.10	0.90	−0.20	1.16	0.75	−0.40
1.34	1.09	0.89	−0.20	1.15	0.73	−0.41
1.36	1.08	0.88	−0.20	1.14	0.72	−0.42
1.38	1.07	0.87	−0.20	1.13	0.70	−0.43
1.40	1.06	0.86	−0.21	1.12	0.69	−0.43
1.42	1.05	0.84	−0.21	1.11	0.67	−0.44
1.44	1.04	0.83	−0.21	1.10	0.66	−0.45
1.46	1.04	0.82	−0.21	1.10	0.64	−0.45
1.48	1.03	0.81	−0.22	1.09	0.63	−0.46
1.50	1.02	0.80	−0.22	1.08	0.61	−0.47
1.52	1.01	0.79	−0.22	1.08	0.60	−0.48
1.54	1.00	0.78	−0.22	1.07	0.59	−0.48
1.56	1.00	0.77	−0.22	1.06	0.57	−0.49
1.58	0.99	0.76	−0.23	1.06	0.56	−0.50
1.60	0.98	0.75	−0.23	1.05	0.55	−0.51
1.62	0.98	0.74	−0.23	1.05	0.53	−0.51
1.64	0.97	0.74	−0.23	1.04	0.52	−0.52
1.66	0.96	0.73	−0.24	1.04	0.51	−0.53
1.68	0.96	0.72	−0.24	1.03	0.49	−0.54
1.70	0.95	0.71	−0.24	1.03	0.48	−0.54

TABLE 8.7 (continued)

	Objective			Projective		
k^2	f/R	z_m/R	z_p/R	f/R	z_m/R	z_p/R
1.72	0.94	0.70	−0.24	1.02	0.47	−0.55
1.74	0.94	0.69	−0.24	1.02	0.46	−0.56
1.76	0.93	0.69	−0.25	1.01	0.45	−0.57
1.78	0.93	0.68	−0.25	1.01	0.43	−0.58
1.80	0.92	0.67	−0.25	1.00	0.42	−0.58
1.82	0.92	0.66	−0.25	1.00	0.41	−0.59
1.84	0.91	0.66	−0.25	1.00	0.40	−0.60
1.86	0.91	0.65	−0.26	0.99	0.39	−0.61
1.88	0.90	0.64	−0.26	0.99	0.38	−0.62
1.90	0.90	0.64	−0.26	0.99	0.36	−0.62
1.92	0.89	0.63	−0.26	0.98	0.35	−0.63
1.94	0.89	0.62	−0.27	0.98	0.34	−0.64
1.96	0.88	0.62	−0.27	0.98	0.33	−0.65
1.98	0.88	0.61	−0.27	0.98	0.32	−0.66
2.00	0.87	0.60	−0.27	0.97	0.31	−0.66
2.02	0.87	0.60	−0.27	0.97	0.30	−0.67
2.04	0.87	0.59	−0.28	0.97	0.29	−0.68
2.06	0.86	0.59	−0.28	0.97	0.28	−0.69
2.08	0.86	0.58	−0.28	0.96	0.27	−0.70
2.10	0.85	0.57	−0.28	0.96	0.26	−0.71
2.12	0.85	0.57	−0.28	0.96	0.24	−0.72
2.14	0.85	0.56	−0.29	0.96	0.23	−0.72
2.16	0.84	0.56	−0.29	0.96	0.22	−0.73
2.18	0.84	0.55	−0.29	0.95	0.21	−0.74
2.20	0.84	0.55	−0.29	0.95	0.20	−0.75
2.22	0.83	0.54	−0.29	0.95	0.19	−0.76
2.24	0.83	0.54	−0.30	0.95	0.18	−0.77
2.26	0.83	0.53	−0.30	0.95	0.17	−0.78
2.28	0.82	0.53	−0.30	0.95	0.16	−0.79
2.30	0.82	0.52	−0.30	0.95	0.15	−0.79
2.32	0.82	0.52	−0.30	0.95	0.14	−0.80
2.34	0.82	0.51	−0.30	0.95	0.13	−0.81
2.36	0.81	0.51	−0.31	0.94	0.12	−0.82
2.38	0.81	0.50	−0.31	0.94	0.11	−0.83
2.40	0.81	0.50	−0.31	0.94	0.10	−0.84
2.42	0.80	0.49	−0.31	0.94	0.09	−0.85
2.44	0.80	0.49	−0.31	0.94	0.08	−0.86
2.46	0.80	0.48	−0.32	0.94	0.07	−0.87
2.48	0.80	0.48	−0.32	0.94	0.06	−0.88
2.50	0.79	0.47	−0.32	0.94	0.05	−0.89

TABLE 8.8
RATIO OF GAP WIDTH TO BORE DIAMETER $S/D = 0.06$

	Objective			Projective		
k^2	f/R	z_m/R	z_p/R	f/R	z_m/R	z_p/R
0.02	50.07	50.07	−0.00	50.07	50.07	−0.00
0.04	25.18	25.17	−0.01	25.18	25.17	−0.01
0.06	16.88	16.87	−0.01	16.88	16.87	−0.01
0.08	12.73	12.71	−0.01	12.73	12.71	−0.01
0.10	10.24	10.22	−0.02	10.24	10.22	−0.02
0.12	8.58	8.56	−0.02	8.58	8.56	−0.02
0.14	7.40	7.37	−0.03	7.40	7.37	−0.03
0.16	6.51	6.48	−0.03	6.51	6.48	−0.03
0.18	5.82	5.78	−0.03	5.82	5.78	−0.03
0.20	5.26	5.23	−0.04	5.26	5.23	−0.04
0.22	4.81	4.77	−0.04	4.81	4.77	−0.04
0.24	4.44	4.39	−0.04	4.44	4.39	−0.04
0.26	4.12	4.07	−0.04	4.12	4.07	−0.05
0.28	3.84	3.80	−0.05	3.85	3.80	−0.05
0.30	3.61	3.56	−0.05	3.61	3.56	−0.05
0.32	3.40	3.35	−0.05	3.40	3.34	−0.06
0.34	3.22	3.16	−0.05	3.22	3.16	−0.06
0.36	3.05	3.00	−0.06	3.06	2.99	−0.07
0.38	2.91	2.85	−0.06	2.92	2.85	−0.07
0.40	2.78	2.72	−0.06	2.79	2.71	−0.07
0.42	2.66	2.60	−0.06	2.67	2.59	−0.08
0.44	2.55	2.49	−0.07	2.56	2.48	−0.08
0.46	2.46	2.39	−0.07	2.46	2.38	−0.09
0.48	2.37	2.29	−0.07	2.37	2.29	−0.09
0.50	2.28	2.21	−0.07	2.29	2.20	−0.09
0.52	2.21	2.13	−0.08	2.22	2.12	−0.10
0.54	2.14	2.06	−0.08	2.15	2.05	−0.10
0.56	2.07	1.99	−0.08	2.08	1.98	−0.11
0.58	2.01	1.93	−0.08	2.02	1.91	−0.11
0.60	1.95	1.87	−0.08	1.97	1.85	−0.11
0.62	1.90	1.81	−0.09	1.91	1.80	−0.12
0.64	1.85	1.76	−0.09	1.86	1.74	−0.12
0.66	1.81	1.71	−0.09	1.82	1.69	−0.13
0.68	1.76	1.67	−0.09	1.78	1.65	−0.13
0.70	1.72	1.63	−0.10	1.74	1.60	−0.13
0.72	1.68	1.58	−0.10	1.70	1.56	−0.14
0.74	1.65	1.55	−0.10	1.66	1.52	−0.14
0.76	1.61	1.51	−0.10	1.63	1.48	−0.15
0.78	1.58	1.47	−0.10	1.59	1.44	−0.15
0.80	1.55	1.44	−0.11	1.56	1.41	−0.15
0.82	1.52	1.41	−0.11	1.53	1.38	−0.16
0.84	1.49	1.38	−0.11	1.51	1.34	−0.16

TABLE 8.8 (continued)

	Objective			Projective		
k^2	f/R	z_m/R	z_p/R	f/R	z_m/R	z_p/R
0.86	1.46	1.35	−0.11	1.48	1.31	−0.17
0.88	1.44	1.32	−0.12	1.45	1.28	−0.17
0.90	1.41	1.29	−0.12	1.43	1.26	−0.17
0.92	1.39	1.27	−0.12	1.41	1.23	−0.18
0.94	1.37	1.24	−0.12	1.39	1.20	−0.18
0.96	1.35	1.22	−0.13	1.37	1.18	−0.19
0.98	1.33	1.20	−0.13	1.35	1.15	−0.19
1.00	1.31	1.18	−0.13	1.33	1.13	−0.20
1.02	1.29	1.16	−0.13	1.31	1.11	−0.20
1.04	1.27	1.14	−0.13	1.29	1.09	−0.20
1.06	1.25	1.12	−0.14	1.27	1.06	−0.21
1.08	1.24	1.10	−0.14	1.26	1.04	−0.21
1.10	1.22	1.08	−0.14	1.24	1.02	−0.22
1.12	1.20	1.06	−0.14	1.23	1.01	−0.22
1.14	1.19	1.04	−0.15	1.21	0.99	−0.23
1.16	1.18	1.03	−0.15	1.20	0.97	−0.23
1.18	1.16	1.01	−0.15	1.19	0.95	−0.24
1.20	1.15	1.00	−0.15	1.17	0.93	−0.24
1.22	1.14	0.98	−0.15	1.16	0.92	−0.24
1.24	1.12	0.97	−0.16	1.15	0.90	−0.25
1.26	1.11	0.95	−0.16	1.14	0.88	−0.25
1.28	1.10	0.94	−0.16	1.13	0.87	−0.26
1.30	1.09	0.93	−0.16	1.12	0.85	−0.26
1.32	1.08	0.91	−0.17	1.11	0.84	−0.27
1.34	1.07	0.90	−0.17	1.10	0.82	−0.27
1.36	1.06	0.89	−0.17	1.09	0.81	−0.28
1.38	1.05	0.88	−0.17	1.08	0.80	−0.28
1.40	1.04	0.86	−0.17	1.07	0.78	−0.29
1.42	1.03	0.85	−0.18	1.06	0.77	−0.29
1.44	1.02	0.84	−0.18	1.05	0.76	−0.30
1.46	1.01	0.83	−0.18	1.04	0.74	−0.30
1.48	1.00	0.82	−0.18	1.04	0.73	−0.31
1.50	0.99	0.81	−0.18	1.03	0.72	−0.31
1.52	0.99	0.80	−0.19	1.02	0.71	−0.32
1.54	0.98	0.79	−0.19	1.02	0.69	−0.32
1.56	0.97	0.78	−0.19	1.01	0.68	−0.33
1.58	0.96	0.77	−0.19	1.00	0.67	−0.33
1.60	0.96	0.76	−0.19	1.00	0.66	−0.34
1.62	0.95	0.75	−0.20	0.99	0.65	−0.34
1.64	0.94	0.74	−0.20	0.98	0.64	−0.35
1.66	0.94	0.73	−0.20	0.98	0.63	−0.35
1.68	0.93	0.73	−0.20	0.97	0.62	−0.36
1.70	0.92	0.72	−0.20	0.97	0.61	−0.36

TABLE 8.8 (continued)

	Objective			Projective		
k^2	f/R	z_m/R	z_p/R	f/R	z_m/R	z_p/R
1.72	0.92	0.71	−0.21	0.96	0.60	−0.37
1.74	0.91	0.70	−0.21	0.96	0.59	−0.37
1.76	0.90	0.69	−0.21	0.95	0.58	−0.38
1.78	0.90	0.69	−0.21	0.95	0.57	−0.38
1.80	0.89	0.68	−0.21	0.94	0.56	−0.39
1.82	0.89	0.67	−0.22	0.94	0.55	−0.39
1.84	0.88	0.66	−0.22	0.93	0.54	−0.40
1.86	0.88	0.66	−0.22	0.93	0.53	−0.40
1.88	0.87	0.65	−0.22	0.93	0.52	−0.41
1.90	0.87	0.64	−0.22	0.92	0.51	−0.41
1.92	0.86	0.64	−0.23	0.92	0.50	−0.42
1.94	0.86	0.63	−0.23	0.91	0.49	−0.42
1.96	0.85	0.62	−0.23	0.91	0.48	−0.43
1.98	0.85	0.62	−0.23	0.91	0.47	−0.43
2.00	0.84	0.61	−0.23	0.90	0.47	−0.44
2.02	0.84	0.60	−0.24	0.90	0.46	−0.44
2.04	0.84	0.60	−0.24	0.90	0.45	−0.45
2.06	0.83	0.59	−0.24	0.90	0.44	−0.46
2.08	0.83	0.59	−0.24	0.89	0.43	−0.46
2.10	0.82	0.58	−0.24	0.89	0.42	−0.47
2.12	0.82	0.58	−0.24	0.89	0.41	−0.47
2.14	0.82	0.57	−0.25	0.89	0.41	−0.48
2.16	0.81	0.56	−0.25	0.88	0.40	−0.48
2.18	0.81	0.56	−0.25	0.88	0.39	−0.49
2.20	0.80	0.55	−0.25	0.88	0.38	−0.50
2.22	0.80	0.55	−0.25	0.88	0.37	−0.50
2.24	0.80	0.54	−0.25	0.87	0.37	−0.51
2.26	0.79	0.54	−0.26	0.87	0.36	−0.51
2.28	0.79	0.53	−0.26	0.87	0.35	−0.52
2.30	0.79	0.53	−0.26	0.87	0.34	−0.53
2.32	0.78	0.52	−0.26	0.87	0.34	−0.53
2.34	0.78	0.52	−0.26	0.87	0.33	−0.54
2.36	0.78	0.51	−0.27	0.86	0.32	−0.54
2.38	0.78	0.51	−0.27	0.86	0.31	−0.55
2.40	0.77	0.50	−0.27	0.86	0.31	−0.56
2.42	0.77	0.50	−0.27	0.86	0.30	−0.56
2.44	0.77	0.49	−0.27	0.86	0.29	−0.57
2.46	0.76	0.49	−0.27	0.86	0.28	−0.57
2.48	0.76	0.49	−0.27	0.86	0.28	−0.58
2.50	0.76	0.48	−0.28	0.85	0.27	−0.59

TABLE 8.9
RATIO OF GAP WIDTH TO BORE DIAMETER $S/D = 0.07$

	Objective			Projective		
k^2	f/R	z_m/R	z_p/R	f/R	z_m/R	z_p/R
0.02	50.02	50.02	−0.00	50.02	50.02	−0.00
0.04	25.15	25.14	−0.01	25.15	25.14	−0.01
0.06	16.86	16.85	−0.01	16.86	16.85	−0.01
0.08	12.72	12.70	−0.01	12.72	12.70	−0.01
0.10	10.23	10.21	−0.02	10.23	10.21	−0.02
0.12	8.57	8.55	−0.02	8.57	8.55	−0.02
0.14	7.39	7.36	−0.03	7.39	7.36	−0.03
0.16	6.50	6.47	−0.03	6.50	6.47	−0.03
0.18	5.81	5.78	−0.03	5.81	5.78	−0.03
0.20	5.26	5.22	−0.04	5.26	5.22	−0.04
0.22	4.81	4.77	−0.04	4.81	4.77	−0.04
0.24	4.43	4.39	−0.04	4.43	4.39	−0.04
0.26	4.11	4.07	−0.04	4.12	4.07	−0.05
0.28	3.84	3.79	−0.05	3.84	3.79	−0.05
0.30	3.60	3.55	−0.05	3.61	3.55	−0.06
0.32	3.40	3.34	−0.05	3.40	3.34	−0.06
0.34	3.21	3.16	−0.05	3.22	3.16	−0.06
0.36	3.05	2.99	−0.06	3.06	2.99	−0.07
0.38	2.91	2.85	−0.06	2.91	2.84	−0.07
0.40	2.78	2.71	−0.06	2.78	2.71	−0.07
0.42	2.66	2.59	−0.06	2.67	2.59	−0.08
0.44	2.55	2.48	−0.07	2.56	2.48	−0.08
0.46	2.45	2.38	−0.07	2.46	2.38	−0.09
0.48	2.36	2.29	−0.07	2.37	2.28	−0.09
0.50	2.28	2.21	−0.07	2.29	2.20	−0.09
0.52	2.21	2.13	−0.08	2.22	2.12	−0.10
0.54	2.13	2.06	−0.08	2.15	2.04	−0.10
0.56	2.07	1.99	−0.08	2.08	1.97	−0.11
0.58	2.01	1.93	−0.08	2.02	1.91	−0.11
0.60	1.95	1.87	−0.08	1.97	1.85	−0.11
0.62	1.90	1.81	−0.09	1.91	1.79	−0.12
0.64	1.85	1.76	−0.09	1.86	1.74	−0.12
0.66	1.80	1.71	−0.09	1.82	1.69	−0.13
0.68	1.76	1.67	−0.09	1.77	1.64	−0.13
0.70	1.72	1.62	−0.10	1.73	1.60	−0.14
0.72	1.68	1.58	−0.10	1.70	1.56	−0.14
0.74	1.64	1.54	−0.10	1.66	1.52	−0.14
0.76	1.61	1.51	−0.10	1.63	1.48	−0.15
0.78	1.58	1.47	−0.10	1.59	1.44	−0.15
0.80	1.55	1.44	−0.11	1.56	1.41	−0.16
0.82	1.52	1.41	−0.11	1.53	1.37	−0.16
0.84	1.49	1.38	−0.11	1.51	1.34	−0.16

TABLE 8.9 (continued)

	Objective			Projective		
k^2	f/R	z_m/R	z_p/R	f/R	z_m/R	z_p/R
0.86	1.46	1.35	−0.11	1.48	1.31	−0.17
0.88	1.44	1.32	−0.12	1.45	1.28	−0.17
0.90	1.41	1.29	−0.12	1.43	1.25	−0.18
0.92	1.39	1.27	−0.12	1.41	1.23	−0.18
0.94	1.37	1.24	−0.12	1.39	1.20	−0.19
0.96	1.35	1.22	−0.13	1.36	1.17	−0.19
0.98	1.32	1.20	−0.13	1.34	1.15	−0.19
1.00	1.31	1.18	−0.13	1.33	1.13	−0.20
1.02	1.29	1.15	−0.13	1.31	1.10	−0.20
1.04	1.27	1.13	−0.13	1.29	1.08	−0.21
1.06	1.25	1.11	−0.14	1.27	1.06	−0.21
1.08	1.23	1.10	−0.14	1.26	1.04	−0.22
1.10	1.22	1.08	−0.14	1.24	1.02	−0.22
1.12	1.20	1.06	−0.14	1.23	1.00	−0.23
1.14	1.19	1.04	−0.15	1.21	0.98	−0.23
1.16	1.17	1.03	−0.15	1.20	0.96	−0.23
1.18	1.16	1.01	−0.15	1.19	0.95	−0.24
1.20	1.15	1.00	−0.15	1.17	0.93	−0.24
1.22	1.14	0.98	−0.15	1.16	0.91	−0.25
1.24	1.12	0.97	−0.16	1.15	0.90	−0.25
1.26	1.11	0.95	−0.16	1.14	0.88	−0.26
1.28	1.10	0.94	−0.16	1.13	0.87	−0.26
1.30	1.09	0.92	−0.16	1.12	0.85	−0.27
1.32	1.08	0.91	−0.17	1.11	0.84	−0.27
1.34	1.07	0.90	−0.17	1.10	0.82	−0.28
1.36	1.06	0.89	−0.17	1.09	0.81	−0.28
1.38	1.05	0.87	−0.17	1.08	0.79	−0.29
1.40	1.04	0.86	−0.17	1.07	0.78	−0.29
1.42	1.03	0.85	−0.18	1.06	0.77	−0.30
1.44	1.02	0.84	−0.18	1.05	0.75	−0.30
1.46	1.01	0.83	−0.18	1.04	0.74	−0.30
1.48	1.00	0.82	−0.18	1.04	0.73	−0.31
1.50	0.99	0.81	−0.18	1.03	0.71	−0.31
1.52	0.99	0.80	−0.19	1.02	0.70	−0.32
1.54	0.98	0.79	−0.19	1.02	0.69	−0.32
1.56	0.97	0.78	−0.19	1.01	0.68	−0.33
1.58	0.96	0.77	−0.19	1.00	0.67	−0.33
1.60	0.96	0.76	−0.20	1.00	0.66	−0.34
1.62	0.95	0.75	−0.20	0.99	0.64	−0.34
1.64	0.94	0.74	−0.20	0.98	0.63	−0.35
1.66	0.94	0.73	−0.20	0.98	0.62	−0.35
1.68	0.93	0.73	−0.20	0.97	0.61	−0.36
1.70	0.92	0.72	−0.21	0.97	0.60	−0.36

TABLE 8.9 (continued)

	Objective			Projective		
k^2	f/R	z_m/R	z_p/R	f/R	z_m/R	z_p/R
1.72	0.92	0.71	−0.21	0.96	0.59	−0.37
1.74	0.91	0.70	−0.21	0.96	0.58	−0.38
1.76	0.90	0.69	−0.21	0.95	0.57	−0.38
1.78	0.90	0.69	−0.21	0.95	0.56	−0.39
1.80	0.89	0.68	−0.22	0.94	0.55	−0.39
1.82	0.89	0.67	−0.22	0.94	0.54	−0.40
1.84	0.88	0.66	−0.22	0.93	0.53	−0.40
1.86	0.88	0.66	−0.22	0.93	0.52	−0.41
1.88	0.87	0.65	−0.22	0.93	0.51	−0.41
1.90	0.87	0.64	−0.22	0.92	0.50	−0.42
1.92	0.86	0·64	−0.23	0.92	0.50	−0.42
1.94	0.86	0.63	−0.23	0.91	0.49	−0.43
1.96	0.85	0.62	−0.23	0.91	0.48	−0.43
1.98	0.85	0.62	−0.23	0.91	0.47	−0.44
2.00	0.84	0.61	−0.23	0.90	0.46	−0.44
2.02	0.84	0.60	−0.24	0.90	0.45	−0.45
2.04	0.84	0.60	−0.24	0.90	0.44	−0.46
2.06	0.83	0.59	−0.24	0.90	0.43	−0.46
2.08	0.83	0.59	−0.24	0.89	0.43	−0.47
2.10	0.82	0.58	−0.24	0.89	0.42	−0.47
2.12	0.82	0.57	−0.24	0.89	0.41	−0.48
2.14	0.82	0.57	−0.25	0.89	0.40	−0.48
2.16	0.81	0.56	−0.25	0.88	0.39	−0.49
2.18	0.81	0.56	−0.25	0.88	0.38	−0.50
2.20	0.80	0.55	−0.25	0.88	0.39	−0.50
2.22	0.80	0.55	−0.25	0.88	0.37	−0.51
2.24	0.80	0.54	−0.26	0.87	0.36	−0.51
2.26	0.79	0.54	−0.26	0.87	0.35	−0.52
2.28	0.79	0.53	−0.26	0.87	0.35	−0.53
2.30	0.79	0.53	−0.26	0.87	0.34	−0.53
2.32	0.78	0.52	−0.26	0.87	0.33	−0.54
2.34	0.78	0.52	−0.26	0.87	0.32	−0.54
2.36	0.78	0.51	−0.27	0.86	0.31	−0.55
2.38	0.78	0.51	−0.27	0.86	0.31	−0.56
2.40	0.77	0.50	−0.27	0.86	0.30	−0.56
2.42	0.77	0.50	−0.27	0.86	0.29	−0.57
2.44	0.77	0.49	−0.27	0.86	0.28	−0.57
2.46	0.76	0.49	−0.27	0.86	0.28	−0.58
2.48	0.76	0.49	−0.28	0.86	0.27	−0.59
2.50	0.76	0.48	−0.28	0.86	0.26	−0.59

TABLE 8.10
RATIO OF GAP WIDTH TO BORE DIAMETER $S/D = 0.08$

	Objective			Projective		
k^2	f/R	z_m/R	z_p/R	f/R	z_m/R	z_p/R
0.02	50.53	50.53	−0.00	50.53	50.53	−0.00
0.04	25.39	25.38	−0.00	25.39	25.38	−0.00
0.06	17.01	17.00	−0.01	17.01	17.00	−0.01
0.08	12.82	12.81	−0.01	12.82	12.81	−0.01
0.10	10.30	10.29	−0.01	10.30	10.29	−0.01
0.12	8.63	8.62	−0.01	8.63	8.62	−0.01
0.14	7.43	7.42	−0.01	7.43	7.42	−0.01
0.16	6.53	6.52	−0.02	6.53	6.52	−0.02
0.18	5.84	5.82	−0.02	5.84	5.82	−0.02
0.20	5.28	5.26	−0.02	5.28	5.26	−0.02
0.22	4.82	4.80	−0.02	4.82	4.80	−0.02
0.24	4.44	4.42	−0.02	4.44	4.42	−0.02
0.26	4.12	4.09	−0.02	4.12	4.09	−0.02
0.28	3.84	3.82	−0.03	3.84	3.82	−0.03
0.30	3.60	3.58	−0.03	3.60	3.58	−0.03
0.32	3.40	3.36	−0.03	3.40	3.36	−0.03
0.34	3.21	3.18	−0.03	3.21	3.18	−0.03
0.36	3.05	3.01	−0.03	3.05	3.01	−0.03
0.38	2.90	2.87	−0.04	2.90	2.87	−0.04
0.40	2.77	2.73	−0.04	2.77	2.73	−0.04
0.42	2.65	2.61	−0.04	2.65	2.61	−0.04
0.44	2.54	2.50	−0.04	2.54	2.50	−0.04
0.46	2.44	2.40	−0.04	2.44	2.40	−0.04
0.48	2.35	2.31	−0.05	2.35	2.31	−0.05
0.50	2.27	2.22	−0.05	2.27	2.22	−0.05
0.52	2.20	2.14	−0.05	2.20	2.14	−0.05
0.54	2.12	2.07	−0.05	2.12	2.07	−0.05
0.56	2.06	2.00	−0.06	2.06	2.00	−0.06
0.58	2.00	1.94	−0.06	2.00	1.94	−0.06
0.60	1.94	1.88	−0.06	1.94	1.88	−0.06
0.62	1.89	1.83	−0.06	1.89	1.83	−0.06
0.64	1.84	1.77	−0.06	1.84	1.77	−0.06
0.66	1.79	1.72	0.07	1.79	1.72	−0.07
0.68	1.75	1.68	−0.07	1.75	1.68	−0.07
0.70	1.71	1.64	−0.07	1.71	1.64	−0.07
0.72	1.67	1.59	−0.07	1.67	1.59	−0.07
0.74	1.63	1.55	−0.07	1.63	1.55	−0.07
0.76	1.59	1.52	−0.08	1.59	1.52	−0.08
0.78	1.56	1.48	−0.08	1.56	1.48	−0.08
0.80	1.53	1.45	−0.08	1.53	1.45	−0.08
0.82	1.50	1.42	−0.08	1.50	1.42	−0.08
0.84	1.47	1.39	−0.09	1.47	1.39	−0.09

TABLE 8.10 (continued)

	Objective			Projective		
k^2	f/R	z_m/R	z_p/R	f/R	z_m/R	z_p/R
0.86	1.44	1.36	−0.09	1.45	1.36	−0.09
0.88	1.42	1.33	−0.09	1.42	1.33	−0.09
0.90	1.39	1.30	−0.09	1.40	1.30	−0.09
0.92	1.37	1.28	−0.09	1.37	1.28	−0.10
0.94	1.35	1.25	−0.10	1.35	1.25	−0.10
0.96	1.33	1.23	−0.10	1.33	1.23	−0.10
0.98	1.31	1.21	−0.10	1.31	1.21	−0.10
1.00	1.29	1.18	−0.10	1.29	1.18	−0.10
1.02	1.27	1.16	−0.10	1.27	1.16	−0.11
1.04	1.25	1.14	−0.11	1.25	1.14	−0.11
1.06	1.23	1.12	−0.11	1.23	1.12	−0.11
1.08	1.22	1.10	−0.11	1.22	1.10	−0.11
1.10	1.20	1.09	−0.11	1.20	1.09	−0.12
1.12	1.18	1.07	−0.12	1.19	1.07	−0.12
1.14	1.17	1.05	−0.12	1.17	1.05	−0.12
1.16	1.15	1.04	−0.12	1.16	1.03	−0.12
1.18	1.14	1.02	−0.12	1.14	1.02	−0.13
1.20	1.13	1.00	−0.12	1.13	1.00	−0.13
1.22	1.11	0.99	−0.13	1.12	0.99	−0.13
1.24	1.10	0.97	−0.13	1.11	0.97	−0.13
1.26	1.09	0.96	−0.13	1.09	0.96	−0.14
1.28	1.08	0.95	−0.13	1.08	0.94	−0.14
1.30	1.07	0.93	−0.13	1.07	0.93	−0.14
1.32	1.06	0.92	−0.14	1.06	0.92	−0.14
1.34	1.04	0.91	−0.14	1.05	0.90	−0.15
1.36	1.03	0.89	−0.14	1.04	0.89	−0.15
1.38	1.02	0.88	−0.14	1.03	0.88	−0.15
1.40	1.01	0.87	−0.14	1.02	0.87	−0.15
1.42	1.01	0.86	−0.15	1.01	0.86	−0.16
1.44	1.00	0.85	−0.15	1.00	0.84	−0.16
1.46	0.99	0.84	−0.15	1.00	0.83	−0.16
1.48	0.98	0.83	−0.15	0.99	0.82	−0.17
1.50	0.97	0.82	−0.15	0.98	0.81	−0.17
1.52	0.96	0.81	−0.16	0.97	0.80	−0.17
1.54	0.95	0.80	−0.16	0.96	0.79	−0.17
1.56	0.95	0.79	−0.16	0.96	0.78	−0.18
1.58	0.94	0.78	−0.16	0.95	0.77	−0.18
1.60	0.93	0.77	−0.16	0.94	0.76	−0.18
1.62	0.92	0.76	−0.16	0.94	0.75	−0.18
1.64	0.92	0.75	−0.17	0.93	0.74	−0.19
1.66	0.91	0.74	−0.17	0.92	0.73	−0.19
1.68	0.90	0.73	−0.17	0.92	0.73	−0.19
1.70	0.90	0.72	−0.17	0.91	0.72	−0.20

TABLE 8.10 (continued)

	Objective			Projective		
k^2	f/R	z_m/R	z_p/R	f/R	z_m/R	z_p/R
1.72	0.89	0.72	−0.17	0.91	0.71	−0.20
1.74	0.88	0.71	−0.18	0.90	0.70	−0.20
1.76	0.88	0.70	−0.18	0.90	0.69	−0.20
1.78	0.87	0.69	−0.18	0.89	0.68	−0.21
1.80	0.87	0.69	−0.18	0.89	0.68	−0.21
1.82	0.86	0.68	−0.18	0.88	0.67	−0.21
1.84	0.86	0.67	−0.18	0.88	0.66	−0.22
1.86	0.85	0.66	−0.19	0.87	0.65	−0.22
1.88	0.85	0.66	−0.19	0.87	0.65	−0.22
1.90	0.84	0.65	−0.19	0.86	0.64	−0.23
1.92	0.84	0.64	−0.19	0.86	0.63	−0.23
1.94	0.83	0.64	−0.19	0.86	0.62	−0.23
1.96	0.83	0.63	−0.20	0.85	0.62	−0.23
1.98	0.82	0.62	−0.20	0.85	0.61	−0.24
2.00	0.82	0.62	−0.20	0.84	0.60	−0.24
2.02	0.81	0.61	−0.20	0.84	0.60	−0.24
2.04	0.81	0.61	−0.20	0.84	0.59	−0.25
2.06	0.80	0.60	−0.20	0.83	0.58	−0.25
2.08	0.80	0.59	−0.21	0.83	0.58	−0.25
2.10	0.79	0.59	−0.21	0.83	0.57	−0.26
2.12	0.79	0.58	−0.21	0.82	0.56	−0.26
2.14	0.79	0.58	−0.21	0.82	0.56	−0.26
2.16	0.78	0.57	−0.21	0.82	0.55	−0.27
2.18	0.78	0.57	−0.21	0.81	0.55	−0.27
2.20	0.77	0.56	−0.21	0.81	0.54	−0.27
2.22	0.77	0.55	−0.22	0.81	0.53	−0.28
2.24	0.77	0.55	−0.22	0.81	0.53	−0.28
2.26	0.76	0.54	−0.22	0.80	0.52	−0.28
2.28	0.76	0.54	−0.22	0.80	0.52	−0.29
2.30	0.76	0.53	−0.22	0.80	0.51	−0.29
2.32	0.75	0.53	−0.22	0.80	0.50	−0.29
2.34	0.75	0.52	−0.23	0.79	0.50	−0.30
2.36	0.75	0.52	−0.23	0.79	0.49	−0.30
2.38	0.74	0.52	−0.23	0.79	0.49	−0.30
2.40	0.74	0.51	−0.23	0.79	0.48	−0.31
2.42	0.74	0.51	−0.23	0.79	0.48	−0.31
2.44	0.73	0.50	−0.23	0.79	0.47	−0.31
2.46	0.73	0.50	−0.23	0.78	0.47	−0.32
2.48	0.73	0.49	−0.24	0.78	0.46	−0.32
2.50	0.73	0.49	−0.24	0.78	0.45	−0.33

TABLE 8.11
RATIO OF GAP WIDTH TO BORE DIAMETER $S/D = 0.09$

	Objective			Projective		
k^2	f/R	z_m/R	z_p/R	f/R	z_m/R	z_p/R
0.02	50.52	50.52	−0.00	50.52	50.52	−0.00
0.04	25.38	25.38	−0.00	25.38	25.38	−0.00
0.06	17.00	17.00	−0.01	17.00	17.00	−0.01
0.08	12.81	12.81	−0.01	12.81	12.81	−0.01
0.10	10.30	10.29	−0.01	10.30	10.29	−0.01
0.12	8.62	8.61	−0.01	8.62	8.61	−0.01
0.14	7.43	7.42	−0.01	7.43	7.42	−0.01
0.16	6.53	6.52	−0.02	6.53	6.52	−0.02
0.18	5.83	5.82	−0.02	5.83	5.82	−0.02
0.20	5.28	5.26	−0.02	5.28	5.26	−0.02
0.22	4.82	4.80	−0.02	4.82	4.80	−0.02
0.24	4.44	4.42	−0.02	4.44	4.42	−0.02
0.26	4.12	4.09	−0.02	4.12	4.09	−0.02
0.28	3.84	3.82	−0.03	3.84	3.82	−0.03
0.30	3.60	3.58	−0.03	3.60	3.58	−0.03
0.32	3.40	3.36	−0.03	3.40	3.36	−0.03
0.34	3.21	3.18	−0.03	3.21	3.18	−0.03
0.36	3.05	3.01	−0.03	3.05	3.01	−0.03
0.38	2.90	2.86	−0.04	2.90	2.86	−0.04
0.40	2.77	2.73	−0.04	2.77	2.73	−0.04
0.42	2.65	2.61	−0.04	2.65	2.61	−0.04
0.44	2.54	2.50	−0.04	2.54	2.50	−0.C4
0.46	2.44	2.40	−0.05	2.44	2.40	−0.05
0.48	2.35	2.31	−0.05	2.35	2.31	−0.05
0.50	2.27	2.22	−0.05	2.27	2.22	−0.05
0.52	2.20	2.14	−0.05	2.20	2.14	−0.05
0.54	2.12	2.07	−0.05	2.12	2.07	−0.05
0.56	2.06	2.00	−0.06	2.06	2.00	−0.06
0.58	2.00	1.94	−0.06	2.00	1.94	−0.06
0.60	1.94	1.88	−0.06	1.94	1.88	−0.06
0.62	1.89	1.83	−0.06	1.89	1.83	−0.06
0.64	1.84	1.77	−0.06	1.84	1.77	−0.06
0.66	1.79	1.72	−0.07	1.79	1.72	−0.07
0.68	1.75	1.68	−0.07	1.75	1.68	−0.07
0.70	1.71	1.64	−0.07	1.71	1.63	−0.07
0.72	1.67	1.59	−0.07	1.67	1.59	−0.07
0.74	1.63	1.55	−0.07	1.63	1.55	−0.08
0.76	1.59	1.52	−0.08	1.60	1.52	−0.08
0.78	1.56	1.48	−0.08	1.56	1.48	−0.08
0.80	1.53	1.45	−0.08	1.53	1.45	−0.08
0.82	1.50	1.42	−0.08	1.50	1.42	−0.08
0.84	1.47	1.39	−0.09	1.47	1.39	−0.09

TABLE 8.11 (continued)

	Objective			Projective		
k^2	f/R	z_m/R	z_p/R	f/R	z_m/R	z_p/R
0.86	1.45	1.36	−0.09	1.45	1.36	−0.09
0.88	1.42	1.33	−0.09	1.42	1.33	−0.09
0.90	1.39	1.30	−0.09	1.40	1.30	−0.09
0.92	1.37	1.28	−0.09	1.37	1.28	−0.10
0.94	1.35	1.25	−0.10	1.35	1.25	−0.10
0.96	1.33	1.23	−0.10	1.33	1.23	−0.10
0.98	1.31	1.21	−0.10	1.31	1.21	−0.10
1.00	1.29	1.18	−0.10	1.29	1.18	−0.10
1.02	1.27	1.16	−0.11	1.27	1.16	−0.11
1.04	1.25	1.14	−0.11	1.25	1.14	−0.11
1.06	1.23	1.12	−0.11	1.24	1.12	−0.11
1.08	1.22	1.10	−0.11	1.22	1.10	−0.11
1.10	1.20	1.09	−0.11	1.20	1.09	−0.12
1.12	1.18	1.07	−0.12	1.19	1.07	−0.12
1.14	1.17	1.05	−0.12	1.17	1.05	−0.12
1.16	1.16	1.04	−0.12	1.16	1.03	−0.12
1.18	1.14	1.02	−0.12	1.14	1.02	−0.13
1.20	1.13	1.00	−0.12	1.13	1.00	−0.13
1.22	1.11	0.99	−0.13	1.12	0.99	−0.13
1.24	1.10	0.97	−0.13	1.11	0.97	−0.13
1.26	1.09	0.96	−0.13	1.09	0.96	−0.14
1.28	1.08	0.94	−0.13	1.08	0.94	−0.14
1.30	1.07	0.93	−0.13	1.07	0.93	−0.14
1.32	1.05	0.92	−0.14	1.06	0.92	−0.15
1.34	1.04	0.90	−0.14	1.05	0.90	−0.15
1.36	1.03	0.89	−0.14	1.04	0.89	−0.15
1.38	1.02	0.88	−0.14	1.03	0.88	−0.15
1.40	1.01	0.87	−0.14	1.02	0.87	−0.16
1.42	1.00	0.86	−0.15	1.01	0.85	−0.16
1.44	1.00	0.85	−0.15	1.00	0.84	−0.16
1.46	0.99	0.84	−0.15	1.00	0.83	−0.16
1.48	0.98	0.82	−0.15	0.99	0.82	−0.17
1.50	0.97	0.81	−0.15	0.98	0.81	−0.17
1.52	0.96	0.80	−0.16	0.97	0.80	−0.17
1.54	0.95	0.79	−0.16	0.96	0.79	−0.17
1.56	0.95	0.78	−0.16	0.96	0.78	−0.18
1.58	0.94	0.78	−0.16	0.95	0.77	−0.18
1.60	0.93	0.77	−0.16	0.94	0.76	−0.18
1.62	0.92	0.76	−0.17	0.94	0.75	−0.19
1.64	0.92	0.75	−0.17	0.93	0.74	−0.19
1.66	0.91	0.74	−0.17	0.92	0.73	−0.19
1.68	0.90	0.73	−0.17	0.92	0.72	−0.19
1.70	0.90	0.72	−0.17	0.91	0.72	−0.20

TABLE 8.11 (continued)

	Objective			Projective		
k^2	f/R	z_m/R	z_p/R	f/R	z_m/R	z_p/R
1.72	0.89	0.71	−0.18	0.91	0.71	−0.20
1.74	0.88	0.71	−0.18	0.90	0.70	−0.20
1.76	0.88	0.70	−0.18	0.90	0.69	−0.21
1.78	0.87	0.69	−0.18	0.89	0.68	−0.21
1.80	0.87	0.68	−0.18	0.89	0.67	−0.21
1.82	0.86	0.68	−0.18	0.88	0.67	−0.21
1.84	0.86	0.67	−0.19	0.88	0.66	−0.22
1.86	0.85	0.66	−0.19	0.87	0.65	−0.22
1.88	0.84	0.66	−0.19	0.87	0.64	−0.22
1.90	0.84	0.65	−0.19	0.86	0.64	−0.23
1.92	0.83	0.64	−0.19	0.86	0.63	−0.23
1.94	0.83	0.64	−0.19	0.86	0.62	−0.23
1.96	0.83	0.63	−0.20	0.85	0.61	−0.24
1.98	0.82	0.62	−0.20	0.85	0.61	−0.24
2.00	0.82	0.62	−0.20	0.84	0.60	−0.24
2.02	0.81	0.61	−0.20	0.84	0.59	−0.25
2.04	0.81	0.60	−0.20	0.84	0.59	−0.25
2.06	0.80	0.60	−0.20	0.83	0.58	−0.25
2.08	0.80	0.59	−0.21	0.83	0.57	−0.26
2.10	0.79	0.59	−0.21	0.83	0.57	−0.26
2.12	0.79	0.58	−0.21	0.82	0.56	−0.26
2.14	0.79	0.57	−0.21	0.82	0.56	−0.27
2.16	0.78	0.57	−0.21	0.82	0.55	−0.27
2.18	0.78	0.56	−0.21	0.81	0.54	−0.27
2.20	0.77	0.56	−0.22	0.81	0.54	−0.28
2.22	0.77	0.55	−0.22	0.81	0.53	−0.28
2.24	0.77	0.55	−0.22	0.81	0.52	−0.28
2.26	0.76	0.54	−0.22	0.80	0.52	−0.29
2.28	0.76	0.54	−0.22	0.80	0.51	−0.29
2.30	0.76	0.53	−0.22	0.80	0.51	−0.29
2.32	0.75	0.53	−0.23	0.80	0.50	−0.30
2.34	0.75	0.52	−0.23	0.80	0.50	−0.30
2.36	0.75	0.52	−0.23	0.79	0.49	−0.30
2.38	0.74	0.51	−0.23	0.79	0.49	−0.31
2.40	0.74	0.51	−0.23	0.79	0.48	−0.31
2.42	0.74	0.51	−0.23	0.79	0.48	−0.31
2.44	0.74	0.50	−0.23	0.79	0.47	−0.32
2.46	0.73	0.50	−0.24	0.78	0.46	−0.32
2.48	0.73	0.49	−0.24	0.78	0.46	−0.32
2.50	0.73	0.49	−0.24	0.78	0.45	−0.33

TABLE 8.12

RATIO OF GAP WIDTH TO BORE DIAMETER $S/D = 0.1$

	Objective			Projective		
k^2	f/R	z_m/R	z_p/R	f/R	z_m/R	z_p/R
0.02	50.40	50.39	−0.00	50.40	50.39	−0.00
0.04	25.32	25.32	−0.00	25.32	25.32	−0.00
0.06	16.96	16.96	−0.01	16.96	16.96	−0.01
0.08	12.78	12.77	−0.01	12.78	12.77	−0.01
0.10	10.28	10.27	−0.01	10.28	10.27	−0.01
0.12	8.60	8.59	−0.01	8.60	8.59	−0.01
0.14	7.41	7.40	−0.01	7.41	7.40	−0.01
0.16	6.52	6.50	−0.02	6.52	6.50	−0.02
0.18	5.82	5.80	−0.02	5.82	5.80	−0.02
0.20	5.26	5.24	−0.02	5.26	5.24	−0.02
0.22	4.81	4.79	−0.02	4.81	4.79	−0.02
0.24	4.43	4.41	−0.02	4.43	4.41	−0.02
0.26	4.11	4.08	−0.03	4.11	4.08	−0.03
0.28	3.83	3.81	−0.03	3.83	3.81	−0.03
0.30	3.60	3.57	−0.03	3.60	3.57	−0.03
0.32	3.39	3.36	−0.03	3.39	3.36	−0.03
0.34	3.20	3.17	−0.03	3.20	3.17	−0.03
0.36	3.04	3.01	−0.04	3.04	3.01	−0.04
0.38	2.90	2.86	−0.04	2.90	2.86	−0.04
0.40	2.76	2.72	−0.04	2.76	2.72	−0.04
0.42	2.65	2.60	−0.04	2.65	2.60	−0.04
0.44	2.54	2.49	−0.04	2.54	2.49	−0.04
0.46	2.44	2.39	−0.05	2.44	2.39	−0.05
0.48	2.35	2.30	−0.05	2.35	2.30	−0.05
0.50	2.27	2.22	−0.05	2.27	2.22	−0.05
0.52	2.19	2.14	−0.05	2.19	2.14	−0.05
0.54	2.12	2.07	−0.05	2.12	2.07	−0.05
0.56	2.05	2.00	−0.06	2.05	2.00	−0.06
0.58	1.99	1.94	−0.06	1.99	1.94	−0.06
0.60	1.94	1.88	−0.06	1.94	1.88	−0.06
0.62	1.88	1.82	−0.06	1.88	1.82	0.06
0.64	1.83	1.77	−0.06	1.83	1.77	−0.06
0.66	1.79	1.72	−0.07	1.79	1.72	−0.07
0.68	1.74	1.67	−0.07	1.74	1.67	−0.07
0.70	1.70	1.63	−0.07	1.70	1.63	−0.07
0.72	1.66	1.59	−0.07	1.66	1.59	−0.07
0.74	1.63	1.55	−0.08	1.63	1.55	−0.08
0.76	1.59	1.51	−0.08	1.59	1.51	−0.08
0.78	1.56	1.48	−0.08	1.56	1.48	−0.08
0.80	1.53	1.45	−0.08	1.53	1.45	−0.08
0.82	1.50	1.41	−0.08	1.50	1.41	−0.08
0.84	1.47	1.38	−0.09	1.47	1.38	−0.09

TABLE 8.12 (continued)

	Objective			Projective		
k^2	f/R	z_m/R	z_p/R	f/R	z_m/R	z_p/R
0.86	1.44	1.35	−0.09	1.44	1.35	−0.09
0.88	1.42	1.33	−0.09	1.42	1.33	−0.09
0.90	1.39	1.30	−0.09	1.39	1.30	−0.09
0.92	1.37	1.27	−0.09	1.37	1.27	−0.10
0.94	1.35	1.25	−0.10	1.35	1.25	−0.10
0.96	1.33	1.23	−0.10	1.33	1.23	−0.10
0.98	1.30	1.20	−0.10	1.31	1.20	−0.10
1.00	1.29	1.18	−0.10	1.29	1.18	−0.11
1.02	1.27	1.16	−0.11	1.27	1.16	−0.11
1.04	1.25	1.14	−0.11	1.25	1.14	−0.11
1.06	1.23	1.12	−0.11	1.23	1.12	−0.11
1.08	1.21	1.10	−0.11	1.22	1.10	−0.12
1.10	1.20	1.08	−0.11	1.20	1.08	−0.12
1.12	1.18	1.07	−0.12	1.19	1.07	−0.12
1.14	1.17	1.05	−0.12	1.17	1.05	−0.12
1.16	1.15	1.03	−0.12	1.16	1.03	−0.12
1.18	1.14	1.02	−0.12	1.14	1.02	−0.13
1.20	1.13	1.00	−0.12	1.13	1.00	−0.13
1.22	1.11	0.99	−0.13	1.12	0.98	−0.13
1.24	1.10	0.97	−0.13	1.10	0.97	−0.13
1.26	1.09	0.96	−0.13	1.09	0.96	−0.14
1.28	1.08	0.94	−0.13	1.08	0.94	−0.14
1.30	1.07	0.93	−0.13	1.07	0.93	−0.14
1.32	1.05	0.92	−0.14	1.06	0.92	−0.15
1.34	1.04	0.90	−0.14	1.05	0.90	−0.15
1.36	1.03	0.89	−0.14	1.04	0.89	−0.15
1.38	1.02	0.88	−0.14	1.03	0.88	−0.15
1.40	1.01	0.87	−0.14	1.02	0.87	−0.16
1.42	1.00	0.86	−0.15	1.01	0.85	−0.16
1.44	1.00	0.85	−0.15	1.00	0.84	−0.16
1.46	0.99	0.84	−0.15	1.00	0.83	−0.16
1.48	0.98	0.82	−0.15	0.99	0.82	−0.17
1.50	0.97	0.81	−0.15	0.98	0.81	−0.17
1.52	0.96	0.80	−0.16	0.97	0.80	−0.17
1.54	0.95	0.79	−0.16	0.96	0.79	−0.17
1.56	0.95	0.78	−0.16	0.96	0.78	−0.18
1.58	0.94	0.78	−0.16	0.95	0.77	−0.18
1.60	0.93	0.77	−0.16	0.94	0.76	−0.18
1.62	0.92	0.76	−0.17	0.94	0.75	−0.19
1.64	0.92	0.75	−0.17	0.93	0.74	−0.19
1.66	0.91	0.74	−0.17	0.92	0.73	−0.19
1.68	0.90	0.73	−0.17	0.92	0.72	−0.19
1.70	0.90	0.72	−0.17	0.91	0.72	−0.20

TABLE 8.12 (continued)

	Objective			Projective		
k^2	f/R	z_m/R	z_p/R	f/R	z_m/R	z_p/R
1.72	0.89	0.71	−0.18	0.91	0.71	−0.20
1.74	0.88	0.71	−0.18	0.90	0.70	−0.20
1.76	0.88	0.70	−0.18	0.90	0.69	−0.21
1.78	0.87	0.69	−0.18	0.89	0.68	−0.21
1.80	0.87	0.68	−0.18	0.89	0.67	−0.21
1.82	0.86	0.68	−0.18	0.88	0.67	−0.21
1.84	0.86	0.67	−0.19	0.88	0.66	−0.22
1.86	0.85	0.66	−0.19	0.87	0.65	−0.22
1.88	0.84	0.66	−0.19	0.87	0.64	−0.22
1.90	0.84	0.65	−0.19	0.86	0.64	−0.23
1.92	0.83	0.64	−0.19	0.86	0.63	−0.23
1.94	0.83	0.64	−0.19	0.86	0.62	−0.23
1.96	0.83	0.63	−0.20	0.85	0.61	−0.24
1.98	0.82	0.62	−0.20	0.85	0.61	−0.24
2.00	0.82	0.62	−0.20	0.84	0.60	−0.24
2.02	0.81	0.61	−0.20	0.84	0.59	−0.25
2.04	0.81	0.60	−0.20	0.84	0.59	−0.25
2.06	0.80	0.60	−0.20	0.83	0.58	−0.25
2.08	0.80	0.59	−0.21	0.83	0.57	−0.26
2.10	0.79	0.59	−0.21	0.83	0.57	−0.26
2.12	0.79	0.58	−0.21	0.82	0.56	−0.26
2.14	0.79	0.57	−0.21	0.82	0.56	−0.27
2.16	0.78	0.57	−0.21	0.82	0.55	−0.27
2.18	0.78	0.56	−0.21	0.81	0.54	−0.27
2.20	0.77	0.56	−0.22	0.81	0.54	−0.28
2.22	0.77	0.55	−0.22	0.81	0.53	−0.28
2.24	0.77	0.55	−0.22	0.81	0.52	−0.28
2.26	0.76	0.54	−0.22	0.80	0.52	−0.29
2.28	0.76	0.54	−0.22	0.80	0.51	−0.29
2.30	0.76	0.53	−0.22	0.80	0.51	−0.29
2.32	0.75	0.53	−0.23	0.80	0.50	−0.30
2.34	0.75	0.52	−0.23	0.80	0.50	−0.30
2.36	0.75	0.52	−0.23	0.79	0.49	−0.30
2.38	0.74	0.51	−0.23	0.79	0.48	−0.31
2.40	0.74	0.51	−0.23	0.79	0.48	−0.31
2.42	0.74	0.50	−0.23	0.79	0.47	−0.31
2.44	0.73	0.50	−0.23	0.79	0.47	−0.32
2.46	0.73	0.50	−0.24	0.78	0.46	−0.32
2.48	0.73	0.49	−0.24	0.78	0.46	−0.32
2.50	0.73	0.49	−0.24	0.78	0.45	−0.33

TABLE 8.13

RATIO OF GAP WIDTH TO BORE DIAMETER $S/D = 0.2$

	Objective			Projective		
k^2	f/R	z_m/R	z_p/R	f/R	z_m/R	z_p/R
0.01	98.20	98.19	−0.00	98.20	98.19	−0.00
0.02	49.22	49.22	−0.00	49.22	49.22	−0.00
0.04	24.74	24.73	−0.00	24.74	24.73	−0.00
0.06	16.57	16.57	−0.01	16.57	16.57	−0.01
0.08	12.49	12.48	−0.01	12.49	12.48	−0.01
0.10	10.04	10.03	−0.01	10.04	10.03	−0.01
0.12	8.41	8.40	−0.01	8.41	8.40	−0.01
0.14	7.25	7.23	−0.01	7.25	7.23	−0.01
0.16	6.37	6.36	−0.02	6.37	6.36	−0.02
0.18	5.69	5.68	−0.02	5.69	5.68	−0.02
0.20	5.15	5.13	−0.02	5.15	5.13	−0.02
0.22	4.71	4.68	−0.02	4.71	4.68	−0.02
0.24	4.34	4.31	−0.02	4.34	4.31	−0.02
0.26	4.02	4.00	−0.03	4.02	4.00	−0.03
0.28	3.75	3.73	−0.03	3.75	3.73	−0.03
0.30	3.52	3.49	−0.03	3.52	3.49	−0.03
0.32	3.32	3.29	−0.03	3.32	3.29	−0.03
0.34	3.14	3.10	−0.03	3.14	3.10	−0.03
0.36	2.98	2.94	−0.04	2.98	2.94	−0.04
0.38	2.84	2.80	−0.04	2.84	2.80	−0.04
0.40	2.71	2.67	−0.04	2.71	2.67	−0.04
0.42	2.59	2.55	−0.04	2.59	2.55	−0.04
0.44	2.49	2.44	−0.05	2.49	2.44	−0.05
0.46	2.39	2.35	−0.05	2.39	2.35	−0.05
0.48	2.31	2.26	−0.05	2.31	2.26	−0.05
0.50	2.23	2.17	−0.05	2.23	2 17	−0 05
0.52	2.15	2.10	−0.05	2.15	2.10	−00.5
0.54	2.08	2.02	−0.06	2.08	2.02	−0.06
0.56	2.02	1.96	−0.06	2.02	1.96	−0.06
0.58	1.96	1.90	−0.06	1.96	1.90	−0.06
0.60	1.90	1.84	−0.06	1.90	1.84	−0.06
0.62	1.85	1.79	−0.07	1.85	1.79	−0.07
0.64	1.80	1.73	−0.07	1.80	1.73	−0.07
0.66	1.76	1.69	−0.07	1.76	1.69	−0.07
0.68	1.71	1.64	−0.07	1.71	1.64	−0.07
0.70	1.67	1.60	−0.08	1.67	1.60	−0.08
0.72	1.64	1.56	−0.08	1.64	1.56	−0.08
0.74	1.60	1.52	−0.08	1.60	1.52	−0.08
0.76	1.57	1.48	−0.08	1.57	1.48	−0.08
0.78	1.53	1.45	−0.08	1.54	1.45	−0.08
0.80	1.50	1.42	−0.09	1.50	1.42	−0.09
0.82	1.48	1.39	−0.09	1.48	1.39	−0.09

TABLE 8.13 (continued)

	Objective			Projective		
k^2	f/R	z_m/R	z_p/R	f/R	z_m/R	z_p/R
0.84	1.45	1.36	−0.09	1.45	1.36	−0.09
0.86	1.42	1.33	−0.09	1.42	1.33	−0.09
0.88	1.40	1.30	−0.10	1.40	1.30	−0.10
0.90	1.37	1.27	−0.10	1.37	1.27	−0.10
0.92	1.35	1.25	−0.10	1.35	1.25	−0.10
0.94	1.33	1.23	−0.10	1.33	1.23	−0.10
0.96	1.31	1.20	−0.10	1.31	1.20	−0.11
0.98	1.29	1.18	−0.11	1.29	1.18	−0.11
1.00	1.27	1.16	−0.11	1.27	1.16	−0.11
1.02	1.25	1.14	−0.11	1.25	1.14	−0.11
1.04	1.23	1.12	−0.11	1.23	1.12	−0.12
1.06	1.22	1.10	−0.12	1.22	1.10	−0.12
1.08	1.20	1.08	−0.12	1.20	1.08	−0.12
1.10	1.18	1.06	−0.12	1.19	1.06	−0.12
1.12	1.17	1.05	−0.12	1.17	1.04	−0.13
1.14	1.15	1.03	−0.13	1.16	1.03	−0.13
1.16	1.14	1.01	−0.13	1.14	1.01	−0.13
1.18	1.13	1.00	−0.13	1.13	1.00	−0.14
1.20	1.11	0.98	−0.13	1.12	0.98	−0.14
1.22	1.10	0.97	−0.13	1.11	0.96	−0.14
1.24	1.09	0.95	−0.14	1.09	0.95	−0.14
1.26	1.08	0.94	−0.14	1.08	0.94	−0.15
1.28	1.07	0.92	−0.14	1.07	0.92	−0.15
1.30	1.05	0.91	−0.14	1.06	0.91	−0.15
1.32	1.04	0.90	−0.14	1.05	0.90	−0.15
1.34	1.03	0.89	−0.15	1.04	0.88	−0.16
1.36	1.02	0.87	−0.15	1.03	0.87	−0.16
1.38	1.01	0.86	−0.15	1.02	0.86	−0.16
1.40	1.00	0.85	−0.15	1.01	0.85	−0.17
1.42	0.99	0.84	−0.16	1.00	0.84	−0.17
1.44	0.99	0.83	−0.16	1.00	0.82	−0.17
1.46	0.98	0.82	−0.16	0.99	0.81	−0.17
1.48	0.97	0.81	−0.16	0.98	0.80	−0.18
1.50	0.96	0.80	−0.16	0.97	0.79	−0.18
1.52	0.95	0.79	−0.17	0.97	0.78	−0.18
1.54	0.95	0.78	−0.17	0.96	0.77	−0.19
1.56	0.94	0.77	−0.17	0.95	0.76	−0.19
1.58	0.93	0.76	−0.17	0.94	0.75	−0.19
1.60	0.92	0.75	−0.17	0.94	0.74	−0.20
1.62	0.92	0.74	−0.18	0.93	0.73	−0.20
1.64	0.91	0.73	−0.18	0.93	0.72	−0.20
1.66	0.90	0.72	−0.18	0.92	0.72	−0.20
1.68	0.90	0.72	−0.18	0.91	0.71	−0.21

TABLE 8.13 (continued)

	Objective			Projective		
k^2	f/R	z_m/R	z_p/R	f/R	z_m/R	z_p/R
1.70	0.89	0.71	−0.18	0.91	0.70	−0.21
1.72	0.88	0.70	−0.19	0.90	0.69	−0.21
1.74	0.88	0.69	−0.19	0.90	0.68	−0.22
1.76	0.87	0.68	−0.19	0.89	0.67	−0.22
1.78	0.87	0.68	−0.19	0.89	0.66	−0.22
1.80	0.86	0.67	−0.19	0.88	0.66	−0.23
1.82	0.86	0.66	−0.19	0.88	0.65	−0.23
1.84	0.85	0.65	−0.20	0.87	0.64	−0.23
1.86	0.85	0.65	−0.20	0.87	0.63	−0.24
1.88	0.84	0.64	−0.20	0.87	0.63	−0.24
1.90	0.84	0.63	−0.20	0.86	0.62	−0.24
1.92	0.83	0.63	−0.20	0.86	0.61	−0.25
1.94	0.83	0.62	−0.21	0.85	0.60	−0.25
1.96	0.82	0.61	−0.21	0.85	0.60	−0.25
1.98	0.82	0.61	−0.21	0.85	0.59	−0.26
2.00	0.81	0.60	−0.21	0.84	0.58	−0.26
2.02	0.81	0.59	−0.21	0.84	0.58	−0.26
2.04	0.80	0.59	−0.21	0.84	0.57	−0.27
2.06	0.80	0.58	−0.22	0.83	0.56	−0.27
2.08	0.79	0.58	−0.22	0.83	0.56	−0.27
2.10	0.79	0.57	−0.22	0.83	0.55	−0.28
2.12	0.79	0.57	−0.22	0.82	0.54	−0.28
2.14	0.78	0.56	−0.22	0.82	0.54	−0.28
2.16	0.78	0.55	−0.22	0.82	0.53	−0.29
2.18	0.78	0.55	−0.23	0.82	0.53	−0.29
2.20	0.77	0.54	−0.23	0.81	0.52	−0.30
2.22	0.77	0.54	−0.23	0.81	0.51	−0.30
2.24	0.76	0.53	−0.23	0.81	0.51	−0.30
2.26	0.76	0.53	−0.23	0.81	0.50	−0.31
2.28	0.76	0.52	−0.23	0.81	0.50	−0.31
2.30	0.75	0.52	−0.24	0.80	0.49	−0.31
2.32	0.75	0.51	−0.24	0.80	0.48	−0.32
2.34	0.75	0.51	−0.24	0.80	0.48	−0.32
2.36	0.74	0.50	−0.24	0.80	0.47	−0.33
2.38	0.74	0.50	−0.24	0.80	0.47	−0.33
2.40	0.74	0.49	−0.24	0.79	0.46	−0.33
2.42	0.74	0.49	−0.24	0.79	0.46	−0.34
2.44	0.73	0.49	−0.25	0.79	0.45	−0.34
2.46	0.73	0.48	−0.25	0.79	0.44	−0.35
2.48	0.73	0.48	−0.25	0.79	0.44	−0.35
2.50	0.72	0.47	−0.25	0.79	0.43	−0.35
3.00	0.67	0.38	−0.29	0.77	0.31	−0.47
3.50	0.62	0.31	−0.32	0.79	0.18	−0.60

TABLE 8.13 (continued)

k^2	Objective f/R	Objective z_m/R	Objective z_p/R	Projective f/R	Projective z_m/R	Projective z_p/R
4.00	0.59	0.25	−0.34	0.83	0.06	−0.77
4.50	0.57	0.20	−0.37	0.91	−0.09	−1.00
5.00	0.55	0.16	−0.39	1.03	−0.27	−1.29
5.50	0.53	0.12	−0.41	1.21	−0.50	−1.72
6.00	0.52	0.09	−0.43	1.52	−0.85	−2.36
6.50	0.51	0.06	−0.45	2.06	−1.43	−3.49
7.00	0.50	0.04	−0.46	3.25	−2.67	−5.93
7.50	0.49	0.02	−0.48	7.89	−7.35	−15.22
8.00	0.48	−0.01	−0.49	−19.22	19.63	38.96
8.50	0.48	−0.03	−0.50	−4.41	4.90	9.31
9.00	0.47	−0.04	−0.51	−2.54	2.99	5.53
9.50	0.46	−0.06	−0.53	−1.81	2.23	4.04
10.00	0.46	−0.08	−0.54	−1.43	1.81	3.24
15.00	0.43	−0.20	−0.62	−0.70	0.65	1.35
20.00	0.41	−0.28	−0.60	−1.10	−0.15	0.95
30.00	0.39	−0.38	−0.77	0.92	1.40	0.48
40.00	0.38	−0.45	−0.83	0.79	0.43	−0.37
50.00	0.37	−0.51	−0.88	−6.15	6.99	13.19
60.00	0.37	−0.55	−0.92	−0.74	1.23	1.97
70.00	0.37	−0.59	−0.95	−0.77	0.57	1.34
80.00	0.36	−0.62	−0.98	−3.92	−2.91	1.03
90.00	0.36	−0.64	−1.00	1.09	1.86	0.77
100.00	0.36	−0.67	−1.03	0.67	1.10	0.43

TABLE 8.14

RATIO OF GAP WIDTH TO BORE DIAMETER $S/D = 0.3$

k^2	Objective f/R	Objective z_m/R	Objective z_p/R	Projective f/R	Projective z_m/R	Projective z_p/R
0.01	94.69	94.69	−0.00	94.69	94.69	−0.00
0.02	47.47	47.47	−0.00	47.47	47.47	−0.00
0.04	23.86	23.86	−0.00	23.86	23.86	−0.00
0.06	15.99	15.99	−0.01	15.99	15.99	−0.01
0.08	12.06	12.05	−0.01	12.06	12.05	−0.01
0.10	9.70	9.69	−0.01	9.70	9.69	−0.01
0.12	8.13	8.11	−0.01	8.13	8.11	−0.01
0.14	7.00	6.99	−0.02	7.00	6.99	−0.02
0.16	6.16	6.14	−0.02	6.16	6.14	−0.02
0.18	5.51	5.49	−0.02	5.51	5.49	−0.02

TABLE 8.14 (continued)

	Objective			Projective		
k^2	f/R	z_m/R	z_p/R	f/R	z_m/R	z_p/R
0.20	4.98	4.96	−0.02	4.98	4.96	−0.02
0.22	4.55	4.53	−0.02	4.55	4.53	−0.02
0.24	4.20	4.17	−0.03	4.20	4.17	−0.03
0.26	3.90	3.87	−0.03	3.90	3.87	−0.03
0.28	3.64	3.61	−0.03	3.64	3.61	−0.03
0.30	3.41	3.38	−0.03	3.41	3.38	−0.03
0.32	3.22	3.18	−0.04	3.22	3.18	−0.04
0.34	3.04	3.01	−0.04	3.04	3.01	−0.04
0.36	2.89	2.85	−0.04	2.89	2.85	−0.04
0.38	2.75	2.71	−0.04	2.75	2.71	−0.04
0.40	2.63	2.59	−0.05	2.63	2.59	−0.05
0.42	2.52	2.47	−0.05	2.52	2.47	−0.05
0.44	2.42	2.37	−0.05	2.42	2.37	−0.05
0.46	2.33	2.27	−0.05	2.33	2.27	−0.05
0.48	2.24	2.19	−0.06	2.24	2.19	−0.06
0.50	2.16	2.11	−0.06	2.16	2.11	−0.06
0.52	2.09	2.03	−0.06	2.09	2.03	−0.06
0.54	2.03	1.96	−0.06	2.03	1.96	−0.06
0.56	1.96	1.90	−0.07	1.96	1.90	−0.07
0.58	1.91	1.84	−0.07	1.91	1.84	−0.07
0.60	1.85	1.78	−0.07	1.85	1.78	−0.07
0.62	1.80	1.73	−0.07	1.80	1.73	−0.07
0.64	1.76	1.68	−0.08	1.76	1.68	−0.08
0.66	1.71	1.64	−0.08	1.71	1.64	−0.08
0.68	1.67	1.59	−0.08	1.67	1.59	−0.08
0.70	1.63	1.55	−0.08	1.63	1.55	−0.08
0.72	1.60	1.51	−0.09	1.60	1.51	−0.09
0.74	1.56	1.48	−0.09	1.56	1.48	−0.09
0.76	1.53	1.44	−0.09	1.53	1.44	−0.09
0.78	1.50	1.41	−0.09	1.50	1.41	−0.09
0.80	1.47	1.38	−0.10	1.47	1.38	−0.10
0.82	1.44	1.35	−0.10	1.44	1.35	−0.10
0.84	1.42	1.32	−0.10	1.42	1.32	−0.10
0.86	1.39	1.29	−0.10	1.39	1.29	−0.10
0.88	1.37	1.26	−0.11	1.37	1.26	−0.11
0.90	1.34	1.24	−0.11	1.35	1.24	−0.11
0.92	1.32	1.21	−0.11	1.32	1.21	−0.11
0.94	1.30	1.19	−0.11	1.30	1.19	−0.12
0.96	1.28	1.17	−0.12	1.28	1.17	−0.12
0.98	1.26	1.15	−0.12	1.27	1.14	−0.12
1.00	1.24	1.12	−0.12	1.25	1.12	−0.12
1.02	1.23	1.10	−0.12	1.23	1.10	−0.13
1.04	1.21	1.08	−0.13	1.21	1.08	−0.13

TABLE 8.14 (continued)

	Objective			Projective		
k^2	f/R	z_m/R	z_p/R	f/R	z_m/R	z_p/R
1.06	1.19	1.07	−0.13	1.20	1.06	−0.13
1.08	1.18	1.05	−0.13	1.18	1.05	−0.14
1.10	1.16	1.03	−0.13	1.17	1.03	−0.14
1.12	1.15	1.01	−0.14	1.15	1.01	−0.14
1.14	1.13	1.00	−0.14	1.14	1.00	−0.14
1.16	1.12	0.98	−0.14	1.13	0.98	−0.15
1.18	1.11	0.97	−0.14	1.11	0.96	−0.15
1.20	1.10	0.95	−0.14	1.10	0.95	−0.15
1.22	1.08	0.94	−0.15	1.09	0.93	−0.16
1.24	1.07	0.92	−0.15	1.08	0.92	−0.16
1.26	1.06	0.91	−0.15	1.07	0.91	−0.16
1.28	1.05	0.90	−0.15	1.06	0.89	−0.17
1.30	1.04	0.88	−0.16	1.05	0.88	−0.17
1.32	1.03	0.87	−0.16	1.04	0.87	−0.17
1.34	1.02	0.86	−0.16	1.03	0.85	−0.17
1.36	1.01	0.85	−0.16	1.02	0.84	−0.18
1.38	1.00	0.83	−0.17	1.01	0.83	−0.18
1.40	0.99	0.82	−0.17	1.00	0.82	−0.18
1.42	0.98	0.81	−0.17	0.99	0.81	−0.19
1.44	0.97	0.80	−0.17	0.99	0.80	−0.19
1.46	0.97	0.79	−0.17	0.98	0.79	−0.19
1.48	0.96	0.78	−0.18	0.97	0.77	−0.20
1.50	0.95	0.77	−0.18	0.97	0.76	−0.20
1.52	0.94	0.76	−0.18	0.96	0.75	−0.20
1.54	0.94	0.75	−0.18	0.95	0.74	−0.21
1.56	0.93	0.74	−0.19	0.95	0.73	−0.21
1.58	0.92	0.73	−0.19	0.94	0.72	−0.21
1.60	0.91	0.72	−0.19	0.93	0.72	−0.22
1.62	0.91	0.72	−0.19	0.93	0.71	−0.22
1.64	0.90	0.71	−0.19	0.92	0.70	−0.22
1.66	0.89	0.70	−0.20	0.92	0.69	−0.23
1.68	0.89	0.69	−0.20	0.91	0.68	−0.23
1.70	0.88	0.68	−0.20	0.91	0.67	−0.24
1.72	0.88	0.67	−0.20	0.90	0.66	−0.24
1.74	0.87	0.67	−0.20	0.90	0.65	−0.24
1.76	0.87	0.66	−0.21	0.89	0.65	−0.25
1.78	0.86	0.65	−0.21	0.89	0.64	−0.25
1.80	0.85	0.64	−0.21	0.88	0.63	−0.25
1.82	0.85	0.64	−0.21	0.88	0.62	−0.26
1.84	0.84	0.63	−0.21	0.87	0.61	−0.26
1.86	0.84	0.62	−0.22	0.87	0.61	−0.26
1.88	0.83	0.62	−0.22	0.87	0.60	−0.27
1.90	0.83	0.61	−0.22	0.86	0.59	−0.27

TABLE 8.14 (continued)

k^2	Objective			Projective		
	f/R	z_m/R	z_p/R	f/R	z_m/R	z_p/R
1.92	0.82	0.60	−0.22	0.86	0.58	−0.28
1.94	0.82	0.60	−0.22	0.86	0.58	−0.28
1.96	0.82	0.59	−0.23	0.85	0.57	−0.28
1.98	0.81	0.58	−0.23	0.85	0.56	−0.29
2.00	0.81	0.58	−0.23	0.85	0.56	−0.29
2.02	0.80	0.57	−0.23	0.84	0.55	−0.30
2.04	0.80	0.57	−0.23	0.84	0.54	−0.30
2.06	0.79	0.56	−0.23	0.84	0.54	−0.30
2.08	0.79	0.55	−0.24	0.84	0.53	−0.31
2.10	0.79	0.55	−0.24	0.83	0.52	−0.31
2.12	0.78	0.54	−0.24	0.83	0.52	−0.32
2.14	0.78	0.54	−0.24	0.83	0.51	−0.32
2.16	0.78	0.53	−0.24	0.83	0.50	−0.32
2.18	0.77	0.53	−0.25	0.82	0.50	−0.33
2.20	0.77	0.52	−0.25	0.82	0.49	−0.33
2.22	0.77	0.52	−0.25	0.82	0.48	−0.34
2.24	0.76	0.51	−0.25	0.82	0.48	−0.34
2.26	0.76	0.51	−0.25	0.82	0.47	−0.35
2.28	0.76	0.50	−0.25	0.81	0.47	−0.35
2.30	0.75	0.50	−0.26	0.81	0.46	−0.35
2.32	0.75	0.49	−0.26	0.81	0.45	−0.36
2.34	0.75	0.49	−0.26	0.81	0.45	−0.36
2.36	0.74	0.48	−0.26	0.81	0.44	−0.37
2.38	0.74	0.48	−0.26	0.81	0.44	−0.37
2.40	0.74	0.47	−0.26	0.81	0.43	−0.38
2.42	0.73	0.47	−0.27	0.80	0.42	−0.38
2.44	0.73	0.46	−0.27	0.80	0.42	−0.39
2.46	0.73	0.46	−0.27	0.80	0.41	−0.39
2.48	0.73	0.46	−0.27	0.80	0.41	−0.39
2.50	0.72	0.45	−0.27	0.80	0.40	−0.40
3.00	0.67	0.36	−0.31	0.80	0.27	−0.53
3.50	0.63	0.29	−0.34	0.83	0.13	−0.70
4.00	0.60	0.23	−0.37	0.90	−0.02	−0.91
4.50	0.57	0.18	−0.39	1.01	−0.20	−1.20
5.00	0.55	0.14	−0.42	1.18	−0.43	−1.61
5.50	0.54	0.10	−0.44	1.47	−0.77	−2.24
6.00	0.53	0.07	−0.45	2.00	−1.35	−3.35
6.50	0.51	0.04	−0.47	3.19	−2.58	−5.77
7.00	0.50	0.02	−0.49	8.01	−7.44	−15.45
7.50	0.50	−0.01	−0.50	−16.14	16.67	32.81
8.00	0.49	−0.03	−0.52	−4.10	4.60	8.70
8.50	0.48	−0.05	−0.53	−2.39	2.85	5.25
9.00	0.48	−0.07	−0.54	−1.72	2.14	3.86

TABLE 8.14 (continued)

k^2	Objective f/R	z_m/R	z_p/R	Projective f/R	z_m/R	z_p/R
9.50	0.47	−0.08	−0.55	−1.36	1.75	3.11
10.00	0.46	−0.10	−0.57	−1.15	1.50	2.64
15.00	0.43	−0.22	−0.65	−0.72	0.55	1.26
20.00	0.41	−0.30	−0.71	−1.55	−0.65	0.90
30.00	0.39	−0.41	−0.80	0.77	1.17	0.40
40.00	0.38	−0.48	−0.86	1.02	0.10	−0.92
50.00	0.38	−0.53	−0.91	−1.58	2.34	3.92
60.00	0.37	−0.58	−0.95	−0.68	1.00	1.67
70.00	0.37	−0.61	−0.98	−1.10	0.12	1.21
80.00	0.36	−0.65	−1.01	2.72	3.65	0.92
90.00	0.36	−0.67	−1.04	0.75	1.37	0.62
100.00	0.36	−0.70	−1.06	0.71	0.81	0.10

TABLE 8.15

RATIO OF GAP WIDTH TO BORE DIAMETER $S/D = 0.4$

k^2	Objective f/R	z_m/R	z_p/R	Projective f/R	z_m/R	z_p/R
0.01	90.23	90.23	−0.00	90.23	90.23	−0.00
0.02	45.25	45.25	−0.00	45.25	45.25	−0.00
0.04	22.76	22.75	−0.00	22.76	22.75	−0.00
0.06	15.26	15.25	−0.01	15.26	15.25	−0.01
0.08	11.51	11.50	−0.01	11.51	11.50	−0.01
0.10	9.26	9.25	−0.01	9.26	9.25	−0.01
0.12	7.76	7.75	−0.01	7.76	7.75	−0.01
0.14	6.69	6.68	−0.02	6.69	6.68	−0.02
0.16	5.89	5.87	−0.02	5.89	5.87	−0.02
0.18	5.27	5.25	−0.02	5.27	5.25	−0.02
0.20	4.77	4.74	−0.03	4.77	4.74	−0.03
0.22	4.36	4.33	0.03	4.36	4.33	−0.03
0.24	4.02	3.99	−0.03	4.02	3.99	−0.03
0.26	3.73	3.70	−0.03	3.73	3.70	−0.03
0.28	3.49	3.45	−0.04	3.49	3.45	−0.04
0.30	3.28	3.24	−0.04	3.28	3.24	−0.04
0.32	3.09	3.05	−0.04	3.09	3.05	−0.04
0.34	2.93	2.88	−0.04	2.93	2.88	−0.04
0.36	2.78	2.73	−0.05	2.78	2.73	−0.05
0.38	2.65	2.60	−0.05	2.65	2.60	−0.05
0.40	2.53	2.48	−0.05	2.53	2.48	−0.05
0.42	2.43	2.37	−0.05	2.43	2.37	−0.05

TABLE 8.15 (continued)

	Objective			Projective		
k^2	f/R	z_m/R	z_p/R	f/R	z_m/R	z_p/R
0.44	2.33	2.27	−0.06	2.33	2.27	−0.06
0.46	2.24	2.18	−0.06	2.24	2.18	−0.06
0.48	2.16	2.10	−0.06	2.16	2.10	−0.06
0.50	2.09	2.02	−0.07	2.09	2.02	−0.07
0.52	2.02	1.95	−0.07	2.02	1.95	−0.07
0.54	1.96	1.89	−0.07	1.96	1.89	−0.07
0.56	1.90	1.82	−0.07	1.90	1.82	−0.07
0.58	1.84	1.77	−0.08	1.84	1.77	−0.08
0.60	1.79	1.71	−0.08	1.79	1.71	−0.08
0.62	1.75	1.66	−0.08	1.75	1.66	−0.08
0.64	1.70	1.62	−0.09	1.70	1.62	−0.09
0.66	1.66	1.57	−0.09	1.66	1.57	−0.09
0.68	1.62	1.53	−0.09	1.62	1.53	−0.09
0.70	1.58	1.49	−0.09	1.59	1.49	−0.09
0.72	1.55	1.45	−0.10	1.55	1.45	−0.10
0.74	1.52	1.42	−0.10	1.52	1.42	−0.10
0.76	1.49	1.38	−0.10	1.49	1.38	−0.10
0.78	1.46	1.35	−0.11	1.46	1.35	−0.11
0.80	1.43	1.32	−0.11	1.43	1.32	−0.11
0.82	1.40	1.29	−0.11	1.41	1.29	−0.11
0.84	1.38	1.26	−0.11	1.38	1.26	−0.12
0.86	1.35	1.24	−0.12	1.36	1.24	−0.12
0.88	1.33	1.21	−0.12	1.33	1.21	−0.12
0.90	1.31	1.19	−0.12	1.31	1.19	−0.13
0.92	1.29	1.16	−0.13	1.29	1.16	−0.13
0.94	1.27	1.14	−0.13	1.27	1.14	−0.13
0.96	1.25	1.12	−0.13	1.25	1.12	−0.13
0.98	1.23	1.10	−0.13	1.24	1.10	−0.14
1.00	1.22	1.08	−0.14	1.22	1.08	−0.14
1.02	1.20	1.06	−0.14	1.20	1.06	−0.14
1.04	1.18	1.04	−0.14	1.19	1.04	−0.15
1.06	1.17	1.02	−0.14	1.17	1.02	−0.15
1.08	1.15	1.01	−0.15	1.16	1.00	−0.15
1.10	1.14	0.99	−0.15	1.15	0.99	−0.16
1.12	1.13	0.97	−0.15	1.13	0.97	−0.16
1.14	1.11	0.96	−0.16	1.12	0.95	−0.17
1.16	1.10	0.94	−0.16	1.11	0.94	−0.17
1.18	1.09	0.93	−0.16	1.10	0.92	−0.17
1.20	1.08	0.91	−0.16	1.08	0.91	−0.18
1.22	1.06	0.90	−0.17	1.07	0.89	−0.18
1.24	1.05	0.88	−0.17	1.06	0.88	−0.18
1.26	1.04	0.87	−0.17	1.05	0.87	−0.19
1.28	1.03	0.86	−0.17	1.04	0.85	−0.19

TABLE 8.15 (continued)

	Objective			Projective		
k^2	f/R	z_m/R	z_p/R	f/R	z_m/R	z_p/R
1.30	1.02	0.85	−0.18	1.03	0.84	−0.19
1.32	1.01	0.83	−0.18	1.03	0.83	−0.20
1.34	1.00	0.82	−0.18	1.02	0.82	−0.20
1.36	0.99	0.81	−0.18	1.01	0.80	−0.20
1.38	0.99	0.80	−0.19	1.00	0.79	−0.21
1.40	0.98	0.79	−0.19	0.99	0.78	−0.21
1.42	0.97	0.78	−0.19	0.99	0.77	−0.22
1.44	0.96	0.77	−0.19	0.98	0.76	−0.22
1.46	0.95	0.76	−0.20	0.97	0.75	−0.22
1.48	0.95	0.75	−0.20	0.96	0.74	−0.23
1.50	0.94	0.74	−0.20	0.96	0.73	−0.23
1.52	0.93	0.73	−0.20	0.95	0.72	−0.24
1.54	0.92	0.72	−0.21	0.95	0.71	−0.24
1.56	0.92	0.71	−0.21	0.94	0.70	−0.24
1.58	0.91	0.70	−0.21	0.94	0.69	−0.25
1.60	0.90	0.69	−0.21	0.93	0.68	−0.25
1.62	0.90	0.68	−0.22	0.92	0.67	−0.26
1.64	0.89	0.67	−0.22	0.92	0.66	−0.26
1.66	0.89	0.67	−0.22	0.91	0.65	−0.26
1.68	0.88	0.66	−0.22	0.91	0.64	−0.27
1.70	0.87	0.65	−0.22	0.91	0.63	−0.27
1.72	0.87	0.64	−0.23	0.90	0.62	−0.28
1.74	0.86	0.63	−0.23	0.90	0.62	−0.28
1.76	0.86	0.63	−0.23	0.89	0.61	−0.29
1.78	0.85	0.62	−0.23	0.89	0.60	−0.29
1.80	0.85	0.61	−0.24	0.89	0.59	−0.29
1.82	0.84	0.61	−0.24	0.88	0.58	−0.30
1.84	0.84	0.60	−0.24	0.88	0.58	−0.30
1.86	0.83	0.59	−0.24	0.88	0.57	−0.31
1.88	0.83	0.58	−0.24	0.87	0.56	−0.31
1.90	0.82	0.58	−0.25	0.87	0.55	−0.32
1.92	0.82	0.57	−0.25	0.87	0.55	−0.32
1.94	0.82	0.57	−0.25	0.86	0.54	−0.33
1.96	0.81	0.56	−0.25	0.86	0.53	−0.33
1.98	0.81	0.55	−0.25	0.86	0.52	−0.33
2.00	0.80	0.55	−0.26	0.86	0.52	−0.34
2.02	0.80	0.54	−0.26	0.85	0.51	−0.34
2.04	0.79	0.54	−0.26	0.85	0.50	−0.35
2.06	0.79	0.53	−0.26	0.85	0.50	−0.35
2.08	0.79	0.52	−0.26	0.85	0.49	−0.36
2.10	0.78	0.52	−0.27	0.85	0.48	−0.36
2.12	0.78	0.51	−0.27	0.84	0.47	−0.37
2.14	0.78	0.51	−0.27	0.84	0.47	−0.37

TABLE 8.15 (continued)

	Objective			Projective		
k^2	f/R	z_m/R	z_p/R	f/R	z_m/R	z_p/R
2.16	0.77	0.50	−0.27	0.84	0.46	−0.38
2.18	0.77	0.50	−0.27	0.84	0.45	−0.38
2.20	0.77	0.49	−0.27	0.84	0.45	−0.39
2.22	0.76	0.49	−0.28	0.84	0.44	−0.39
2.24	0.76	0.48	−0.28	0.83	0.43	−0.40
2.26	0.76	0.48	−0.28	0.83	0.43	−0.41
2.28	0.75	0.47	−0.28	0.83	0.42	−0.41
2.30	0.75	0.47	−0.28	0.83	0.42	−0.42
2.32	0.75	0.46	−0.29	0.83	0.41	−0.42
2.34	0.74	0.46	−0.29	0.83	0.40	−0.43
2.36	0.74	0.45	−0.29	0.83	0.40	−0.43
2.38	0.74	0.45	−0.29	0.83	0.39	−0.44
2.40	0.74	0.44	−0.29	0.83	0.38	−0.44
2.42	0.73	0.44	−0.29	0.83	0.38	−0.45
2.44	0.73	0.43	−0.30	0.83	0.37	−0.46
2.46	0.73	0.43	−0.30	0.83	0.37	−0.46
2.48	0.73	0.43	−0.30	0.83	0.36	−0.47
2.50	0.72	0.42	−0.30	0.83	0.35	−0.47
3.00	0.67	0.33	−0.34	0.84	0.20	−0.64
3.50	0.63	0.26	−0.37	0.90	0.04	−0.86
4.00	0.60	0.20	−0.40	1.01	−0.15	−1.16
4.50	0.58	0.15	−0.43	1.19	−0.40	−1.59
5.00	0.56	0.11	−0.45	1.51	−0.77	−2.27
5.50	0.54	0.07	−0.47	2.10	−1.42	−3.52
6.00	0.53	0.04	−0.49	3.60	−2.96	−6.56
6.50	0.52	0.01	−0.51	12.78	−12.19	−24.97
7.00	0.51	−0.02	−0.53	−8.40	8.95	17.35
7.50	0.50	−0.04	−0.54	−3.23	3.74	6.97
8.00	0.49	−0.06	−0.56	−2.04	2.51	4.55
8.50	0.49	−0.08	−0.57	−1.52	1.95	3.47
9.00	0.48	−0.10	−0.58	−1.23	1.62	2.85
9.50	0.48	−0.12	−0.59	−1.05	1.40	2.45
10.00	0.47	−0.13	−0.60	−0.94	1.24	2.17
15.00	0.44	−0.26	−0.69	−0.79	0.39	1.18
20.00	0.42	−0.34	−0.76	−4.08	−3.24	0.84
30.00	0.40	−0.45	−0.84	0.70	0.94	0.25
40.00	0.39	−0.52	−0.91	2.18	−1.16	−3.34
50.00	0.38	−0.57	−0.95	−0.87	1.50	2.37
60.00	0.37	−0.62	−0.99	−0.74	0.68	1.43
70.00	0.37	−0.65	−1.02	−5.40	−4.33	1.07
80.00	0.37	−0.69	−1.05	0.92	1.68	0.76
90.00	0.36	−0.71	−1.08	0.68	0.98	0.30
100.00	0.36	−0.74	−1.10	1.21	0.09	−1.12

TABLE 8.16
RATIO OF GAP WIDTH TO BORE DIAMETER $S/D = 0.5$

	Objective			Projective		
k^2	f/R	z_m/R	z_p/R	f/R	z_m/R	z_p/R
0.01	85.16	85.16	−0.00	85.16	85.16	−0.00
0.02	42.72	42.71	−0.00	42.72	42.71	−0.00
0.04	21.50	21.49	−0.01	21.50	21.49	−0.01
0.06	14.42	14.42	−0.01	14.42	14.42	−0.01
0.08	10.89	10.88	−0.01	10.89	10.88	−0.01
0.10	8.77	8.75	−0.01	8.77	8.75	−0.01
0.12	7.35	7.34	−0.02	7.35	7.34	−0.02
0.14	6.34	6.32	−0.02	6.34	6.32	−0.02
0.16	5.59	5.56	−0.02	5.59	5.56	−0.02
0.18	5.00	4.97	−0.03	5.00	4.97	−0.03
0.20	4.53	4.50	−0.03	4.53	4.50	−0.03
0.22	4.14	4.11	−0.03	4.14	4.11	−0.03
0.24	3.82	3.79	−0.03	3.82	3.79	−0.03
0.26	3.55	3.52	−0.04	3.55	3.52	−0.04
0.28	3.32	3.28	−0.04	3.32	3.28	−0.04
0.30	3.12	3.08	−0.04	3.12	3.08	−0.04
0.32	2.95	2.90	−0.05	2.95	2.90	−0.05
0.34	2.79	2.74	−0.05	2.79	2.74	−0.05
0.36	2.65	2.60	−0.05	2.65	2.60	−0.05
0.38	2.53	2.47	−0.06	2.53	2.47	−0.06
0.40	2.42	2.36	−0.06	2.42	2.36	−0.06
0.42	2.32	2.26	−0.06	2.32	2.26	−0.06
0.44	2.23	2.16	−0.07	2.23	2.16	−0.07
0.46	2.15	2.08	−0.07	2.15	2.08	−0.07
0.48	2.07	2.00	−0.07	2.07	2.00	−0.07
0.50	2.00	1.93	−0.08	2.00	1.93	−0.08
0.52	1.94	1.86	−0.08	1.94	1.86	−0.08
0.54	1.88	1.80	−0.08	1.88	1.80	−0.08
0.56	1.82	1.74	−0.09	1.82	1.74	−0.09
0.58	1.77	1.68	−0.09	1.77	1.68	−0.09
0.60	1.73	1.63	−0.09	1.73	1.63	−0.09
0.62	1.68	1.59	−0.10	1.68	1.59	−0.10
0.64	1.64	1.54	−0.10	1.64	1.54	−0.10
0.66	1.60	1.50	−0.10	1.60	1.50	−0.10
0.68	1.57	1.46	−0.11	1.57	1.46	−0.11
0.70	1.53	1.42	−0.11	1.53	1.42	−0.11
0.72	1.50	1.39	−0.11	1.50	1.39	−0.11
0.74	1.47	1.35	−0.12	1.47	1.35	−0.12
0.76	1.44	1.32	−0.12	1.44	1.32	−0.12
0.78	1.41	1.29	−0.12	1.41	1.29	−0.12
0.80	1.39	1.26	−0.13	1.39	1.26	−0.13
0.82	1.36	1.23	−0.13	1.36	1.23	−0.13

TABLE 8.16 (continued)

	Objective			Projective		
k^2	f/R	z_m/R	z_p/R	f/R	z_m/R	z_p/R
0.84	1.34	1.21	−0.13	1.34	1.20	−0.14
0.86	1.32	1.18	−0.14	1.32	1.18	−0.14
0.88	1.29	1.16	−0.14	1.30	1.15	−0.14
0.90	1.27	1.13	−0.14	1.28	1.13	−0.15
0.92	1.26	1.11	−0.15	1.26	1.11	−0.15
0.94	1.24	1.09	−0.15	1.24	1.09	−0.15
0.96	1.22	1.07	−0.15	1.22	1.07	−0.16
0.98	1.20	1.05	−0.16	1.21	1.04	−0.16
1.00	1.19	1.03	−0.16	1.19	1.03	−0.17
1.02	1.17	1.01	−0.16	1.18	1.01	−0.17
1.04	1.16	0.99	−0.16	1.16	0.99	−0.17
1.06	1.14	0.97	−0.17	1.15	0.97	−0.18
1.08	1.13	0.96	−0.17	1.14	0.95	−0.18
1.10	1.11	0.94	−0.17	1.12	0.94	−0.19
1.12	1.10	0.92	−0.18	1.11	0.92	−0.19
1.14	1.09	0.91	−0.18	1.10	0.90	−0.19
1.16	1.08	0.89	−0.18	1.09	0.89	−0.20
1.18	1.07	0.88	−0.19	1.08	0.87	−0.20
1.20	1.06	0.87	−0.19	1.07	0.86	−0.21
1.22	1.04	0.85	−0.19	1.06	0.85	−0.21
1.24	1.03	0.84	−0.20	1.05	0.83	−0.22
1.26	1.02	0.83	−0.20	1.04	0.82	−0.22
1.28	1.01	0.81	−0.20	1.03	0.81	−0.23
1.30	1.01	0.80	−0.20	1.02	0.79	−0.23
1.32	1.00	0.79	−0.21	1.02	0.78	−0.23
1.34	0.99	0.78	−0.21	1.01	0.77	−0.24
1.36	0.98	0.77	−0.21	1.00	0.76	−0.24
1.38	0.97	0.76	−0.22	0.99	0.75	−0.25
1.40	0.96	0.75	−0.22	0.99	0.73	−0.25
1.42	0.96	0.73	−0.22	0.98	0.72	−0.26
1.44	0.95	0.72	−0.22	0.97	0.71	−0.26
1.46	0.94	0.71	−0.23	0.97	0.70	−0.27
1.48	0.93	0.70	−0.23	0.96	0.69	−0.27
1.50	0.93	0.70	−0.23	0.96	0.68	−0.28
1.52	0.92	0.69	−0.23	0.95	0.67	−0.28
1.54	0.91	0.68	−0.24	0.95	0.66	−0.29
1.56	0.91	0.67	−0.24	0.94	0.65	−0.29
1.58	0.90	0.66	−0.24	0.94	0.64	−0.30
1.60	0.89	0.65	−0.24	0.93	0.63	−0.30
1.62	0.89	0.64	−0.25	0.93	0.62	−0.31
1.64	0.88	0.63	−0.25	0.92	0.61	−0.31
1.66	0.88	0.63	−0.25	0.92	0.60	−0.32
1.68	0.87	0.62	−0.25	0.92	0.59	−0.32

TABLE 8.16 (continued)

	Objective			Projective		
k^2	f/R	z_m/R	z_p/R	f/R	z_m/R	z_p/R
1.70	0.87	0.61	−0.26	0.91	0.59	−0.33
1.72	0.86	0.60	−0.26	0.91	0.58	−0.33
1.74	0.86	0.59	−0.26	0.91	0.57	−0.34
1.76	0.85	0.59	−0.26	0.90	0.56	−0.34
1.78	0.85	0.58	−0.27	0.90	0.55	−0.35
1.80	0.84	0.57	−0.27	0.90	0.54	−0.35
1.82	0.84	0.57	−0.27	0.89	0.53	−0.36
1.84	0.83	0.56	−0.27	0.89	0.53	−0.36
1.86	0.83	0.55	−0.28	0.89	0.52	−0.37
1.88	0.82	0.55	−0.28	0.89	0.51	−0.38
1.90	0.82	0.54	−0.28	0.88	0.50	−0.38
1.92	0.82	0.53	−0.28	0.88	0.49	−0.39
1.94	0.81	0.53	−0.28	0.88	0.49	−0.39
1.96	0.81	0.52	−0.29	0.88	0.48	−0.40
1.98	0.80	0.51	−0.29	0.88	0.47	−0.41
2.00	0.80	0.51	−0.29	0.88	0.46	−0.41
2.02	0.80	0.50	−0.29	0.87	0.46	−0.42
2.04	0.79	0.50	−0.30	0.87	0.45	−0.42
2.06	0.79	0.49	−0.30	0.87	0.44	−0.43
2.08	0.79	0.49	−0.30	0.87	0.43	−0.44
2.10	0.78	0.48	−0.30	0.87	0.43	−0.44
2.12	0.78	0.47	−0.30	0.87	0.42	−0.45
2.14	0.77	0.47	−0.31	0.87	0.41	−0.46
2.16	0.77	0.46	−0.31	0.87	0.40	−0.46
2.18	0.77	0.46	−0.31	0.87	0.40	−0.47
2.20	0.77	0.45	−0.31	0.87	0.39	−0.48
2.22	0.76	0.45	−0.31	0.87	0.38	−0.48
2.24	0.76	0.44	−0.32	0.86	0.38	−0.49
2.26	0.76	0.44	−0.32	0.86	0.37	−0.50
2.28	0.75	0.43	−0.32	0.86	0.36	−0.50
2.30	0.75	0.43	−0.32	0.86	0.35	−0.51
2.32	0.75	0.42	−0.32	0.86	0.35	−0.52
2.34	0.74	0.42	−0.33	0.86	0.34	−0.52
2.36	0.74	0.41	−0.33	0.86	0.33	−0.53
2.38	0.74	0.41	−0.33	0.87	0.33	−0.54
2.40	0.74	0.41	−0.33	0.87	0.32	−0.55
2.42	0.73	0.40	−0.33	0.87	0.31	−0.55
2.44	0.73	0.40	−0.33	0.87	0.31	−0.56
2.46	0.73	0.39	−0.34	0.87	0.30	−0.57
2.48	0.73	0.39	−0.34	0.87	0.29	−0.58
2.50	0.72	0.38	−0.34	0.87	0.28	−0.58
3.00	0.67	0.29	−0.38	0.92	0.11	−0.81
3.50	0.64	0.22	−0.42	1.03	−0.10	−1.13

TABLE 8.16 (continued)

	Objective			Projective		
k^2	f/R	z_m/R	z_p/R	f/R	z_m/R	z_p/R
4.00	0.61	0.16	−0.45	1.23	−0.38	−1.60
4.50	0.59	0.11	−0.47	1.58	−0.81	−2.39
5.00	0.57	0.07	−0.50	2.33	−1.61	−3.95
5.50	0.55	0.03	−0.52	4.62	−3.95	−8.58
6.00	0.54	0.00	−0.54	415.32	−414.70	−830.02
6.50	0.53	−0.03	−0.56	−4.83	5.40	10.23
7.00	0.52	−0.06	−0.57	−2.45	2.98	5.43
7.50	0.51	−0.08	−0.59	−1.68	2.16	3.84
8.00	0.50	−0.10	−0.60	−1.31	1.74	3.05
8.50	0.50	−0.12	−0.62	−1.09	1.48	2.57
9.00	0.49	−0.14	−0.63	−0.96	1.29	2.24
9.50	0.48	−0.16	−0.64	−0.86	1.14	2.01
10.00	0.48	−0.18	−0.65	−0.80	1.03	1.84
15.00	0.44	−0.30	−0.74	−0.99	0.10	1.09
20.00	0.42	−0.38	−0.81	3.68	4.45	0.77
30.00	0.40	−0.49	−0.90	0.73	0.68	−0.05
40.00	0.39	−0.57	−0.96	−3.60	4.50	8.10
50.00	0.38	−0.62	−1.00	−0.69	1.08	1.77
60.00	0.38	−0.67	−1.04	−1.37	−0.14	1.23
70.00	0.37	−0.70	−1.08	1.30	2.19	0.89
80.00	0.37	−0.73	−1.11	0.68	1.13	0.45
90.00	0.37	−0.76	−1.13	1.08	0.28	−0.81
100.00	0.37	−0.79	−1.15	−3.17	4.25	7.42

TABLE 8.17
RATIO OF GAP WIDTH TO BORE DIAMETER $S/D = 0.6$

	Objective			Projective		
k^2	f/R	z_m/R	z_p/R	f/R	z_m/R	z_p/R
0.01	79.79	79.79	−0.00	79.79	79.79	−0.00
0.02	40.04	40.04	−0.00	40.04	40.04	−0.00
0.04	20.17	20.16	−0.01	20.17	20.16	−0.01
0.06	13.54	13.53	−0.01	13.54	13.53	−0.01
0.08	10.23	10.22	−0.01	10.23	10.22	−0.01
0.10	8.24	8.23	−0.02	8.24	8.23	−0.02
0.12	6.92	6.90	−0.02	6.92	6.90	−0.02
0.14	5.98	5.95	−0.02	5.98	5.95	−0.02
0.16	5.27	5.24	−0.03	5.27	5.24	−0.03
0.18	4.72	4.69	−0.03	4.72	4.69	−0.03

TABLE 8.17 (continued)

	Objective			Projective		
k^2	f/R	z_m/R	z_p/R	f/R	z_m/R	z_p/R
0.20	4.28	4.24	−0.03	4.28	4.24	−0.03
0.22	3.92	3.88	−0.04	3.92	3.88	−0.04
0.24	3.62	3.58	−0.04	3.62	3.58	−0.04
0.26	3.36	3.32	−0.04	3.36	3.32	−0.04
0.28	3.15	3.10	−0.05	3.15	3.10	−0.05
0.30	2.96	2.91	−0.05	2.96	2.91	−0.05
0.32	2.80	2.74	−0.06	2.80	2.74	−0.06
0.34	2.65	2.59	−0.06	2.65	2.59	−0.06
0.36	2.52	2.46	−0.06	2.52	2.46	−0.06
0.38	2.41	2.34	−0.07	2.41	2.34	−0.07
0.40	2.30	2.23	−0.07	2.30	2.23	−0.07
0.42	2.21	2.14	−0.07	2.21	2.14	−0.07
0.44	2.13	2.05	−0.08	2.13	2.05	−0.08
0.46	2.05	1.97	−0.08	2.05	1.97	−0.08
0.48	1.98	1.89	−0.09	1.98	1.89	−0.09
0.50	1.92	1.82	−0.09	1.92	1.82	−0.09
0.52	1.86	1.76	−0.09	1.86	1.76	−0.09
0.54	1.80	1.70	−0.10	1.80	1.70	−0.10
0.56	1.75	1.65	−0.10	1.75	1.65	−0.10
0.58	1.70	1.60	−0.11	1.70	1.60	−0.11
0.60	1.66	1.55	−0.11	1.66	1.55	−0.11
0.62	1.62	1.50	−0.11	1.62	1.50	−0.11
0.64	1.58	1.46	−0.12	1.58	1.46	−0.12
0.66	1.54	1.42	−0.12	1.54	1.42	−0.12
0.68	1.51	1.38	−0.13	1.51	1.38	−0.13
0.70	1.48	1.35	−0.13	1.48	1.35	−0.13
0.72	1.45	1.31	−0.13	1.45	1.31	−0.14
0.74	1.42	1.28	−0.14	1.42	1.28	−0.14
0.76	1.39	1.25	−0.14	1.39	1.25	−0.15
0.78	1.37	1.22	−0.15	1.37	1.22	−0.15
0.80	1.34	1.19	−0.15	1.35	1.19	−0.15
0.82	1.32	1.17	−0.15	1.32	1.17	−0.16
0.84	1.30	1.14	−0.16	1.30	1.14	−0.16
0.86	1.28	1.12	−0.16	1.28	1.11	−0.17
0.88	1.26	1.09	−0.17	1.26	1.09	−0.17
0.90	1.24	1.07	−0.17	1.25	1.07	−0.18
0.92	1.22	1.05	−0.17	1.23	1.05	−0.18
0.94	1.20	1.03	−0.18	1.21	1.03	−0.19
0.96	1.19	1.01	−0.18	1.20	1.01	−0.19
0.98	1.17	0.99	−0.18	1.18	0.99	−0.20
1.00	1.16	0.97	−0.19	1.17	0.97	−0.20
1.02	1.14	0.95	−0.19	1.15	0.95	−0.21
1.04	1.13	0.93	−0.19	1.14	0.93	−0.21

TABLE 8.17 (continued)

	Objective			Projective		
k^2	f/R	z_m/R	z_p/R	f/R	z_m/R	z_p/R
1.06	1.12	0.92	−0.20	1.13	0.91	−0.22
1.08	1.10	0.90	−0.20	1.12	0.90	−0.22
1.10	1.09	0.89	−0.21	1.11	0.88	−0.23
1.12	1.08	0.87	−0.21	1.10	0.86	−0.23
1.14	1.07	0.86	−0.21	1.09	0.85	−0.24
1.16	1.06	0.84	−0.22	1.08	0.83	−0.24
1.18	1.05	0.83	−0.22	1.07	0.82	−0.25
1.20	1.04	0.81	−0.22	1.06	0.81	−0.25
1.22	1.03	0.80	−0.23	1.05	0.79	−0.26
1.24	1.02	0.79	−0.23	1.04	0.78	−0.26
1.26	1.01	0.78	−0.23	1.03	0.76	−0.27
1.28	1.00	0.76	−0.24	1.03	0.75	−0.27
1.30	0.99	0.75	−0.24	1.02	0.74	−0.28
1.32	0.98	0.74	−0.24	1.01	0.73	−0.29
1.34	0.97	0.73	−0.25	1.01	0.71	−0.29
1.36	0.97	0.72	−0.25	1.00	0.70	−0.30
1.38	0.96	0.71	−0.25	0.99	0.69	−0.30
1.40	0.95	0.70	−0.25	0.99	0.68	−0.31
1.42	0.94	0.69	−0.26	0.98	0.67	−0.31
1.44	0.94	0.68	−0.26	0.98	0.66	−0.32
1.46	0.93	0.67	−0.26	0.97	0.65	−0.33
1.48	0.92	0.66	−0.27	0.97	0.63	−0.33
1.50	0.92	0.65	−0.27	0.96	0.62	−0.34
1.52	0.91	0.64	−0.27	0.96	0.61	−0.35
1.54	0.91	0.63	−0.28	0.96	0.60	−0.35
1.56	0.90	0.62	−0.28	0.95	0.59	−0.36
1.58	0.89	0.61	−0.28	0.95	0.58	−0.36
1.60	0.89	0.60	−0.28	0.94	0.57	−0.37
1.62	0.88	0.60	−0.29	0.94	0.56	−0.38
1.64	0.88	0.59	−0.29	0.94	0.55	−0.38
1.66	0.87	0.58	−0.29	0.94	0.54	−0.39
1.68	0.87	0.57	−0.29	0.93	0.54	−0.40
1.70	0.86	0.56	−0.30	0.93	0.53	−0.40
1.72	0.86	0.56	−0.30	0.93	0.52	−0.41
1.74	0.85	0.55	−0.30	0.93	0.51	−0.42
1.76	0.85	0.54	−0.31	0.92	0.50	−0.43
1.78	0.84	0.54	−0.31	0.92	0.49	−0.43
1.80	0.84	0.53	−0.31	0.92	0.48	−0.44
1.82	0.83	0.52	−0.31	0.92	0.47	−0.45
1.84	0.83	0.51	−0.32	0.92	0.46	−0.45
1.86	0.83	0.51	−0.32	0.92	0.45	−0.46
1.88	0.82	0.50	−0.32	0.91	0.45	−0.47
1.90	0.82	0.49	−0.32	0.91	0.44	−0.48

TABLE 8.17 (continued)

	Objective			Projective		
k^2	f/R	z_m/R	z_p/R	f/R	z_m/R	z_p/R
1.92	0.81	0.49	−0.33	0.91	0.43	−0.48
1.94	0.81	0.48	−0.33	0.91	0.42	−0.49
1.96	0.81	0.48	−0.33	0.91	0.41	−0.50
1.98	0.80	0.47	−0.33	0.91	0.40	−0.51
2.00	0.80	0.46	−0.33	0.91	0.39	−0.52
2.02	0.80	0.46	−0.34	0.91	0.39	−0.52
2.04	0.79	0.45	−0.34	0.91	0.38	−0.53
2.06	0.79	0.45	−0.34	0.91	0.37	−0.54
2.08	0.79	0.44	−0.34	0.91	0.36	−0.55
2.10	0.78	0.44	−0.35	0.91	0.35	−0.56
2.12	0.78	0.43	−0.35	0.91	0.34	−0.57
2.14	0.78	0.43	−0.35	0.91	0.34	−0.58
2.16	0.77	0.42	−0.35	0.91	0.33	−0.59
2.18	0.77	0.41	−0.35	0.91	0.32	−0.59
2.20	0.77	0.41	−0.36	0.91	0.31	−0.60
2.22	0.76	0.40	−0.36	0.92	0.30	−0.61
2.24	0.76	0.40	−0.36	0.92	0.29	−0.62
2.26	0.76	0.39	−0.36	0.92	0.29	−0.63
2.28	0.76	0.39	−0.37	0.92	0.28	−0.64
2.30	0.75	0.38	−0.37	0.92	0.27	−0.65
2.32	0.75	0.38	−0.37	0.92	0.26	−0.66
2.34	0.75	0.38	−0.37	0.92	0.25	−0.67
2.36	0.74	0.37	−0.37	0.93	0.24	−0.68
2.38	0.74	0.37	−0.38	0.93	0.24	−0.69
2.40	0.74	0.36	−0.38	0.93	0.23	−0.70
2.42	0.74	0.36	−0.38	0.93	0.22	−0.71
2.44	0.73	0.35	−0.38	0.93	0.21	−0.72
2.46	0.73	0.35	−0.38	0.94	0.20	−0.73
2.48	0.73	0.34	−0.39	0.94	0.19	−0.74
2.50	0.73	0.34	−0.39	0.94	0.19	−0.76
3.00	0.69	0.25	−0.43	1.05	−0.05	−1.10
3.50	0.64	0.18	−0.47	1.26	−0.36	−1.62
4.00	0.62	0.12	−0.50	1.68	−0.86	−2.54
4.50	0.59	0.07	0.53	2.67	−1.91	−4.59
5.00	0.58	0.02	−0.55	6.93	−6.23	−13.17
5.50	0.56	−0.01	−0.57	−11.58	12.22	23.81
6.00	0.55	−0.05	−0.59	−3.23	3.82	7.05
6.50	0.54	−0.08	−0.61	−1.93	2.46	4.39
7.00	0.53	−0.10	−0.63	−1.41	1.89	3.30
7.50	0.52	−0.13	−0.65	−1.14	1.56	2.70
8.00	0.51	−0.15	−0.66	−0.98	1.35	2.32
8.50	0.50	−0.17	−0.67	−0.88	1.19	2.06
9.00	0.50	−0.19	−0.69	−0.81	1.06	1.87

TABLE 8.17 (continued)

	Objective			Projective		
k^2	f/R	z_m/R	z_p/R	f/R	z_m/R	z_p/R
9.50	0.49	−0.21	−0.70	−0.77	0.95	1.72
10.00	0.49	−0.23	−0.71	−0.75	0.85	1.60
15.00	0.45	−0.35	−0.80	−1.72	−0.72	1.00
20.00	0.43	−0.44	−0.87	1.22	1.88	0.67
30.00	0.41	−0.55	−0.96	1.05	0.19	−0.87
40.00	0.40	−0.62	−1.02	−0.97	1.71	2.67
50.00	0.39	−0.68	−1.07	−0.81	0.64	1.45
60.00	0.38	−0.72	−1.10	3.14	4.17	1.04
70.00	0.38	−0.76	−1.14	0.70	1.30	0.60
80.00	0.37	−0.79	−1.17	0.99	0.45	−0.54
90.00	0.37	−0.82	−1.19	−2.95	4.07	7.02
100.00	0.37	−0.85	−1.21	−0.73	1.51	2.24

TABLE 8.18

RATIO OF GAP WIDTH TO BORE DIAMETER $S/D = 0.7$

	Objective			Projective		
k^2	f/R	z_m/R	z_p/R	f/R	z_m/R	z_p/R
0.01	74.38	74.38	−0.00	74.38	74.38	−0.00
0.02	37.34	37.34	−0.00	37.34	37.34	−0.00
0.04	18.83	18.82	−0.01	18.83	18.82	−0.01
0.06	12.65	12.64	−0.01	12.65	12.64	−0.01
0.08	9.57	9.55	−0.02	9.57	9.55	−0.02
0.10	7.72	7.70	−0.02	7.72	7.70	−0.02
0.12	6.49	6.46	−0.02	6.49	6.46	−0.02
0.14	5.61	5.58	−0.03	5.61	5.58	−0.03
0.16	4.95	4.91	−0.03	4.95	4.91	−0.03
0.18	4.43	4.40	−0.04	4.43	4.40	−0.04
0.20	4.02	3.98	−0.04	4.02	3.98	−0.04
0.22	3.69	3.64	−0.05	3.69	3.64	−0.05
0.24	3.41	3.36	−0.05	3.41	3.36	−0.05
0.26	3.18	3.12	−0.05	3.18	3.12	−0.05
0.28	2.97	2.92	−0.06	2.97	2.92	−0.06
0.30	2.80	2.74	−0.06	2.80	2.74	−0.06
0.32	2.65	2.58	−0.07	2.65	2.58	−0.07
0.34	2.51	2.44	−0.07	2.51	2.44	−0.07
0.36	2.39	2.32	−0.08	2.39	2.32	−0.08
0.38	2.29	2.21	−0.08	2.29	2.21	−0.08
0.40	2.19	2.11	−0.09	2.19	2.11	−0.09

TABLE 8.18 (continued)

	Objective			Projective		
k^2	f/R	z_m/R	z_p/R	f/R	z_m/R	z_p/R
0.42	2.11	2.01	−0.09	2.11	2.01	−0.09
0.44	2.03	1.93	−0.09	2.03	1.93	−0.10
0.46	1.96	1.86	−0.10	1.96	1.86	−0.10
0.48	1.89	1.79	−0.10	1.89	1.79	−0.10
0.50	1.83	1.72	−0.11	1.83	1.72	−0.11
0.52	1.78	1.66	−0.11	1.78	1.66	−0.11
0.54	1.73	1.61	−0.12	1.73	1.61	−0.12
0.56	1.68	1.55	−0.12	1.68	1.55	−0.12
0.58	1.63	1.51	−0.13	1.64	1.51	−0.13
0.60	1.59	1.46	−0.13	1.60	1.46	−0.13
0.62	1.56	1.42	−0.14	1.56	1.42	−0.14
0.64	1.52	1.38	−0.14	1.52	1.38	−0.14
0.66	1.49	1.34	−0.15	1.49	1.34	−0.15
0.68	1.46	1.30	−0.15	1.46	1.30	−0.16
0.70	1.43	1.27	−0.16	1.43	1.27	−0.16
0.72	1.40	1.24	−0.16	1.40	1.24	−0.17
0.74	1.37	1.21	−0.17	1.38	1.21	−0.17
0.76	1.35	1.18	−0.17	1.35	1.18	−0.18
0.78	1.33	1.15	−0.18	1.33	1.15	−0.18
0.80	1.30	1.12	−0.18	1.31	1.12	−0.19
0.82	1.28	1.10	−0.18	1.29	1.10	−0.19
0.84	1.26	1.07	−0.19	1.27	1.07	−0.20
0.86	1.24	1.05	−0.19	1.25	1.05	−0.21
0.88	1.23	1.03	−0.20	1.24	1.02	−0.21
0.90	1.21	1.01	−0.20	1.22	1.00	−0.22
0.92	1.19	0.99	−0.21	1.20	0.98	−0.22
0.94	1.18	0.97	−0.21	1.19	0.96	−0.23
0.96	1.16	0.95	−0.22	1.18	0.94	−0.23
0.98	1.15	0.93	−0.22	1.16	0.92	−0.24
1.00	1.13	0.91	−0.22	1.15	0.90	−0.25
1.02	1.12	0.89	−0.23	1.14	0.88	−0.25
1.04	1.11	0.88	−0.23	1.13	0.87	−0.26
1.06	1.10	0.86	−0.24	1.12	0.85	−0.27
1.08	1.08	0.84	−0.24	1.11	0.83	−0.27
1.10	1.07	0.83	−0.24	1.10	0.82	−0.28
1.12	1.06	0.81	−0.25	1.09	0.80	−0.29
1.14	1.05	0.80	−0.25	1.08	0.79	−0.29
1.16	1.04	0.78	−0.26	1.07	0.77	−0.30
1.18	1.03	0.77	−0.26	1.06	0.76	−0.31
1.20	1.02	0.76	−0.26	1.06	0.74	−0.31
1.22	1.01	0.75	−0.27	1.05	0.73	−0.32
1.24	1.00	0.73	−0.27	1.04	0.72	−0.33
1.26	1.00	0.72	−0.28	1.04	0.70	−0.33

TABLE 8.18 (continued)

	Objective			Projective		
k^2	f/R	z_m/R	z_p/R	f/R	z_m/R	z_p/R
1.28	0.99	0.71	−0.28	1.03	0.69	−0.34
1.30	0.98	0.70	−0.28	1.02	0.68	−0.35
1.32	0.97	0.69	−0.29	1.02	0.66	−0.36
1.34	0.96	0.67	−0.29	1.01	0.65	−0.36
1.36	0.96	0.66	−0.29	1.01	0.64	−0.37
1.38	0.95	0.65	−0.30	1.00	0.63	−0.38
1.40	0.94	0.64	−0.30	1.00	0.61	−0.39
1.42	0.94	0.63	−0.30	1.00	0.60	−0.39
1.44	0.93	0.62	−0.31	0.99	0.59	−0.40
1.46	0.92	0.61	−0.31	0.99	0.58	−0.41
1.48	0.92	0.60	−0.31	0.99	0.57	−0.42
1.50	0.91	0.60	−0.32	0.98	0.56	−0.43
1.52	0.91	0.59	−0.32	0.98	0.55	−0.44
1.54	0.90	0.58	−0.32	0.98	0.53	−0.44
1.56	0.89	0.57	−0.33	0.98	0.52	−0.45
1.58	0.89	0.56	−0.33	0.97	0.51	−0.46
1.60	0.88	0.55	−0.33	0.97	0.50	−0.47
1.62	0.88	0.54	−0.34	0.97	0.49	−0.48
1.64	0.87	0.54	−0.34	0.97	0.48	−0.49
1.66	0.87	0.53	−0.34	0.97	0.47	−0.50
1.68	0.86	0.52	−0.34	0.97	0.46	−0.51
1.70	0.86	0.51	−0.35	0.97	0.45	−0.52
1.72	0.86	0.51	−0.35	0.96	0.44	−0.52
1.74	0.85	0.50	−0.35	0.96	0.43	−0.53
1.76	0.85	0.49	−0.36	0.96	0.42	−0.54
1.78	0.84	0.48	−0.36	0.96	0.41	−0.55
1.80	0.84	0.48	−0.36	0.96	0.40	−0.56
1.82	0.83	0.47	−0.36	0.96	0.39	−0.57
1.84	0.83	0.46	−0.37	0.96	0.38	−0.58
1.86	0.83	0.46	−0.37	0.96	0.37	−0.59
1.88	0.82	0.45	−0.37	0.97	0.36	−0.61
1.90	0.82	0.44	−0.37	0.97	0.35	−0.62
1.92	0.81	0.44	−0.38	0.97	0.34	−0.63
1.94	0.81	0.43	−0.38	0.97	0.33	−0.64
1.96	0.81	0.43	−0.38	0.97	0.32	−0.65
1.98	0.80	0.42	−0.38	0.97	0.31	−0.66
2.00	0.80	0.41	−0.39	0.97	0.30	−0.67
2.02	0.80	0.41	−0.39	0.97	0.29	−0.68
2.04	0.79	0.40	−0.39	0.98	0.28	−0.70
2.06	0.79	0.40	−0.39	0.98	0.27	−0.71
2.08	0.79	0.39	−0.40	0.98	0.26	−0.72
2.10	0.78	0.39	−0.40	0.98	0.25	−0.73
2.12	0.78	0.38	−0.40	0.99	0.24	−0.74

TABLE 8.18 (continued)

	Objective			Projective		
k^2	f/R	z_m/R	z_p/R	f/R	z_m/R	z_p/R
2.14	0.78	0.37	−0.40	0.99	0.23	−0.76
2.16	0.78	0.37	−0.41	0.99	0.22	−0.77
2.18	0.77	0.36	−0.41	0.99	0.21	−0.78
2.20	0.77	0.36	−0.41	1.00	0.20	−0.80
2.22	0.77	0.35	−0.41	1.00	0.19	−0.81
2.24	0.76	0.35	−0.42	1.00	0.18	−0.82
2.26	0.76	0.34	−0.42	1.01	0.17	−0.84
2.28	0.76	0.34	−0.42	1.01	0.16	−0.85
2.30	0.76	0.33	−0.42	1.02	0.15	−0.87
2.32	0.75	0.33	−0.42	1.02	0.14	−0.88
2.34	0.75	0.33	−0.43	1.02	0.13	−0.90
2.36	0.75	0.32	−0.43	1.03	0.12	−0.91
2.38	0.75	0.32	−0.43	1.03	0.10	−0.93
2.40	0.74	0.31	−0.43	1.04	0.09	−0.94
2.42	0.74	0.31	−0.43	1.04	0.08	−0.96
2.44	0.74	0.30	−0.44	1.05	0.07	−0.98
2.46	0.74	0.30	−0.44	1.05	0.06	−0.99
2.48	0.73	0.29	−0.44	1.06	0.05	−1.01
2.50	0.73	0.29	−0.44	1.07	0.04	−1.03
3.00	0.69	0.20	−0.49	1.29	−0.31	−1.60
3.50	0.65	0.12	−0.53	1.77	−0.88	−2.65
4.00	0.62	0.06	−0.56	3.06	−2.25	−5.31
4.50	0.60	0.01	−0.59	13.07	−12.34	−25.41
5.00	0.58	−0.03	−0.61	−5.78	6.45	12.22
5.50	0.57	−0.07	−0.64	−2.43	3.04	5.47
6.00	0.56	−0.10	−0.66	−1.59	2.13	3.72
6.50	0.54	−0.13	−0.68	−1.22	1.70	2.91
7.00	0.53	−0.16	−0.69	−1.01	1.43	2.44
7.50	0.53	−0.18	−0.71	−0.89	1.24	2.14
8.00	0.52	−0.21	−0.72	−0.82	1.10	1.92
8.50	0.51	−0.23	−0.74	−0.78	0.97	1.75
9.00	0.50	−0.25	−0.75	−0.76	0.86	1.62
9.50	0.50	−0.27	−0.76	−0.76	0.76	1.52
10.00	0.49	−0.28	−0.78	−0.78	0.65	1.43
15.00	0.46	0.41	−0.87	−128.51	−127.59	0.92
20.00	0.44	−0.50	−0.93	0.80	1.32	0.52
30.00	0.41	−0.61	−1.02	5.63	−4.54	−10.17
40.00	0.40	−0.69	−1.09	−0.70	1.16	1.86
50.00	0.39	−0.74	−1.13	−3.01	−1.78	1.23
60.00	0.38	−0.79	−1.17	0.78	1.56	0.78
70.00	0.38	−0.83	−1.21	0.88	0.66	−0.21
80.00	0.38	−0.86	−1.23	−3.16	4.33	7.49
90.00	0.37	−0.89	−1.26	−0.71	1.48	2.19
100.00	0.37	−0.91	−1.28	−0.86	0.74	1.61

TABLE 8.19
RATIO OF GAP WIDTH TO BORE DIAMETER $S/D = 0.8$

	Objective			Projective		
k^2	f/R	z_m/R	z_p/R	f/R	z_m/R	z_p/R
0.01	69.12	69.12	−0.00	69.12	69.12	−0.00
0.02	34.72	34.72	−0.00	34.72	34.72	−0.00
0.04	17.52	17.52	−0.01	17.52	17.52	−0.01
0.06	11.79	11.78	−0.01	11.79	11.78	−0.01
0.08	8.93	8.91	−0.02	8.93	8.91	−0.02
0.10	7.21	7.19	−0.02	7.21	7.19	−0.02
0.12	6.07	6.04	−0.03	6.07	6.04	−0.03
0.14	5.25	5.22	−0.03	5.25	5.22	−0.03
0.16	4.64	4.60	−0.04	4.64	4.60	−0.04
0.18	4.16	4.12	−0.4	4.16	4.12	−0.04
0.20	3.78	3.73	−0.05	3.78	3.73	−0.05
0.22	3.47	3.42	−0.05	3.47	3.42	−0.05
0.24	3.21	3.15	−0.06	3.21	3.15	−0.06
0.26	3.00	2.93	−0.07	3.00	2.93	−0.07
0.28	2.81	2.74	−0.07	2.81	2.74	−0.07
0.30	2.65	2.57	−0.08	2.65	2.57	−0.08
0.32	2.51	2.42	−0.08	2.51	2.42	−0.08
0.34	2.38	2.29	−0.09	2.38	2.29	−0.09
0.36	2.27	2.18	−0.09	2.27	2.18	−0.09
0.38	2.17	2.08	−0.10	2.17	2.08	−0.10
0.40	2.09	1.98	−0.10	2.09	1.98	−0.10
0.42	2.01	1.90	−0.11	2.01	1.90	−0.11
0.44	1.93	1.82	−0.12	1.93	1.82	−0.12
0.46	1.87	1.75	−0.12	1.87	1.75	−0.12
0.48	1.81	1.68	−0.13	1.81	1.68	−0.13
0.50	1.75	1.62	−0.13	1.75	1.62	−0.13
0.52	1.70	1.56	−0.14	1.70	1.56	−0.14
0.54	1.66	1.51	−0.14	1.66	1.51	−0.15
0.56	1.61	1.46	−0.15	1.62	1.46	−0.15
0.58	1.57	1.42	−0.16	1.58	1.42	−0.16
0.60	1.54	1.37	−0.16	1.54	1.37	−0.17
0.62	1.50	1.33	−0.17	1.50	1.33	−0.17
0.64	1.47	1.30	−0.17	1.47	1.29	−0.18
0.66	1.44	1.26	−0.18	1.44	1.26	−0.18
0.68	1.41	1.23	−0.19	1.42	1.22	−0.19
0.70	1.38	1.19	−0.19	1.39	1.19	−0.20
0.72	1.36	1.16	−0.20	1.37	1.16	−0.21
0.74	1.33	1.13	−0.20	1.34	1.13	−0.21
0.76	1.31	1.10	−0.21	1.32	1.10	−0.22
0.78	1.29	1.08	−0.21	1.30	1.07	−0.23
0.80	1.27	1.05	−0.22	1.28	1.05	−0.23
0.82	1.25	1.03	−0.22	1.26	1.02	−0.24

TABLE 8.19 (continued)

	Objective			Projective		
k^2	f/R	z_m/R	z_p/R	f/R	z_m/R	z_p/R
0.84	1.23	1.00	−0.23	1.25	1.00	−0.25
0.86	1.22	0.98	−0.23	1.23	0.98	−0.26
0.88	1.20	0.96	−0.24	1.22	0.95	−0.26
0.90	1.18	0.94	−0.24	1.20	0.93	−0.27
0.92	1.17	0.92	−0.25	1.19	0.91	−0.28
0.94	1.15	0.90	−0.25	1.18	0.89	−0.29
0.96	1.14	0.88	−0.26	1.17	0.87	−0.29
0.98	1.13	0.86	−0.26	1.15	0.85	−0.30
1.00	1.12	0.85	−0.27	1.14	0.83	−0.31
1.02	1.10	0.83	−0.27	1.13	0.82	−0.32
1.04	1.09	0.81	−0.28	1.12	0.80	−0.33
1.06	1.08	0.80	−0.28	1.12	0.78	−0.33
1.08	1.07	0.78	−0.29	1.11	0.76	−0.34
1.10	1.06	0.77	−0.29	1.10	0.75	−0.35
1.12	1.05	0.75	−0.30	1.09	0.73	−0.36
1.14	1.04	0.74	−0.30	1.09	0.72	−0.37
1.16	1.03	0.72	−0.31	1.08	0.70	−0.38
1.18	1.02	0.71	−0.31	1.07	0.69	−0.39
1.20	1.01	0.70	−0.31	1.07	0.67	−0.40
1.22	1.00	0.69	−0.32	1.06	0.66	−0.41
1.24	1.00	0.67	−0.32	1.06	0.64	−0.42
1.26	0.99	0.66	−0.33	1.05	0.63	−0.43
1.28	0.98	0.65	−0.33	1.05	0.61	−0.43
1.30	0.97	0.64	−0.33	1.05	0.60	−0.44
1.32	0.97	0.63	−0.34	1.04	0.59	−0.45
1.34	0.96	0.62	−0.34	1.04	0.57	−0.46
1.36	0.95	0.61	−0.35	1.04	0.56	−0.48
1.38	0.95	0.60	−0.35	1.03	0.55	−0.49
1.40	0.94	0.59	−0.35	1.03	0.54	−0.50
1.42	0.93	0.58	−0.36	1.03	0.52	−0.51
1.44	0.93	0.57	−0.36	1.03	0.51	−0.52
1.46	0.92	0.56	−0.37	1.03	0.50	−0.53
1.48	0.92	0.55	−0.37	1.03	0.48	−0.54
1.50	0.91	0.54	−0.37	1.02	0.47	−0.55
1.52	0.90	0.53	−0.38	1.02	0.46	−0.56
1.54	0.90	0.52	−0.38	1.02	0.45	−0.58
1.56	0.89	0.51	−0.38	1.02	0.44	−0.59
1.58	0.89	0.50	−0.39	1.02	0.42	−0.60
1.60	0.88	0.50	−0.39	1.02	0.41	−0.61
1.62	0.88	0.49	−0.39	1.02	0.40	−0.62
1.64	0.87	0.48	−0.40	1.02	0.39	−0.64
1.66	0.87	0.47	−0.40	1.03	0.38	−0.65
1.68	0.87	0.46	−0.40	1.03	0.36	−0.66

TABLE 8.19 (continued)

	Objective			Projective		
k^2	f/R	z_m/R	z_p/R	f/R	z_m/R	z_p/R
1.70	0.86	0.46	−0.40	1.03	0.35	−0.68
1.72	0.86	0.45	−0.41	1.03	0.34	−0.69
1.74	0.85	0.44	−0.41	1.03	0.33	−0.70
1.76	0.85	0.43	−0.41	1.03	0.32	−0.72
1.78	0.84	0.43	−0.42	1.04	0.30	−0.73
1.80	0.84	0.42	−0.42	1.04	0.29	−0.75
1.82	0.84	0.41	−0.42	1.04	0.28	−0.76
1.84	0.83	0.41	−0.43	1.04	0.27	−0.78
1.86	0.83	0.40	−0.43	1.05	0.25	−0.79
1.88	0.83	0.39	−0.43	1.05	0.24	−0.81
1.90	0.82	0.39	−0.43	1.05	0.23	−0.82
1.92	0.82	0.38	−0.44	1.06	0.22	−0.84
1.94	0.81	0.38	−0.44	1.06	0.20	−0.86
1.96	0.81	0.37	−0.44	1.07	0.19	−0.87
1.98	0.81	0.36	−0.45	1.07	0.18	−0.89
2.00	0.80	0.36	−0.45	1.08	0.17	−0.91
2.02	0.80	0.35	−0.45	1.08	0.15	−0.93
2.04	0.80	0.35	−0.45	1.09	0.14	−0.94
2.06	0.80	0.34	−0.46	1.09	0.13	−0.96
2.08	0.79	0.33	−0.46	1.10	0.11	−0.98
2.10	0.79	0.33	−0.46	1.10	0.10	−1.00
2.12	0.79	0.32	−0.46	1.11	0.09	−1.02
2.14	0.78	0.32	−0.47	1.12	0.07	−1.04
2.16	0.78	0.31	−0.47	1.12	0.06	−1.06
2.18	0.78	0.31	−0.47	1.13	0.05	−1.08
2.20	0.78	0.30	−0.47	1.14	0.03	−1.10
2.22	0.77	0.30	−0.48	1.14	0.02	−1.13
2.24	0.77	0.29	−0.48	1.15	0.00	−1.15
2.26	0.77	0.29	−0.48	1.16	−0.01	−1.17
2.28	0.76	0.28	−0.48	1.17	−0.03	−1.19
2.30	0.76	0.28	−0.48	1.18	−0.04	−1.22
2.32	0.76	0.27	−0.49	1.19	−0.06	−1.24
2.34	0.76	0.27	−0.49	1.20	−0.07	−1.27
2.36	0.75	0.26	−0.49	1.21	−0.09	−1.29
2.38	0.75	0.26	−0.49	1.22	−0.10	−1.32
2.40	0.75	0.25	−0.50	1.23	−0.12	−1.35
2.42	0.75	0.25	−0.50	1.24	−0.14	−1.37
2.44	0.75	0.25	−0.50	1.25	−0.15	−1.40
2.46	0.74	0.24	−0.50	1.26	−0.17	−1.43
2.48	0.74	0.24	−0.50	1.27	−0.19	−1.46
2.50	0.74	0.23	−0.51	1.28	−0.21	−1.49
3.00	0.69	0.14	−0.55	1.79	−0.83	−2.62
3.50	0.66	0.07	−0.59	3.33	−2.47	−5.79

TABLE 8.19 (continued)

k^2	Objective f/R	z_m/R	z_p/R	Projective f/R	z_m/R	z_p/R
4.00	0.63	0.01	−0.63	37.53	−36.76	−74.29
4.50	0.61	−0.05	−0.66	−4.09	4.79	8.88
5.00	0.59	−0.09	−0.68	−2.01	2.63	4.64
5.50	0.58	−0.13	−0.71	−1.38	1.93	3.31
6.00	0.56	−0.16	−0.73	−1.09	1.57	2.66
6.50	0.55	−0.19	−0.75	−0.93	1.33	2.27
7.00	0.54	−0.22	−0.76	−0.84	1.16	2.00
7.50	0.53	−0.25	−0.78	−0.80	1.02	1.81
8.00	0.53	−0.27	−0.80	−0.78	0.89	1.67
8.50	0.52	−0.29	−0.81	−0.78	0.77	1.55
9.00	0.51	−0.31	−0.82	−0.81	0.65	1.46
9.50	0.51	−0.33	−0.84	−0.86	0.51	1.37
10.00	0.50	−0.35	−0.85	−0.95	0.36	1.30
15.00	0.46	−0.48	−0.94	1.63	2.45	0.82
20.00	0.44	−0.56	−1.01	0.72	0.98	0.26
30.00	0.42	−0.68	−1.10	−1.37	2.30	3.68
40.00	0.40	−0.76	−1.16	−0.90	0.62	1.51
50.00	0.39	−0.81	−1.21	1.18	2.18	1.00
60.00	0.39	−0.86	−1.25	0.75	0.95	0.20
70.00	0.38	−0.90	−1.28	−5.90	7.15	13.05
80.00	0.38	−0.93	−1.31	−0.71	1.51	2.22
90.00	0.38	−0.96	−1.33	−1.01	0.60	1.60
100.00	0.37	−0.98	−1.36	2.25	3.50	1.25

TABLE 8.20
RATIO OF GAP WIDTH TO BORE DIAMETER $S/D = 0.9$

k^2	Objective f/R	z_m/R	z_p/R	Projective f/R	z_m/R	z_p/R
0.01	64.12	64.12	−0.00	64.12	64.12	−0.00
0.02	32.23	32.23	−0.01	32.23	32.23	−0.01
0.04	16.29	16.28	−0.01	16.29	16.28	−0.01
0.06	10.98	10.96	−0.02	10.98	10.96	−0.02
0.08	8.32	8.30	−0.02	8.32	8.30	−0.02
0.10	6.73	6.70	−0.03	6.73	6.70	−0.03
0.12	5.67	5.64	−0.04	5.67	5.64	−0.04
0.14	4.91	4.87	−0.04	4.91	4.87	−0.04
0.16	4.35	4.30	−0.05	4.35	4.30	−0.05
0.18	3.91	3.85	−0.05	3.91	3.85	−0.05

TABLE 8.20 (continued)

	Objective			Projective		
k^2	f/R	z_m/R	z_p/R	f/R	z_m/R	z_p/R
0.20	3.56	3.50	−0.06	3.56	3.50	−0.06
0.22	3.27	3.20	−0.07	3.27	3.20	−0.07
0.24	3.03	2.96	−0.07	3.03	2.96	−0.07
0.26	2.83	2.75	−0.08	2.83	2.75	−0.08
0.28	2.66	2.57	−0.09	2.66	2.57	−0.09
0.30	2.51	2.41	−0.09	2.51	2.41	−0.09
0.32	2.38	2.28	−0.10	2.38	2.28	−0.10
0.34	2.26	2.16	−0.11	2.26	2.16	−0.11
0.36	2.16	2.05	−0.11	2.16	2.05	−0.11
0.38	2.07	1.95	−0.12	2.07	1.95	−0.12
0.40	1.99	1.86	−0.13	1.99	1.86	−0.13
0.42	1.92	1.78	−0.14	1.92	1.78	−0.14
0.44	1.85	1.71	−0.14	1.85	1.71	−0.14
0.46	1.79	1.64	−0.15	1.79	1.64	−0.15
0.48	1.74	1.58	−0.16	1.74	1.58	−0.16
0.50	1.69	1.52	−0.16	1.69	1.52	−0.17
0.52	1.64	1.47	−0.17	1.64	1.47	−0.17
0.54	1.60	1.42	−0.18	1.60	1.42	−0.18
0.56	1.56	1.37	−0.18	1.56	1.37	−0.19
0.58	1.52	1.33	−0.19	1.53	1.33	−0.20
0.60	1.49	1.29	−0.20	1.49	1.29	−0.21
0.62	1.46	1.25	−0.21	1.46	1.25	−0.21
0.64	1.43	1.21	−0.21	1.43	1.21	−0.22
0.66	1.40	1.18	−0.22	1.41	1.18	−0.23
0.68	1.37	1.15	−0.23	1.38	1.14	−0.24
0.70	1.35	1.12	−0.23	1.36	1.11	−0.25
0.72	1.33	1.09	−0.24	1.34	1.08	−0.26
0.74	1.30	1.06	−0.25	1.32	1.05	−0.27
0.76	1.28	1.03	−0.25	1.30	1.03	−0.27
0.78	1.26	1.01	−0.26	1.28	1.00	−0.28
0.80	1.25	0.98	−0.27	1.27	0.97	−0.29
0.82	1.23	0.96	−0.27	1.25	0.95	−0.30
0.84	1.21	0.93	−0.28	1.24	0.92	−0.31
0.86	1.20	0.91	−0.28	1.22	0.90	−0.32
0.88	1.18	0.89	−0.29	1.21	0.88	−0.33
0.90	1.17	0.87	−0.30	1.20	0.86	−0.34
0.92	1.15	0.85	−0.30	1.19	0.84	−0.35
0.94	1.14	0.83	−0.31	1.18	0.82	−0.36
0.96	1.13	0.81	−0.31	1.17	0.80	−0.37
0.98	1.11	0.80	−0.32	1.16	0.78	−0.38
1.00	1.10	0.78	−0.32	1.15	0.76	−0.39
1.02	1.09	0.76	−0.33	1.14	0.74	−0.40
1.04	1.08	0.75	−0.33	1.14	0.72	−0.42

TABLE 8.20 (continued)

	Objective			Projective		
k^2	f/R	z_m/R	z_p/R	f/R	z_m/R	z_p/R
1.06	1.07	0.73	−0.34	1.13	0.70	−0.43
1.08	1.06	0.72	−0.34	1.12	0.69	−0.44
1.10	1.05	0.70	−0.35	1.12	0.67	−0.45
1.12	1.04	0.69	−0.35	1.11	0.65	−0.46
1.14	1.03	0.67	−0.36	1.11	0.64	−0.47
1.16	1.02	0.66	−0.36	1.11	0.62	−0.49
1.18	1.02	0.65	−0.37	1.10	0.60	−0.50
1.20	1.01	0.63	−0.37	1.10	0.59	−0.51
1.22	1.00	0.62	−0.38	1.10	0.57	−0.52
1.24	0.99	0.61	−0.38	1.09	0.56	−0.54
1.26	0.99	0.60	−0.39	1.09	0.54	−0.55
1.28	0.98	0.59	−0.39	1.09	0.53	−0.56
1.30	0.97	0.58	−0.40	1.09	0.51	−0.58
1.32	0.96	0.56	−0.40	1.09	0.50	−0.59
1.34	0.96	0.55	−0.40	1.09	0.48	−0.61
1.36	0.95	0.54	−0.41	1.09	0.47	−0.62
1.38	0.95	0.53	−0.41	1.09	0.45	−0.64
1.40	0.94	0.52	−0.42	1.09	0.44	−0.65
1.42	0.93	0.51	−0.42	1.09	0.42	−0.67
1.44	0.93	0.50	−0.42	1.09	0.41	−0.68
1.46	0.92	0.49	−0.43	1.09	0.39	−0.70
1.48	0.92	0.49	−0.43	1.09	0.38	−0.72
1.50	0.91	0.48	−0.44	1.09	0.36	−0.73
1.52	0.91	0.47	−0.44	1.10	0.35	−0.75
1.54	0.90	0.46	−0.44	1.10	0.33	−0.77
1.56	0.90	0.45	−0.45	1.10	0.32	−0.78
1.58	0.89	0.44	−0.45	1.11	0.30	−0.80
1.60	0.89	0.43	−0.45	1.11	0.29	−0.82
1.62	0.88	0.43	−0.46	1.11	0.27	−0.84
1.64	0.88	0.42	−0.46	1.12	0.26	−0.86
1.66	0.87	0.41	−0.46	1.12	0.24	−0.88
1.68	0.87	0.40	−0.47	1.13	0.23	−0.90
1.70	0.86	0.39	−0.47	1.13	0.21	−0.92
1.72	0.86	0.39	−0.47	1.14	0.20	−0.94
1.74	0.86	0.38	−0.48	1.14	0.18	−0.96
1.76	0.85	0.37	−0.48	1.15	0.17	−0.98
1.78	0.85	0.37	−0.48	1.16	0.15	−1.00
1.80	0.84	0.36	−0.49	1.16	0.14	−1.03
1.82	0.84	0.35	−0.49	1.17	0.12	−1.05
1.84	0.84	0.35	−0.49	1.18	0.10	−1.07
1.86	0.83	0.34	−0.50	1.18	0.09	−1.10
1.88	0.83	0.33	−0.50	1.19	0.07	−1.12
1.90	0.83	0.33	−0.50	1.20	0.05	−1.15

TABLE 8.20 (continued)

	Objective			Projective		
k^2	f/R	z_m/R	z_p/R	f/R	z_m/R	z_p/R
1.92	0.82	0.32	−0.50	1.21	0.04	−1.18
1.94	0.82	0.31	−0.51	1.22	0.02	−1.20
1.96	0.82	0.31	−0.51	1.23	0.00	−1.23
1.98	0.81	0.30	−0.51	1.24	−0.02	−1.26
2.00	0.81	0.30	−0.52	1.25	−0.04	−1.29
2.02	0.81	0.29	−0.52	1.26	−0.06	−1.32
2.04	0.80	0.28	−0.52	1.27	−0.07	−1.35
2.06	0.80	0.28	−0.52	1.29	−0.09	−1.38
2.08	0.80	0.27	−0.53	1.30	−0.11	−1.41
2.10	0.80	0.27	−0.53	1.31	−0.13	−1.45
2.12	0.79	0.26	−0.53	1.33	−0.15	−1.48
2.14	0.79	0.26	−0.53	1.34	−0.18	−1.51
2.16	0.79	0.25	−0.54	1.35	−0.20	−1.55
2.18	0.78	0.25	−0.54	1.37	−0.22	−1.59
2.20	0.78	0.24	−0.54	1.39	−0.24	−1.63
2.22	0.78	0.24	−0.54	1.40	−0.26	−1.67
2.24	0.78	0.23	−0.55	1.42	−0.29	−1.71
2.26	0.77	0.23	−0.55	1.44	−0.31	−1.75
2.28	0.77	0.22	−0.55	1.46	−0.34	−1.79
2.30	0.77	0.22	−0.55	1.47	−0.36	−1.84
2.32	0.77	0.21	−0.56	1.50	−0.39	−1.88
2.34	0.76	0.21	−0.56	1.52	−0.42	−1.93
2.36	0.76	0.20	−0.56	1.54	−0.44	−1.98
2.38	0.76	0.20	−0.56	1.56	−0.47	−2.03
2.40	0.76	0.19	−0.57	1.58	−0.50	−2.09
2.42	0.76	0.19	−0.57	1.61	−0.53	−2.14
2.44	0.75	0.18	−0.57	1.63	−0.56	−2.20
2.46	0.75	0.18	−0.57	1.66	−0.60	−2.26
2.48	0.75	0.17	−0.57	1.69	−0.63	−2.32
2.50	0.75	0.17	−0.58	1.72	−0.66	−2.38
3.00	0.70	0.08	−0.63	3.27	−2.34	−5.61
3.50	0.67	0.00	−0.67	148.52	−147.69	−296.22
4.00	0.64	−0.06	−0.70	−3.46	4.20	7.66
4.50	0.62	−0.11	−0.73	−1.78	2.44	4.22
5.00	0.60	−0.16	−0.76	−1.25	1.82	3.08
5.50	0.59	−0.20	−0.78	−1.01	1.49	2.50
6.00	0.57	−0.23	−0.80	−0.89	1.27	2.16
6.50	0.56	−0.26	−0.82	−0.82	1.10	1.92
7.00	0.55	−0.29	−0.84	−0.80	0.95	1.75
7.50	0.54	−0.32	−0.86	−0.80	0.81	1.61
8.00	0.53	−0.34	−0.87	−0.84	0.66	1.50
8.50	0.53	−0.36	−0.89	−0.91	0.50	1.41
9.00	0.52	−0.38	−0.90	−1.03	0.30	1.33

TABLE 8.20 (continued)

k^2	Objective			Projective		
	f/R	z_m/R	z_p/R	f/R	z_m/R	z_p/R
9.50	0.51	−0.40	−0.91	−1.23	0.04	1.26
10.00	0.51	−0.42	−0.93	−1.57	−0.37	1.20
15.00	0.47	−0.55	−1.02	0.89	1.57	0.68
20.00	0.45	−0.64	−1.08	0.94	0.54	−0.40
30.00	0.42	−0.75	−1.18	−0.74	1.43	2.17
40.00	0.41	−0.83	−1.24	−12.01	−10.73	1.28
50.00	0.40	−0.89	−1.29	0.69	1.32	0.62
60.00	0.39	−0.94	−1.33	5.30	−3.92	−9.22
70.00	0.38	−0.97	−1.36	−0.73	1.62	2.36
80.00	0.38	−1.01	−1.39	−1.09	0.54	1.64
90.00	0.38	−1.04	−1.41	1.41	2.66	1.24
100.00	0.37	−1.06	−1.44	0.68	1.41	0.73

TABLE 8.21
RATIO OF GAP WIDTH TO BORE DIAMETER $S/D = 1.0$

k^2	Objective			Projective		
	f/R	z_m/R	z_p/R	f/R	z_m/R	z_p/R
0.01	59.46	59.46	−0.00	59.46	59.46	−0.00
0.02	29.91	29.91	−0.01	29.91	29.91	−0.01
0.04	15.14	15.13	−0.01	15.14	15.13	−0.01
0.06	10.22	10.20	−0.02	10.22	10.20	−0.02
0.08	7.76	7.73	−0.03	7.76	7.73	−0.03
0.10	6.29	6.25	−0.04	6.29	6.25	−0.04
0.12	5.31	5.26	−0.04	5.31	5.26	−0.04
0.14	4.61	4.55	−0.05	4.61	4.55	−0.05
0.16	4.08	4.02	−0.06	4.08	4.02	−0.06
0.18	3.67	3.61	−0.07	3.67	3.61	−0.07
0.20	3.35	3.28	−0.07	3.35	3.28	−0.07
0.22	3.08	3.00	−0.08	3.08	3.00	−0.08
0.24	2.86	2.77	−0.09	2.86	2.77	−0.09
0.26	2.68	2.58	−0.10	2.68	2.58	−0.10
0.28	2.52	2.41	−0.11	2.52	2.41	−0.11
0.30	2.38	2.27	−0.11	2.38	2.27	−0.11
0.32	2.26	2.14	−0.12	2.26	2.14	−0.12
0.34	2.16	2.03	−0.13	2.16	2.03	−0.13
0.36	2.06	1.92	−0.14	2.06	1.92	−0.14
0.38	1.98	1.83	−0.15	1.98	1.83	−0.15
0.40	1.91	1.75	−0.16	1.91	1.75	−0.16

TABLE 8.21 (continued)

	Objective			Projective		
k^2	f/R	z_m/R	z_p/R	f/R	z_m/R	z_p/R
0.42	1.84	1.67	−0.17	1.84	1.67	−0.17
0.44	1.78	1.61	−0.18	1.78	1.61	−0.18
0.46	1.73	1.54	−0.18	1.73	1.54	−0.19
0.48	1.68	1.48	−0.19	1.68	1.48	−0.20
0.50	1.63	1.43	−0.20	1.63	1.43	−0.21
0.52	1.59	1.38	−0.21	1.59	1.38	−0.22
0.54	1.55	1.33	−0.22	1.55	1.33	−0.23
0.56	1.51	1.29	−0.23	1.52	1.28	−0.24
0.58	1.48	1.24	−0.24	1.49	1.24	−0.25
0.60	1.45	1.21	−0.24	1.46	1.20	−0.26
0.62	1.42	1.17	−0.25	1.43	1.16	−0.27
0.64	1.39	1.13	−0.26	1.41	1.13	−0.28
0.66	1.37	1.10	−0.27	1.38	1.10	−0.29
0.68	1.35	1.07	−0.28	1.36	1.06	−0.30
0.70	1.32	1.04	−0.28	1.34	1.03	−0.31
0.72	1.30	1.01	−0.29	1.32	1.00	−0.32
0.74	1.28	0.98	−0.30	1.31	0.97	−0.33
0.76	1.26	0.96	−0.31	1.29	0.95	−0.35
0.78	1.25	0.93	−0.31	1.28	0.92	−0.36
0.80	1.23	0.91	−0.32	1.26	0.89	−0.37
0.82	1.21	0.88	−0.33	1.25	0.87	−0.38
0.84	1.20	0.86	−0.33	1.24	0.84	−0.40
0.86	1.18	0.84	−0.34	1.23	0.82	−0.41
0.88	1.17	0.82	−0.35	1.22	0.80	−0.42
0.90	1.16	0.80	−0.35	1.21	0.78	−0.44
0.92	1.14	0.78	−0.36	1.20	0.75	−0.45
0.94	1.13	0.76	−0.37	1.20	0.73	−0.46
0.96	1.12	0.74	−0.37	1.19	0.71	−0.48
0.98	1.11	0.73	−0.38	1.18	0.69	−0.49
1.00	1.10	0.71	−0.39	1.18	0.67	−0.51
1.02	1.09	0.69	−0.39	1.17	0.65	−0.52
1.04	1.08	0.68	−0.40	1.17	0.63	−0.54
1.06	1.07	0.66	−0.40	1.17	0.61	−0.55
1.08	1.06	0.65	−0.41	1.16	0.59	−0.57
1.10	1.05	0.63	−0.41	1.16	0.57	−0.59
1.12	1.04	0.62	−0.42	1.16	0.56	−0.60
1.14	1.03	0.61	−0.42	1.16	0.54	−0.62
1.16	1.02	0.59	−0.43	1.16	0.52	−0.64
1.18	1.01	0.58	−0.43	1.16	0.50	−0.66
1.20	1.01	0.57	−0.44	1.16	0.48	−0.67
1.22	1.00	0.56	−0.44	1.16	0.47	−0.69
1.24	0.99	0.54	−0.45	1.16	0.45	−0.71
1.26	0.99	0.53	−0.45	1.16	0.43	−0.73

TABLE 8.21 (continued)

k^2	Objective			Projective		
	f/R	z_m/R	z_p/R	f/R	z_m/R	z_p/R
1.28	0.98	0.52	−0.46	1.16	0.41	−0.75
1.30	0.97	0.51	−0.46	1.16	0.39	−0.77
1.32	0.97	0.50	−0.47	1.17	0.38	−0.79
1.34	0.96	0.49	−0.47	1.17	0.36	−0.81
1.36	0.95	0.48	−0.48	1.17	0.34	−0.83
1.38	0.95	0.47	−0.48	1.18	0.32	−0.86
1.40	0.94	0.46	−0.49	1.18	0.30	−0.88
1.42	0.94	0.45	−0.49	1.19	0.29	−0.90
1.44	0.93	0.44	−0.49	1.19	0.27	−0.92
1.46	0.93	0.43	−0.50	1.20	0.25	−0.95
1.48	0.92	0.42	−0.50	1.20	0.23	−0.97
1.50	0.92	0.41	−0.51	1.21	0.21	−1.00
1.52	0.91	0.40	−0.51	1.22	0.19	−1.03
1.54	0.91	0.39	−0.51	1.23	0.17	−1.05
1.56	0.90	0.38	−0.52	1.23	0.15	−1.08
1.58	0.90	0.37	−0.52	1.24	0.13	−1.11
1.60	0.89	0.37	−0.52	1.25	0.11	−1.14
1.62	0.89	0.36	−0.53	1.26	0.09	−1.17
1.64	0.88	0.35	−0.53	1.27	0.07	−1.20
1.66	0.88	0.34	−0.54	1.28	0.05	−1.23
1.68	0.87	0.34	−0.54	1.29	0.03	−1.26
1.70	0.87	0.33	−0.54	1.31	−0.01	−1.30
1.72	0.87	0.32	−0.55	1.32	−0.01	−1.33
1.74	0.86	0.31	−0.55	1.33	−0.04	−1.37
1.76	0.86	0.31	−0.55	1.35	−0.06	−1.40
1.78	0.85	0.30	−0.56	1.36	−0.08	−1.44
1.80	0.85	0.29	−0.56	1.38	−0.11	−1.48
1.82	0.85	0.28	−0.56	1.39	−0.13	−1.52
1.84	0.84	0.28	−0.57	1.41	−0.16	−1.56
1.86	0.84	0.27	−0.57	1.43	−0.18	−1.61
1.88	0.84	0.27	−0.57	1.44	−0.21	−1.65
1.90	0.83	0.26	−0.57	1.46	−0.23	−1.70
1.92	0.83	0.25	−0.58	1.48	−0.26	−1.74
1.94	0.83	0.25	−0.58	1.50	−0.29	−1.79
1.96	0.82	0.24	−0.58	1.53	−0.32	−1.85
1.98	0.82	0.23	−0.59	1.55	−0.35	−1.90
2.00	0.82	0.23	−0.59	1.57	−0.38	−1.95
2.02	0.81	0.22	−0.59	1.60	−0.41	−2.01
2.04	0.81	0.22	−0.60	1.62	−0.45	−2.07
2.06	0.81	0.21	−0.60	1.65	−0.48	−2.13
2.08	0.81	0.21	−0.60	1.68	−0.52	−2.20
2.10	0.80	0.20	−0.60	1.71	−0.56	−2.27
2.12	0.80	0.19	−0.61	1.74	−0.59	−2.34

TABLE 8.21 (continued)

	Objective			Projective		
k^2	f/R	z_m/R	z_p/R	f/R	z_m/R	z_p/R
2.14	0.80	0.19	−0.61	1.77	−0.63	−2.41
2.16	0.79	0.18	−0.61	1.81	−0.68	−2.49
2.18	0.79	0.18	−0.61	1.85	−0.72	−2.57
2.20	0.79	0.17	−0.62	1.88	−0.76	−2.65
2.22	0.79	0.17	−0.62	1.93	−0.81	−2.74
2.24	0.78	0.16	−0.62	1.97	−0.86	−2.83
2.26	0.78	0.16	−0.62	2.01	−0.91	−2.93
2.28	0.78	0.15	−0.63	2.06	−0.97	−3.03
2.30	0.78	0.15	−0.63	2.11	−1.02	−3.14
2.32	0.77	0.14	−0.63	2.17	−1.08	−3.25
2.34	0.77	0.14	−0.63	2.22	−1.15	−3.37
2.36	0.77	0.13	−0.64	2.28	−1.21	−3.50
2.38	0.77	0.13	−0.64	2.35	−1.28	−3.63
2.40	0.77	0.12	−0.64	2.42	−1.36	−3.78
2.42	0.76	0.12	−0.64	2.49	−1.44	−3.93
2.44	0.76	0.12	−0.65	2.57	−1.52	−4.09
2.46	0.76	0.11	−0.65	2.65	−1.61	−4.26
2.48	0.76	0.11	−0.65	2.74	−1.71	−4.45
2.50	0.75	0.10	−0.65	2.84	−1.81	−4.65
3.00	0.71	0.01	−0.70	36.31	−35.40	−71.70
3.50	0.68	−0.07	−0.74	−3.37	4.16	7.53
4.00	0.65	−0.13	−0.78	−1.69	2.38	4.07
4.50	0.63	−0.18	−0.81	−1.19	1.78	2.97
5.00	0.61	−0.23	−0.84	−0.97	1.46	2.43
5.50	0.59	−0.27	−0.86	−0.86	1.23	2.10
6.00	0.58	−0.30	−0.88	−0.82	1.05	1.87
6.50	0.57	−0.34	−0.90	−0.82	0.89	1.71
7.00	0.56	−0.36	−0.92	−0.86	0.72	1.58
7.50	0.55	−0.39	−0.94	−0.95	0.52	1.47
8.00	0.54	−0.41	−0.95	−1.10	0.28	1.38
8.50	0.53	−0.44	−0.97	−1.38	−0.07	1.31
9.00	0.53	−0.46	−0.98	−1.92	−0.68	1.24
9.50	0.52	−0.48	−1.00	−3.31	−2.14	1.17
10.00	0.51	−0.49	−1.01	−13.21	−12.10	1.11
15.00	0.47	−0.63	−1.10	0.74	1.19	0.44
20.00	0.45	−0.72	−1.17	2.91	−1.60	−4.51
30.00	0.42	−0.83	−1.26	−0.80	0.93	1.73
40.00	0.41	−0.91	−1.32	0.95	1.98	1.03
50.00	0.40	−0.97	−1.37	1.24	0.36	−0.89
60.00	0.39	−1.02	−1.41	−0.84	1.91	2.75
70.00	0.39	−1.06	−1.44	−1.01	0.72	1.72
80.00	0.38	−1.09	−1.47	1.24	2.52	1.28
90.00	0.38	−1.12	−1.50	0.71	1.32	0.61
100.00	0.38	−1.14	−1.52	5.68	−4.12	−9.80

TABLE 8.22
RATIO OF GAP WIDTH TO BORE DIAMETER $S/D = 1.1$

	Objective			Projective		
k^2	f/R	z_m/R	z_p/R	f/R	z_m/R	z_p/R
0.01	55.17	55.16	−0.00	55.17	55.16	−0.00
0.02	27.78	27.77	−0.01	27.78	27.77	−0.01
0.04	14.09	14.07	−0.02	14.09	14.07	−0.02
0.06	9.53	9.50	−0.03	9.53	9.50	−0.03
0.08	7.25	7.21	−0.03	7.25	7.21	−0.03
0.10	5.88	5.84	−0.04	5.88	5.84	−0.04
0.12	4.97	4.92	−0.05	4.97	4.92	−0.05
0.14	4.32	4.26	−0.06	4.32	4.26	−0.06
0.16	3.84	3.77	−0.07	3.84	3.77	−0.07
0.18	3.46	3.38	−0.08	3.46	3.38	−0.08
0.20	3.16	3.07	−0.09	3.16	3.07	−0.09
0.22	2.92	2.82	−0.10	2.92	2.82	−0.10
0.24	2.72	2.60	−0.11	2.72	2.60	−0.11
0.26	2.54	2.42	−0.12	2.54	2.42	−0.12
0.28	2.40	2.27	−0.13	2.40	2.27	−0.13
0.30	2.27	2.13	−0.14	2.27	2.13	−0.14
0.32	2.16	2.01	−0.15	2.16	2.01	−0.15
0.34	2.07	1.90	−0.16	2.07	1.90	−0.16
0.36	1.98	1.81	−0.17	1.98	1.81	−0.17
0.38	1.91	1.72	−0.18	1.91	1.72	−0.19
0.40	1.84	1.64	−0.19	1.84	1.64	−0.20
0.42	1.78	1.57	−0.21	1.78	1.57	−0.21
0.44	1.72	1.51	−0.22	1.73	1.51	−0.22
0.46	1.67	1.45	−0.23	1.68	1.45	−0.23
0.48	1.63	1.39	−0.24	1.63	1.39	−0.24
0.50	1.59	1.34	−0.25	1.59	1.34	−0.26
0.52	1.55	1.29	−0.26	1.56	1.29	−0.27
0.54	1.51	1.24	−0.27	1.52	1.24	−0.28
0.56	1.48	1.20	−0.28	1.49	1.20	−0.30
0.58	1.45	1.16	−0.29	1.47	1.16	−0.31
0.60	1.42	1.12	−0.30	1.44	1.12	−0.32
0.62	1.39	1.09	−0.31	1.42	1.08	−0.34
0.64	1.37	1.05	−0.32	1.40	1.04	−0.35
0.66	1.35	1.02	−0.33	1.38	1.01	−0.37
0.68	1.33	0.99	−0.33	1.36	0.98	−0.38
0.70	1.31	0.96	−0.34	1.34	0.95	−0.40
0.72	1.29	0.93	−0.35	1.33	0.92	−0.41
0.74	1.27	0.91	−0.36	1.31	0.89	−0.43
0.76	1.25	0.88	−0.37	1.30	0.86	−0.44
0.78	1.23	0.86	−0.38	1.29	0.83	−0.46
0.80	1.22	0.83	−0.39	1.28	0.81	−0.48
0.82	1.20	0.81	−0.39	1.27	0.78	−0.49

TABLE 8.22 (continued)

	Objective			Projective		
k^2	f/R	z_m/R	z_p/R	f/R	z_m/R	z_p/R
0.84	1.19	0.79	−0.40	1.26	0.75	−0.51
0.86	1.18	0.77	−0.41	1.26	0.73	−0.53
0.88	1.16	0.75	−0.42	1.25	0.71	−0.55
0.90	1.15	0.73	−0.42	1.25	0.68	−0.56
0.92	1.14	0.71	−0.43	1.24	0.66	−0.58
0.94	1.13	0.69	−0.44	1.24	0.63	−0.60
0.96	1.12	0.67	−0.44	1.23	0.61	−0.62
0.98	1.10	0.66	−0.45	1.23	0.59	−0.64
1.00	1.09	0.64	−0.46	1.23	0.57	−0.66
1.02	1.08	0.62	−0.46	1.23	0.54	−0.69
1.04	1.08	0.61	−0.47	1.23	0.52	−0.71
1.06	1.07	0.59	−0.47	1.23	0.50	−0.73
1.08	1.06	0.58	−0.48	1.23	0.48	−0.75
1.10	1.05	0.56	−0.49	1.23	0.46	−0.78
1.12	1.04	0.55	−0.49	1.24	0.44	−0.80
1.14	1.03	0.54	−0.50	1.24	0.41	−0.83
1.16	1.02	0.52	−0.50	1.24	0.39	−0.85
1.18	1.02	0.51	−0.51	1.25	0.37	−0.88
1.20	1.01	0.50	−0.51	1.25	0.35	−0.90
1.22	1.00	0.48	−0.52	1.26	0.33	−0.93
1.24	1.00	0.47	−0.52	1.26	0.30	−0.96
1.26	0.99	0.46	−0.53	1.27	0.28	−0.99
1.28	0.98	0.45	−0.53	1.28	0.26	−1.02
1.30	0.98	0.44	−0.54	1.29	0.24	−1.05
1.32	0.97	0.43	−0.54	1.30	0.21	−1.08
1.34	0.96	0.42	−0.55	1.30	0.19	−1.12
1.36	0.96	0.41	−0.55	1.31	0.16	−1.15
1.38	0.95	0.40	−0.56	1.33	0.14	−1.18
1.40	0.95	0.39	−0.56	1.34	0.12	−1.22
1.42	0.94	0.38	−0.57	1.35	0.09	−1.26
1.44	0.94	0.37	−0.57	1.36	0.07	−1.30
1.46	0.93	0.36	−0.57	1.38	0.04	−1.34
1.48	0.93	0.35	−0.58	1.39	0.01	−1.38
1.50	0.92	0.34	−0.58	1.41	−0.01	−1.42
1.52	0.92	0.33	−0.59	1.42	−0.04	−1.46
1.54	0.91	0.32	−0.59	1.44	−0.07	−1.51
1.56	0.91	0.31	−0.59	1.46	−0.10	−1.56
1.58	0.90	0.30	−0.60	1.48	−0.13	−1.61
1.60	0.90	0.30	−0.60	1.50	−0.16	−1.66
1.62	0.89	0.29	−0.61	1.52	−0.19	−1.71
1.64	0.89	0.28	−0.61	1.54	−0.22	−1.77
1.66	0.88	0.27	−0.61	1.57	−0.26	−1.82
1.68	0.88	0.26	−0.62	1.59	−0.29	−1.88

TABLE 8.22 (continued)

	Objective			Projective		
k^2	f/R	z_m/R	z_p/R	f/R	z_m/R	z_p/R
1.70	0.88	0.26	−0.62	1.62	−0.33	−1.95
1.72	0.87	0.25	−0.62	1.65	−0.36	−2.01
1.74	0.87	0.24	−0.63	1.67	−0.40	−2.08
1.76	0.87	0.23	−0.63	1.71	−0.44	−2.15
1.78	0.86	0.23	−0.63	1.74	−0.48	−2.22
1.80	0.86	0.22	−0.64	1.77	−0.53	−2.30
1.82	0.85	0.21	−0.64	1.81	−0.57	−2.38
1.84	0.85	0.21	−0.64	1.85	−0.62	−2.47
1.86	0.85	0.20	−0.65	1.89	−0.67	−2.56
1.88	0.84	0.19	−0.65	1.93	−0.72	−2.65
1.90	0.84	0.19	−0.65	1.98	−0.77	−2.75
1.92	0.84	0.18	−0.66	2.03	−0.83	−2.86
1.94	0.83	0.17	−0.66	2.08	−0.89	−2.97
1.96	0.83	0.17	−0.66	2.13	−0.95	−3.09
1.98	0.83	0.16	−0.67	2.19	−1.02	−3.21
2.00	0.82	0.16	−0.67	2.26	−1.09	−3.35
2.02	0.82	0.15	−0.67	2.32	−1.16	−3.49
2.04	0.82	0.14	−0.67	2.39	−1.24	−3.64
2.06	0.82	0.14	−0.68	2.47	−1.33	−3.80
2.08	0.81	0.13	−0.68	2.55	−1.42	−3.97
2.10	0.81	0.13	−0.68	2.64	−1.51	−4.16
2.12	0.81	0.12	−0.69	2.74	−1.62	−4.36
2.14	0.80	0.12	−0.69	2.85	−1.73	−4.57
2.16	0.80	0.11	−0.69	2.96	−1.85	−4.81
2.18	0.80	0.11	−0.69	3.08	−1.98	−5.06
2.20	0.80	0.10	−0.70	3.22	−2.12	−5.34
2.22	0.79	0.09	−0.70	3.37	−2.28	−5.65
2.24	0.79	0.09	−0.70	3.53	−2.45	−5.98
2.26	0.79	0.08	−0.70	3.71	−2.64	−6.35
2.28	0.79	0.08	−0.71	3.92	−2.85	−6.76
2.30	0.78	0.07	−0.71	4.15	−3.08	−7.23
2.32	0.78	0.07	−0.71	4.40	−3.34	−7.75
2.34	0.78	0.07	−0.71	4.69	−3.64	−8.34
2.36	0.78	0.06	−0.72	5.03	−3.98	9.01
2.38	0.78	0.06	−0.72	5.41	−4.37	−9.79
2.40	0.77	0.05	−0.72	5.87	−4.83	−10.70
2.42	0.77	0.05	−0.72	6.40	−5.38	−11.78
2.44	0.77	0.04	−0.73	7.05	−6.03	−13.08
2.46	0.77	0.04	−0.73	7.84	−6.83	−14.67
2.48	0.76	0.03	−0.73	8.84	−7.83	−16.67
2.50	0.76	0.03	−0.73	10.13	−9.13	−19.26
3.00	0.72	−0.07	−0.78	−3.86	4.73	8.60
3.50	0.68	−0.14	−0.83	−1.71	2.46	4.16

TABLE 8.22 (continued)

	Objective			Projective		
k^2	f/R	z_m/R	z_p/R	f/R	z_m/R	z_p/R
4.00	0.66	−0.21	−0.86	−1.17	1.80	2.97
4.50	0.63	−0.26	−0.89	−0.95	1.46	2.41
5.00	0.62	−0.31	−0.92	−0.86	1.22	2.08
5.50	0.60	−0.35	−0.95	−0.84	1.02	1.86
6.00	0.59	−0.38	−0.97	−0.87	0.82	1.69
6.50	0.57	−0.41	−0.99	−0.96	0.61	1.56
7.00	0.56	−0.44	−1.01	−1.13	0.33	1.46
7.50	0.55	−0.47	−1.02	−1.46	−0.09	1.37
8.00	0.55	−0.49	−1.04	−2.20	−0.91	1.29
8.50	0.54	−0.52	−1.05	−4.81	−3.60	1.22
9.00	0.53	−0.54	−1.07	22.71	23.85	1.15
9.50	0.52	−0.56	−1.08	3.44	4.52	1.08
10.00	0.52	−0.57	−1.09	1.91	2.92	1.01
15.00	0.48	−0.71	−1.19	0.87	0.80	−0.08
20.00	0.45	−0.80	−1.25	−1.80	2.95	4.75
30.00	0.43	−0.92	−1.34	−2.52	−1.04	1.48
40.00	0.41	−1.00	−1.41	0.72	1.28	0.57
50.00	0.40	−1.06	−1.46	−1.56	2.87	4.43
60.00	0.39	−1.10	−1.50	−0.81	1.09	1.90
70.00	0.39	−1.14	−1.53	1.44	2.81	1.37
80.00	0.38	−1.17	−1.56	0.73	1.32	0.59
90.00	0.38	−1.20	−1.58	−6.34	7.89	14.23
100.00	0.38	−1.23	−1.61	−0.69	1.75	2.45

TABLE 8.23

RATIO OF GAP WIDTH TO BORE DIAMETER $S/D = 1.2$

	Objective			Projective		
k^2	f/R	z_m/R	z_p/R	f/R	z_m/R	z_p/R
0.01	51.25	51.24	−0.00	51.25	51.24	−0.00
0.02	25.83	25.82	−0.01	25.83	25.82	−0.01
0.04	13.13	13.11	−0.02	13.13	13.11	−0.02
0.06	8.89	8.86	−0.03	8.89	8.86	−0.03
0.08	6.78	6.74	−0.04	6.78	6.74	−0.04
0.10	5.52	5.46	−0.05	5.52	5.46	−0.05
0.12	4.67	4.61	−0.06	4.67	4.61	−0.06
0.14	4.07	4.00	−0.08	4.07	4.00	−0.08
0.16	3.62	3.54	−0.09	3.62	3.54	−0.09
0.18	3.28	3.18	−0.10	3.28	3.18	−0.10

TABLE 8.23 (continued)

	Objective			Projective		
k^2	f/R	z_m/R	z_p/R	f/R	z_m/R	z_p/R
0.20	3.00	2.89	−0.11	3.00	2.89	−0.11
0.22	2.77	2.65	−0.12	2.77	2.65	−0.12
0.24	2.59	2.45	−0.14	2.59	2.45	−0.14
0.26	2.43	2.28	−0.15	2.43	2.28	−0.15
0.28	2.29	2.13	−0.16	2.29	2.13	−0.16
0.30	2.18	2.01	−0.17	2.18	2.00	−0.17
0.32	2.08	1.89	−0.19	2.08	1.89	−0.19
0.34	1.99	1.79	−0.20	1.99	1.79	−0.20
0.36	1.91	1.70	−0.21	1.92	1.70	−0.22
0.38	1.84	1.62	−0.23	1.85	1.62	−0.23
0.40	1.78	1.54	−0.24	1.79	1.54	−0.24
0.42	1.73	1.48	−0.25	1.73	1.47	−0.26
0.44	1.68	1.41	−0.27	1.68	1.41	−0.27
0.46	1.63	1.35	−0.28	1.64	1.35	−0.29
0.48	1.59	1.30	−0.29	1.60	1.30	−0.31
0.50	1.55	1.25	−0.30	1.57	1.25	−0.32
0.52	1.52	1.20	−0.31	1.54	1.20	−0.34
0.54	1.49	1.16	−0.33	1.51	1.15	−0.36
0.56	1.46	1.12	−0.34	1.48	1.11	−0.37
0.58	1.43	1.08	−0.35	1.46	1.07	−0.39
0.60	1.40	1.04	−0.36	1.44	1.03	−0.41
0.62	1.38	1.01	−0.37	1.42	0.99	−0.43
0.64	1.36	0.97	−0.38	1.40	0.96	−0.45
0.66	1.33	0.94	−0.39	1.39	0.92	−0.46
0.68	1.31	0.91	−0.40	1.37	0.89	−0.48
0.70	1.29	0.88	−0.41	1.36	0.86	−0.51
0.72	1.28	0.86	−0.42	1.35	0.82	−0.53
0.74	1.26	0.83	−0.43	1.34	0.79	−0.55
0.76	1.24	0.80	−0.44	1.33	0.76	−0.57
0.78	1.23	0.78	−0.45	1.33	0.73	−0.59
0.80	1.21	0.76	−0.46	1.32	0.71	−0.61
0.82	1.20	0.73	−0.47	1.32	0.68	−0.64
0.84	1.19	0.71	−0.47	1.31	0.65	−0.66
0.86	1.17	0.69	−0.48	1.31	0.62	−0.69
0.88	1.16	0.67	−0.49	1.31	0.59	−0.71
0.90	1.15	0.65	−0.50	1.31	0.57	−0.74
0.92	1.14	0.63	−0.50	1.31	0.54	−0.77
0.94	1.13	0.61	−0.51	1.31	0.51	−0.79
0.96	1.12	0.60	−0.52	1.31	0.49	−0.82
0.98	1.11	0.58	−0.53	1.31	0.46	−0.85
1.00	1.10	0.56	−0.53	1.32	0.44	−0.88
1.02	1.09	0.55	−0.54	1.32	0.41	−0.91
1.04	1.08	0.53	−0.55	1.33	0.38	−0.95

TABLE 8.23 (continued)

	Objective			Projective		
k^2	f/R	z_m/R	z_p/R	f/R	z_m/R	z_p/R
1.06	1.07	0.52	−0.55	1.34	0.35	−0.98
1.08	1.06	0.50	−0.56	1.34	0.33	−1.02
1.10	1.05	0.49	−0.56	1.35	0.30	−1.05
1.12	1.04	0.47	−0.57	1.36	0.27	−1.09
1.14	1.04	0.46	−0.58	1.37	0.24	−1.13
1.16	1.03	0.45	−0.58	1.38	0.22	−1.17
1.18	1.02	0.43	−0.59	1.39	0.19	−1.21
1.20	1.01	0.42	−0.59	1.41	0.16	−1.25
1.22	1.01	0.41	−0.60	1.42	0.13	−1.29
1.24	1.00	0.40	−0.60	1.44	0.10	−1.34
1.26	0.99	0.39	−0.61	1.45	0.07	−1.39
1.28	0.99	0.37	−0.61	1.47	0.03	−1.44
1.30	0.98	0.36	−0.62	1.49	0.00	−1.49
1.32	0.97	0.35	−0.62	1.51	−0.03	−1.54
1.34	0.97	0.34	−0.63	1.53	−0.07	−1.60
1.36	0.96	0.33	−0.63	1.55	−0.10	−1.65
1.38	0.96	0.32	−0.64	1.58	−0.14	−1.71
1.40	0.95	0.31	−0.64	1.60	−0.18	−1.78
1.42	0.95	0.30	−0.65	1.63	−0.21	−1.84
1.44	0.94	0.29	−0.65	1.66	−0.26	−1.91
1.46	0.94	0.28	−0.66	1.69	−0.30	−1.98
1.48	0.93	0.27	−0.66	1.72	−0.34	−2.06
1.50	0.93	0.26	−0.66	1.75	−0.39	−2.14
1.52	0.92	0.25	−0.67	1.79	−0.43	−2.22
1.54	0.92	0.24	−0.67	1.83	−0.48	−2.31
1.56	0.91	0.24	−0.68	1.87	−0.54	−2.40
1.58	0.91	0.23	−0.68	1.91	−0.59	−2.50
1.60	0.90	0.22	−0.68	1.96	−0.65	−2.61
1.62	0.90	0.21	−0.69	2.01	−0.71	−2.72
1.64	0.90	0.20	−0.69	2.06	−0.77	−2.83
1.66	0.89	0.20	−0.70	2.12	−0.84	−2.96
1.68	0.89	0.19	−0.70	2.18	−0.91	−3.09
1.70	0.88	0.18	−0.70	2.25	−0.98	−3.23
1.72	0.88	0.17	−0.71	2.32	−1.06	−3.38
1.74	0.88	0.17	−0.71	2.40	−1.15	−3.54
1.76	0.87	0.16	−0.71	2.48	−1.24	−3.72
1.78	0.87	0.15	−0.72	2.57	−1.34	−3.91
1.80	0.86	0.14	−0.72	2.67	−1.44	−4.11
1.82	0.86	0.14	−0.72	2.77	−1.56	−4.33
1.84	0.86	0.13	−0.73	2.89	−1.68	−4.57
1.86	0.85	0.12	−0.73	3.01	−1.82	−4.83
1.88	0.85	0.12	−0.73	3.15	−1.97	−5.12
1.90	0.85	0.11	−0.74	3.31	−2.13	−5.44

TABLE 8.23 (continued)

	Objective			Projective		
k^2	f/R	z_m/R	z_p/R	f/R	z_m/R	z_p/R
1.92	0.84	0.10	−0.74	3.48	−2.31	−5.79
1.94	0.84	0.10	−0.74	3.67	−2.51	−6.18
1.96	0.84	0.09	−0.75	3.89	−2.73	−6.62
1.98	0.83	0.09	−0.75	4.13	−2.98	−7.11
2.00	0.83	0.08	−0.75	4.41	−3.27	−7.68
2.02	0.83	0.07	−0.76	4.73	−3.60	−8.32
2.04	0.83	0.07	−0.76	5.10	−3.97	−9.07
2.06	0.82	0.06	−0.76	5.54	−4.42	−9.96
2.08	0.82	0.06	−0.76	6.06	−4.95	−11.00
2.10	0.82	0.05	−0.77	6.69	−5.58	−12.27
2.12	0.81	0.04	−0.77	7.47	−6.37	−13.84
2.14	0.81	0.04	−0.77	8.46	−7.37	−15.82
2.16	0.81	0.03	−0.78	9.75	−8.67	−18.42
2.18	0.81	0.03	−0.78	11.51	−10.44	−21.95
2.20	0.80	0.02	−0.78	14.06	−12.99	−27.06
2.22	0.80	0.02	−0.78	18.07	−17.01	−35.08
2.24	0.80	0.01	−0.79	25.28	−24.22	−49.50
2.26	0.80	0.01	−0.79	42.07	−41.02	−83.08
2.28	0.79	0.00	−0.79	125.40	−124.35	−249.75
2.30	0.79	−0.00	−0.79	−127.80	128.84	256.64
2.32	0.79	−0.01	−0.80	−42.33	43.36	85.70
2.34	0.79	−0.01	−0.80	−25.37	26.40	51.77
2.36	0.79	−0.02	−0.80	−18.12	19.14	37.25
2.38	0.78	−0.02	−0.80	−14.09	15.10	29.20
2.40	0.78	−0.03	−0.81	−11.53	12.54	24.07
2.42	0.78	−0.03	−0.81	−9.76	10.76	20.53
2.44	0.78	−0.04	−0.81	−8.47	9.46	17.92
2.46	0.77	−0.04	−0.81	−7.47	8.46	15.94
2.48	0.77	−0.04	−0.82	−6.69	7.67	14.37
2.50	0.77	−0.05	−0.82	6.06	7.03	13.09
3.00	0.72	−0.15	−0.87	−1.88	2.71	4.58
3.50	0.69	−0.22	−0.91	−1.20	1.89	3.09
4.00	0.66	−0.29	−0.95	−0.96	1.50	2.46
4.50	0.64	−0.34	0.98	−0.87	1.23	2.10
5.00	0.62	−0.39	−1.01	−0.87	1.00	1.87
5.50	0.61	−0.43	−1.03	−0.93	0.77	1.70
6.00	0.59	−0.46	−1.06	−1.09	0.48	1.57
6.50	0.58	−0.50	−1.08	−1.42	0.04	1.46
7.00	0.57	−0.52	−1.09	−2.20	−0.83	1.37
7.50	0.56	−0.55	−1.11	−5.50	−4.22	1.28
8.00	0.55	−0.58	−1.13	10.30	11.50	1.20
8.50	0.54	−0.60	−1.14	2.72	3.84	1.12
9.00	0.54	−0.62	−1.16	1.62	2.67	1.05

TABLE 8.23 (continued)

	Objective			Projective		
k^2	f/R	z_m/R	z_p/R	f/R	z_m/R	z_p/R
9.50	0.53	−0.64	−1.17	1.20	2.16	0.97
10.00	0.52	−0.66	−1.18	0.98	1.86	0.88
15.00	0.48	−0.79	−1.28	1.86	−0.36	−2.22
20.00	0.46	−0.89	−1.34	−0.82	1.75	2.56
30.00	0.43	−1.00	−1.43	1.32	2.58	1.26
40.00	0.41	−1.08	−1.50	1.94	−0.29	−2.23
50.00	0.40	−1.14	−1.55	−0.70	1.57	2.26
60.00	0.40	−1.19	−1.59	3.72	5.26	1.54
70.00	0.39	−1.23	−1.62	0.71	1.47	0.76
80.00	0.39	−1.26	−1.65	−3.97	5.57	9.53
90.00	0.38	−1.29	−1.67	−0.68	1.67	2.35
100.00	0.38	−1.32	−1.70	−3.31	−1.53	1.77

TABLE 8.24

RATIO OF GAP WIDTH TO BORE DIAMETER $S/D = 1.3$

	Objective			Projective		
k^2	f/R	z_m/R	z_p/R	f/R	z_m/R	z_p/R
0.01	47.69	47.69	−0.01	47.69	47.68	−0.01
0.02	24.07	24.05	−0.01	24.07	24.05	−0.01
0.04	12.26	12.23	−0.02	12.26	12.23	−0.02
0.06	8.33	8.29	−0.04	8.33	8.29	−0.04
0.08	6.36	6.31	−0.05	6.36	6.31	−0.05
0.10	5.19	5.12	−0.06	5.19	5.12	−0.06
0.12	4.41	4.33	−0.08	4.41	4.33	−0.08
0.14	3.85	3.76	−0.09	3.85	3.76	−0.09
0.16	3.43	3.33	−0.11	3.43	3.33	−0.11
0.18	3.11	2.99	−0.12	3.11	2.99	−0.12
0.20	2.86	2.72	−0.14	2.86	2.72	−0.14
0.22	2.65	2.50	−0.15	2.65	2.50	−0.15
0.24	2.47	2.31	−0.17	2.48	2.31	−0.17
0.26	2.33	2.15	−0.18	2.33	2.15	−0.18
0.28	2.21	2.01	−0.20	2.21	2.01	−0.20
0.30	2.10	1.89	−0.21	2.10	1.89	−0.21
0.32	2.01	1.78	−0.23	2.01	1.78	−0.23
0.34	1.93	1.68	−0.25	1.93	1.68	−0.25
0.36	1.86	1.60	−0.26	1.86	1.60	−0.27
0.38	1.80	1.52	−0.28	1.80	1.52	−0.29
0.40	1.74	1.45	−0.29	1.75	1.45	−0.30

TABLE 8.24 (continued)

	Objective			Projective		
k^2	f/R	z_m/R	z_p/R	f/R	z_m/R	z_p/R
0.42	1.69	1.38	−0.31	1.70	1.38	−0.32
0.44	1.64	1.32	−0.32	1.66	1.32	−0.34
0.46	1.60	1.26	−0.34	1.62	1.26	−0.36
0.48	1.56	1.21	−0.35	1.59	1.21	−0.38
0.50	1.53	1.16	−0.37	1.56	1.15	−0.40
0.52	1.50	1.12	−0.38	1.53	1.11	−0.43
0.54	1.47	1.07	−0.39	1.51	1.06	−0.45
0.56	1.44	1.03	−0.41	1.49	1.02	−0.47
0.58	1.41	1.00	−0.42	1.47	0.97	−0.49
0.60	1.39	0.96	−0.43	1.45	0.93	−0.52
0.62	1.37	0.92	−0.44	1.44	0.90	−0.54
0.64	1.35	0.89	−0.45	1.43	0.86	−0.57
0.66	1.33	0.86	−0.47	1.42	0.82	−0.60
0.68	1.31	0.83	−0.48	1.41	0.79	−0.62
0.70	1.29	0.80	−0.49	1.40	0.75	−0.65
0.72	1.27	0.78	−0.50	1.40	0.72	−0.68
0.74	1.26	0.75	−0.51	1.39	0.68	−0.71
0.76	1.24	0.72	−0.52	1.39	0.65	−0.74
0.78	1.23	0.70	−0.53	1.39	0.62	−0.77
0.80	1.21	0.68	−0.53	1.39	0.59	−0.80
0.82	1.20	0.65	−0.54	1.39	0.55	−0.84
0.84	1.19	0.63	−0.55	1.39	0.52	−0.87
0.86	1.17	0.61	−0.56	1.40	0.49	−0.91
0.88	1.16	0.59	−0.57	1.40	0.46	−0.95
0.90	1.15	0.57	−0.58	1.41	0.42	−0.98
0.92	1.14	0.55	−0.58	1.42	0.39	−1.02
0.94	1.13	0.54	−0.59	1.42	0.36	−1.07
0.96	1.12	0.52	−0.60	1.43	0.33	−1.11
0.98	1.11	0.50	−0.61	1.45	0.29	−1.15
1.00	1.10	0.48	−0.61	1.46	0.26	−1.20
1.02	1.09	0.47	−0.62	1.47	0.22	−1.25
1.04	1.08	0.45	−0.63	1.49	0.19	−1.30
1.06	1.07	0.44	0.63	1.50	0.15	1.35
1.08	1.06	0.42	−0.64	1.52	0.12	−1.41
1.10	1.06	0.41	−0.65	1.54	0.08	−1.46
1.12	1.05	0.39	−0.65	1.56	0.04	−1.52
1.14	1.04	0.38	−0.66	1.59	−0.00	−1.59
1.16	1.03	0.37	−0.66	1.61	−0.04	−1.65
1.18	1.03	0.35	−0.67	1.64	−0.08	−1.72
1.20	1.02	0.34	−0.68	1.67	−0.13	−1.79
1.22	1.01	0.33	−0.68	1.70	−0.17	−1.87
1.24	1.01	0.32	−0.69	1.73	−0.22	−1.95
1.26	1.00	0.31	−0.69	1.76	−0.27	−2.03

TABLE 8.24 (continued)

	Objective			Projective		
k^2	f/R	z_m/R	z_p/R	f/R	z_m/R	z_p/R
1.28	0.99	0.29	−0.70	1.80	−0.32	−2.12
1.30	0.99	0.28	−0.70	1.84	−0.38	−2.22
1.32	0.98	0.27	−0.71	1.88	−0.43	−2.32
1.34	0.97	0.26	−0.71	1.93	−0.49	−2.42
1.36	0.97	0.25	−0.72	1.98	−0.56	−2.54
1.38	0.96	0.24	−0.72	2.03	−0.62	−2.65
1.40	0.96	0.23	−0.73	2.09	−0.69	−2.78
1.42	0.95	0.22	−0.73	2.15	−0.77	−2.92
1.44	0.95	0.21	−0.74	2.22	−0.84	−3.06
1.46	0.94	0.20	−0.74	2.29	−0.93	−3.22
1.48	0.94	0.19	−0.75	2.37	−1.02	−3.39
1.50	0.93	0.18	−0.75	2.46	−1.12	−3.57
1.52	0.93	0.17	−0.75	2.55	−1.22	−3.77
1.54	0.92	0.17	−0.76	2.65	−1.33	−3.99
1.56	0.92	0.16	−0.76	2.77	−1.46	−4.22
1.58	0.91	0.15	−0.77	2.89	−1.59	−4.48
1.60	0.91	0.14	−0.77	3.03	−1.74	−4.77
1.62	0.91	0.13	−0.77	3.18	−1.90	−5.08
1.64	0.90	0.12	−0.78	3.35	−2.08	−5.43
1.66	0.90	0.12	−0.78	3.54	−2.28	−5.82
1.68	0.89	0.11	−0.79	3.76	−2.50	−6.26
1.70	0.89	0.10	−0.79	4.00	−2.76	−6.76
1.72	0.89	0.09	−0.79	4.28	−3.05	−7.33
1.74	0.88	0.09	−0.80	4.61	−3.39	−8.00
1.76	0.88	0.08	−0.80	4.99	−3.78	−8.77
1.78	0.87	0.07	−0.80	5.45	−4.24	−9.69
1.80	0.87	0.06	−0.81	6.00	−4.80	−10.80
1.82	0.87	0.06	−0.81	6.68	−5.49	−12.16
1.84	0.86	0.05	−0.81	7.53	−6.35	−13.88
1.86	0.86	0.04	−0.82	8.63	−7.46	−16.10
1.88	0.86	0.04	−0.82	10.13	−8.96	−19.09
1.90	0.85	0.03	−0.82	12.25	−11.09	−23.34
1.92	0.85	0.02	−0.83	15.50	−14.36	−29.86
1.94	0.85	0.02	−0.83	21.13	−19.99	−41.12
1.96	0.84	0.01	−0.83	33.20	−32.06	−65.26
1.98	0.84	0.00	−0.84	77.74	−76.35	−153.82
2.00	0.84	−0.00	−0.84	−231.54	232.65	464.19
2.02	0.84	−0.01	−0.84	−46.40	47.51	93.92
2.04	0.83	−0.01	−0.85	−25.79	26.89	52.67
2.06	0.83	−0.02	−0.85	−17.86	18.95	36.81
2.08	0.83	−0.03	−0.85	−13.66	14.75	28.40
2.10	0.82	−0.03	−0.85	−11.06	12.14	23.20
2.12	0.82	−0.04	−0.86	−9.30	10.37	19.67

TABLE 8.24 (continued)

k^2	Objective			Projective		
	f/R	z_m/R	z_p/R	f/R	z_m/R	z_p/R
2.14	0.82	−0.04	−0.86	−8.02	9.08	17.10
2.16	0.82	−0.05	−0.86	−7.05	8.11	15.16
2.18	0.81	−0.05	−0.87	−6.30	7.34	13.64
2.20	0.81	−0.06	−0.87	−5.69	6.73	12.42
2.22	0.81	−0.06	−0.87	−5.19	6.22	11.41
2.24	0.81	−0.07	−0.87	−4.77	5.80	10.57
2.26	0.80	−0.07	−0.88	−4.41	5.43	9.85
2.28	0.80	−0.08	−0.88	−4.11	5.12	9.23
2.30	0.80	−0.08	−0.88	−3.85	4.85	8.70
2.32	0.80	−0.09	−0.88	−3.62	4.61	8.23
2.34	0.79	−0.09	−0.89	−3.41	4.40	7.81
2.36	0.79	−0.10	−0.89	−3.23	4.22	7.45
2.38	0.79	−0.10	−0.89	−3.07	4.05	7.11
2.40	0.79	−0.11	−0.89	−2.92	3.89	6.82
2.42	0.78	−0.11	−0.90	−2.79	3.75	6.55
2.44	0.78	−0.12	−0.90	−2.67	3.63	6.30
2.46	0.78	−0.12	−0.90	−2.56	3.51	6.07
2.48	0.78	−0.13	−0.90	−2.46	3.41	5.87
2.50	0.78	−0.13	−0.91	−2.37	3.31	5.68
3.00	0.73	−0.23	−0.96	−1.30	2.07	3.37
3.50	0.70	−0.31	−1.00	−0.99	1.59	2.58
4.00	0.67	−0.37	−1.04	−0.89	1.28	2.17
4.50	0.65	−0.42	−1.07	−0.90	1.01	1.91
5.00	0.63	−0.47	−1.10	−1.00	0.73	1.73
5.50	0.61	−0.51	−1.12	−1.26	0.34	1.59
6.00	0.60	−0.55	−1.15	−1.87	−0.39	1.47
6.50	0.58	−0.58	−1.17	−4.17	−2.80	1.37
7.00	0.57	−0.61	−1.18	14.59	15.87	1.28
7.50	0.56	−0.64	−1.20	2.71	3.90	1.19
8.00	0.55	−0.66	−1.22	1.56	2.66	1.10
8.50	0.55	−0.69	−1.23	1.14	2.15	1.01
9.00	0.54	−0.71	−1.25	0.94	1.85	0.91
9.50	0.53	−0.73	−1.26	0.84	1.64	0.81
10.00	0.53	−0.74	−1.27	0.79	1.47	0.68
15.00	0.48	−0.88	−1.37	−3.47	4.82	8.28
20.00	0.46	−0.97	−1.43	−0.77	1.27	2.04
30.00	0.43	−1.09	−1.53	0.72	1.64	0.92
40.00	0.42	−1.17	−1.59	−1.15	2.52	3.67
50.00	0.41	−1.23	−1.64	−1.72	0.09	1.81
60.00	0.40	−1.28	−1.68	0.70	1.78	1.08
70.00	0.39	−1.32	−1.71	−24.54	26.34	50.80
80.00	0.39	−1.35	−1.74	−0.69	1.71	2.40
90.00	0.38	−1.38	−1.77	−146.08	−144.30	1.78
100.00	0.38	−1.41	−1.79	0.69	1.90	1.21

TABLE 8.25
RATIO OF GAP WIDTH TO BORE DIAMETER $S/D = 1.4$

	Objective			Projective		
k^2	f/R	z_m/R	z_p/R	f/R	z_m/R	z_p/R
0.01	44.48	44.47	−0.01	44.48	44.47	−0.01
0.02	22.47	22.46	−0.01	22.47	22.46	−0.01
0.04	11.48	11.45	−0.03	11.48	11.45	−0.03
0.06	7.81	7.77	−0.05	7.81	7.77	−0.05
0.08	5.99	5.93	−0.06	5.99	5.93	−0.06
0.10	4.89	4.82	−0.08	4.89	4.82	−0.08
0.12	4.17	4.07	−0.09	4.17	4.07	−0.09
0.14	3.65	3.54	−0.11	3.65	3.54	−0.11
0.16	3.27	3.14	−0.13	3.27	3.14	−0.13
0.18	2.97	2.82	−0.15	2.97	2.82	−0.15
0.20	2.73	2.57	−0.17	2.73	2.57	−0.17
0.22	2.54	2.36	−0.18	2.54	2.36	−0.18
0.24	2.38	2.18	−0.20	2.38	2.18	−0.20
0.26	2.25	2.03	−0.22	2.25	2.03	−0.22
0.28	2.14	1.90	−0.24	2.14	1.89	−0.24
0.30	2.04	1.78	−0.26	2.04	1.78	−0.26
0.32	1.96	1.68	−0.28	1.96	1.68	−0.29
0.34	1.88	1.58	−0.30	1.89	1.58	−0.31
0.36	1.82	1.50	−0.32	1.83	1.50	−0.33
0.38	1.76	1.42	−0.34	1.77	1.42	−0.35
0.40	1.71	1.35	−0.36	1.73	1.35	−0.38
0.42	1.66	1.29	−0.37	1.69	1.28	−0.40
0.44	1.62	1.23	−0.39	1.65	1.22	−0.43
0.46	1.58	1.18	−0.41	1.62	1.16	−0.45
0.48	1.55	1.13	−0.42	1.59	1.11	−0.48
0.50	1.52	1.08	−0.44	1.57	1.06	−0.51
0.52	1.49	1.03	−0.45	1.55	1.01	−0.54
0.54	1.46	0.99	−0.47	1.53	0.96	−0.57
0.56	1.43	0.95	−0.48	1.51	0.92	−0.60
0.58	1.41	0.91	−0.50	1.50	0.87	−0.63
0.60	1.38	0.88	−0.51	1.49	0.83	−0.66
0.62	1.36	0.84	−0.52	1.48	0.79	−0.70
0.64	1.34	0.81	−0.53	1.48	0.75	−0.73
0.66	1.32	0.78	−0.55	1.47	0.71	−0.77
0.68	1.31	0.75	−0.56	1.47	0.67	−0.81
0.70	1.29	0.72	−0.57	1.47	0.63	−0.84
0.72	1.27	0.69	−0.58	1.47	0.59	−0.89
0.74	1.26	0.67	−0.59	1.48	0.55	−0.93
0.76	1.24	0.64	−0.60	1.48	0.51	−0.97
0.78	1.23	0.62	−0.61	1.49	0.47	−1.02
0.80	1.21	0.60	−0.62	1.50	0.43	−1.06
0.82	1.20	0.57	−0.63	1.51	0.39	−1.11

TABLE 8.25 (continued)

	Objective			Projective		
k^2	f/R	z_m/R	z_p/R	f/R	z_m/R	z_p/R
0.84	1.19	0.55	−0.64	1.52	0.35	−1.17
0.86	1.18	0.53	−0.64	1.53	0.31	−1.22
0.88	1.16	0.51	−0.65	1.55	0.27	−1.28
0.90	1.15	0.49	−0.66	1.57	0.23	−1.34
0.92	1.14	0.47	−0.67	1.59	0.19	−1.40
0.94	1.13	0.45	−0.68	1.61	0.14	−1.46
0.96	1.12	0.44	−0.68	1.63	0.10	−1.53
0.98	1.11	0.42	−0.69	1.66	0.05	−1.60
1.00	1.10	0.40	−0.70	1.68	0.00	−1.68
1.02	1.09	0.39	−0.71	1.71	−0.05	−1.76
1.04	1.08	0.37	−0.71	1.75	−0.10	−1.84
1.06	1.08	0.36	−0.72	1.78	−0.15	−1.93
1.08	1.07	0.34	−0.73	1.82	−0.21	−2.03
1.10	1.06	0.33	−0.73	1.86	−0.27	−2.13
1.12	1.05	0.31	−0.74	1.91	−0.33	−2.24
1.14	1.05	0.30	−0.75	1.95	−0.40	−2.35
1.16	1.04	0.29	−0.75	2.01	−0.46	−2.47
1.18	1.03	0.27	−0.76	2.06	−0.54	−2.60
1.20	1.02	0.26	−0.76	2.13	−0.62	−2.74
1.22	1.02	0.25	−0.77	2.19	−0.70	−2.89
1.24	1.01	0.24	−0.77	2.27	−0.79	−3.05
1.26	1.00	0.22	−0.78	2.35	−0.88	−3.23
1.28	1.00	0.21	−0.79	2.44	−0.98	−3.42
1.30	0.99	0.20	−0.79	2.53	−1.09	−3.63
1.32	0.99	0.19	−0.80	2.64	−1.21	−3.86
1.34	0.98	0.18	−0.80	2.76	−1.35	−4.11
1.36	0.97	0.17	−0.81	2.89	−1.49	−4.38
1.38	0.97	0.16	−0.81	3.04	−1.65	−4.69
1.40	0.96	0.15	−0.82	3.20	−1.83	−5.03
1.42	0.96	0.14	−0.82	3.39	−2.02	−5.41
1.44	0.95	0.13	−0.83	3.60	−2.25	−5.85
1.46	0.95	0.12	−0.83	3.84	−2.50	−6.34
1.48	0.94	0.11	−0.83	4.12	−2.79	−6.91
1.50	0.94	0.10	−0.84	4.45	−3.13	−7.57
1.52	0.93	0.09	−0.84	4.83	−3.52	−8.35
1.54	0.93	0.08	−0.85	5.29	−3.99	−9.28
1.56	0.92	0.07	−0.85	5.85	−4.57	−10.42
1.58	0.92	0.06	−0.86	6.55	−5.28	−11.83
1.60	0.92	0.06	−0.86	7.45	−6.18	−13.63
1.62	0.91	0.05	−0.86	8.64	−7.38	−16.02
1.64	0.91	0.04	−0.87	10.28	−9.03	−19.31
1.66	0.90	0.03	−0.87	12.71	−11.47	−24.17
1.68	0.90	0.02	−0.88	16.64	−15.42	−32.06

TABLE 8.25 (continued)

	Objective			Projective		
k^2	f/R	z_m/R	z_p/R	f/R	z_m/R	z_p/R
1.70	0.90	0.02	−0.88	24.14	−22.92	−47.07
1.72	0.89	0.01	−0.88	44.01	−42.80	−86.81
1.74	0.89	0.00	−0.89	250.10	−248.90	−499.01
1.76	0.88	−0.01	−0.89	−67.84	69.03	136.88
1.78	0.88	−0.01	−0.89	−29.86	31.05	60.91
1.80	0.88	−0.02	−0.90	−19.15	20.32	39.47
1.82	0.87	−0.03	−0.90	−14.09	15.26	29.35
1.84	0.87	−0.03	−0.90	−11.15	12.31	23.46
1.86	0.87	−0.04	−0.91	−9.23	10.37	19.60
1.88	0.86	−0.05	−0.91	−7.87	9.01	16.88
1.90	0.86	−0.05	−0.91	−6.87	7.99	14.86
1.92	0.86	−0.06	−0.92	−6.09	7.21	13.20
1.94	0.85	−0.07	−0.92	−5.47	6.58	12.06
1.96	0.85	−0.07	−0.92	−4.97	6.07	11.05
1.98	0.85	−0.08	−0.93	−4.56	5.65	10.21
2.00	0.84	−0.09	−0.93	−4.21	5.29	9.50
2.02	0.84	−0.09	−0.93	−3.91	4.98	8.89
2.04	0.84	−0.10	−0.94	−3.65	4.72	8.37
2.06	0.84	−0.10	−0.94	−3.42	4.48	7.91
2.08	0.83	−0.11	−0.94	−3.23	4.28	7.50
2.10	0.83	−0.12	−0.94	−3.05	4.10	7.15
2.12	0.83	−0.12	−0.95	−2.89	3.93	6.83
2.14	0.82	−0.13	−0.95	−2.75	3.78	6.54
2.16	0.82	−0.13	−0.95	−2.63	3.65	6.28
2.18	0.82	−0.14	−0.96	−2.51	3.53	6.04
2.20	0.82	−0.14	−0.96	−2.41	3.41	5.82
2.22	0.81	−0.15	−0.96	−2.31	3.31	5.63
2.24	0.81	−0.15	−0.96	−2.23	3.22	5.44
2.26	0.81	−0.16	−0.97	−2.15	3.13	5.28
2.28	0.81	−0.16	−0.97	−2.07	3.05	5.12
2.30	0.80	−0.17	−0.97	−2.01	2.97	4.98
2.32	0.80	−0.17	−0.98	−1.94	2.90	4.84
2.34	0.80	−0.18	−0.98	−1.88	2.83	4.72
2.36	0.80	−0.18	−0.98	−1.83	2.77	4.60
2.38	0.79	−0.19	−0.98	−1.78	2.71	4.49
2.40	0.79	−0.19	−0.99	−1.73	2.66	4.39
2.42	0.79	−0.20	−0.99	−1.68	2.60	4.29
2.44	0.79	−0.20	−0.99	−1.64	2.55	4.20
2.46	0.79	−0.21	−0.99	−1.60	2.51	4.11
2.48	0.78	−0.21	−0.99	−1.57	2.46	4.03
2.50	0.78	−0.22	−1.00	−1.53	2.42	3.95
3.00	0.74	−0.31	−1.05	−1.06	1.74	2.80
3.50	0.70	−0.39	−1.10	−0.92	1.37	2.29

TABLE 8.25 (continued)

	Objective			Projective		
k^2	f/R	z_m/R	z_p/R	f/R	z_m/R	z_p/R
4.00	0.67	−0.46	−1.13	−0.93	1.06	1.99
4.50	0.65	−0.51	−1.16	−1.07	0.72	1.79
5.00	0.63	−0.56	−1.19	−1.43	0.20	1.63
5.50	0.61	−0.60	−1.22	−2.55	−1.04	1.50
6.00	0.60	−0.64	−1.24	−17.59	−16.20	1.39
6.50	0.59	−0.67	−1.26	3.58	4.87	1.29
7.00	0.58	−0.70	−1.28	1.70	2.88	1.19
7.50	0.57	−0.73	−1.29	1.17	2.25	1.08
8.00	0.56	−0.75	−1.31	0.95	1.91	0.97
8.50	0.55	−0.77	−1.33	0.84	1.68	0.84
9.00	0.54	−0.80	−1.34	0.80	1.49	0.69
9.50	0.54	−0.82	−1.35	0.80	1.31	0.51
10.00	0.53	−0.83	−1.36	0.85	1.12	0.27
15.00	0.49	−0.97	−1.46	−0.98	2.16	3.14
20.00	0.46	−1.06	−1.53	−1.45	0.33	1.78
30.00	0.43	−1.19	−1.62	1.05	0.89	−0.17
40.00	0.42	−1.27	−1.68	−0.72	1.60	2.31
50.00	0.41	−1.33	−1.73	1.00	2.47	1.47
60.00	0.40	−1.37	−1.77	1.84	0.09	−1.75
70.00	0.39	−1.41	−1.81	−0.70	1.91	2.61
80.00	0.39	−1.45	−1.83	−149.26	−147.41	1.85
90.00	0.38	−1.48	−1.86	0.68	1.85	1.17
100.00	0.38	−1.50	−1.88	18.40	−16.51	−34.92

TABLE 8.26

RATIO OF GAP WIDTH TO BORE DIAMETER $S/D = 1.5$

	Objective			Projective		
k^2	f/R	z_m/R	z_p/R	f/R	z_m/R	z_p/R
0.01	41.58	41.57	−0.01	41.58	41.57	−0.01
0.02	21.04	21.02	−0.02	21.04	21.02	0.02
0.04	10.77	10.74	−0.04	10.77	10.74	−0.04
0.06	7.36	7.30	−0.05	7.36	7.30	−0.05
0.08	5.65	5.58	−0.07	5.65	5.58	−0.07
0.10	4.64	4.54	−0.09	4.64	4.54	−0.09
0.12	3.96	3.85	−0.11	3.96	3.85	−0.11
0.14	3.48	3.34	−0.13	3.48	3.34	−0.13
0.16	3.12	2.97	−0.16	3.12	2.97	−0.16
0.18	2.85	2.67	−0.18	2.85	2.67	−0.18

TABLE 8.26 (continued)

	Objective			Projective		
k^2	f/R	z_m/R	z_p/R	f/R	z_m/R	z_p/R
0.20	2.63	2.43	−0.20	2.63	2.43	−0.20
0.22	2.45	2.23	−0.22	2.45	2.23	−0.22
0.24	2.31	2.06	−0.25	2.31	2.06	−0.25
0.26	2.18	1.91	−0.27	2.19	1.91	−0.27
0.28	2.08	1.79	−0.29	2.09	1.79	−0.30
0.30	1.99	1.68	−0.32	2.00	1.67	−0.32
0.32	1.92	1.58	−0.34	1.93	1.57	−0.35
0.34	1.85	1.49	−0.36	1.86	1.48	−0.38
0.36	1.79	1.41	−0.38	1.81	1.40	−0.41
0.38	1.74	1.33	−0.41	1.76	1.32	−0.44
0.40	1.69	1.26	−0.43	1.72	1.25	−0.47
0.42	1.65	1.20	−0.45	1.69	1.19	−0.50
0.44	1.61	1.14	−0.46	1.66	1.12	−0.54
0.46	1.57	1.09	−0.48	1.64	1.07	−0.57
0.48	1.54	1.04	−0.50	1.62	1.01	−0.61
0.50	1.51	0.99	−0.52	1.60	0.96	−0.64
0.52	1.48	0.95	−0.53	1.58	0.90	−0.68
0.54	1.45	0.90	−0.55	1.57	0.85	−0.72
0.56	1.43	0.86	−0.56	1.57	0.80	−0.76
0.58	1.40	0.83	−0.58	1.56	0.75	−0.81
0.60	1.38	0.79	−0.59	1.56	0.71	−0.85
0.62	1.36	0.76	−0.60	1.56	0.66	−0.90
0.64	1.34	0.72	−0.62	1.56	0.61	−0.95
0.66	1.32	0.69	−0.63	1.57	0.57	−1.00
0.68	1.31	0.66	−0.64	1.57	0.52	−1.05
0.70	1.29	0.64	−0.65	1.58	0.47	−1.11
0.72	1.27	0.61	−0.66	1.60	0.43	−1.17
0.74	1.26	0.58	−0.68	1.61	0.38	−1.23
0.76	1.24	0.56	−0.69	1.63	0.33	−1.30
0.78	1.23	0.53	−0.70	1.65	0.28	−1.37
0.80	1.22	0.51	−0.71	1.67	0.23	−1.44
0.82	1.20	0.49	−0.71	1.69	0.18	−1.52
0.84	1.19	0.47	−0.72	1.72	0.12	−1.60
0.86	1.18	0.45	−0.73	1.75	0.07	−1.68
0.88	1.17	0.43	−0.74	1.78	0.01	−1.77
0.90	1.16	0.41	−0.75	1.82	−0.05	−1.87
0.92	1.15	0.39	−0.76	1.86	−0.11	−1.97
0.94	1.14	0.37	−0.77	1.90	−0.18	−2.08
0.96	1.13	0.35	−0.77	1.95	−0.25	−2.20
0.98	1.12	0.34	−0.78	2.00	−0.32	−2.33
1.00	1.11	0.32	−0.79	2.06	−0.40	−2.46
1.02	1.10	0.30	−0.80	2.12	−0.48	−2.61
1.04	1.09	0.29	−0.80	2.19	−0.57	−2.77

TABLE 8.26 (continued)

	Objective			Projective		
k^2	f/R	z_m/R	z_p/R	f/R	z_m/R	z_p/R
1.06	1.08	0.27	−0.81	2.27	−0.67	−2.94
1.08	1.07	0.26	−0.82	2.35	−0.77	−3.12
1.10	1.07	0.24	−0.82	2.45	−0.88	−3.33
1.12	1.06	0.23	−0.83	2.55	−1.00	−3.55
1.14	1.05	0.21	−0.84	2.66	−1.13	−3.79
1.16	1.04	0.20	−0.84	2.79	−1.27	−4.07
1.18	1.04	0.19	−0.85	2.94	−1.43	−4.37
1.20	1.03	0.18	−0.85	3.10	−1.61	−4.71
1.22	1.02	0.16	−0.86	3.28	−1.81	−5.09
1.24	1.02	0.15	−0.87	3.49	−2.03	−5.53
1.26	1.01	0.14	−0.87	3.74	−2.29	−6.03
1.28	1.00	0.13	−0.88	4.02	−2.59	−6.60
1.30	1.00	0.12	−0.88	4.35	−2.93	−7.28
1.32	0.99	0.10	−0.89	4.75	−3.34	−8.09
1.34	0.99	0.09	−0.89	5.23	−3.84	−9.07
1.36	0.98	0.08	−0.90	5.82	−4.44	−10.27
1.38	0.97	0.07	−0.90	6.58	−5.21	−11.79
1.40	0.97	0.06	−0.91	7.56	−6.21	−13.77
1.42	0.96	0.05	−0.91	8.91	−7.56	−16.47
1.44	0.96	0.04	−0.92	10.84	−9.51	−20.35
1.46	0.95	0.03	−0.92	13.86	−12.54	−26.41
1.48	0.95	0.02	−0.93	19.25	−17.94	−37.18
1.50	0.94	0.01	−0.93	31.52	−30.22	−61.74
1.52	0.94	0.01	−0.93	87.36	−86.07	−173.43
1.54	0.93	−0.00	−0.94	−112.84	114.12	226.96
1.56	0.93	−0.01	−0.94	−34.26	25.52	69.78
1.58	0.93	−0.02	−0.95	−20.19	21.45	41.64
1.60	0.92	−0.03	−0.95	−14.32	15.56	29.87
1.62	0.92	−0.04	−0.96	−11.09	12.32	23.41
1.64	0.91	−0.05	−0.96	−9.05	10.28	19.33
1.66	0.91	−0.05	−0.96	−7.65	8.86	16.51
1.68	0.90	−0.06	−0.97	−6.63	7.83	14.46
1.70	0.90	−0.07	−0.97	−5.85	7.04	12.88
1.72	0.90	−0.08	−0.97	−5.23	6.41	11.65
1.74	0.89	−0.08	−0.98	−4.74	5.91	10.65
1.76	0.89	−0.09	−0.98	−4.33	5.49	9.82
1.78	0.89	−0.10	−0.99	−3.99	5.14	9.13
1.80	0.88	−0.11	−0.99	−3.70	4.84	8.54
1.82	0.88	−0.11	−0.99	−3.45	4.58	8.03
1.84	0.88	−0.12	−1.00	−3.23	4.35	7.58
1.86	0.87	−0.13	−1.00	−3.04	4.15	7.20
1.88	0.87	−0.13	−1.00	−2.87	3.98	6.85
1.90	0.87	−0.14	−1.01	−2.72	3.82	6.54

TABLE 8.26 (continued)

	Objective			Projective		
k^2	f/R	z_m/R	z_p/R	f/R	z_m/R	z_p/R
1.92	0.86	−0.15	−1.01	−2.59	3.68	6.27
1.94	0.86	−0.15	−1.01	−2.47	3.55	6.02
1.96	0.86	−0.16	−1.02	−2.36	3.43	5.79
1.98	0.85	−0.17	−1.02	−2.27	3.32	5.59
2.00	0.85	−0.17	−1.02	−2.18	3.22	5.40
2.02	0.85	−0.18	−1.03	−2.09	3.13	5.23
2.04	0.84	−0.18	−1.03	−2.02	3.05	5.07
2.06	0.84	−0.19	−1.03	−1.95	2.97	4.92
2.08	0.84	−0.20	−1.03	−1.89	2.90	4.78
2.10	0.84	−0.20	−1.04	−1.83	2.83	4.66
2.12	0.83	−0.21	−1.04	−1.77	2.76	4.54
2.14	0.83	−0.21	−1.04	−1.72	2.70	4.43
2.16	0.83	−0.22	−1.05	−1.67	2.65	4.32
2.18	0.82	−0.22	−1.05	−1.63	2.59	4.22
2.20	0.82	−0.23	−1.05	−1.59	2.54	4.13
2.22	0.82	−0.24	−1.05	−1.55	2.50	4.05
2.24	0.82	−0.24	−1.06	−1.51	2.45	3.96
2.26	0.81	−0.25	−1.06	−1.48	2.41	3.89
2.28	0.81	−0.25	−1.06	−1.45	2.37	3.81
2.30	0.81	−0.26	−1.07	−1.42	2.33	3.74
2.32	0.81	−0.26	−1.07	−1.39	2.29	3.68
2.34	0.80	−0.27	−1.07	−1.36	2.25	3.61
2.36	0.80	−0.27	−1.07	−1.34	2.22	3.55
2.38	0.80	−0.28	−1.08	−1.31	2.19	3.50
2.40	0.80	−0.28	−1.08	−1.29	2.15	3.44
2.42	0.80	−0.29	−1.08	−1.27	2.12	3.39
2.44	0.79	−0.29	−1.08	−1.25	2.09	3.34
2.46	0.79	−0.29	−1.09	−1.23	2.07	3.29
2.48	0.79	−0.30	−1.09	−1.21	2.04	3.25
2.50	0.79	−0.30	−1.09	−1.19	2.01	3.20
3.00	0.74	−0.40	−1.14	−0.96	1.52	2.48
3.50	0.71	−0.48	−1.19	−0.95	1.16	2.11
4.00	0.68	−0.55	−1.22	−1.10	0.77	1.87
4.50	0.65	−0.60	−1.26	−1.57	0.12	1.69
5.00	0.63	−0.65	−1.28	−3.43	−1.88	1.55
5.50	0.62	−0.69	−1.31	12.33	13.75	1.43
6.00	0.60	−0.73	−1.33	2.27	3.58	1.31
6.50	0.59	−0.76	−1.35	1.33	2.51	1.19
7.00	0.58	−0.79	−1.37	1.00	2.06	1.06
7.50	0.57	−0.82	−1.39	0.86	1.78	0.92
8.00	0.56	−0.84	−1.40	0.81	1.56	0.75
8.50	0.55	−0.87	−1.42	0.82	1.36	0.53
9.00	0.55	−0.89	−1.43	0.89	1.14	0.24

TABLE 8.26 (continued)

k^2	Objective f/R	Objective z_m/R	Objective z_p/R	Projective f/R	Projective z_m/R	Projective z_p/R
9.50	0.54	−0.91	−1.45	1.05	0.86	−0.19
10.00	0.53	−0.93	−1.46	1.36	0.44	−0.92
15.00	0.49	−1.07	−1.56	−0.76	1.64	2.40
20.00	0.46	−1.16	−1.62	3.07	4.66	1.59
30.00	0.44	−1.28	−1.71	−3.61	5.29	8.91
40.00	0.42	−1.36	−1.78	−2.26	−0.34	1.92
50.00	0.41	−1.42	−1.83	0.77	1.53	0.76
60.00	0.40	−1.47	−1.87	−0.85	2.38	3.23
70.00	0.39	−1.51	−1.90	−2.74	−0.74	2.00
80.00	0.39	−1.54	−1.93	0.69	1.96	1.27
90.00	0.38	−1.57	−1.95	−7.27	9.21	16.48
100.00	0.38	−1.60	−1.98	−0.69	1.90	2.59

TABLE 8.27

RATIO OF GAP WIDTH TO BORE DIAMETER $S/D = 1.6$

k^2	Objective f/R	Objective z_m/R	Objective z_p/R	Projective f/R	Projective z_m/R	Projective z_p/R
0.01	38.97	38.96	−0.01	38.97	38.96	−0.01
0.02	19.75	19.73	−0.02	19.75	19.73	−0.02
0.04	10.14	10.10	−0.04	10.14	10.10	−0.04
0.06	6.95	6.88	−0.07	6.95	6.88	−0.07
0.08	5.36	5.27	−0.09	5.36	5.27	−0.09
0.10	4.41	4.29	−0.11	4.41	4.29	−0.11
0.12	3.78	3.64	−0.14	3.78	3.64	−0.14
0.14	3.33	3.17	−0.16	3.33	3.17	−0.16
0.16	3.00	2.81	−0.19	3.00	2.81	−0.19
0.18	2.74	2.53	−0.22	2.74	2.53	−0.22
0.20	2.54	2.30	−0.24	2.54	2.30	−0.24
0.22	2.38	2.11	−0.27	2.38	2.11	−0.27
0.24	2.25	1.95	−0.30	2.25	1.95	−0.30
0.26	2.14	1.81	−0.33	2.14	1.81	−0.33
0.28	2.04	1.69	−0.36	2.05	1.68	−0.37
0.30	1.96	1.58	−0.38	1.97	1.57	−0.40
0.32	1.89	1.48	−0.41	1.91	1.48	−0.43
0.34	1.83	1.39	−0.43	1.85	1.39	−0.47
0.36	1.77	1.31	−0.46	1.81	1.30	−0.51
0.38	1.72	1.24	−0.48	1.77	1.23	−0.55
0.40	1.68	1.17	−0.50	1.74	1.15	−0.59

TABLE 8.27 (continued)

	Objective			Projective		
k^2	f/R	z_m/R	z_p/R	f/R	z_m/R	z_p/R
0.42	1.64	1.11	−0.53	1.71	1.08	−0.63
0.44	1.60	1.05	−0.55	1.69	1.02	−0.67
0.46	1.57	1.00	−0.57	1.67	0.96	−0.72
0.48	1.53	0.95	−0.58	1.66	0.90	−0.77
0.50	1.50	0.90	−0.60	1.65	0.84	−0.82
0.52	1.48	0.86	−0.62	1.65	0.78	−0.87
0.54	1.45	0.82	−0.63	1.65	0.72	−0.92
0.56	1.43	0.78	−0.65	1.65	0.67	−0.98
0.58	1.40	0.74	−0.66	1.65	0.61	−1.04
0.60	1.38	0.70	−0.68	1.66	0.56	−1.11
0.62	1.36	0.67	−0.69	1.67	0.50	−1.17
0.64	1.34	0.64	−0.71	1.69	0.44	−1.25
0.66	1.33	0.61	−0.72	1.71	0.39	−1.32
0.68	1.31	0.58	−0.73	1.73	0.33	−1.40
0.70	1.29	0.55	−0.74	1.75	0.27	−1.49
0.72	1.28	0.52	−0.75	1.78	0.21	−1.58
0.74	1.26	0.50	−0.76	1.81	0.14	−1.67
0.76	1.25	0.47	−0.78	1.85	0.08	−1.77
0.78	1.23	0.45	−0.79	1.89	0.01	−1.88
0.80	1.22	0.42	−0.80	1.93	−0.07	−2.00
0.82	1.21	0.40	−0.81	1.98	−0.14	−2.13
0.84	1.19	0.38	−0.81	2.04	−0.22	−2.26
0.86	1.18	0.36	−0.82	2.10	−0.31	−2.41
0.88	1.17	0.34	−0.83	2.17	−0.40	−2.57
0.90	1.16	0.32	−0.84	2.24	−0.50	−2.74
0.92	1.15	0.30	−0.85	2.32	−0.61	−2.93
0.94	1.14	0.28	−0.86	2.42	−0.72	−3.14
0.96	1.13	0.27	−0.87	2.52	−0.85	−3.37
0.98	1.12	0.25	−0.87	2.64	−0.98	−3.62
1.00	1.11	0.23	−0.88	2.77	−1.13	−3.90
1.02	1.10	0.22	−0.89	2.92	−1.30	−4.22
1.04	1.09	0.20	−0.89	3.09	−1.49	−4.58
1.06	1.09	0.18	−0.90	3.28	−1.70	−4.98
1.08	1.08	0.17	−0.91	3.51	−1.94	−5.45
1.10	1.07	0.15	−0.92	3.77	−2.22	−5.99
1.12	1.06	0.14	−0.92	4.08	−2.55	−6.62
1.14	1.05	0.13	−0.93	4.45	−2.93	−7.38
1.16	1.05	0.11	−0.93	4.90	−3.40	−8.29
1.18	1.04	0.10	−0.94	5.45	−3.97	−9.42
1.20	1.03	0.09	−0.95	6.16	−4.69	−10.85
1.22	1.03	0.07	−0.95	7.09	−5.63	−12.72
1.24	1.02	0.06	−0.96	8.35	−6.91	−15.26

TABLE 8.27 (continued)

	Objective			Projective		
k^2	f/R	z_m/R	z_p/R	f/R	z_m/R	z_p/R
1.26	1.01	0.05	−0.96	10.18	−8.76	−18.94
1.28	1.01	0.04	−0.97	13.06	−11.65	−24.72
1.30	1.00	0.03	−0.97	18.25	−16.85	−35.10
1.32	1.00	0.02	−0.98	30.34	−28.96	−59.31
1.34	0.99	0.01	−0.98	90.49	−89.12	−179.61
1.36	0.98	−0.01	−0.99	−91.75	93.11	184.86
1.38	0.98	−0.02	−0.99	−30.41	31.76	62.17
1.40	0.97	−0.03	−1.00	−18.22	19.56	37.78
1.42	0.97	−0.04	−1.00	−13.01	14.33	27.34
1.44	0.96	−0.05	−1.01	−10.12	11.43	21.54
1.46	0.96	−0.06	−1.01	−8.28	9.58	17.86
1.48	0.95	−0.07	−1.02	−7.01	8.29	15.30
1.50	0.95	−0.07	−1.02	−6.08	7.35	13.43
1.52	0.94	−0.08	−1.03	−5.37	6.63	12.00
1.54	0.94	−0.09	−1.03	−4.81	6.06	10.87
1.56	0.93	−0.10	−1.04	−4.36	5.59	9.95
1.58	0.93	−0.11	−1.04	−3.99	5.21	9.20
1.60	0.93	−0.12	−1.04	−3.67	4.89	8.56
1.62	0.92	−0.13	−1.05	−3.41	4.61	8.02
1.64	0.92	−0.13	−1.05	−3.18	4.37	7.55
1.66	0.91	−0.14	−1.06	−2.98	4.16	7.15
1.68	0.91	−0.15	−1.06	−2.81	3.98	6.79
1.70	0.91	−0.16	−1.06	−2.66	3.82	6.47
1.72	0.90	−0.17	−1.07	−2.52	3.67	6.19
1.74	0.90	−0.17	−1.07	−2.40	3.54	5.94
1.76	0.89	−0.18	−1.08	−2.29	3.42	5.71
1.78	0.89	−0.19	−1.08	−2.19	3.31	5.50
1.80	0.89	−0.20	−1.08	−2.10	3.21	5.31
1.82	0.88	−0.20	−1.09	−2.02	3.11	5.14
1.84	0.88	−0.21	−1.09	−1.95	3.03	4.98
1.86	0.88	−0.22	−1.09	−1.88	2.95	4.83
1.88	0.87	−0.22	−1.10	−1.82	2.88	4.69
1.90	0.87	−0.23	−1.10	−1.76	2.81	4.57
1.92	0.87	−0.24	−1.10	−1.70	2.74	4.45
1.94	0.86	−0.24	−1.11	−1.65	2.68	4.34
1.96	0.86	−0.25	−1.11	−1.61	2.63	4.24
1.98	0.86	−0.26	−1.11	−1.57	2.57	4.14
2.00	0.85	−0.26	−1.12	−1.53	2.52	4.05
2.02	0.85	−0.27	−1.12	−1.49	2.48	3.97
2.04	0.85	−0.27	−1.12	−1.45	2.43	3.89
2.06	0.84	−0.28	−1.13	−1.42	2.39	3.81
2.08	0.84	−0.29	−1.13	−1.39	2.35	3.74
2.10	0.84	−0.29	−1.13	−1.36	2.31	3.67
2.12	0.84	−0.30	−1.13	−1.34	2.27	3.61

TABLE 8.27 (continued)

	Objective			Projective		
k^2	f/R	z_m/R	z_p/R	f/R	z_m/R	z_p/R
2.14	0.83	−0.30	−1.14	−1.31	2.23	3.55
2.16	0.83	−0.31	−1.14	−1.29	2.20	3.49
2.18	0.83	−0.31	−1.14	−1.26	2.17	3.43
2.20	0.83	−0.32	−1.15	−1.24	2.13	3.38
2.22	0.82	−0.33	−1.15	−1.22	2.10	3.33
2.24	0.82	−0.33	−1.15	−1.21	2.07	3.28
2.26	0.82	−0.34	−1.15	−1.19	2.05	3.23
2.28	0.82	−0.34	−1.16	−1.17	2.02	3.19
2.30	0.81	−0.35	−1.16	−1.16	1.99	3.15
2.32	0.81	−0.35	−1.16	−1.14	1.96	3.11
2.34	0.81	−0.36	−1.16	−1.13	1.94	3.07
2.36	0.81	−0.36	−1.17	−1.11	1.91	3.03
2.38	0.80	−0.37	−1.17	−1.10	1.89	2.99
2.40	0.80	−0.37	−1.17	−1.09	1.87	2.96
2.42	0.80	−0.38	−1.17	−1.08	1.84	2.92
2.44	0.80	−0.38	−1.18	−1.07	1.82	2.89
2.46	0.79	−0.38	−1.18	−1.06	1.80	2.86
2.48	0.79	−0.39	−1.18	−1.05	1.78	2.83
2.50	0.79	−0.39	−1.18	−1.04	1.76	2.80
3.00	0.74	−0.49	−1.24	−0.96	1.32	2.28
3.50	0.71	−0.57	−1.28	−1.10	0.88	1.99
4.00	0.68	−0.64	−1.32	−1.61	0.17	1.78
4.50	0.66	−0.69	−1.35	−4.14	−2.52	1.62
5.00	0.64	−0.74	−1.38	6.02	7.50	1.47
5.50	0.62	−0.78	−1.40	1.82	3.16	1.34
6.00	0.61	−0.82	−1.43	1.16	2.36	1.20
6.50	0.59	−0.85	−1.45	0.92	1.97	1.05
7.00	0.58	−0.88	−1.47	0.83	1.70	0.87
7.50	0.57	−0.91	−1.48	0.83	1.47	0.64
8.00	0.56	−0.94	−1.50	0.90	1.22	0.32
8.50	0.56	−0.96	−1.51	1.08	0.91	−0.17
9.00	0.55	−0.98	−1.53	1.47	0.39	−1.08
9.50	0.54	−1.00	−1.54	2.60	−0.84	−3.44
10.00	0.53	−1.02	−1.55	14.79	−13.12	−27.92
15.00	0.49	−1.16	−1.65	−0.96	1.12	2.09
20.00	0.47	−1.25	−1.72	0.86	2.21	1.35
30.00	0.44	−1.37	−1.81	−0.78	2.15	2.93
40.00	0.42	−1.45	−1.87	0.98	2.58	1.60
50.00	0.41	−1.51	−1.92	−6.25	8.16	14.41
60.00	0.40	−1.56	−1.96	−0.96	1.32	2.28
70.00	0.39	−1.60	−2.00	0.73	2.26	1.52
80.00	0.39	−1.64	−2.02	67.43	−65.38	−132.82
90.00	0.39	−1.66	−2.05	−0.70	1.92	2.62
100.00	0.38	−1.69	−2.07	2.13	4.10	1.97

TABLE 8.28
RATIO OF GAP WIDTH TO BORE DIAMETER $S/D = 1.7$

	Objective			Projective		
k^2	f/R	z_m/R	z_p/R	f/R	z_m/R	z_p/R
0.01	36.61	36.60	−0.01	36.61	36.60	−0.01
0.02	18.58	18.56	−0.03	18.58	18.56	−0.03
0.04	9.58	9.53	−0.05	9.58	9.53	−0.05
0.06	6.59	6.51	−0.08	6.59	6.51	−0.08
0.08	5.09	4.99	−0.11	5.09	4.99	−0.11
0.10	4.21	4.07	−0.13	4.21	4.07	−0.13
0.12	3.62	3.45	−0.16	3.62	3.45	−0.16
0.14	3.20	3.01	−0.19	3.20	3.01	−0.19
0.16	2.89	2.67	−0.23	2.89	2.67	−0.23
0.18	2.66	2.40	−0.26	2.66	2.40	−0.26
0.20	2.47	2.18	−0.29	2.47	2.18	−0.29
0.22	2.32	2.00	−0.33	2.33	2.00	−0.33
0.24	2.20	1.84	−0.36	2.21	1.84	−0.37
0.26	2.10	1.71	−0.39	2.11	1.70	−0.41
0.28	2.01	1.59	−0.42	2.03	1.58	−0.45
0.30	1.94	1.48	−0.46	1.96	1.47	−0.49
0.32	1.87	1.39	−0.49	1.91	1.38	−0.53
0.34	1.81	1.30	−0.51	1.86	1.28	−0.58
0.36	1.76	1.22	−0.54	1.83	1.20	−0.63
0.38	1.71	1.15	−0.56	1.80	1.12	−0.68
0.40	1.67	1.08	−0.59	1.78	1.04	−0.73
0.42	1.63	1.02	−0.61	1.76	0.97	−0.79
0.44	1.60	0.96	−0.63	1.75	0.90	−0.85
0.46	1.56	0.91	−0.65	1.74	0.83	−0.91
0.48	1.53	0.86	−0.67	1.74	0.77	−0.97
0.50	1.50	0.81	−0.69	1.74	0.70	−1.04
0.52	1.48	0.77	−0.71	1.75	0.63	−1.12
0.54	1.45	0.73	−0.72	1.76	0.57	−1.19
0.56	1.43	0.69	−0.74	1.78	0.50	−1.28
0.58	1.41	0.65	−0.76	1.79	0.43	−1.36
0.60	1.39	0.62	−0.77	1.82	0.36	−1.46
0.62	1.37	0.58	−0.78	1.85	0.29	−1.56
0.64	1.35	0.55	−0.80	1.88	0.22	−1.66
0.66	1.33	0.52	−0.81	1.92	0.14	1.78
0.68	1.31	0.49	−0.82	1.96	0.06	−1.90
0.70	1.29	0.46	−0.83	2.01	−0.02	−2.03
0.72	1.28	0.43	−0.85	2.07	−0.11	−2.18
0.74	1.26	0.41	−0.86	2.13	−0.21	−2.34
0.76	1.25	0.38	−0.87	2.20	−0.31	−2.51
0.78	1.24	0.36	−0.88	2.28	−0.41	−2.70
0.80	1.22	0.34	−0.89	2.37	−0.53	−2.90
0.82	1.21	0.31	−0.90	2.47	−0.66	−3.13

TABLE 8.28 (continued)

	Objective			Projective		
k^2	f/R	z_m/R	z_p/R	f/R	z_m/R	z_p/R
0.84	1.20	0.29	−0.91	2.59	−0.80	−3.39
0.86	1.19	0.27	−0.92	2.72	−0.95	−3.67
0.88	1.18	0.25	−0.93	2.87	−1.13	−3.99
0.90	1.16	0.23	−0.93	3.04	−1.32	−4.36
0.92	1.15	0.21	−0.94	3.23	−1.54	−4.77
0.94	1.14	0.19	−0.95	3.46	−1.79	−5.25
0.96	1.13	0.18	−0.96	3.74	−2.08	−5.82
0.98	1.12	0.16	−0.97	4.06	−2.42	−6.48
1.00	1.12	0.14	−0.97	4.45	−2.84	−7.29
1.02	1.11	0.13	−0.98	4.94	−3.34	−8.27
1.04	1.10	0.11	−0.99	5.55	−3.97	−9.51
1.06	1.09	0.09	−1.00	6.34	−4.78	−11.12
1.08	1.08	0.08	−1.00	7.41	−5.87	−13.28
1.10	1.07	0.06	−1.01	8.94	−7.41	−16.35
1.12	1.07	0.05	−1.02	11.27	−9.76	−21.03
1.14	1.06	0.04	−1.02	15.29	−13.80	−29.09
1.16	1.05	0.02	−1.03	23.85	−22.37	−46.22
1.18	1.04	0.01	−1.03	54.43	−52.97	−107.40
1.20	1.04	−0.00	−1.04	−190.14	191.58	381.72
1.22	1.03	−0.02	−1.05	−34.55	35.98	70.54
1.24	1.02	−0.03	−1.05	−18.99	20.41	39.40
1.26	1.02	−0.04	−1.06	−13.09	14.50	27.59
1.28	1.01	−0.05	−1.06	−9.99	11.38	21.37
1.30	1.01	−0.06	−1.07	−8.08	9.46	17.54
1.32	1.00	−0.07	−1.07	−6.79	8.15	14.93
1.34	0.99	−0.08	−1.08	−5.85	7.20	13.05
1.36	0.99	−0.10	−1.08	−5.15	6.48	11.62
1.38	0.98	−0.11	−1.09	−4.60	5.91	10.51
1.40	0.98	−0.12	−1.09	−4.15	5.46	9.61
1.42	0.97	−0.13	−1.10	−3.79	5.08	8.87
1.44	0.97	−0.14	−1.10	−3.49	4.77	8.25
1.46	0.96	−0.15	−1.11	−3.23	4.50	7.73
1.48	0.96	−0.16	−1.11	−3.01	4.27	7.28
1.50	0.95	−0.17	−1.12	−2.82	4.06	6.89
1.52	0.95	−0.17	−1.12	−2.66	3.88	6.54
1.54	0.94	−0.18	−1.13	−2.51	3.73	6.24
1.56	0.94	−0.19	−1.13	−2.38	3.58	5.97
1.58	0.93	−0.20	−1.14	−2.27	3.46	5.72
1.60	0.93	−0.21	−1.14	−2.16	3.34	5.51
1.62	0.93	−0.22	−1.14	−2.07	3.24	5.31
1.64	0.92	−0.23	−1.15	−1.99	3.14	5.13
1.66	0.92	−0.23	−1.15	−1.91	3.05	4.96
1.68	0.91	−0.24	−1.16	−1.84	2.97	4.81

TABLE 8.28 (continued)

	Objective			Projective		
k^2	f/R	z_m/R	z_p/R	f/R	z_m/R	z_p/R
1.70	0.91	−0.25	−1.16	−1.78	2.89	4.67
1.72	0.91	−0.26	−1.16	−1.72	2.82	4.54
1.74	0.90	−0.27	−1.17	−1.67	2.76	4.42
1.76	0.90	−0.27	−1.17	−1.62	2.69	4.31
1.78	0.89	−0.28	−1.17	−1.57	2.64	4.21
1.80	0.89	−0.29	−1.18	−1.53	2.58	4.11
1.82	0.89	−0.29	−1.18	−1.49	2.53	4.02
1.84	0.88	−0.30	−1.18	−1.45	2.48	3.93
1.86	0.88	−0.31	−1.19	−1.42	2.43	3.85
1.88	0.88	−0.31	−1.19	−1.39	2.39	3.78
1.90	0.87	−0.32	−1.19	−1.36	2.35	3.70
1.92	0.87	−0.33	−1.20	−1.33	2.31	3.64
1.94	0.87	−0.33	−1.20	−1.30	2.27	3.57
1.96	0.86	−0.34	−1.20	−1.28	2.23	3.51
1.98	0.86	−0.35	−1.21	−1.26	2.20	3.45
2.00	0.86	−0.35	−1.21	−1.23	2.16	3.40
2.02	0.85	−0.36	−1.21	−1.21	2.13	3.35
2.04	0.85	−0.37	−1.22	−1.20	2.10	3.30
2.06	0.85	−0.37	−1.22	−1.18	2.07	3.25
2.08	0.85	−0.38	−1.22	−1.16	2.04	3.20
2.10	0.84	−0.38	−1.23	−1.15	2.01	3.16
2.12	0.84	−0.39	−1.23	−1.13	1.98	3.12
2.14	0.84	−0.39	−1.23	−1.12	1.96	3.08
2.16	0.83	−0.40	−1.24	−1.11	1.93	3.04
2.18	0.83	−0.41	−1.24	−1.10	1.91	3.00
2.20	0.83	−0.41	−1.24	−1.08	1.88	2.96
2.22	0.83	−0.42	−1.24	−1.07	1.86	2.93
2.24	0.82	−0.42	−1.25	−1.06	1.83	2.90
2.26	0.82	−0.43	−1.25	−1.06	1.81	2.87
2.28	0.82	−0.43	−1.25	−1.05	1.79	2.83
2.30	0.82	−0.44	−1.25	−1.04	1.76	2.80
2.32	0.81	−0.44	−1.26	−1.03	1.74	2.78
2.34	0.81	−0.45	−1.26	−1.03	1.72	2.75
2.36	0.81	−0.45	−1.26	−1.02	1.70	2.72
2.38	0.81	−0.46	−1.26	−1.02	1.68	2.69
2.40	0.80	−0.46	−1.27	−1.01	1.66	2.67
2.42	0.80	−0.47	−1.27	−1.01	1.64	2.64
2.44	0.80	−0.47	−1.27	−1.00	1.62	2.62
2.46	0.80	−0.48	−1.27	−1.00	1.60	2.60
2.48	0.80	−0.48	−1.28	−1.00	1.58	2.58
2.50	0.79	−0.49	−1.28	−0.99	1.56	2.55
3.00	0.75	−0.59	−1.33	−1.07	1.08	2.15
3.50	0.71	−0.67	−1.38	−1.52	0.37	1.90

TABLE 8.28 (continued)

	Objective			Projective		
k^2	f/R	z_m/R	z_p/R	f/R	z_m/R	z_p/R
4.00	0.68	−0.73	−1.41	−3.98	−2.27	1.70
4.50	0.66	−0.79	−1.45	5.20	6.74	1.54
5.00	0.64	−0.83	−1.47	1.66	3.05	1.38
5.50	0.62	−0.88	−1.50	1.08	2.31	1.23
6.00	0.61	−0.91	−1.52	0.89	1.93	1.04
6.50	0.60	−0.95	−1.54	0.84	1.65	0.82
7.00	0.58	−0.98	−1.56	0.88	1.39	0.51
7.50	0.57	−1.01	−1.58	1.04	1.06	0.02
8.00	0.57	−1.03	−1.60	1.43	0.54	−0.89
8.50	0.56	−1.05	−1.61	2.61	−0.77	−3.38
9.00	0.55	−1.08	−1.63	28.35	−26.61	−54.96
9.50	0.54	−1.10	−1.64	−3.22	4.86	8.08
10.00	0.54	−1.11	−1.65	−1.59	3.13	4.72
15.00	0.49	−1.26	−1.75	−3.77	−1.88	1.90
20.00	0.47	−1.35	−1.81	0.77	1.66	0.89
30.00	0.44	−1.47	−1.91	−0.87	1.46	2.33
40.00	0.42	−1.55	−1.97	0.76	1.72	0.95
50.00	0.41	−1.61	−2.02	−0.71	2.17	2.88
60.00	0.40	−1.66	−2.06	1.25	3.12	1.87
70.00	0.39	−1.70	−2.09	1.79	0.46	−1.33
80.00	0.39	−1.73	−2.12	−0.69	2.11	2.80
90.00	0.39	−1.76	−2.15	1.99	4.02	2.04
100.00	0.38	−1.79	−2.17	0.81	1.74	0.93

TABLE 8.29
RATIO OF GAP WIDTH TO BORE DIAMETER $S/D = 1.8$

	Objective			Projective		
k^2	f/R	z_m/R	z_p/R	f/R	z_m/R	z_p/R
0.01	34.49	34.47	−0.01	34.49	34.47	−0.01
0.02	17.54	17.51	−0.03	17.54	17.51	−0.03
0.04	9.07	9.01	−0.06	9.07	9.01	−0.06
0.06	6.26	6.17	−0.09	6.26	6.17	−0.09
0.08	4.86	4.74	−0.12	4.86	4.74	−0.12
0.10	4.03	3.87	−0.16	4.03	3.87	−0.16
0.12	3.48	3.29	−0.20	3.48	3.29	−0.20
0.14	3.09	2.86	−0.23	3.09	2.86	−0.23
0.16	2.81	2.54	−0.27	2.81	2.54	−0.27
0.18	2.59	2.28	−0.31	2.59	2.28	−0.31

TABLE 8.29 (continued)

	Objective			Projective		
k^2	f/R	z_m/R	z_p/R	f/R	z_m/R	z_p/R
0.20	2.42	2.07	−0.35	2.42	2.07	−0.35
0.22	2.28	1.89	−0.39	2.29	1.89	−0.40
0.24	2.17	1.74	−0.43	2.18	1.74	−0.45
0.26	2.07	1.61	−0.47	2.10	1.60	−0.49
0.28	1.99	1.49	−0.50	2.03	1.48	−0.55
0.30	1.92	1.39	−0.54	1.97	1.37	−0.60
0.32	1.86	1.29	−0.57	1.93	1.27	−0.66
0.34	1.80	1.21	−0.60	1.89	1.18	−0.72
0.36	1.75	1.13	−0.63	1.87	1.09	−0.78
0.38	1.71	1.06	−0.65	1.85	1.00	−0.85
0.40	1.67	0.99	−0.68	1.84	0.92	−0.92
0.42	1.63	0.93	−0.70	1.83	0.84	−0.99
0.44	1.60	0.87	−0.72	1.83	0.76	−1.07
0.46	1.56	0.82	−0.74	1.84	0.68	−1.16
0.48	1.53	0.77	−0.76	1.85	0.61	−1.25
0.50	1.50	0.72	−0.78	1.87	0.53	−1.34
0.52	1.48	0.68	−0.80	1.89	0.45	−1.45
0.54	1.45	0.64	−0.82	1.92	0.36	−1.56
0.56	1.43	0.60	−0.83	1.96	0.28	−1.68
0.58	1.41	0.56	−0.85	2.00	0.19	−1.81
0.60	1.39	0.52	−0.86	2.05	0.10	−1.95
0.62	1.37	0.49	−0.88	2.11	0.00	−2.11
0.64	1.35	0.46	−0.89	2.17	−0.10	−2.28
0.66	1.33	0.43	−0.90	2.25	−0.21	−2.46
0.68	1.31	0.40	−0.92	2.33	−0.33	−2.67
0.70	1.30	0.37	−0.93	2.43	−0.46	−2.89
0.72	1.28	0.34	−0.94	2.54	−0.61	−3.15
0.74	1.27	0.32	−0.95	2.67	−0.76	−3.43
0.76	1.25	0.29	−0.96	2.81	−0.94	−3.76
0.78	1.24	0.27	−0.97	2.98	−1.14	−4.12
0.80	1.23	0.24	−0.98	3.18	−1.36	−4.55
0.82	1.21	0.22	−0.99	3.42	−1.62	−5.04
0.84	1.20	0.20	−1.00	3.69	−1.93	−5.62
0.86	1.19	0.18	−1.01	4.03	−2.28	−6.31
0.88	1.18	0.16	−1.02	4.44	−2.72	−7.16
0.90	1.17	0.14	−1.03	4.96	−3.26	−8.21
0.92	1.16	0.12	−1.04	5.62	−3.94	−9.56
0.94	1.15	0.10	−1.05	6.50	−4.84	−11.34
0.96	1.14	0.08	−1.05	7.72	−6.09	−13.81
0.98	1.13	0.07	−1.06	9.54	−7.92	−17.46
1.00	1.12	0.05	−1.07	12.51	−10.91	−23.42
1.02	1.11	0.03	−1.08	18.22	−16.64	−34.86
1.04	1.10	0.02	−1.08	33.71	−32.15	−65.86

TABLE 8.29 (continued)

	Objective			Projective		
k^2	f/R	z_m/R	z_p/R	f/R	z_m/R	z_p/R
1.06	1.09	0.00	−1.09	230.98	−229.44	−460.42
1.08	1.09	−0.01	−1.10	−47.41	48.94	96.35
1.10	1.08	−0.03	−1.10	−21.48	22.98	44.46
1.12	1.07	−0.04	−1.11	−13.88	15.37	29.24
1.14	1.06	−0.06	−1.12	−10.25	11.72	21.97
1.16	1.06	−0.07	−1.12	−8.13	9.58	17.71
1.18	1.05	−0.08	−1.13	−6.74	8.18	14.92
1.20	1.04	−0.09	−1.14	−5.76	7.18	12.94
1.22	1.03	−0.11	−1.14	−5.03	6.43	11.46
1.24	1.03	−0.12	−1.15	−4.47	5.86	10.32
1.26	1.02	−0.13	−1.15	−4.02	5.39	9.41
1.28	1.02	−0.14	−1.16	−3.66	5.02	8.67
1.30	1.01	−0.16	−1.16	−3.36	4.70	8.06
1.32	1.00	−0.17	−1.17	−3.10	4.43	7.54
1.34	1.00	−0.18	−1.17	−2.89	4.20	7.09
1.36	0.99	−0.19	−1.18	−2.70	4.00	6.70
1.38	0.99	−0.20	−1.19	−2.54	3.83	6.37
1.40	0.98	−0.21	−1.19	−2.40	3.67	6.07
1.42	0.98	−0.22	−1.19	−2.27	3.53	5.81
1.44	0.97	−0.23	−1.20	−2.16	3.41	5.57
1.46	0.97	−0.24	−1.20	−2.06	3.29	5.36
1.48	0.96	−0.25	−1.21	−1.98	3.19	5.17
1.50	0.96	−0.26	−1.21	−1.90	3.10	4.99
1.52	0.95	−0.27	−1.22	−1.82	3.01	4.83
1.54	0.95	−0.28	−1.22	−1.76	2.93	4.69
1.56	0.94	−0.28	−1.23	−1.70	2.85	4.55
1.58	0.94	−0.29	−1.23	−1.64	2.78	4.43
1.60	0.93	−0.30	−1.24	−1.59	2.72	4.31
1.62	0.93	−0.31	−1.24	−1.54	2.66	4.20
1.64	0.92	−0.32	−1.24	−1.50	2.60	4.10
1.66	0.92	−0.33	−1.25	−1.46	2.55	4.01
1.68	0.92	−0.33	−1.25	−1.43	2.50	3.92
1.70	0.91	−0.34	−1.26	−1.39	2.45	3.84
1.72	0.91	−0.35	−1.26	−1.36	2.40	3.77
1.74	0.90	−0.36	−1.26	−1.33	2.36	3.69
1.76	0.90	−0.37	−1.27	−1.30	2.32	3.62
1.78	0.90	−0.37	−1.27	−1.28	2.28	3.56
1.80	0.89	−0.38	−1.27	−1.26	2.24	3.50
1.82	0.89	−0.39	−1.28	−1.23	2.21	3.44
1.84	0.89	−0.39	−1.28	−1.21	2.17	3.38
1.86	0.88	−0.40	−1.28	−1.20	2.14	3.33
1.88	0.88	−0.41	−1.29	−1.18	2.10	3.28
1.90	0.88	−0.41	−1.29	−1.16	2.07	3.23

TABLE 8.29 (continued)

k^2	Objective			Projective		
	f/R	z_m/R	z_p/R	f/R	z_m/R	z_p/R
1.92	0.87	−0.42	−1.29	−1.15	2.04	3.19
1.94	0.87	−0.43	−1.30	−1.13	2.01	3.15
1.96	0.87	−0.43	−1.30	−1.12	1.98	3.10
1.98	0.86	−0.44	−1.30	−1.11	1.96	3.06
2.00	0.86	−0.45	−1.31	−1.10	1.93	3.02
2.02	0.86	−0.45	−1.31	−1.09	1.90	2.99
2.04	0.85	−0.46	−1.31	−1.08	1.88	2.95
2.06	0.85	−0.47	−1.32	−1.07	1.85	2.92
2.08	0.85	−0.47	−1.32	−1.06	1.82	2.89
2.10	0.85	−0.48	−1.32	−1.05	1.80	2.85
2.12	0.84	−0.48	−1.33	−1.05	1.78	2.82
2.14	0.84	−0.49	−1.33	−1.04	1.75	2.79
2.16	0.84	−0.49	−1.33	−1.04	1.73	2.77
2.18	0.83	−0.50	−1.33	−1.03	1.71	2.74
2.20	0.83	−0.51	−1.34	−1.03	1.68	2.71
2.22	0.83	−0.51	−1.34	−1.02	1.66	2.69
2.24	0.83	−0.52	−1.34	−1.02	1.64	2.66
2.26	0.82	−0.52	−1.35	−1.02	1.62	2.64
2.28	0.82	−0.53	−1.35	−1.02	1.60	2.61
2.30	0.82	−0.53	−1.35	−1.02	1.57	2.59
2.32	0.82	−0.54	−1.35	−1.01	1.55	2.57
2.34	0.81	−0.54	−1.36	−1.01	1.53	2.55
2.36	0.81	−0.55	−1.36	−1.02	1.51	2.52
2.38	0.81	−0.55	−1.36	−1.02	1.49	2.50
2.40	0.81	−0.56	−1.36	−1.02	1.47	2.48
2.42	0.81	−0.56	−1.37	−1.02	1.45	2.47
2.44	0.80	−0.57	−1.37	−1.02	1.42	2.45
2.46	0.80	−0.57	−1.37	−1.02	1.40	2.43
2.48	0.80	−0.58	−1.37	−1.03	1.38	2.41
2.50	0.80	−0.58	−1.38	−1.03	1.36	2.39
3.00	0.75	−0.68	−1.43	−1.35	0.70	2.06
3.50	0.71	−0.76	−1.47	−3.07	−1.25	1.82
4.00	0.69	−0.83	−1.51	6.46	8.09	1.64
4.50	0.66	−0.88	−1.54	1.66	3.12	1.46
5.00	0.64	−0.93	−1.57	1.06	2.34	1.28
5.50	0.63	−0.97	−1.60	0.88	1.94	1.06
6.00	0.61	−1.01	−1.62	0.86	1.64	0.78
6.50	0.60	−1.04	−1.64	0.95	1.32	0.37
7.00	0.59	−1.07	−1.66	1.24	0.87	−0.37
7.50	0.58	−1.10	−1.68	2.09	−0.13	−2.22
8.00	0.57	−1.13	−1.69	9.60	−7.76	−17.36
8.50	0.56	−1.15	−1.71	−3.63	5.36	8.98
9.00	0.55	−1.17	−1.72	−1.60	3.21	4.81

TABLE 8.29 (continued)

k^2	Objective			Projective		
	f/R	z_m/R	z_p/R	f/R	z_m/R	z_p/R
9.50	0.54	−1.19	−1.74	−1.09	2.58	3.67
10.00	0.54	−1.21	−1.74	−0.89	2.24	3.13
15.00	0.49	−1.35	−1.86	1.55	3.27	1.72
20.00	0.47	−1.44	−1.91	1.64	0.50	−1.14
30.00	0.44	−1.57	−2.00	21.92	23.95	2.03
40.00	0.42	−1.65	−2.07	26.83	−24.72	−51.55
50.00	0.41	−1.71	−2.12	−1.62	0.68	2.30
60.00	0.40	−1.76	−2.16	0.77	1.84	1.07
70.00	0.40	−1.80	−2.19	−0.76	2.52	3.28
80.00	0.39	−1.83	−2.22	6.07	8.26	2.19
90.00	0.39	−1.86	−2.24	0.79	1.85	1.06
100.00	0.38	−1.88	−2.27	−0.92	2.89	3.82

TABLE 8.30

RATIO OF GAP WIDTH TO BORE DIAMETER $S/D = 1.9$

k^2	Objective			Projective		
	f/R	z_m/R	z_p/R	f/R	z_m/R	z_p/R
0.01	32.57	32.55	−0.02	32.57	32.55	−0.02
0.02	16.59	16.56	−0.03	16.59	16.56	−0.03
0.04	8.62	8.55	−0.07	8.62	8.55	−0.07
0.06	5.97	5.86	−0.11	5.97	5.86	−0.11
0.08	4.66	4.51	−0.15	4.66	4.51	−0.15
0.10	3.88	3.69	−0.19	3.88	3.69	−0.19
0.12	3.36	3.13	−0.23	3.36	3.13	−0.23
0.14	3.00	2.73	−0.28	3.00	2.73	−0.28
0.16	2.74	2.42	−0.32	2.74	2.42	−0.32
0.18	2.54	2.17	−0.37	2.54	2.17	−0.37
0.20	2.38	1.96	−0.42	2.39	1.96	−0.43
0.22	2.25	1.79	−0.46	2.27	1.79	−0.48
0.24	2.15	1.64	−0.51	2.17	1.63	−0.54
0.26	2.06	1.51	−0.55	2.10	1.50	−0.60
0.28	1.98	1.40	−0.59	2.04	1.38	−0.66
0.30	1.91	1.29	−0.62	2.00	1.26	−0.73
0.32	1.85	1.20	−0.66	1.97	1.16	−0.81
0.34	1.80	1.11	−0.69	1.94	1.06	−0.88
0.36	1.75	1.04	−0.72	1.93	0.96	−0.97
0.38	1.71	0.96	−0.74	1.93	0.87	−1.06
0.40	1.67	0.90	−0.77	1.93	0.78	−1.15
0.42	1.63	0.84	−0.79	1.94	0.69	−1.25

TABLE 8.30 (continued)

	Objective			Projective		
k^2	f/R	z_m/R	z_p/R	f/R	z_m/R	z_p/R
0.44	1.60	0.78	−0.81	1.96	0.59	−1.36
0.46	1.56	0.73	−0.84	1.98	0.50	−1.48
0.48	1.53	0.68	−0.86	2.02	0.40	−1.61
0.50	1.51	0.63	−0.88	2.06	0.30	−1.75
0.52	1.48	0.59	−0.89	2.11	0.20	−1.91
0.54	1.46	0.55	−0.91	2.17	0.09	−2.07
0.56	1.43	0.51	−0.93	2.24	−0.02	−2.26
0.58	1.41	0.47	−0.94	2.32	−0.15	−2.47
0.60	1.39	0.43	−0.96	2.41	−0.29	−2.69
0.62	1.37	0.40	−0.97	2.52	−0.43	−2.95
0.64	1.35	0.37	−0.99	2.64	−0.60	−3.24
0.66	1.33	0.34	−1.00	2.79	−0.78	−3.57
0.68	1.32	0.31	−1.01	2.96	−0.98	−3.94
0.70	1.30	0.28	−1.02	3.16	−1.22	−4.38
0.72	1.29	0.25	−1.04	3.40	−1.49	−4.89
0.74	1.27	0.22	−1.05	3.69	−1.81	−5.51
0.76	1.26	0.20	−1.06	4.05	−2.19	−6.24
0.78	1.24	0.17	−1.07	4.49	−2.66	−7.16
0.80	1.23	0.15	−1.08	5.06	−3.26	−8.31
0.82	1.22	0.13	−1.09	5.80	−4.02	−9.82
0.84	1.21	0.11	−1.10	6.81	−5.06	−11.88
0.86	1.19	0.09	−1.11	8.29	−6.56	−14.85
0.88	1.18	0.07	−1.12	10.61	−8.91	−19.52
0.90	1.17	0.05	−1.13	14.80	−13.12	−27.91
0.92	1.16	0.03	−1.13	24.57	−22.92	−47.49
0.94	1.15	0.01	−1.14	73.36	−71.72	−145.08
0.96	1.14	−0.01	−1.15	−73.76	75.38	149.13
0.98	1.13	−0.03	−1.16	−24.49	26.09	50.58
1.00	1.12	−0.04	−1.17	−14.67	16.25	30.92
1.02	1.11	−0.06	−1.17	−10.47	12.03	22.50
1.04	1.10	−0.08	−1.18	−8.15	9.68	17.83
1.06	1.10	−0.09	−1.19	−6.67	8.19	14.85
1.08	1.09	−0.11	−1.19	−5.65	7.15	12.79
1.10	1.08	−0.12	−1.20	−4.90	6.38	11.28
1.12	1.07	−0.13	−1.21	−4.33	5.79	10.12
1.14	1.07	−0.15	−1.21	−3.88	5.33	9.21
1.16	1.06	−0.16	−1.22	−3.52	4.95	8.47
1.18	1.05	−0.18	−1.23	−3.23	4.64	7.86
1.20	1.04	−0.19	−1.23	−2.98	4.37	7.35
1.22	1.04	−0.20	−1.24	−2.77	4.14	6.91
1.24	1.03	−0.21	−1.24	−2.59	3.95	6.53
1.26	1.02	−0.23	−1.25	−2.43	3.77	6.20
1.28	1.02	−0.24	−1.26	−2.29	3.62	5.91

TABLE 8.30 (continued)

	Objective			Projective		
k^2	f/R	z_m/R	z_p/R	f/R	z_m/R	z_p/R
1.30	1.01	−0.25	−1.26	−2.17	3.48	5.66
1.32	1.01	−0.26	−1.27	−2.07	3.36	5.43
1.34	1.00	−0.27	−1.27	−1.97	3.25	5.22
1.36	0.99	−0.28	−1.28	−1.89	3.15	5.04
1.38	0.99	−0.29	−1.28	−1.81	3.06	4.87
1.40	0.98	−0.30	−1.29	−1.74	2.97	4.72
1.42	0.98	−0.31	−1.29	−1.68	2.89	4.57
1.44	0.97	−0.32	−1.30	−1.62	2.82	4.44
1.46	0.97	−0.33	−1.30	−1.57	2.75	4.33
1.48	0.96	−0.34	−1.31	−1.53	2.69	4.21
1.50	0.96	−0.35	−1.31	−1.48	2.63	4.11
1.52	0.95	−0.36	−1.31	−1.44	2.57	4.02
1.54	0.95	−0.37	−1.32	−1.41	2.52	3.93
1.56	0.94	−0.38	−1.32	−1.37	2.47	3.84
1.58	0.94	−0.39	−1.33	−1.34	2.42	3.77
1.60	0.94	−0.40	−1.33	−1.31	2.38	3.69
1.62	0.93	−0.40	−1.34	−1.29	2.33	3.62
1.64	0.93	−0.41	−1.34	−1.26	2.29	3.56
1.66	0.92	−0.42	−1.34	−1.24	2.25	3.49
1.68	0.92	−0.43	−1.35	−1.22	2.22	3.44
1.70	0.92	−0.44	−1.35	−1.20	2.18	3.38
1.72	0.91	−0.44	−1.36	−1.18	2.14	3.33
1.74	0.91	−0.45	−1.36	−1.17	2.11	3.28
1.76	0.90	−0.46	−1.36	−1.15	2.08	3.23
1.78	0.90	−0.47	−1.37	−1.14	2.05	3.18
1.80	0.90	−0.47	−1.37	−1.13	2.01	3.14
1.82	0.89	−0.48	−1.37	−1.11	1.98	3.10
1.84	0.89	−0.49	−1.38	−1.10	1.95	3.06
1.86	0.89	−0.50	−1.38	−1.09	1.93	3.02
1.88	0.88	−0.50	−1.38	−1.09	1.90	2.98
1.90	0.88	−0.51	−1.39	−1.08	1.87	2.95
1.92	0.88	−0.52	−1.39	−1.07	1.84	2.91
1.94	0.87	−0.52	−1.39	−1.06	1.82	2.88
1.96	0.87	−0.53	−1.40	−1.06	1.79	2.85
1.98	0.87	−0.54	−1.40	−1.05	1.76	2.82
2.00	0.86	−0.54	−1.40	−1.05	1.74	2.79
2.02	0.86	−0.55	−1.41	−1.05	1.71	2.76
2.04	0.86	−0.55	−1.41	−1.05	1.69	2.73
2.06	0.85	−0.56	−1.41	−1.04	1.66	2.71
2.08	0.85	−0.57	−1.42	−1.04	1.64	2.68
2.10	0.85	−0.57	−1.42	−1.04	1.61	2.66
2.12	0.85	−0.58	−1.42	−1.04	1.59	2.63
2.14	0.84	−0.58	−1.43	−1.04	1.57	2.61

TABLE 8.30 (continued)

	Objective			Projective		
k^2	f/R	z_m/R	z_p/R	f/R	z_m/R	z_p/R
2.16	0.84	−0.59	−1.43	−1.05	1.54	2.59
2.18	0.84	−0.59	−1.43	−1.05	1.52	2.57
2.20	0.83	−0.60	−1.43	−1.05	1.49	2.54
2.22	0.83	−0.61	−1.44	−1.05	1.47	2.52
2.24	0.83	−0.61	−1.44	−1.06	1.44	2.50
2.26	0.83	−0.62	−1.44	−1.06	1.42	2.48
2.28	0.82	−0.62	−1.45	−1.07	1.40	2.46
2.30	0.82	−0.63	−1.45	−1.08	1.37	2.45
2.32	0.82	−0.63	−1.45	−1.08	1.34	2.43
2.34	0.82	−0.64	−1.45	−1.09	1.32	2.41
2.36	0.81	−0.64	−1.46	−1.10	1.29	2.39
2.38	0.81	−0.65	−1.46	−1.11	1.27	2.38
2.40	0.81	−0.65	−1.46	−1.12	1.24	2.36
2.42	0.81	−0.66	−1.46	−1.13	1.21	2.34
2.44	0.81	−0.66	−1.47	−1.14	1.19	2.33
2.46	0.80	−0.67	−1.47	−1.15	1.16	2.31
2.48	0.80	−0.67	−1.47	−1.17	1.13	2.30
2.50	0.80	−0.67	−1.47	−1.18	1.10	2.28
3.00	0.75	−0.78	−1.53	−2.16	−0.17	1.98
3.50	0.72	−0.86	−1.57	21.40	23.16	1.76
4.00	0.69	−0.92	−1.61	1.84	3.40	1.56
4.50	0.66	−0.98	−1.64	1.08	2.43	1.35
5.00	0.64	−1.03	−1.67	0.89	2.00	1.11
5.50	0.63	−1.07	−1.69	0.88	1.66	0.78
6.00	0.61	−1.11	−1.72	1.03	1.28	0.24
6.50	0.60	−1.14	−1.74	1.49	0.64	−0.86
7.00	0.59	−1.17	−1.76	3.53	−1.54	−5.07
7.50	0.58	−1.20	−1.77	−7.57	9.42	16.99
8.00	0.57	−1.22	−1.79	−1.90	3.62	5.51
8.50	0.56	−1.25	−1.81	−1.16	2.75	3.91
9.00	0.55	−1.27	−1.82	−0.91	2.35	3.26
9.50	0.54	−1.29	−1.83	−0.81	2.08	2.89
10.00	0.54	−1.31	−1.85	−0.80	1.84	2.64
15.00	0.49	−1.45	−1.94	0.81	2.28	1.47
20.00	0.47	−1.54	−2.01	−2.07	3.99	6.06
30.00	0.44	−1.66	−2.10	0.83	2.55	1.72
40.00	0.42	−1.74	−2.17	−0.78	2.51	3.29
50.00	0.41	−1.80	−2.22	0.92	2.83	1.91
60.00	0.41	−1.85	−2.25	−2.20	4.37	6.57
70.00	0.40	−1.89	−2.29	−1.64	0.82	2.46
80.00	0.39	−1.93	−2.32	0.70	2.18	1.48
90.00	0.39	−1.96	−2.34	−1.03	3.12	4.15
100.00	0.38	−1.98	−2.37	−1.61	0.92	2.53

TABLE 8.31
RATIO OF GAP WIDTH TO BORE DIAMETER $S/D = 2.0$

	Objective			Projective		
k^2	f/R	z_m/R	z_p/R	f/R	z_m/R	z_p/R
0.01	30.84	30.82	−0.02	30.84	30.82	−0.02
0.02	15.74	15.70	−0.04	15.74	15.70	−0.04
0.04	8.21	8.13	−0.08	8.21	8.13	−0.08
0.06	5.72	5.59	−0.13	5.72	5.59	−0.13
0.08	4.48	4.31	−0.17	4.48	4.31	−0.17
0.10	3.75	3.52	−0.22	3.75	3.52	−0.22
0.12	3.27	2.99	−0.27	3.27	2.99	−0.27
0.14	2.93	2.60	−0.33	2.93	2.60	−0.33
0.16	2.68	2.30	−0.38	2.69	2.30	−0.38
0.18	2.50	2.06	−0.44	2.51	2.06	−0.45
0.20	2.35	1.86	−0.49	2.37	1.86	−0.51
0.22	2.23	1.69	−0.54	2.26	1.68	−0.58
0.24	2.13	1.54	−0.59	2.18	1.53	−0.65
0.26	2.05	1.41	−0.63	2.12	1.39	−0.73
0.28	1.97	1.30	−0.67	2.08	1.27	−0.81
0.30	1.91	1.20	−0.71	2.05	1.15	−0.90
0.32	1.85	1.10	−0.75	2.03	1.03	−0.99
0.34	1.80	1.02	−0.78	2.02	0.92	−1.10
0.36	1.75	0.94	−0.81	2.03	0.82	−1.21
0.38	1.71	0.87	−0.84	2.04	0.71	−1.33
0.40	1.67	0.81	−0.86	2.06	0.60	−1.46
0.42	1.63	0.74	−0.89	2.09	0.49	−1.60
0.44	1.60	0.69	−0.91	2.14	0.38	−1.76
0.46	1.57	0.63	−0.93	2.19	0.26	−1.93
0.48	1.54	0.58	−0.95	2.25	0.14	−2.12
0.50	1.51	0.54	−0.97	2.33	0.00	−2.33
0.52	1.48	0.49	−0.99	2.42	−0.14	−2.57
0.54	1.46	0.45	−1.01	2.53	−0.30	−2.83
0.56	1.43	0.41	−1.02	2.66	−0.48	−3.14
0.58	1.41	0.37	−1.04	2.81	−0.67	−3.49
0.60	1.39	0.34	−1.05	2.99	−0.89	−3.89
0.62	1.37	0.31	−1.07	3.21	−1.15	−4.36
0.64	1.35	0.27	−1.08	3.47	−1.45	−4.92
0.66	1.34	0.24	−1.09	3.79	−1.80	−5.60
0.68	1.32	0.21	−1.11	4.19	−2.24	−6.43
0.70	1.30	0.18	−1.12	4.70	−2.78	−7.48
0.72	1.29	0.16	−1.13	5.37	−3.48	−8.85
0.74	1.27	0.13	−1.14	6.28	−4.42	−10.70
0.76	1.26	0.10	−1.15	7.60	−5.76	−13.36
0.78	1.25	0.08	−1.16	9.64	−7.83	−17.48
0.80	1.23	0.06	−1.18	13.26	−11.48	−24.74
0.82	1.22	0.03	−1.19	21.36	−19.61	−40.97

TABLE 8.31 (continued)

	Objective			Projective		
k^2	f/R	z_m/R	z_p/R	f/R	z_m/R	z_p/R
0.84	1.21	0.01	−1.19	55.61	−53.88	−109.49
0.86	1.20	−0.01	−1.20	−90.73	92.43	183.16
0.88	1.18	−0.03	−1.21	−24.91	26.59	51.50
0.90	1.17	−0.05	−1.22	−14.42	16.08	30.51
0.92	1.16	−0.07	−1.23	−10.15	11.79	21.94
0.94	1.15	−0.09	−1.24	−7.83	9.45	17.28
0.96	1.14	−0.10	−1.25	−6.38	7.97	14.35
0.98	1.13	−0.12	−1.25	−5.38	6.96	12.34
1.00	1.13	−0.14	−1.26	−4.66	6.21	10.88
1.02	1.12	−0.15	−1.27	−4.11	5.65	9.76
1.04	1.11	−0.17	−1.28	−3.68	5.20	8.88
1.06	1.10	−0.19	−1.28	−3.34	4.83	8.17
1.08	1.09	−0.20	−1.29	−3.06	4.53	7.59
1.10	1.08	−0.22	−1.30	−2.82	4.27	7.10
1.12	1.08	−0.23	−1.30	−2.62	4.06	6.68
1.14	1.07	−0.24	−1.31	−2.45	3.87	6.32
1.16	1.06	−0.26	−1.32	−2.30	3.70	6.00
1.18	1.05	−0.27	−1.32	−2.18	3.55	5.73
1.20	1.05	−0.28	−1.33	−2.06	3.42	5.48
1.22	1.04	−0.30	−1.34	−1.96	3.30	5.27
1.24	1.03	−0.31	−1.34	−1.88	3.20	5.07
1.26	1.03	−0.32	−1.35	−1.80	3.10	4.90
1.28	1.02	−0.33	−1.35	−1.73	3.01	4.73
1.30	1.01	−0.34	−1.36	−1.66	2.93	4.59
1.32	1.01	−0.36	−1.36	−1.60	2.85	4.46
1.34	1.00	−0.37	−1.37	−1.55	2.78	4.33
1.36	1.00	−0.38	−1.37	−1.51	2.71	4.22
1.38	0.99	−0.39	−1.38	−1.46	2.65	4.11
1.40	0.99	−0.40	−1.38	−1.42	2.59	4.02
1.42	0.98	−0.41	−1.39	−1.39	2.54	3.93
1.44	0.98	−0.42	−1.39	−1.35	2.49	3.84
1.46	0.97	−0.43	−1.40	−1.32	2.44	3.76
1.48	0.97	−0.44	−1.40	−1.30	2.39	3.69
1.50	0.96	−0.45	−1.41	−1.27	2.35	3.62
1.52	0.96	−0.46	−1.41	−1.25	2.30	3.55
1.54	0.95	−0.47	−1.42	−1.23	2.26	3.49
1.56	0.95	−0.47	−1.42	−1.21	2.22	3.43
1.58	0.94	−0.48	−1.43	−1.19	2.19	3.38
1.60	0.94	−0.49	−1.43	−1.17	2.15	3.32
1.62	0.93	−0.50	−1.43	−1.16	2.11	3.27
1.64	0.93	−0.51	−1.44	−1.15	2.08	3.22
1.66	0.93	−0.52	−1.44	−1.13	2.04	3.18
1.68	0.92	−0.52	−1.45	−1.12	2.01	3.14

TABLE 8.31 (continued)

	Objective			Projective		
k^2	f/R	z_m/R	z_p/R	f/R	z_m/R	z_p/R
1.70	0.92	−0.53	−1.45	−1.11	1.98	3.09
1.72	0.91	−0.54	−1.45	−1.11	1.95	3.05
1.74	0.91	−0.55	−1.46	−1.10	1.92	3.02
1.76	0.91	−0.56	−1.46	−1.09	1.89	2.98
1.78	0.90	−0.56	−1.46	−1.09	1.86	2.94
1.80	0.90	−0.57	−1.47	−1.08	1.83	2.91
1.82	0.89	−0.58	−1.47	−1.08	1.80	2.88
1.84	0.89	−0.58	−1.48	−1.07	1.77	2.85
1.86	0.89	−0.59	−1.48	−1.07	1.74	2.82
1.88	0.88	−0.60	−1.48	−1.07	1.72	2.79
1.90	0.88	−0.60	−1.49	−1.07	1.69	2.76
1.92	0.88	−0.61	−1.49	−1.07	1.66	2.73
1.94	0.87	−0.62	−1.49	−1.07	1.63	2.71
1.96	0.87	−0.62	−1.50	−1.08	1.61	2.68
1.98	0.87	−0.63	−1.50	−1.08	1.58	2.66
2.00	0.86	−0.64	−1.50	−1.08	1.55	2.63
2.02	0.86	−0.64	−1.51	−1.09	1.52	2.61
2.04	0.86	−0.65	−1.51	−1.09	1.50	2.59
2.06	0.86	−0.66	−1.51	−1.10	1.47	2.57
2.08	0.85	−0.66	−1.51	−1.11	1.44	2.54
2.10	0.85	−0.67	−1.52	−1.11	1.41	2.52
2.12	0.85	−0.67	−1.52	−1.12	1.38	2.50
2.14	0.84	−0.68	−1.52	−1.13	1.35	2.48
2.16	0.84	−0.68	−1.53	−1.14	1.32	2.47
2.18	0.84	−0.69	−1.53	−1.16	1.29	2.45
2.20	0.84	−0.70	−1.53	−1.17	1.26	2.43
2.22	0.83	−0.70	−1.53	−1.18	1.23	2.41
2.24	0.83	−0.71	−1.54	−1.20	1.20	2.39
2.26	0.83	−0.71	−1.54	−1.21	1.16	2.38
2.28	0.83	−0.72	−1.54	−1.23	1.13	2.36
2.30	0.82	−0.72	−1.55	−1.25	1.09	2.34
2.32	0.82	−0.73	−1.55	−1.27	1.06	2.33
2.34	0.82	−0.73	−1.55	−1.29	1.02	2.31
2.36	0.82	−0.74	−1.55	−1.32	0.98	2.30
2.38	0.81	−0.74	−1.56	−1.34	0.94	2.28
2.40	0.81	−0.75	−1.56	−1.37	0.90	2.27
2.42	0.81	−0.75	−1.56	−1.40	0.85	2.25
2.44	0.81	−0.76	−1.56	−1.43	0.81	2.24
2.46	0.80	−0.76	−1.57	−1.46	0.76	2.23
2.48	0.80	−0.77	−1.57	−1.50	0.71	2.21
2.50	0.80	−0.77	−1.57	−1.54	0.66	2.20
3.00	0.75	−0.87	−1.63	−7.26	−5.34	1.92
3.50	0.72	−0.95	−1.67	2.40	4.09	1.69

TABLE 8.31 (continued)

	Objective			Projective		
k^2	f/R	z_m/R	z_p/R	f/R	z_m/R	z_p/R
4.00	0.69	−1.02	−1.71	1.17	2.63	1.46
4.50	0.66	−1.07	−1.74	0.91	2.10	1.19
5.00	0.64	−1.12	−1.77	0.90	1.72	0.82
5.50	0.63	−1.16	−1.79	1.09	1.28	0.19
6.00	0.61	−1.20	−1.82	1.76	0.42	−1.34
6.50	0.60	−1.24	−1.84	7.42	−5.40	−12.93
7.00	0.59	−1.27	−1.86	−3.19	5.07	8.26
7.50	0.58	−1.29	−1.87	−1.40	3.13	4.53
8.00	0.57	−1.32	−1.89	−0.98	2.55	3.54
8.50	0.56	−1.34	−1.90	−0.84	2.22	3.06
9.00	0.55	−1.37	−1.92	−0.81	1.96	2.76
9.50	0.55	−1.39	−1.93	−0.87	1.69	2.56
10.00	0.54	−1.40	−1.94	−1.06	1.34	2.40
15.00	0.50	−1.55	−2.04	0.83	1.79	0.97
20.00	0.47	−1.64	−2.11	−0.79	2.44	3.24
30.00	0.44	−1.76	−2.20	0.83	1.83	1.00
40.00	0.42	−1.84	−2.26	−1.00	1.59	2.59
50.00	0.41	−1.90	−2.31	0.87	1.82	0.94
60.00	0.40	−1.95	−2.35	−0.71	2.22	2.93
70.00	0.40	−1.99	−2.39	0.76	2.72	1.96
80.00	0.39	−2.02	−2.42	−2.12	4.43	6.55
90.00	0.39	−2.05	−2.44	−1.18	1.49	2.67
100.00	0.38	−2.08	−2.46	0.69	2.56	1.87

References

In English

COSSLETT, V. E. (1946). "Introduction to Electron Optics," 2nd ed. Oxford Univ. Press, London and New York.

GABOR, D. (1945). "The Electron Microscope." Hutton, London.

GRIVET, P. (1965). "Electron Optics." Pergamon Press, New York.

HAINE, M. (1961). "The Electron Microscope—The Present State of the Art." Spon, London.

HALL, C. (1966). "Introduction to Electron Microscopy." McGraw-Hill, New York.

JACOB, L. (1950). "An Introduction to Electron Optics." Methuen, London.

KLEMPERER, O. (1953). "Electron Optics." Cambridge Univ. Press, London and New York.

MALOFF, I. G., and EPSTEIN, D. W. (1938). "Electron Optics in Television." McGraw-Hill, New York.

MEYER, L. M. (1939). "Electron Optics." Chapman & Hall, London.

MOSS, H. (1968). "Narrow-Angle Electron Gun and Cathode-Ray Tubes." Academic Press, New York.

PIERCE, J. R. (1954). "Theory and Design of Electron Beams." Van Nostrand, Princeton New Jersey.

SEPTIER, A. (1967). "Focusing of Charged Particles." Academic Press, New York.

STURROCK, P. A. (1955). "Static and Dynamic Electron Optics." Cambridge Univ. Press, London and New York.

TSUKKERMAN, I. I. (1961). "Electron Optics in Television." Pergamon Press, New York.

ZWORYKIN, V. K., MORTON, G. A., RAMBERG, E. G., VANCE, A. W., and HILLIER, J. (1945). "Electron Optics and the Electron Microscope." Wiley, New York.

In French

DE BROGLIE, L. (1950). "Optique Electronique et Corpusculaire." Hermann, Paris.

DE BROGLIE, L., MAGNAN, C., *et al.* (1946). "L'Optique Electronique." Revue d'Optique, Paris.

DUPOUY, G. (1952). "Elements d'Optique Electronique." Armand Colin, Paris.

In German

Busch, H., and Brüche, E. (1937). "Beiträge zur Elektronenoptik," Barth, Leipzig.
Brüche, E., and Scherzer, O. (1934). "Geometrische Elektronenoptik." Springer, Berlin.
Glaser, W. (1952). "Grundlagen der Elektronenoptik." Springer, Vienna.
Ollendorff, F. (1955). "Elektronik des Einzelelektrons." Springer, Berlin.
Picht, J. (1966). "Einführung in die Theorie der Elektronenmikroskopie." Verlag Technik. H. Leins, Berlin.
Von Ardenne, M. (1940). "Elektronen Übermikroskopie." Springer, Berlin.

Index